Isolation, Identification and Characterization of Allelochemicals/Natural Products

Isolation, Identification and Characterization of Allelochemicals/Natural Products

Editors

DIEGO A. SAMPIETRO
Instituto de Estudios Vegetales *"Dr. A. R. Sampietro"*
Universidad Nacional de Tucumán, Tucumán
Argentina

CESAR A. N. CATALAN
Instituto de Química Orgánica
Universidad Nacional de Tucumán, Tucumán
Argentina

MARTA A. VATTUONE
Instituto de Estudios Vegetales *"Dr. A. R. Sampietro"*
Universidad Nacional de Tucumán, Tucumán
Argentina

Series Editor

S. S. NARWAL
Haryana Agricultural University
Hisar, India

Science Publishers

Enfield (NH)   Jersey   Plymouth

Science Publishers   *www.scipub.net*

234 May Street
Post Office Box 699
Enfield, New Hampshire 03748
United States of America

General enquiries : *info@scipub.net*
Editorial enquiries : *editor@scipub.net*
Sales enquiries   : *sales@scipub.net*

Published by Science Publishers, Enfield, NH, USA
An imprint of Edenbridge Ltd., British Channel Islands
Printed in India

© 2009 reserved

ISBN: 978-1-57808-577-4

Library of Congress Cataloging-in-Publication Data

Isolation, identification and characterization of allelo-
chemicals/natural products/editors, Diego A. Sampietro, Cesar
A. N. Catalan, Marta A. Vattuone.
   p. cm.
 Includes bibliographical references and index.
 ISBN 978-1-57808-577-4 (hardcover)
 1. Allelochemicals. 2. Natural products. I. Sampietro, Diego
A. II. Catalan, Cesar A. N. III. Vattuone, Marta A.
  QK898.A43I86 2009
  571.9'2--dc22
                                              2008048397

Preface

Allelopathy is a newly emerging multidisciplinary field of agricultural research. Till now, a lot of allelopathy research work has been done in various fields of Agriculture and Plant Sciences. However, no standard methods are being used by various workers due to lack of compendium on the techniques, hence, the results obtained are not easily comparable with each other. This situation has caused a lot of problems to researchers working in underdeveloped/third world countries in small towns, where library and research facilities are not available. Therefore, to make available the standard methods for conducting allelopathy research work, this multi-volume book has been planned, with one volume each for each discipline. In all the conferences held worldwide since 1990's, a need has always been felt for a Manual of Allelopathy Research Methods. Hence, Prof. S.S. Narwal has planned this multivolume book **Research Methods in Plant Sciences : Allelopathy.** The objective of this book series is to provide research workers, this information about various methods, so that they can conduct research independently. The methods have been described in a simple manner.

Discovery of new bioactive compounds from natural products/secondary metabolites has played a major role in the development of organic chemistry. The ecological role of these allelochemicals/natural products as herbicides, insecticides and fungicides for ecological/organic sustainable agriculture has recently drawn great attention, due to increasing public concern against synthetic pesticides. Current researches on natural products emphasize not only the need of structural elucidation of the isolated compounds, but also of their natural biological function and possible utility for human purposes. These concerns require reliable protocols for such studies, which sometimes are not available worldwide. In allelopathy research, identification of allelochemicals is always required. Majority of researchers in under developed/third world countries, do not identify such allelochemicals, mainly because of non-compilation of such methods in book form.

This book is divided into three sections:

Section I. Sample Collection, Handling and Storage: It includes three chapters: Allelopathy and Allelochemicals, Soil, Air and Water Samples and Plant Sampling and Sample Preparation.

Section II. Isolation, Identification and Structural Elucidation : It has eight chapters: Colorimetric Reactions, TLC and PC, HPLC and LC- MS, Chromatography and Spectroscopy of Alkaloids, GC and GC-MS of Terpenoids, GC and GC-MS for Non-volatile Compounds, Spectrometry: Ultraviolet and Visible Spectra, Spectrometry: Infrared Spectra.

Section III. Biological Activity of Natural Products : It (has six Chapters: Bioassays with whole plants and plant organs, Bioassays on Plants: Plant cells and organelles, Bioassays on Microorganisms: Antifungal and antibacterial activities, Bioassays for Antioxidant, Genotoxic, Mutagenic and Cytotoxic activities, Bioassays: Inhibitors of insect chitin-degrading enzymes and Bioassays: insect behaviour, development and survival.

Section IV. Appendices: Abbreviations, Chemical Formulae and Molecular Weight of Solvents and Reagents, Molecular Weight of Organic Compounds.

This book will serve as a ready reference in the laboratory or class room and probably provide solutions to many problems of isolation, identification and characterization of allelochemicals/natural products. It will be particularly useful for UG and PG students pursing this field, as well as for organic chemists.

We are indebted to all the contributors, who have actually used all these methods in their fields of specialization for the last 10 to 25 years and by accepting the challenging task of presenting the procedures of various methods in a very simple language, easily understood by beginners.

We would appreciate receiveing valuable suggestions from students and researchers, to enable us to make further improvements in future editions of this book, so that it is more useful and meaningful.

May 28, 2008

Editors

Contents

SECTION III. Biological Activity of Natural Products

SECTION IV. Appendices

Sample Collection,
Handling and Storage

Allelopathy and Allelochemicals

S.S. Narwal[1][*] and D.A. Sampietro[2]

1. INTRODUCTION

Allelopathy refers to any process involving secondary metabolites produced by plants, microorganisms, viruses and fungi that influence the growth and development of agricultural and biological systems (Narwal, 1994). It has been established that allelopathy offers great potential to (a) increase agricultural production (food grains, vegetables, fruits and forestry), (b) decrease harmful effects of modern agricultural practices [multiple cropping, leaching losses from N fertilizers, indiscriminate use of pesticides (weedicides, fungicides, insecticides and nematicides), tolerant/resistant biotypes in pests] on soil health/productivity and on environment and (c) maintain soil productivity and a pollution-free environment for our future generations. It is likely that in the near future allelopathy will be used in crop production, crop protection, agroforestry and agrohorticultural practices in developed and developing countries. Allelopathy may become one of the strategic sciences to reduce the environmental pollution and to increase agricultural production in sustainable agriculture of the 21st century. Allelopathy provides the basis for sustainable agriculture, hence, currently allelopathy research is being done in most countries worldwide and is receiving more attention from agricultural and bioscientists.

Allelopathy is a new field of science and the term 'Allelopathy' was coined by Prof. Hans Molisch, a German plant physiologist in 1937. Therefore, till now

————————————————————
Authors' addresses: [1]Department of Agronomy, CCS Haryana Agricultural University, Hisar 125 004, India.
[2]Instituto de Estudios Vegetales *"Dr. A. R. Sampietro"*, España 2903, 4000, San Miguel de Tucumán, Tucumán, Argentina.
Corresponding author: E-mail: narwal_1947@yahoo.com

there is no comprehensive book on Methodology of Allelopathy Research and is causing problems to researchers working in underdeveloped/third world countries, in small towns without library facilities. Therefore, to make available the standard methods for conducting allelopathy research independently, this multivolume book has been planned. Since allelopathy is a multi-disciplinary area of research, individual volumes have been planned for each discipline.

2. ALLELOCHEMICALS

Allelochemicals are secondary substances, biosynthesized from the metabolism of carbohydrates, fats and amino acids and arise from acetate or the shikimic acid pathway. These are biosynthesized and stored in the plant cells and do not affect the cell activities. However, after their release from the plant cells (through volatilization, leaching root exudates and decomposition of biomass), these allelochemicals start influencing the organisms (plants, pathogens, insect pests, etc.), when they come in contact. Rice (1984) indicated that allelopathic agents influence the plant growth through following physiological processes: (a) cell division and cell elongation, (b) phytohormone induced growth, (c) membrane permeability, (d) mineral uptake, (e) availability of soil phosphorus and potash, (f) stomatal opening and photosynthesis, (g) respiration, (h) protein synthesis and changes in lipid and organic acid metabolism, (i) inhibition of porphyrin synthesis, (j) inhibition or stimulation of specific enzymes, (k) corking and clogging of xylem elements, stem conductance of water and internal water relations and (l) miscellaneous mechanisms. Most of these physiological and biochemical processes are responses of cells to various allelochemicals.

2.1   Nature of Allelochemicals

Rice (1984) has divided these compounds into 14 chemical categories: (a) cinnamic acid derivatives, (b) coumarins, (c) simple phenols, benzoic acid derivatives, gallic acid and protocatechuicacid, (d) flavonoids, (e) condensed and hydrolysable tannins, (f) terpenoids and steroids, (g) water soluble organic acids, straight chain alcohols, aliphatic aldehydes and ketones, (h) simple unsaturated lactones, (i) long chain fatty acids, (j) naphthoquinones, anthraquinones and complex quinones, (k) amino acids and polypeptides, (l) alkaloids and cyanohydrins, (m) sulfides and mustard oil glycosides and (n) purines and nucleotides. However, Putnam and Tang (1986) grouped these chemicals into 11 classes: (a) toxic gases, (b) organic acids and aldehydes, (c) aromatic acids, (d) simple unsaturated lactones, (e) coumarins, (f) quinines, (g) flavonoids, (h) tannins, (i) alkaloids, (j) terpenoids and steroids, and (k) miscellaneous and unknown.

Rice (1984) outlined the following factors which affect the amount of allelochemicals produced: (a) radiation, (b) mineral deficiencies, (c) water stress, (d) temperature, (e) allelopathic agents, (f) age of plant organs, (g) genetics, and (h) pathogens and predators.

3. ISOLATION, IDENTIFICATION AND CHARACTERIZATION OF ALLELOCHEMICALS

Allelochemicals are an integral part of allelopathy research. Hence, allelopathy research is not complete till the allelochemicals present in the experimental conditions are isolated, identified and characterized. Hence, the knowledge of chemistry methods is required. The preceding section 2.1 shows that allelochemicals belong to several distinct chemical classes and consequently the techniques for their isolation, identification and characterization are also different.

Discovery of new bioactive compounds from natural products/secondary metabolites has played a major role in the development of organic chemistry. The ecological role of these allelochemicals/natural products as herbicides, insecticides and fungicides for ecological/organic sustainable agriculture has recently drawn great attention, due to increasing public concern against the use of synthetic pesticides. Current researches on natural products emphasize not only the need for structural elucidation of the isolated compounds, but also for their natural biological function and possible utility for human purposes. These concerns require reliable protocols for such studies, which sometimes are not available worldwide. Hence, this book has been written to provide consolidated information about the isolation, identification and characterization of allelochemicals/natural products.

SUGGESTED READINGS

Narwal, S.S. (1994). *Allelopathy in Crop Production*. Scientific Publishers, Jodhpur, India.
Putnam, A. R. and Tang, C.S. (1986). *The Science of Allelopathy*. Wiley Interscience, New York.
Rice E.L. (1984). *Allelopathy*. 2nd edition. Academic Press, New York.

2

Soil, Air and Water Samples

**Diego A. Sampietro*, Jose R. Soberón, Melina A. Sgariglia,
Emma N. Quiroga and Marta A. Vattuone**

1. INTRODUCTION

Chemical analysis of natural products usually involves the following sequence: defining purpose of analysis, sample collection, sample preparation for chromatography, analytical chromatography, data processing and results of chemical analysis (Fig. 1). Sampling, or sample collection, is the second operation of this process and consists of obtaining the representative samples using adequate sampling techniques, as per sample nature and the purpose of chemical analysis (Moldoveanu and Victor, 2002). Improper sampling, sample storage or sample

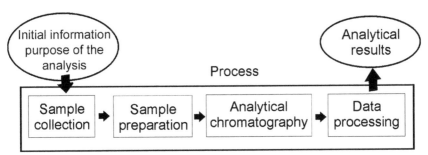

Figure 1   Steps in the chemical analysis of natural products (Source: Adapted from Modolveanu and Victor, 2002).

Authors' address: Instituto de Estudios Vegetales *"Dr. A. R. Sampietro"*, Facultad de Bioquímica, Química y Farmacia, Universidad Nacional de Tucumán, España 2903, 4000, San Miguel de Tucumán, Tucumán, Argentina.
*Corresponding author: E-mail: dasampietro2006@yahoo.com.ar

handling often lead to experimental errors and drastically reduces the reliability of analytical results. This chapter provides general techniques applied for collection, handling and storage of soil, air and water samples in natural products research.

1.1 Sample Collection

Sample collection consists of taking small representative portions of a larger amount of material to be analyzed. Sampling should be reproducible, timely, economical and safe. It should not disturb the analyzed system. Samples must be collected following a specific sampling design. Selection of sampling design depends on the nature of the bulk material, the system investigated and the purpose of analysis. Sampling design usually follows one of the following possibilities (Lohr, 1999):

(i) **Random sampling:** It is the most common sampling design. It assumes that the bulk material is made of a large number of equally sized and discrete portions. A random sampling is possible only if each portion has an equal chance of being selected.

(ii) **Systematic sampling:** It consists of collecting samples at locations and times according to spatial or temporal patterns. The most common systematic sampling approach is to overlay a grid on an area and then to collect a sample at each grid node.

(iii) **Stratified sampling:** It is achieved by separating the sample population into similar non-overlapping groups, called strata, and then selecting either a simple random or systematic sample from each strata. In many types of ecological sampling, stratification is believed to increase the reliability of the data relative to unstratified studies with a similar cost. The success of a stratified design is only possible when each stratum is homogeneous and different from others.

(iv) **Composite samples:** It is a common sampling method that simplifies sample analyses. Composite samples are obtained by mixing several samples collected from different points of the bulk sample, at different locations or at different times. A composite sample is expected to be more representative, homogeneous and descriptive of the bulk sample rather than single samples. After preparation, the composite sample is re-sampled by taking only a portion for analysis.

1.2 General Precautions for Sample Handling and Storage

The following general precautions should be taken to preserve organic molecules of a sample during handling and storage:

(i) **Timely sample analysis:** The delays in handling, sample processing, and analysis should be avoided because the sample can suffer physical and chemical changes, leading to the loss of analytes. However, laboratories often have a limited capacity for fast sample processing. In this case, preservation procedures must be applied according to the nature of samples (i.e. storage in darkness at low temperature, addition of preservatives and antioxidants, or adjustment of pH).

(ii) **Analysis of samples with compounds in low concentrations:** In this case, samples should be stored in physical locations, different from those of analytical standards or any other material that may contain a high concentration of the analyte under analysis.

(iii) **Freezing:** It is often applied for sample preservation. Samples that cannot be frozen, or those that do not need to be frozen, are usually stored at 0-5 °C.

(iv) **Relative humidity:** Some samples need storage under specific relative humidity. A specific value of relative humidity can be achieved using saturated salt solutions in a confined space. Salts such as NaCl, KI and $K_3C_6H_5O_7$ [potassium citrate] are often used to regulate relative humidity.

(v) **Drying:** It is a common alternative to preserve samples after collection and sample preparation. Some samples can be dried in an oven at 40-50°C (i.e. most plant samples) or at room temperature (i.e. soil samples) according to thermal stability of the analytes of interest.

(vi) **Addition of organic solvents or standard compounds:** Addition of organic solvents prevents the adsorption of hydrophobic constituents to the container's surfaces and ensures their solubility. Solvents used should be pure, to prevent sample contamination. Significant amounts of volatile organic compounds (VOCs) can be lost during storage. Losses of VOCs from a sample can be estimated by adding a known amount of a standard VOC to a sample.

(vii) **Contamination of samples:** The containers are the main source of sample contaminants. Plastic containers, for example, often contaminate samples with phthalates. Glass containers are preferred for organic analytes. Containers must be washed with sulfate-free detergents followed by thorough rinsing. Standard solutions used for calibration procedures must be stored in inert containers. Reagents used as additives could also introduce detectable levels of contaminant ions.

2. AIR SAMPLING

VOCs are substances normally found in vapour phase at room temperature. The physicochemical constraints on volatility restrict plant VOCs to low-molecular and largely lipophilic products belonging to the classes of terpenes and non-terpene aliphatics (including nitrogen- and sulfur-containing compounds), phenylpropanoids and benzenoids. The concentration of VOCs in air fluctuates in both time and space, and measurement techniques must be able to detect these fluctuations. Common preconcentration techniques used for the headspace analysis in GC [Gas chromatography] are often applied in chemical ecology for identification and quantification of VOCs, after modifications according to the needs of research. Dynamic headspace analysis is more often employed than static headspace analysis.

2.1 Dynamic Headspace Techniques

Dynamic headspace analysis consists in the collection of VOCs from a sample source confined in an entrainment chamber. A carrier gas (usually purified air) is passed over the sample. The sample's VOCs released are carried by the gas to a solid trap, usually a porous organic polymer such as Porapak Q, Tenax TA, or activated charcoal, where the analytes are adsorbed and preconcentrated. Dynamic headspace analysis is also performed in outdoor environments. In this case, collection of VOCs is achieved using small battery-powered pumps attached to the belts of workers. The inlet of the pump is connected by flexible tubing to a tube containing a sorbent material attached to the lapel of the workers' shirts. The pumps pull air at a fixed, calibrated flow-rate through the sorbent tube. At the end of a designated sampling period, the pump is switched off and the start and stop time and flow-rate are recorded. Typically, the flow-rate is calibrated at the beginning and end of the sampling period, if not more often, so the total volume of air sampled can be recorded. The sorbent tube is then capped and sent to the laboratory for analysis, after which the total mass of each VOC is divided by the total pumped air volume to give the concentration in mass per unit volume. The following are examples of classic devices developed for collection of VOCs through the dynamic headspace system.

Experiment 1: Closed Air Recirculating System

The sample source of VOCs confined in an entrapped chamber (i.e. a desiccator) is exposed to recirculating airflow. The air circulates through a solid trap, usually a porous organic polymer such as activated charcoal, where the analytes are adsorbed and reconcentrated (Fig. 2).

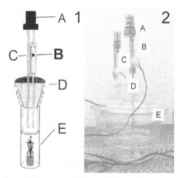

Figure 2 Schematic drawing (1) and picture (2) of the suggested VOCs-collecting device: A) circulating pump, (B) stainless steel housing containing the activated charcoal trap (1.5 mg charcoal), (C) stainless steel tubes (i.e. 1.0 mm), (D) Teflon stopper, (E) glass container, adequate to cover the material source of VOCs (Source: Donath and Boland, 1995).

Materials and equipments required

Small circulation pump; stainless steel housing containing the charcoal trap [polish the stainless steel housing to fit a precision glass liner containing a pad of charcoal (the pad should be filled with at least 1.5 mg charcoal)]; glass liner; stainless steel tubes; Teflon stopper; glass desiccator (capacity: 3 -1 L); plant material (a plant or a plant part).

Procedure

(i) Attach the Teflon stopper to the upper part of the desiccator.
(ii) Connect the circulating pump with the stainless steel housing and the stainless steel tube to the Teflon stopper.
(iii) Attach the circulation pump to the upper part of desiccator.
(iv) Close the desiccator. Switch on the pump and let it work for few minutes. If air from outside the desiccator is not aspirated, switch off the pump and put the source of VOCs (i.e. a plant or a plant part in a beaker with distilled water) inside the desiccator.
(v) Switch on the pump. Let air to circulate for 20 hours.
(vi) Switch off the pump. Open the desiccator and detach the stainless steel housing.
(vii) Extract VOCs from the activated charcoal pad with a volume of CH_2Cl_2 (i.e. 0.2 mL) and analyze by GC-MS [gas chromatography – mass spectrometry].

Calculations

Amount of VOCs released by the source material can be estimated, after GC-MS analysis, as under:

$$VOCs(\mu g/L) = \frac{\text{Total VOCs in } CH_2Cl_2 (\mu g)}{\text{Time of trapping (20 h)} \times \text{airflow (L/h)}}$$

Statistical analysis

Differences between samples in concentrations of VOCs are detected using a one-way analysis of variance at 0.01 or 0.05 level. Significant differences between the mean of VOCs concentrations is determined by using Dunnet T3 test.

Observations

If the amount of VOCs is less for GC-MS analyses, repeat the experiment increasing the time for VOCs trapping. The volume of CH_2Cl_2 used to extract VOCs also can be reduced. The suggested trapping system is suitable for collecting the low VOCs levels, because it minimizes the air contaminants collected in open systems. A control consisting in trapping air components through the system with the empty glass container in the same conditions performed for plant material is advised, to detect unexpected air contaminants. Make at least three independent experiments (N = 3). Each treatment should have at least three replications.

Experiment 2: Closed No-Recirculating Air System

A source of VOCs (i.e. a plant material) confined in an entrapped chamber (i.e. a glass carboy) is exposed to air aspirated through a solvent where the analytes are solubilized and concentrated (Fig. 3).

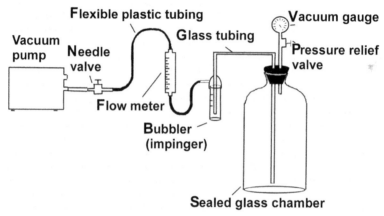

Figure 3   Air sampling apparatus used for collection of VOCs. The source material of VOCs is placed in the sealed glass chamber (Source: Wilt *et al.*, 1993).

Materials and equipments required

Pyrex glass carboy (9.5 L); rubber stopper; flow meter; flexible plastic tubing; glass tubing; bubbler; needle valve; vacuum pump; vacuum gauge; liquid nitrogen; plant material (a plant or a plant part).

Procedure

(i) Fit the rubber stopper to the pyrex glass carboy.

(ii) Fit two glass tubes to the stopper and seal with a water-based glue.

(iii) Attach a vacuum gauge to one of the protruding tubes. Connect the other tube with a glass tubing to a 50 mL impinger.

(iv) Attach a flow meter with a vacuum line to the bubbler, which is connected to a vacuum pump through another length of tubing.

(v) Put a needle valve on-line between the air flow meter and the vacuum pump so that the air flow rate could be accurately controlled.

(vi) On the following day, place an appropriate amount of the source of VOCs into the carboy. The analyst should simulate a bulk density similar to that observed in natural conditions to obtain meaningful values during VOCs analysis.

(vii) Carefully shake the source of VOCs when loading it into the carboy, to avoid breaking and shattering.

(viii) Seal off the carboy with an aluminum oil lined stopper and the glass tubes with a water-based glue.

(ix) Let the carboy content stabilize in an oven at 38°C for some time (hours or days). Then, connect the carboy to the vacuum pump.

(x) Fill the bubbler with 30 mL of hexane. Apply vacuum adjusting the air flow to 150 mL/min. Air is sampled until the vacuum gauge indicates that exactly 0.5 atm of air is remaining in the carboy.

(xi) Switch off the pump, fresh air is slowly allowed back into the system to re-establish the atmospheric pressure.

(xii) Transfer the contents of the bubbler into glass containers. Then, seal and refrigerate them. Disconnect the carboy from the vacuum pump and place back in the oven at 38°C for 1 hour. Repeat this procedure as many times as needed with the samples to trap enough VOCs for analysis. Concentrate hexane aliquot to 3 mL each adding liquid nitrogen, while kept cold in an ice bath. Take little volumes of these aliquots for GC-MS analysis.

Calculations

Record the volume of air pumped according to pumping time. Calculate the VOCs concentration, after GC-MS analysis, as follows:

$$\text{VOCs } (\mu g/L) = \frac{\text{Total VOCs in hexane } (\mu g)}{\text{Time of trapping (h)} \times \text{airflow (L/h)}}$$

where, h = hour and L = liter.

Observations

Between tests, the carboy and the flow meter should be thoroughly washed with detergent and then rinsed with hexane. A control consisting of pumping air through the system without addition of plant material to the glass container is advisable, to detect unexpected air contamination. Make at least three independent experiments (N = 3). Each treatment should have at least three replications.

Statistical analysis

Data are analyzed as indicated in Statistical analysis of Experiment 1.

Precautions

Liquid nitrogen is extremely cold and skin contact causes severe frostbite. Safety glasses, gloves and special containers should be used as a precaution. Before use, take complete instructions for the use of this liquid from your supplier.

2.2   Static Headspace Analysis of Air VOCs

In static headspace analysis, the sample is tightly closed in a vessel, where it comes into equilibrium with its vapour at a predetermined temperature (Tholl et al., 2006). The headspace can be sampled using a syringe or a similar device and directly injected into the gas chromatograph. The sample can be outdoor air. Collection of outdoor air samples consists of entrapping the air volumes in canisters provided with Teflon-coated silicone septum lids that allow a hermetic seal. Then, a gas syringe is inserted through the septum and a volume of the confined air is drawn for injection in a GC or GC-MS.

Experiment 3: Outdoor Air Samples

Air confined in a canister with the source of VOCs is sampled using a gas glass syringe.

Materials and equipments required

Canisters provided with Teflon-coated silicone septum lids; gas glass syringe; a sample source of VOCs (i.e. a plant or a plant part).

Procedure

(i)   Place the sample source of VOCs in the glass canister (in case of outdoor air sampling, open the canister for several minutes and close).

(ii)   Seal the canister with a Teflon-coated silicone septum and leave for 2 hours for equilibration.

(iii)   Using the gas glass syringe, draw 5 mL of the canister space and immediately inject in a GC-MS.

Calculations

Calculations should be done after the GC-MS analysis, as follows:

$$\text{VOCs (ppm)} = \frac{\text{Total VOCs in 5 mL of injected air (µg)}}{\text{Total air volume (5 mL)}}$$

Observations

This technique cannot be successfully applied for VOCs analysis, when their concentration is low in the air. In these cases, the use of dynamic space techniques may be more appropriate. Make at least three independent experiments ($N = 3$). Each treatment should have at least three replications.

Statistical analysis

Data are analyzed as indicated in Statistical analysis of Experiment 1.

3. SOIL SAMPLING

Concentration of natural products in soil is often needed in chemical ecology research. In allelopathy, for example, analysis of quantitative and qualitative changes in soil content of allelochemicals at different times may suggest that these substances participate in plant-plant interactions (Blum *et al.*, 1992). Before sampling, quadrants of the total sampling area and number of samples per quadrant are established as per the selected sampling design. Sometimes, equal amounts of samples collected in the same quadrant are combined to produce a composite representative sample. The best tool for soil sampling is the soil sampling tube or auger that is punched into the soil at the proper depth to extract a small core of soil (Fig. 4). When augers are not available, soil sampling may be performed with garden spades or shovels.

Experiment 4: Use of a Bucket Auger

Materials and equipments required

Stainless steel soil auger (7 cm diameter bucket); soil auger extensions prelabelled in 14 cm increments (enough to reach desired sampling depth); soil auger handle; 9 cm diameter heavy walled PVC sleeve cut into 10 cm length (1 or more); rubber mallet (1 or more); 5 × 10 cm wood sheet cut into 2.3 cm length (1 or

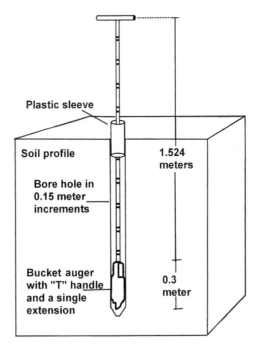

Figure 4 Soil sampling with a bucket auger.

more); 40 × 50 cm plastic bags (1 per sample); 5 gallon plastic buckets with handles, filled with clean water; 5 gallon bucket filled with soapy water; water sprayer; wash brushes; double distilled water; isopropyl alcohol; 500 mL wash bottle; disposable gloves; plastic bags (label containers or plastic bags before soil sampling; record date, location, number and depth of each soil sample).

Procedure

(i) Clear the soil top area where the auger will be punched. Place the PVC sleeve over the site to be sampled and drive it into the ground approximately 5 cm using the wood and the rubber mallet.

(ii) Place a clean auger in the sleeve. Turn the handle to advance the auger until it is filled with soil. Pull the auger up and tap all the soil into a 40 × 50 cm plastic bag. Return the auger to the sleeve and continue turning until the top of the auger is levelled with the soil surface. Again pull the auger up and load all the soil into the same bag. Surface soils tend to be compact, hence, fill the auger two or more times to go from a depth of 0 to 10 cm.

(iii) If deep soil samples are needed, drive the auger as indicated in step 2 but discard the soil extracted until the desired depth is reached. Collect the samples in bags as indicated in step 2 at the desired depth.

(iv) Put on the gloves and hold the top of the plastic bag tightly closed while shaking the soil to mix it thoroughly.

(v) Clean the auger by scrubbing it with a brush in a 5-gallon bucket filled with soapy water. Then, rinse in a 5-gallon bucket filled with clear water. Then, rinse again spraying with double distilled water. Give a final rinse with isopropyl alcohol dispensed from a clean bottle.

(vi) Transport the samples to the laboratory on dry ice and store at −20°C.

(vii) Prevent contamination from the 0 to 10 cm soil, cleaning the PVC sleeve with a gloved hand and scooping it out of the bottom of the hole before taking the second sample.

(viii) Label and transport the samples to the laboratory on dry ice.

3.1 Examples of Soil Collection

Collection of soil rhizosphere from a plant producing allelochemicals (Fujii et al., 2005): Grow the plants from a selected species in plastic pots in controlled conditions for a suitable time (i.e. *Medicago sativa* in a greenhouse at 25°C for four weeks). Take out the plants from the plastic pot without disturbance. Then, gently shake the plant roots and collect the root-zone soil in plastic bags (Fig. 5A). Then, remove and collect the soil adhering to the surroundings of roots (rhizosphere soil) in plastic bags. Sieve soil samples through 1 mm mesh. Discard root residues of rhizosphere soils. Place in plastic bags and freeze (−20°C) before extraction of soil allelochemicals. These samples should be processed during the following three months after collection.

Collection of soil samples in a walnut–corn cropping system (Jose and Gillespie, 1998): Take the soil cores (0-10 cm depth) at four different distances from the tree rows: 0, 0.9, 2.5 and 4 m from the row (Fig. 5B). Collect and combine 10 soil cores at each plot and collection distance. Place each soil sample in a plastic bag and immediately transport to the laboratory. Refrigerate for no longer than 24 hours before allelochemical extraction. Determine gravimetric soil water for each sample using subsamples. Collect on each sampling date, a composite sample of 10 soil cores from a nearby cornfield, which serves as a control for recovery experiments. Take samples as mentioned at different times during the year to evaluate fluctuations of soil allelochemical content.

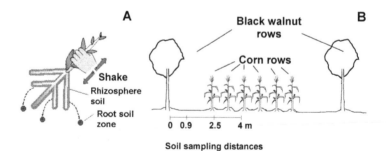

Figure 5 (A) Separation between root-zone soil and rhizosphere soil from a plant producing allelochemicals (Source: Fujii *et al.*, 2005). (B) Walnut–corn cropping system (Adapted from Jose and Gillespie, 1998).

Collection of soil samples from crop plots subjected to conventional and no-till cropping systems (Blum *et al.*, 1992): Assuming plots of 30 × 8 m with eight rows subjected to conventional and no-till systems, divide each plot into four sections (15 × 4 m) for sampling. Take two soil cores (5.5 cm in diameter; 0-10 cm depth) per section and combine them. Also take additional soil cores (0 to 2.5 cm depth) adjacent to the previous sampling locations. The 0 to 2.5 cm core samples are taken because allelochemicals released from plant residues are often more concentrated in top soil, making it possible to detect vertical gradients in the soil. Take samples, as mentioned, at several times after crop harvest to evaluate changes in soil allelochemical content.

3.2 Handling Soil Samples

Air-dry composite soil samples by spreading out on paper or any other suitable material in the laboratory. After five or six weeks, store the dry samples in labelled paper bags at room temperature. Several compounds are irreversibly adsorbed to soil particles during the drying process. Hence, these substances will not be available when soil is re-hydrated for chemical analysis. This problem is overcome by freezing the fresh soil samples at −20°C upto the time of chemical analysis.

3.3 Extraction of Soil Allelochemicals

Experiment 5: Extraction of Phenolic Acids

Soil is extracted with water or ethylenediamine tetraacetic acid (EDTA). Water extraction recovers primarily phenolic acids in soil solution, while the difference between EDTA and water extraction provides an estimate of the reversibly bound phenolic acids weakly sorbed to soil particles (Blum *et al.*, 1992).

(a) Water Extraction

Materials and equipments required

Soil samples (collected according to Experiment 4): drying oven; balance with 0.1 g sensitivity and at least 400 g capacity; thermometer capable of measuring upto 110°C; containers suitable for the drying oven; Whatman # 1 filter paper; Erlenmeyer flasks; glass funnel; separating funnel with a Teflon stopper; diethyl ether; methanol; 2 N HCl [transfer 200 mL of concentrated HCl (37% w/v, 10 M) to a beaker and dissolve in 800 mL of distilled water]; refrigerator; shaker; centrifuge capable of 900 × g; rotary evaporator; pH-meter.

Procedure

(i) **For dry soil samples:** Suspend the soil samples (100 g dry weight) in 200 mL of double distilled water.

For frozen soil samples:

(a) Thaw samples overnight at 10°C.
(b) **Determination of water content:** Weigh 10-15 g soil sample and dry it at 105°C in an oven until constant weight.
(c) Suspend soil sample (100 g fresh weigh) in 200 mL of double distilled water.

(ii) Shake for 18 hours (in darkness, 10°C). Centrifugate at 900 rpm and filter the supernatant with Whatman # 1 paper using a glass funnel.
(iii) Acidify the filtrate in an Erlenmeyer flask with 2 N HCl to pH 3.
(iv) Extract the phenolic acids from the filtrates with diethyl ether in a ratio 1:2 (v/v, water:diethyl ether). Add the acid filtrate to the separation funnel and then the diethyl ether. Insert the stopper, invert the separation funnel and shake vigorously for 3 minutes. Place the separating funnel with the stopper in the upside and leave for 10 minutes to separate in two phases. Carefully recover the upper phase (diethyl ether) transferring it to an Erlenmeyer flask.
(v) Evaporate the diethyl ether in a rotary evaporator, dissolve the residue in 4 mL of methanol and analyze by HPLC (Blum *et al.*, 1992).

Calculations

Step 1: Calculate the amount of a phenolic compound (A_{Phc}) extracted with water, considering concentration of the phenolic compound (C_{Phc}) determined after HPLC analysis and total volume of the methanol fraction (V_{MeOH}):

$$A_{Phc} \text{ (μmol)} = C_{Phc} \text{ (μmol/mL)} \times V_{MeOH} \text{ (4 mL)}$$

Step 2: Express content of a phenolic compound extracted with water (Cw) per g of dry soil:

$$Cw = A_{Phc} \text{ (μmol)}/\text{dry weight (g)}$$

For example, if A_{Phc} is 10 μmol and dry weight is 20 g, then:

$$Cw = 10 \text{ μmol/20 g}   \text{ Hence, } Cw = 0.5 \text{ μmol/g}$$

(b) Extraction with Neutral EDTA

Reagents and materials required

EDTA extractant (add 374 g of disodium ethylenediamine tetraacetic acid to 500 mL water in a beaker; add NaOH pellets very slowly as the EDTA-water suspension is stirred rapidly, until the EDTA is dissolved; the resulting solution has a pH of approximately 8.5. Adjust to pH 7 with concentrated HCl. Make up the final volume to 2 L); materials and equipments as indicated for water extraction.

Procedure

(i) **For dry soil samples:** Suspend soil samples (10 g dry weight) in 40 mL of neutral EDTA extractant.

For frozen soil samples:

(a) Thaw samples overnight at 10°C.
(b) **Determination of water content:** Weigh 10-15 g of the soil sample and dry it at 105°C in an oven until constant weight.
(c) Suspend soil sample (10 g fresh weigh) in 40 mL of neutral EDTA extractant.

(ii) Shake for 2.5 hours (in darkness, 10°C). Centrifuge for 15 minutes at 900 rpm and filter the supernatant through Whatman # 1 paper.
(iii) Proceed as indicated in steps 4 and 5 for water extraction.

Calculations

Step 1: Calculate the amount of a phenolic compound (A_{Phc}) extracted with neutral EDTA, considering concentration of phenolic compound (C_{Phc}) determined after HPLC analysis and total volume of methanol fraction (V_{MeOH}):

$$A_{Phc} \ (\mu mol) = C_{Phc} \ (\mu mol/mL) \times V_{MeOH} \ (4 \ mL)$$

Step 2: Express content of a phenolic compound extracted with EDTA (C_{EDTA}) per g of dry soil:

$$C_{EDTA} = A_{Phc} \ (\mu mol)/dry \ weight \ (g)$$

For example, if A_{Phc} is 100 μmol and dry weight is 20 g, then

$$C_{EDTA} = 100 \ \mu mol/20 \ g \quad Hence, \ C_{EDTA} = 5 \ \mu mol/g$$

Step 3: Calculate reversibly bound phenolic acids weakly sorbed on soil particles (Crb), subtracting the amount of phenolic compounds extracted with water (Cw) from that extracted with EDTA (C_{EDTA}):

$$Crb \ (\mu mol/g) = C_{EDTA} \ (\mu mol/g) - Cw \ (\mu mol/g)$$

Observations

Total phenolic compounds can be determined in soil extracts with Folin Ciocalteu reagent before the extraction with diethyl ether or in the methanolic extract as a preliminary measure of phenolic compounds.

Phenolic compounds can also be extracted from the dry/fresh soil sample with methanol in the ratio 5:20 (w/v) and then filtered for further analysis.

Statistical analysis

Differences between samples in concentrations of phenolic compounds are determined using a one-way analysis of variance at 0.01 or 0.05 level. Significant differences between the means of VOCs concentrations are determined using Dunnet T3 test.

Precautions

During the partitioning of aqueous phase with diethyl ether, excessive pressure should be dissipated through the separating funnel tap when it is inverted. Dissipation of pressure should be done in a fume hood.

3.4 Extraction of VOCs

Soil VOCs reside in vapours, dissolved in liquid and sorbed phases of the edaphic environment. VOCs in vapour and liquid phases are readily lost during sampling, whereas they are extracted with difficulty from the sorbed phase. Total soil VOCs are often extracted with methanol at room temperature, heat water-miscible solvents and soxhlet technique. Methanol extraction is an inexpensive method widely used and is described in the following experiment.

Experiment 6: Methanol Extraction

Soil samples collected for VOCs analysis must be handled minimizing losses due to volatilization and biodegradation. Field extraction/preservation of soil samples with methanol overcome these problems. The ratio of methanol to soil is 2.5:1 (v/w). A portion of methanol extract is combined with double distilled water and analyzed by a purge and trap GC or GC-MS system (Ramstad and Nestrick, 1981).

Materials and equipments required

Sample containers: 60 mL wide mouth packer bottles with open-top screw caps and silicone rubbers coated with Teflon septas; methanol; small core sampler (i.e. auger) having a diameter less than the sample container.

Procedure

(i) Fill a sample container with 25 mL of methanol and immediately cap the container tightly. Weigh and register each labelled sample container with 25 mL of methanol.

(ii) Store sample containers at 4°C ± 2°C until use.

(iii) Collect the samples with the core sampler. It must deliver the sample directly into the sample container.

(iv) Tare the weight of the small diameter core sampler.

(v) Once the sampling interval has been selected, trim off the surface soil to expose a fresh soil surface. The loss of volatile organics from the surface soils will occur if the soil is exposed for a short period of time. The removal of the surface soils can be accomplished by scraping the soil surface using a decontaminated spatula. The sampling procedure must begin immediately, once a fresh soil surface has been exposed.

(vi) Collect 12 g of sample (wet weight) with the core sampler.

(vii) Wipe the outside of the core sampler to remove any adherent soil. Then, insert the core sampler into the soil. Depending upon the soil texture, sampling depth and moisture content, the required sample weight can be obtained after one or more insertions into the soil.

(viii) Quickly weigh the sample while contained in the small diameter core sampler. Excess soil sample can be removed from the coring device by extruding a small portion of the core and cleaning away with a spatula.

(ix) Immediately open the sample container and slowly extrude the soil core into the preweighed and prenumbered sample container. Avoid splashing methanol out of the sample container. Do not insert the core sampler into the container's mouth or immerse the sampler device into the methanol.

(x) Use a clean brush or paper towel to remove the particles off the threads. The presence of soil particles prevents the container from being seal-proof, resulting in the loss of methanol which may invalidate the sample.

(xi) Secure the lid of the sample container. Gently swirl the sample to mix and break up soil aggregates until the soil is covered with methanol. **Do not shake.**

(xii) After sample collection, immediately return the containers to an iced cooler in an upright position. Sample containers can be placed in separated ziplock bags to protect other containers in case of leakage during transport. The laboratory sample number or field sample identification number may be pasted/written on the bag.

Calculations

(i) Calculate the weight of each soil sample as under:

Soil sample (g) = [Soil + container + methanol] (g) – [container + methanol] (g)

(ii) To report the sample results on a dry weight basis, collect one duplicate sample **not preserved with methanol** from each sample location for moisture determination (see extraction with neutral EDTA). Tightly seal the container to prevent the loss of soil moisture.

(iii) Weigh a 5 g portion of the wet sample in a tared container. Dry at 105°C in an oven. Allow to cool in a desiccator and weigh again until constant weight.

(iv) Determine percent dry weight by the following formula:

$$\text{Dry weight (\%)} = [\text{SFW} - \text{SDW}]\ (g)/\text{SFW}\ (g)$$

where SFW is soil fresh weight (5 g) and SDW is soil dry weight.

(v) Calculate the sample concentration on a dry weight basis.

Statistical analysis

Data are analyzed as indicated in Statistical analysis of Experiment 5.

Experiment 7: Terpenoids Extraction (Refer Chapter 8, Section 9.1, Experiments 1 and 2)

4. WATER SAMPLING

Throughfall, stemflow and decaying litter leachates can be directly collected from the tree or shrub species growing in the field. Before collection of natural water samples, you should:

(i) Define the sampling area location and register the specimens of plant species of your interest.

(ii) Choose a proper number of trees or shrubs under study with similar diameters and heights. Decide the number of samples to be collected, according to the selected sampling design and statistical analysis.

(iii) Wash collectors with distilled water before each collection period.
(iv) Understand that a collection period of a natural water sample usually comprises 24 to 48 hours.

4.1 Natural Water Samples

4.1.1 Throughfall

It is the rainwater that falls through the plant foliage (Molina *et al.*, 1991). Throughfall is collected using common pluviometers or funnel collectors made of noncontaminating polyethylene plastic (Fig. 6A). These collectors are placed in the canopy of the tree or shrub specimens under study. Control consists of precipitations collected with the same device in open fields close to the trees or shrubs under study. In funnel collectors, an inert screen on the funnel prevents contamination with outside elements. The funnel and the collector bottle are connected with a tube which has a loop to reduce evaporation. A layer of glass wool is inserted in the neck of the funnel to prevent splashing and to retain organic debris.

4.1.2 Stemflow

It is the rainwater running down the bole of trees and shrubs. It can accumulate a significant amount of natural products which are transported to the tree floor and into the soil. Stemflow collars are often used to capture stemflow for chemical analysis (Fig. 6B). The collars are constructed from a polyurethane foam applied around the bole of a tree or shrub. The foam is tightly sealed against the bole creating an effective water barrier. After placing the foam, a trough is dug in the top of the collar and a funnel and a bottle collection apparatus is positioned, under a low point in the trough where the water will drip out. During precipitation, stemflow is intercepted by the collar and drains into the funnel and bottle that is used for analysis.

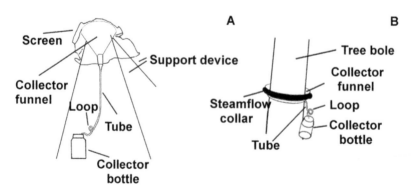

Figure 6 (A) Polyurethane stemflow (Adapted from Likens and Eaton, 1970), and (B) throughfall collectors (Adapted from Hart and Parent, 1974).

4.1.3 *Leachates from Decaying Litter*

The natural litter is made of leaves, bark, twigs and other tree or shrub residues produced by natural abscissions. Several studies require the preparation of litter leachates during litter decaying process. General guidelines for preparation of these leachates are given below:

(i) Divide the sampling area in plots of 1 m². Define the number and location of each plot according to the sampling design and statistical analysis.

(ii) Collect more than 200 g litter per plot in the course of a few days. Litter collected from the same plot should be thoroughly mixed to obtain homogeneous samples.

(iii) Prepare a composite sample mixing the litter samples.

(iv) Put the composite sample in nylon bags of 2 mm mesh. In each bag add 30 g of litter. Prepare several bags (i.e. 30) and place them in groups (i.e. 5 per group) at random points on the sampling site. Place the bags on the soil surface. Remove a bag from each group after 1, 15, 30, 60 days.

(v) Prepare litter leachates saturating the collected material with a known volume of distilled water, allowing it to soak for 24 hours in darkness at 15°C. Filter and use the preparation before the chromatographic process.

Experiment 8: Pluviotron

Principle

This technique simulates the rain leaching from plant materials washing them with water using a recirculating system (Fig. 7). This allows collection of artificial leachate samples.

Materials and equipments required

Steel frame with three layers (A, B, C); two plastic trays (55 × 40 × 15 cm) with needle sized holes (2 cm between holes); plastic tray without holes; pump; metal or plastic tubings; plant material; Whatman # 1 paper.

Procedure

(i) Put plastic trays with holes in layers A and B of the steel frame (Fig. 6). In layer C place the plastic tray without holes.

(ii) Connect the pump to layer C and with the tubing system as shown in Fig. 6.

(iii) Place an amount of chopped plant material in the tray of layer B.

(iv) Fill the tray in layer A with 3 L of tap water. The tap water will drip through the holes like artificial rain and will pass through layer B to layer C.

(v) Return the leachate collected in tray C to tray A and recirculate for 48 hours. Filter the leachate with Whatman # 1 and store in a cold room at 5°C prior to chemical analysis by HPLC.

Observations

Replace plant material every week. The leachate is filtered through a Whatman # 1 filter paper and used for bioassay or chemical analyses. Leachate volumes are large enough to be used for greenhouse pot assays. Make at least three independent leachate collections (N = 3) for each plant species.

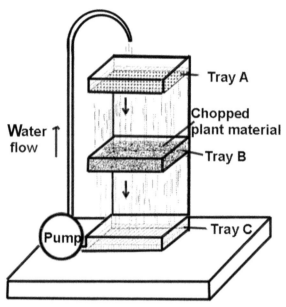

Figure 7 Pluviotron to simulate leaching of plant materials. Water drips through holes at the bottom of tray A to tray B, simulating a raindrip. Tray B contains chopped plant material. Water drips from tray B to no-holed tray C. The pump recirculates water from tray C to tray A (Source: Chou, 1989).

Statistical analysis

Differences between samples in concentrations of organic compounds analyzed by HPLC are detected using a one-way analysis of variance at 0.01 or 0.05 level. Significant differences between means of VOCs concentrations are determined using Dunnet T3 test.

4.2 Root Exudates Trapping Systems

Experiment 9: Agar Trapping System

Principle

Plant seedlings are grown in beakers filled with agar. Root exudates released are extracted with diethyl ether from the agar medium (Huang *et al.*, 2003).

Materials and equipments required

Glass beakers (500 mL), Erlenmeyer flask (1,000 mL); agar; wheat seeds (or other plant species): weigh a known number of seeds. Calculate the number of seeds to be used according to number of beakers prepared and multiply the calculated weight by 1.5; PVC cling film roll; autoclave; white paper sheets; aluminum foil roll; white cotton cord roll (2 mm diameter); cotton plugs; laminar flow; diethyl ether.

Procedure

I. Root Exudates Trapping

(i) Suspend 3 g of agar powder in 1 L of distilled water and heat to boiling until the medium is completely dissolved.

(ii) Transfer warm agar media to 500 mL glass beakers, adding 30 mL to each beaker. Seal the mouth of each beaker with paper. Then, cover the paper seal with aluminum foil. Adjust the cover (aluminum + paper) to the glass walls of the beakers with cotton cord. Add 500 mL of distilled water to an Erlenmeyer flask and close it with a cotton plug.

(iii) Sterilize the beakers and the Erlenmeyer in an autoclave at 1 Atm and 121°C for 20 minutes.

(iv) Go to the laminar flow. Sterilize a weight of wheat seeds with 1% sodium hypochlorite for 15 minutes, considering that 100 mL of hypochlorite solution is needed to sterilize 0.5 g wheat seeds. Finally, rinse with sterilized distilled water.

(v) Open a sterilized beaker and place 24 sterile wheat seeds. Wrap the beaker with PVC film. Repeat the same steps with the next beaker. Beakers without seeds are to be used as controls.

(vi) Place the beakers in a controlled environment (i.e. growth cabinet with a photoperiod of 13/11 hours at 25/13°C) for 6 days. Then, carefully remove wheat seedlings from the agar.

II. Root Exudates Extraction

(i) Uproot wheat seedlings from the agar medium.

(ii) Rinse roots of each beaker twice with 5 mL distilled water and return these washings to their corresponding agar medium.

(iii) Collect agar-water from each beaker, adjust to pH 3 with drops of 0.01 M HCl and stir thoroughly. Extract agar medium three times, each one with 60 mL of diethyl ether. Combine ether layers and evaporate in a rotary evaporator under reduced pressure at 35°C until 2 mL.

(iv) Transfer diethyl ether aliquots into 2 mL minivials. Completely evaporate the diethyl ether in an oven at 30°C. Dissolve in an appropriate solvent and analyze chemical composition.

Observations

Make at least three independent experiments (N = 3). Each treatment should have at least three replications.

Statistical analysis

Data are analyzed as indicated in Statistical analysis of Experiment 8.

Experiment 10: Double-Pot Trapping System

Principle

A plant species is grown in a brown glass container filled with an inert medium and watered with Hoagland's solution (Fig. 8). After proper development of the

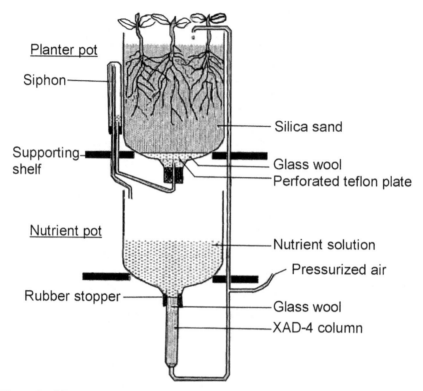

Planter pot

Siphon

Supporting shelf

Nutrient pot

Rubber stopper

Silica sand

Glass wool

Perforated teflon plate

Nutrient solution

Pressurized air

Glass wool

XAD-4 column

Figure 8 Schematic diagram of the double-pot continuous root-exudate trapping system used to collect root exudates (Source: Caswell *et al.*, 1991).

root system, a column packed with XAD-4 resin is attached to the bottom of the bottle. Hoagland's solution is recirculated through the pot and column to trap hydrophobic root exudates. This system is appropriate for collection of root exudates from undisturbed root systems (Tang and Young, 1982).

Materials and equipments required

Brown glass bottles (remove the bottom of 1 gallon glass bottles); culture bed (fill each bottle with an equal weight of clean 60 mesh silica sand; seal the mouth of the bottle with a perforated Teflon plate; place glass wool on the plate); pump; air siphon; Amberlite XAD-4 resin column of 12 mL bed volume (put glass wool at top side of the column; prior to use, clean the resin by treatment with hot water, followed by Soxhlet extraction with methanol, acetone, and diethyl ether, each for 24 hours); device for dry sterilization (i.e. oven at 100°C); Hoagland's solution (Table 1); seeds of a plant species (i.e. wheat): seeds sterilize in 1% sodium hypochlorite solution for 15 minutes; receptor pot (close a bottle with a perforated rubber, attach to the XAD-4 column to the rubber); glass tube for recirculating system.

Table 1 Stock solutions and volumes of each one needed to prepare 1 L of concentrated Hoagland's nutrient solution (Source: Hoagland and Arnon, 1950).

Chemical	Stock solutions		Volumes of stock solutions to
	Weight	Distilled water (mL)	prepare nutrient solution (mL)
$NH_4H_2PO_4$	2.3 g	20	1
KNO_3	10.1 g	100	6
$Ca(NO_3)_2$	11.5 g	70	4
$MgSO_4$	3.8 g	40	2
H_3BO_3	57.2 mg		
$MnCl_2 \cdot 4 H_2O$	36.2 mg	20 mL	
$ZnSO_4 \cdot 7 H_2O$	4.4 mg	all	1
$Cu\,SO_4 \cdot 5 H_2O$	1.6 mg	together	
$H_2MoO_4 \cdot H_2O$	0.4 mg		

Procedure

(i) Wrap the culture beds with aluminum foil. Heat sterilize the culture beds for 2 days at 100°C.

(ii) **Growth of seedlings in Petri dishes:** Place 4 wheat seeds on 10 g of silica sand in each Petri dish and irrigate with 20 mL of 1/10 strength Hoagland's solution. Five days after germination, transfer the seedlings to sterilized culture beds.

(iii) Irrigate daily with 100 mL of 1/10 strength Hoagland's solution. You should also prepare culture beds without wheat seeds as controls. Connect the air siphon to each bottle, as shown in Fig. 8.

(iv) After 30 days you should see a well-developed root system. Attach the XAD-4 column to the receptor pot and the circulating system, as shown in Fig. 8.

(v) Circulate 1/10 strength Hoagland's solution (1 l/h) through the pot and the column. Daily replenish the Hoagland's solution.

(vi) After 3 days, detach the column and wash it with 200 mL of water followed by 100 mL of methanol. Combine the eluates and concentrate the eluates *in vacuo* at 40°C using a rotary evaporator. Extract the aqueous exudates with CH_2Cl_2. First extraction can be made adjusting the concentrated aqueous eluates to pH 3 with 1 N HCl (acidic fraction). Then, add 1 N NaOH to reach pH 7 and extract again (neutral fraction). A basic fraction can be obtained extracting with CH_2Cl_2 after addition of NaOH to the aqueous eluates to reach pH 10. These fractions can be concentrated and used for bioassays and chromatographic procedures. Decide number of XAD-4 columns needed for chemical analysis according to the amount of analytes recovered.

Observations

Make at least three independent collections of root exudates (N = 3) from the plant species assayed.

Statistical analysis

Data are analyzed as indicated in *Statistical analysis* of Experiment 8.

SUGGESTED READINGS

Blum, U., Gerig, T.M., Worsham, A.D., Holappa, L.D. and King, L.D. (1992). Allelopathic activity in wheat conventional and wheat-no-till soils: development of soil extract bioassays. *Journal of Chemical Ecology* **18**: 2191-2221.

Caswell, E.P., Tang, C.S., De Frank, J. and Apt, W.J. (1991). The influence of root exudates of *Chloris gayana* and *Tagetes patula* on *Rotylenchulus reniformis*. *Revue Nématologie* **14**: 581-587.

Chou, C.H. (1989). Allelopathic research of subtropical vegetation in Taiwan IV. Comparative phytotoxic nature of leachate from four subtropical grasses. *Journal of Chemical Ecology* **15**: 2149-2159.

Donath, J. and Boland, W. (1995). Biosynthesis of acyclic homoterpenes – enzyme selectivity and absolute configuration of the nerolidol precursor. *Phytochemistry* **39**: 785–790.

Fujii, Y., Furubayashi, A. and Hiradate, S. (2005). Rhizosphere soil method: A new bioassay to evaluate allelopathy in the field. In: *Proceedings of the Fourth World Congress on Allelopathy: Establishing the Scientific Base* (Eds., J.D.I. Haper, M. An and J.H. Kent). Regional Institute Ltd., Wagga Wagga, Australia.

Hart, G.E. and Parent, D.R. (1974). Chemistry of throughfall under Douglas fir and Rocky Mountain juniper. *American Midland Naturalist* **92**: 191-201.

Hoagland, D.R. and Arnon, D.I. (1950). The water-culture method of growing plants without soil. *California Agricultural Experimental Station Circular* No. 347. USDA-ARS.

Huang, Z., Haig, T., Wu, H., An, M. and Pratley, J. (2003). Correlation between phytotoxicity on annual ryegrass (*Lolium rigidum*) and production dynamics of allelochemicals within root exudates of an allelopathic wheat. *Journal of Chemical Ecology* **29**: 2263-2279.

Jose, S. and Gillespie, A.R. (1998). Allelopathy in black walnut (*Juglans nigra* L.) alley cropping. I. Spatio-temporal variation in soil juglone in a black walnut–corn (*Zea mays* L.) alley cropping system in the midwestern USA. *Plant and Soil* **203**: 191–197.

Likens, G.E. and Eaton, J.E. (1970). A polyurethane stemflow collector for trees and shrubs. *Ecology* **51**: 938-939.

Lohr, S. (1999). *Sampling: Design and Analysis*. Duxbury Press, Pacific Grove, CA.

Moldoveanu, S.C. and Victor, D. (2002). Sample preparation in chromatography. *Journal of Chromatography Library Series* **65**. Elsevier Science Publishing Company, New York, USA.

Molina, A., Reigosa, M.J. and Carballeira, A. (1991). Release of allelochemicals agents from litter, throughfall, and topsoil in plantations of *Eucalyptus globulus* labill in Spain. *Journal of Chemical Ecology* **17**: 147-160.

Ramstad, T. and Nestrick, T.J. (1981). A Procedure for Determining Benzene in Soil by the Purge-and-Trap Technique. *Bulletin of Environmental Contamination and Toxicology* **26**: 440-445.

Tang, C.S. and Young, C.C. (1982). Collection and identification of allelopathic compounds from the undisturbed root system of Bigalta Limpograss (*Hemarthria altissima*). *Plant Physiology* **69**: 155-160.

Tholl, D., Boland, W., Hansel, A., Loreto, R. U. and Schnitzler, J. P. (2006). Techniques for molecular analysis: practical approaches to plant volatile analysis. *The Plant Journal* **45**: 540–560.

Wilt, F. M., Miller, G. C. and Everett, R. L. (1993). Measurement of monoterpene hydrocarbon levels in vapor phase surrounding single leaf pinyon (*Pinnus monophylla* Torr & Frem. Pinaceae) understory litter. *Journal of Chemical Ecology* **19**: 1417-1428.

Plant Sampling and Sample Preparation

**Marta A. Vattuone*, Diego A. Sampietro, José R. Soberón,
Melina A. Sgariglia, and Emma N. Quiroga**

1. INTRODUCTION

Sample collection, transportation, storage, handling and extraction are critical steps in chemical analysis of natural products. Sample extraction is required for the separation of sample constituents to make them suitable for further chemical analysis. This process is necessary when direct analysis of the sample is not possible or to improve the yield of the substance of interest. It is worth considering that a sample can be used in two main parts, the analytes and the matrix. The analytes are the parts of the sample under study, and the matrix is the remaining part of the sample, i.e. the part not required for a particular analysis. Moreover, the sample extraction can be done based on the analytes, on the matrix, or on both, to obtain better analytical results. The study of natural plant products is a difficult task. The main problem is that their nature and the required amount of extract in the plant depends on several factors, which must be controlled as far as possible. Factors such as growth state, plant development and environmental stimulus may change the plant metabolic rate and chemical composition. These problems are present before and after plant

Authors' address: *Cátedra de Fitoquímica. Instituto de Estudios Vegetales *"Dr. A. R. Sampietro"*, Facultad de Bioquímica, Química y Farmacia, Universidad Nacional de Tucumán, España 2903, 4000, San Miguel de Tucumán, Tucumán, Argentina.

*Corresponding author: E-mail: instveg@unt.edu.ar

sample collection. It is worth considering that when cells die (the senescent process), the cellular integrity is lost and enzymes and substrates that are normally separated, come in contact resulting in their consequent interactions. In this situation, the oxidation process increases, phenolics are transformed into quinones and the process of polymerization is favoured. Complex formations occur among the phenolic compound derivatives and also in proteins. Thus, before the application of extraction techniques, it is necessary to stop the metabolic transformations occurring in the plant sample. This is generally done by immersing the plant sample in liquid nitrogen or by boiling in alcohol.

Another important step is the solubilization of components of interest, to facilitate their extraction. In practice, simple and complex analytical problems can appear. Very often the complex problems cannot be solved using a single methodology, therefore, combinations of different extraction methods are required. The best extraction technique depends on the (i) texture, (ii) water content of the plant material and (iii) type of compound to be isolated. The extraction is preceded by preliminary preparation of the plant material: bruising the fresh, wilted, or semidesiccated organs, chopping herbaceous plants, pounding root and rhizomes, or turning wood into chips or shavings. The extraction is done by using a specialized apparatus such as static or moving extractors with floating filter and Soxhlet-type online extractors. The selection of solvent depends on technical and economic parameters viz., selectivity, stability, chemical inertness and handling safety (if possible, non-toxic and non-flammable). Solvent boiling points should not be too high, so that the solvent can be completely eliminated, but not too low, to limit losses and control the cost. The solvents most often used are aliphatic hydrocarbons such as petroleum ether, hexane and also propane or liquid butane. Although benzene is a good solvent, its toxicity severely limits its use. Sometimes halogenated solvents (chlorinated and fluorinated derivatives of methane and ethane) as well as ethanol or methanol are also used. After extraction, the solvent is often evaporated. Low boiling point solvents preclude the degradations or modifications that may be induced by nonvolatile solvents (water at acidic pHs) or solvents of slightly higher boiling point (acetone or alcohols). The main drawbacks of solvent extraction are: (i) lack of selectivity, (ii) toxicity of solvents (which leads to restrictive regulations regarding their use) and (iii) residues in the final product. The polarity of the solvents is another parameter to be considered for sample extraction as well as for sample components fractionation applying different technologies. An elutropic series (a series of solvents with an increasing degree of polarity) is shown in Table 1.

This chapter provides an overview of techniques applied for extraction of different groups of secondary metabolites from plant samples, considering the properties of substance/s to be extracted.

Table 1 Eluotropic series

Solvent	Polarity[a]	Solvent	Polarity	Solvent	Polarity
n-Hexane	0.1	Ethoxybenzene	3.3	2-Picoline	4.9
Isooctane	0.1	Diphenyl ether	3.4	Acetona	5.1
Cyclohexane	0.2	Ethylene chloride	3.5	Metanol	5.1
Carbon disulfide	0.3	n-Butanol	3.9	Nitroethane	5.2
n-Decane	0.4	Isopropanol	3.9	Pyridine	5.3
Carbon tetrachloride	1.6	Tetrahydrofuran	4.0	Methoxyethanol	5.5
Triethylamine	1.9	n-Propanol	4.0	Benzyl alcohol	5.7
Di-n-butyl ether	2.1	Chloroform	4.1	Acetonitrile	5.8
Diisopropyl ether	2.4	ter-Butanol	4.1	Acetic acid	6.0
Toluene	2.4	Dibenzyl ether	4.1	Nitromethane	6.0
p-Xylene	2.5	Ethanol	4.3	Methylformamide	6.0
Benzene	2.7	Ethyl acetate	4.4	Anilina	6.3
Chlorobenzene	2.7	Nitrobenzene	4.4	Dimethylformamide	6.4
Bromobenzene	2.7	Cyclohexanone	4.7	Ethylene glycol	6.9
Diethyl ether	2.8	Ethyl methyl ketone	4.7	Dimethyl sulfoxide	7.2
Iodobenzene	2.8	Dioxane	4.8	Tetrafluoropropanol	8.6
Dichloromethane	3.1	Benzonitrile	4.8	Formamide	9.6
Fluorobenzene	3.2	Acetophenone	4.8	Water	10.2

[a] The polarity of a mixture ($P_{mixture}$) of the miscible solvents A and B is calculated as follows:

$P_{mixture} = fv_A \times P_A + fv_B \times P_B$, where fv_A: volumetric fraction of component A; fv_B: volumetric fraction of component B; P_A: polarity of component A and P_B: polarity of component B. The volumetric fraction is calculated as the percentage of each component in the mixture.

2. NITROGEN CONTAINING COMPOUNDS

Amino acids are constituents of structural and enzymatic proteins. They are precursors of many nitrogenous plant compounds: proteins, amines, short-chain acids, glucosinolates, betalains, cyanogenic glycosides, porphyrins, purines, pyrimidines, cytoquinins, alkaloids and after deamination, all phenylpropanoid compounds. The structure and chemical properties of amino acids and peptides, their biosynthetic origin, properties, and functions of proteins are explained in biochemistry textbooks. This chapter is restricted to amino acid derivatives, and among them, to those that are not protein constituents and to compounds derived from them, namely glucosinolates, cyanogenic glycosides, non-protein amino acids, alkaloids, amines and cytoquinins. The characteristic of nitrogen compounds is that they are usually basic, hence, they form salts with mineral acids. Usually, they can be extracted from plant tissues using weak acidic solvents and can be selectively precipitated from such extracts by addition of ammonia. A general view of nitrogen metabolism in plants can be obtained from Dey and Harborne (1966), Beevers (1976) and Miflin (1981).

2.1   Non-protein Amino Acids

Principle
Non-protein amino acids extraction is based on their solubility in 80% ethanol.

2.1.1   Extraction with Perchloric Acid
Homogenize dried powdered seeds, fresh seedlings or germinating seeds in a chilled mortar and pestle in an appropriate volume of 0.2 M $HClO_4$ (0.4-0.8 g fresh weight tissue/mL of $HClO_4$ solution or 120-250 mg seed powder/5 mL of $HClO_4$ solution). After extraction for 1 hour at 0-4 °C, centrifuge the preparation for 10 minutes at 15,000 × g. Collect the supernatant and derivatize immediately in the presence of sodium bicarbonate or store at 0-4 °C (Yan *et al.*, 2005)

2.1.2   Extraction with Aqueous Ethanol

Procedure I
Soak dried seeds in aqueous ethanol (3:7, v/v), shake or sonicate, leave to stand overnight, centrifuge and filter (Yan *et al.*, 2005).

Procedure II
Shake finely ground seeds (1 g) with 75% ethanol (25 mL) overnight. Centrifuge and apply the supernatant to a small column (12 × 0.8 cm) ZeoKarb 225 (H^+ form in 75% ethanol) to retain organic acids. Wash with aqueous ethanol and elute the column with aqueous ethanol containing 2 M NH_4OH (25 mM). Concentrate this eluate and analyze (Dunnill and Fowden, 1965).

2.2   Polyamines

Polyamines are widely found in diverse biological systems and have been associated with many cell processes, including cell proliferation and differentiation, synthesis of nucleic acids and proteins, membrane stability and signal transduction. In higher plants, polyamines are implicated in molecular signalling events in plant pathogen interactions and are also apparently involved in the plant response to microbial symbionts important in plant nutrition. The content of polyamines sharply increases under abiotic stresses and the major shifts in nitrogen and polyamines metabolism can occur. They originate from the decarboxylation of amino acids. Nevertheless, their biosynthesis by transamination of the corresponding aldehyde has been demonstrated as the main pathway for synthesis of aliphatic amines in more than 50 plants. The most widespread plant polyamines are divided into three main groups: aliphatic monoamines, aliphatic polyamines and aromatic amines. Aliphatic amines are volatile and can be extracted by the techniques described above. Diamines and polyamines are less volatile but frequently have a foul smell. Polyamines are extracted by acidification. Extraction with perchloric acid releases soluble, nonconjugated ("free") polyamines. Acid heat hydrolysis of the perchloric acid-soluble fractions in HCl yield soluble, conjugated polyamines. While, acid-heat hydrolysis of perchloric acid-insoluble pelleted cellular material yield insoluble, conjugated ("bound") polyamines.

Procedure

Homogenize 100 mg of dry plant tissue with a Polytron homogenizer in 5 mL of 5-10% $HClO_4$ (v/v) in a 15 mL Corex glass centrifuge tube. Keep the homogenate in an ice bath for 1 hour before centrifugation at 20,000 × g for 20 minutes at 4°C. Separate the supernatant containing free polyamines. Incubate an aliquot of the supernatant with 6 N HCl at 100°C for 16 hours to hydrolyze the soluble conjugated polyamines.

Wash the extraction pellet twice with 10 mL of 5% $HClO_4$ and once with 10 mL of acetone. Incubate the washed pellet in closed vials containing 2 mL of 6 N HCl at 100°C for 16 hours and centrifuge at 20,000 × g for 20 minutes. Mix 1 mL of each of the supernatants containing free polyamines with 1.5 mL of HPLC grade water and benzoylate for HPLC analysis.

2.3 Cyanogenic Glycosides

Cyanogenic glycosides (glycosides of 2-hydroxynitriles) are readily hydrolyzed, at near neutral pH, by specific β-glucosidases which release a monosaccharide and hydrogen cyanide. These compounds are very fragile, hence their extraction and purification is delicate. Cyanogenic glycosides are formed from the structurally related amino acids via aldoximes and nitriles by a five-step pathway (Conn and Butler, 1969). Cyanogenic glycosides appear in young seedlings and probably have a protective function in the plant kingdom, avoiding plant damage by phytophagous organisms.

Procedure

Cyanogenic glycosides can be isolated by general procedures used for other glycosides with previous de-activation of the glycosidases present in the plant tissue (Kofod and Eyjolfsson, 1969; Maher and Hughes, 1971).

Removal of cyanogenic glycosides of flaxseed meal: Extract the meal (100 g) with 80% ethanol (1:10, w/v) at 7°C for 1 hour. Dry the ethanol extract and dissolve the residue in 10 ml of methanol followed by the addition of 20 mL chloroform; recover the precipitates by centrifugation. Evaporate the supernatant to dryness using a rotary evaporator. Redissolve the dried residue in 4 mL of 15% (v/v) HPLC grade methanol in water. Filter the solution through a 0.45 μm nylon filter before continuing with the isolation (Wanasundara *et al.*, 1993).

2.4 Alkaloids

Sample extraction and processing for analysis of alkaloids is described in Chapter 7, Experiments 1 to 6.

2.5 Cytokinins, Purines and Pyrimidines

Plant purines and pyrimidines can be considered in four groups according to structural and functional characteristics.

(i) **Group I:** These are purines and pyrimidines that are common to all living organisms and are the basis of nucleic acids (DNA and RNA), certain nucleotides as adenosine triphosphate (ATP) and uridine diphosphate glucose that play important functions in the plant metabolism. These subjects are out of the scope of this book and can be referred to in biochemistry books.

(ii) **Group II:** These are bases closely related to the nucleic acid bases in structure and characteristics that show unusual structure and are present in plants.

(iii) **Group III:** These contain methylated purines (theobromine and caffeine are most common).

(iv) **Group IV:** These are substituted purines in C6-position, known as cytokinins.

Cytokinins are a class of plant hormones, which in cooperation with auxin, play a unique role in the control of developmental processes in plants such as cell division and differentiation, formation and growth of roots and shoots, apical dominance and senescence. Natural cytokinins are 6-N-substituted purine derivatives. Those which occur in plants as free bases are supposed to be the biologically active compounds. Glycosidic conjugates of cytokinins are transport, storage or inactivated forms of cytokinins; while cytokinin riboside phosphates mainly represent the primary products of cytokinin biosynthesis. More than 40 natural cytokinins are present in plant tissues.

Procedure

Three different extraction solvents: (a) 80% (v/v) MeOH, (b) Bieleski's MCF-7 solvent (MeOH–CHCl$_3$–HCO$_2$H–H$_2$O 12:5:1:2, v/v/v) and (c) modified Bieleski's solvent (MeOH–HCO$_2$H–H$_2$O; 15/1/4, v/v/v) can be used. Homogenize nitrogen frozen plant leaf material (1 g) with a pestle in a ceramic mortar, in liquid nitrogen. Distribute the powder in 50 mL polypropylene centrifuge tubes. Add aliquots of 10 mL of cold extracting solvents together with a mixture of deuterium labelled cytokinin standards, 50 pmol each, in a total volume of 50 mL of 50% MeOH (v/v). After overnight extraction at 20 °C, remove the solids by centrifugation (13,000 × g, 20 min, 4°C) and re-extract with 5 mL of the corresponding extraction solvent by vortexing for 30 seconds, followed by standing at 20°C for 1 hour. Pass the pooled extracts through Sep-Pak Plus C18 to remove pigments and lipids, and evaporate under vacuum at 40°C near to dryness prior purification (Hoyerová *et al.*, 2006).

3. CARBOHYDRATES AND RELATED COMPOUNDS

Carbohydrates are the most abundant class of organic compounds found in living organisms. They originate as products of photosynthesis, an endothermic reductive condensation of carbon dioxide requiring light energy and the pigment chlorophyll. Carbohydrates are a major source of metabolic energy, both for plants and for animals that depend on plants for food. Apart from the simple sugars, they provide a means of storing energy (starch) that meets this vital nutritional role and transport of energy (as sucrose). Carbohydrates also serve as a structural material (hemicellulose, cellulose), as a component of the energy transport compound ATP, recognition sites on cell surfaces, one of three essential components of DNA and RNA and plant glycosides. They also play a number of ecological roles, in plant-animal interactions (flower nectar is mainly sugars), in protection from wounds and infections, and in the detoxification of foreign substances. The major simple sugars present in plants are the monosaccharides glucose and fructose and the disaccharide sucrose (Loewus and Tanner, 1982; Dey, 1990). There are traces of galactose, rhamnose, xylose and sugar phosphates involved in metabolism. Sugar phosphates are very easily hydrolyzed during the manipulation, thus special procedures are required to detect them. The bulk of carbohydrates present in plants are in bound form, mainly as oligo- and polysaccharides, or bound to different aglycones, as plant glycosides. Two hexoses (glucose and galactose), the pentoses (arabinose and xylose) and one methylpentose (rhamnose), are commonly found in glycosides and polysaccharides. Other monosaccharides found in minor quantities are glucuronic and galacturonic acids, and mannose as a component of polysaccharides. Ribose and deoxyribose are present in plants as components of RNA and DNA. Fructose, a keto sugar, is present as a component of oligosaccharides, mainly sucrose, and of fructans, but is rarely found as a component of glycosides. Monosaccharides can be into two isomeric forms in equilibrium but only one form is normally encountered (glucose is usually in the β-D-form and rhamnose as the α-L-isomer). These compounds are naturally present as a cyclic form, pyrano-(six-membered) and furano-(five membered ring), although one form is predominant. Glucose takes the pyrano-configuration, while fructose is usually in the furano form.

Oligosaccharides are saccharide polymers containing a small number (typically from 2 to 10) of monosaccharides. Monosaccharides are connected by either an α-or β-glycosidic linkage. The type of the glycosidic linkage allows a distinction between a non-reducing disaccharide (sucrose) and a reducing disaccharide (maltose). Even when there are only two subunits, these can be joined together by ether links in a number of different ways, i.e. through different hydroxyls and by α and β links. They are found throughout nature in both the free and bound form. Some, such as the raffinose series, occur as storage or transport carbohydrates in plants. Others, such as maltodextrins or cellodextrins, result from the microbial breakdown of larger polysaccharides such as starch or cellulose. They can be also found either O- or N-linked to compatible amino acid side chains in proteins or to lipid moieties. The number of free oligosaccharides that accumulate in plants are relatively few. Sucrose (2-α-glucosylfructose) is the only one which is of universal occurrence. Many oligosaccharides occur in nature as glycosides, such as flavonoid pigments, saponins and steroidal alkaloids. Oligosaccharides resemble monosaccharides in their properties and methods of plant extraction and separation.

Although sugar alcohols are not carbohydrates in the strict sense, the naturally occurring sugar alcohols are so closely related to the carbohydrates that it is necessary to discuss them. The sugar alcohols most widely distributed in nature are sorbitol (identical with D-glucitol), D-mannitol, and D-galactitol (dulcitol). They correspond to the reduction of glucose, mannose and galactose. Reduction of the anomeric carbon atom alters the possibilities for isomerism and the same sugar alcohol may be formed from several reducing sugars, as happens with sorbitol that is produced from glucose and fructose. Dulcitol and sorbitol are the most widely distributed sugar alcohols in the plant kingdom. They are used for energy storage, but mannitol may also be involved in the translocation in phloem in higher plants. Other possible functions include osmoregulation and protection of plants from desiccation and frost damage.

Related to the sugar alcohols are the polyhydroxycyclohexanes (cyclitols), of which the hexahydroxy derivatives, known as inositols, are of most interest. Nine stereoisomeric forms are possible, two of which are optically active and seven of them are *meso* forms. One of the *meso* forms is widely distributed in nature, and is designated as inositol or *meso* inositol. It is found in microorganisms, plants and animals. It occurs both free and combined in many organs. Some cyclitols have important physiological functions in living organisms.

The polysaccharides are high molecular weight substances (30,000 to 14,000,000). They are insoluble in liquids and are readily altered by the acids and alkalis required to catalyze their conversion into soluble derivatives. Hence, their extraction from tissue and purification are extremely difficult. Moreover, even the purified products are not molecularly homogeneous, since the same substance may consist of polymers of varying molecular weight. Nevertheless the polysaccharides are considerably simpler than proteins because frequently only a single type of building unit, for example glucose, is present. Compounds containing a single type of building unit are known as homopolysaccharides. Some polysaccharides are derived from several different types of building units and are known as heteropolysaccharides. Polysaccharides, viz., starch, glycogen, or inulin, are reserve foodstuffs for plants, animals and others, because cellulose of plants and chitin of crustaceans have a structural function; for others, such as gums and mucilages, the function is not well-known.

The main classes of polysaccharides found in higher plants, algae and fungi are enumerated in Table 2 (Harborne, 1998). Most plant polysaccharides, unlike starch and cellulose, are heteropolysaccharides, having more than one type of sugar unit. Traditionally, plant polysaccharides fall into two groups according to whether they are easily soluble in aqueous solutions or not. Those that are soluble include starch, inulin, pectin and various gums and mucilages. The gums which are exuded by plants, sometimes in response to injury or infection, are almost pure polysaccharides. It was suggested that they have a protective function in the plant. The less soluble polysaccharides usually comprise the structural cell wall material and occur in close association with lignin. To this fraction belong cellulose and various hemicelluloses. Furthermore, the hemicelluloses have a variety of sugar components and fall into three main groups: the xylans, glucomannans and arabinogalactans. They are structurally complex and other monosaccharides can be involved in their structure. Types

Table 2 Main classes of polysaccharides in higher plants, algae and fungi (Source: Harborne, 1998)

Class name	Sugar unit(s)	Linkage	Distribution
Higher plants			
Cellulose	Glucose	$\beta1 \rightarrow 4$	Universal as cell wall material
Starch-amylopectin	Glucose	$\alpha1 \rightarrow 4, \alpha1 \rightarrow 6$	Universal as storage
Starch-amylose	Glucose	$\alpha1 \rightarrow 4$	material
Fructan	Fructose (some glucose)	$\beta2 \rightarrow 1$	In artichoke chicory, etc.
Xylan	Xylose (some arabinose and uronic acid)	$\beta1 \rightarrow 4$	Widespread, e.g. in grasses
Glucomannan	Glucose, mannose	$\beta1 \rightarrow 4$	Widespread, but
Arabinogalactan	Arabinose, galactose	$1 \rightarrow 3, 1 \rightarrow 6$	specific in coniferous wood
Pectin	Galacturonic acid (some others)	$\alpha1 \rightarrow 4$	Widespread
Galactomannan	Mannose, galactose, arabinose, rhamnose	$\beta1 \rightarrow 4, \alpha1 \rightarrow 6$	Seed mucilages
Gum	Arabinose, rhamnose, galactose, glucuronic acid	Highly branched species	*Acacia* and *Prunus*
Algae (seaweeds)			
Laminaran	Glucose	$\beta1 \rightarrow 3$	
Polysaccharide sulfate	Fucose (and others)	–	Phaeophyceae (brown algae)
Alginic acid	Mannuronic and guluronic acid	–	
Amylopectin	Glucose	$\alpha1 \rightarrow 4, \alpha1 \rightarrow 6$	
Galactan	Galactose	$1 \rightarrow 3, 1 \rightarrow 4$	Rhodophyceae
Starch	Glucose	$\alpha1 \rightarrow 4, \alpha1 \rightarrow 6$	
Polysaccharide sulfate	Rhamnose, xylose, glucuronic acid	–	Chlorophyceae (green algae)

of storage polysaccharides as well as structural ones do vary in different groups of plants.

3.1 Simple Sugars and Oligosaccharides

Experiment 1: Extraction of Simple Sugars and Oligosaccharides

Simple sugars are very soluble in water. This characteristic is used for their extraction from fresh plant samples with polar solvents (water, ethanol and methanol).

Materials and equipments required

Fresh plant tissue (leaves or other plant part); distilled water or 96% ethanol; vacuum evaporator; Whatman # 1 filter paper or centrifuge; petroleum ether.

Procedure

(a) **Simple sugars:** Boil finely cut fresh leaves (or other plant part) in water for 5 minutes (96% ethanol or methanol can be also used). Filter or centrifuge. Concentrate the extract in the vacuum evaporator until ethanol is completely removed. The obtained aqueous extract can be analyzed directly by chromatography.

(b) **From polysaccharides or other sugar derivatives:** Analysis of the individual sugars of a polysaccharide or a plant glycoside can be made after hydrolysis. Hydrolysis is performed with 1 M H_2SO_4 or 2 M HCl:MeOH (1:1) in a boiling water bath for 30 minutes. Sugar analysis on the hydrolysates must be performed prior to removal of H_2SO_4. It may be removed as $BaSO_4$ after addition of equimolar quantities of $BaCO_3$ solution followed by centrifugation. In plant glycoside, the aglycone moiety can be removed by extraction with ethyl ether or ethyl acetate. Sometimes ion exchange resins (i.e. Amberlite mixed bed) are used to remove the acid under mild conditions. Then, the sample is ready for further analysis.

(c) **Components collected from plant secretions (nectar and honey):** Collect nectar from flowers with glass capillary tubes. Make different dilutions of nectar and honey by adding distilled water and analyze their composition by paper chromatography (PC).

3.2 Sugar Alcohols

Experiment 2: Extraction of Sugar Alcohols

Principle

The same as for simple sugars.

Materials and equipments required

The same as for simple sugars.

Procedure

Extract fresh plant tissue with 80% ethanol under reflux for 2 hours using two or three changes of solvent. Evaporate the solvent under reduced pressure and take up the residue with a small water volume. Filter or centrifuge to remove any solids. De-proteinize with 10% trichloroacetic acid (TCA) if necessary and deionize the extract with a cation-anion exchange resin. Continue the analysis by PC.

3.3 Cyclitols

Experiment 3: Extraction of Cyclitols

Principle

The same as for simple sugars.

Materials and equipments required

The same as for simple sugars.

Procedure

Homogenize 2 g of dried material or 20 g of fresh material in a blender with 25 mL benzene and centrifuge. Homogenize the residue with 70 mL of boiling water for 10 minutes, shake with 150 mg charcoal and 150 mg of celite and centrifuge. Separate the supernatant and treat with a mixed bed (cationic and anionic) exchange resin and centrifuge to clarify the preparation. Concentrate the supernatant in vacuo to 2-3 mL and dry. Dissolve the syrup in 200 µL of water by warming in a water bath. Analyze the presence of cyclitols using two-dimensional TLC on microcrystalline cellulose, in acetone-water (17:3) followed by *n*-butanol-pyridine-water (10:3:3). If the amount of glucose present exceeds that of cyclitols, oxidize the glucose with catalase and glucose axidase to D-gluconic acid, which solidifies and can be eliminated by centrifugation.

3.4 Polysaccharides

Experiment 4: Extraction of Plant Polysaccharides

The extraction of polysaccharides from plant tissues requires numerous and repeated stages, mainly based on a differential solubility and molecular weights of the sample components. The application of this methodology allows the separation of water-soluble polysaccharides that are separately removed. The residue, after several extractions, will be constituted mainly by more or less pure cellulose.

Materials and equipments required

Separation funnel; 96% ethanol; acetone; ethyl ether; Erlenmeyer flasks of different capacity; beakers.

Procedure

3.4.1 General Procedure

Extract the plant material with 96% ethanol. If the tissue is rich in lipids (e.g. seeds), it may be necessary to remove them with acetone and with ethyl ether, discarding the organic fractions. Extract neutral water-soluble polysaccharides with 1% NaCl solution or with boiling water. Recover these polysaccharides from the solution by pouring the extracts into several volumes of 96% ethanol, where they precipitate. Extract the pectins from the residue with 0.5% ammonium oxalate and subsequently precipitate them from the solution by acidification and pouring into 96% alcohol. Remove lignin by extraction with 1% NaCl at 70°C for 1 hour. For tissues rich in lignin, the extraction procedure must be repeated several times. Alternatively, delignification can be achieved by extraction with chloramines-T (sodium *p*-toluene-sulfonchloramide) and ethanolamine. These extracts are discarded and the residue is washed and dried.

The hemicelluloses are removed from the dried residue by keeping with 7-12% NaOH at room temperature for 24 hours. This last procedure must be repeated to obtain thorough separation of the hemicelluloses. Then, the hemicelluloses are recovered by acidification of the alkaline extracts with acetic acid and precipitation with ethanol. The remaining material is washed and dried, and this is the cellulose fraction.

Gums and mucilages can be purified by dissolving in water and precipitating with ethanol. Then, they are collected and dried.

3.4.2 Extraction of Pectin from Apples

Extract 400 g of apple pulp and boil in 96% ethanol for 10 minutes and discard the ethanol. Homogenize the remaining tissue and extract it with 300 mL of boiling water. Neutralize the extract with 1 M NH_4OH to pH 6.5 and evaporate in vacuum to 100-150 mL. Add ethanol to 80% saturation. The formation of precipitate indicates the presence of pectic acid. Collect it and dry. Further purification is done by redissolving in hot water, filtering, precipitating with ethanol in the same experimental conditions and drying.

3.4.3 Extraction of Starch from Potato Tubers

Homogenize 1 Kg of peeled potatoes with 750 mL of 1% NaCl and filter through cheesecloth. Re-extract the residue with 150 mL of 1% NaCl, filter and add to the first filtrate. Leave the preparation to stand, so that starch granules which pass through the cheesecloth will precipitate. Then, discard the supernatant. Wash the wet starch two times with 1% NaCl, once with 0.01 M NaOH and once with water. Then drain water and dry.

3.4.4 Xylan from Corncob

Extract raw corncob (*ca* 5 g) by soaking in 100 mL of 1% H_2SO_4 at 60°C for 12 hours. Then, filter and wash the solid residue with distilled water at pH 6.0. Autoclave for 30 minutes. After cooling, collect the steamed corncob. Add water to the corncob, in a ratio 1:12 and mesh with a waring blender mixer (3,000 rpm, 20 minutes). For chemical analysis, filter the slurry through a gauze. Analyze the filtrate, named as extract hereinafter, for its content of reducing sugars (xylose and xylooligosaccharides), total soluble sugars and furfural. Since cellulose did not degrade during the aqueous extraction of xylan, total soluble (TS) sugar in the filtrate was the soluble xylan and xylooligosaccharides extracted from the corncob.

3.4.5 Xylan from Straw

Principle

Consists of xylan extraction from the previous delignification.

Procedure

Extract 30 g of the powdered dry straw by boiling in 1 L of 3.5% sodium sulfite. Delignify the residue by repeating extractions with 5% sodium hypochlorite at room temperature and with 6% sodium hypochlorite containing 1% H_2SO_4 in a fume cupboard. Filter, wash and dry the residue and then extract with 500 mL of 6% NaOH for 45 minutes at 100°C and filter. Treat the filtrate with 200 mL of Fehling's solution, collect the precipitate of the complex xylan-copper and wash with 80% ethanol. Decompose the complex by suspending it in 96% ethanol and passing in hydrogen chloride gas at 0°C. Collect the residue by centrifugation, wash it by centrifugation and dry.

4. PHENOLIC COMPOUNDS

Phenolic compounds occur ubiquitously in plants (Pridham, 1960; Harborne, 1982; Harborne and Turner, 1984). They are synthesized in response to stress conditions such as infection, wounding and UV radiation, etc. (Nicholson and Hammerschmidt, 1992; Beckman, 1999). These compounds are a very diversified group of phytochemicals derived from phenylalanine and tyrosine (Shahidi, 2000; 2002; Shahidi and Naczk, 2004). Some of them are shown in Fig. 1.

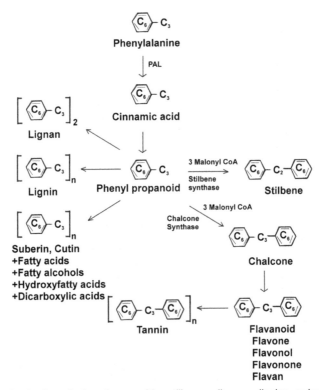

Figure 1   Production of phenylpropanoids, stilbenes, lignans, lignins, suberins, cutins, flavonoids and tannins from phenylalanine (Source: Shahidi and Naczk, 2004).

Plant phenolics include simple phenols, phenolic acids (both benzoic and cinnamic acid derivatives), phenylpropanoids, anthocyanins, coumarins, flavonoids, xanthones, stilbenes and minor flavonoids, hydrolyzable and condensed tannins, lignans, lignins and quinone pigments. In plants, phenolics have several functions such as allelochemicals, phytoalexins, antifeedants, attractants for pollinators, contributors to the plant pigmentation, antioxidants and protective agents against UV light (Shahidi and Naczk, 2004). Phenolics are not uniformly distributed in plant tissues and cellular and subcellular levels. Insoluble phenolics are the components of cell walls, while soluble phenolics are compartmentalized within the plant vacuoles (Pridham, 1960; Bengoechea *et al.*, 1997; Beckman, 1999). Phenolic compounds must first be extracted from their source prior to any analysis, but unfortunately there are no standard procedures for their extraction.

4.1 Extraction Procedures

Extraction of phenolic compounds from plant materials is influenced by their chemical nature, the extraction methods employed, sample particle size, storage time and conditions, as well as presence of interferences. There is a high variability in the chemical nature of plant phenolics. It goes from simple to highly polymerized substances, that is from phenolic acids and phenyl propanoids to highly polymerized substances like tannins. They may also exist as complexes with carbohydrates, proteins and other plant components. Consequently, plant phenolic extracts are always a mixture of different classes of phenolics that are soluble in the solvent system used. Very often, additional steps may be required for the removal of unwanted phenolics and non-phenolic substances such as waxes, fats, terpenes and chlorophylls. Extraction techniques based on acidity are commonly used to remove unwanted phenolics and non-phenolic substances (Robbins, 2003). Solubility of phenolic compounds are governed by the type of solvent (polarity) used, degree of polymerization of phenolics with other plant constituents and formation of insoluble complexes. To date, there is no a satisfactory procedure suitable for extraction of all phenolics or a specific class of phenolic substances in plant materials. Methanol, ethanol, acetone, water, ethyl acetate, and to a lesser extent, propanol, dimethylformamide and their combinations are frequently used for the extraction of phenolics (Antolovich *et al.*, 2000). Extraction periods are another important factor that influence the extraction yield. Usually the contact between solid samples and the extracting solvent is variable, generally between 1 minute and 24 hours. Longer extraction times increase the chance of oxidation of phenolics unless reducing agents are added to the solvent system (Khanna *et al.*, 1968). Other authors found that a two-stage extraction with 70% (v/v) acetone, 1 minute each, using a Polytron homogenizer at 10,000 rpm, was sufficient for the extraction of tannins from commercial canola meals (Naczk and Shahidi, 1991; Naczk *et al.*, 1992). Further extractions only marginally enhanced the yield of extraction of other phenolic compounds. However, it was suggested that the optimum extraction time required for dry bean phenolics was 50-60 minutes. The recovery of polyphenols from plant products is also dependent on the ratio of sample to solvent. In general, this ratio is 1:5 to 1:10 (w/v).

There is no uniform or completely satisfactory procedure suitable for extraction of all phenolics or a specific class of phenolic substances in plant materials. The following paragraphs are an enumeration of the more frequently techniques used for phenolics extraction from diverse sources.

4.2 Extraction of Phenols and Phenolic Acids

Free phenols are relatively rare in plants; hydroquinones are probably the most widely distributed. Acid hydrolysis of plant tissue releases a number of ether-soluble phenolic acids, some of which are widely distributed in plants. These acids can be associated with lignin as ester groups or present in the alcohol-insoluble fraction of plants. They also can be present in the alcohol-soluble fraction as simple glycosides.

Procedure

Extract oil seeds with a mixture of methanol-acetone-water (7:7:6, v/v/v) at room temperature (Krygier *et al.*, 1982). Filtrate or centrifuge the preparation, acidify the supernatant with 2 N HCl and extract free phenolics with diethyl ether. Treat the aqueous fraction with 2 M NaOH under nitrogen for 4 hours to release esterified phenolic acids. The hydrolysate is acidified to pH 2 and the released phenolic acids are extracted with diethyl ether. Extract the residue with a mixture of methanol-acetone-water and then with 4 M NaOH under nitrogen to liberate insoluble-bound phenolic acids. Recently, it was reported that the addition of 1% ascorbic acid and 10 mM ethylendiaminetetraacetic acid (EDTA) prevent the degradation of the phenolic acids during alkaline hydrolysis (Nardini *et al.*, 2002).

4.3 Extraction of Phenolic Acids and Lignins

Phenolic acids extraction is based on their solubility in ether after a mild hydrolysis (0.1 N HCl).

Procedure

Immerse 10 g of sliced fresh leaf samples in 40 mL of 2 M HCl and heat for 30 minutes in a boiling water bath. Cool, extract with ether and separate the ether phase. Concentrate it to dryness. Dissolve in a small volume of 96% ethanol and analyze.

4.4 Extraction of Phenylpropanoids

Phenylpropanoids are biosynthetically derived from the amino acid phenylalanine. They are naturally occurring phenolic compounds, whose general formula is C_6–C_3. Hydroxycinnamic acids are the most widespread. They provide the building blocks of lignin and are related to growth regulation and disease resistance. Hydroxycoumarins, phenylpropenes and lignans also belong to this group of phenolics.

4.4.1 Hydroxycinnamic Acids and Hydroxycoumarins

It is based on solubility of phenylpropanoids in non-polar solvents after acid or alkaline hydrolysis of a plant extract.

Procedure

Subject fresh plant material to an acid or alkaline hydrolysis (as described earlier) for extraction of phenolic acids and lignins. Centrifuge or filter and extract the filtrate with ethyl ether or ethyl acetate. Wash the extract with water, dry with anhydrous Na_2SO_4 and evaporate the solvent under reduced pressure. Analyze the composition of the preparation with TLC and HPLC.

4.4.2 Phenylpropenes

The solubility of phenylpropenes in ethyl ether allows their detection together with essential oils

Procedure

Grind 50 mg of young plant leaves in liquid nitrogen with a mortar and pestle. Soak the powder in 2 mL of methyl *ter*-butyl ether (MTBE) containing an internal standard (0.02 mg of toluene) and extract for 2 hours at room temperature in 5-mL glass vials with tightly sealed rubber septa caps. Remove the MTBE upper layer, which include the volatile oil and place into another vial and concentrate to 200 µL under gentle N_2 gas flow. Analyze by GC-MS (Harborne, 1969).

4.4.3 Furanocoumarins

Furanocoumarins are generally lipid-soluble and occasionally can be found as glycosides. They can be isolated during extraction of dry plant material with ethyl ether without previous acid hydrolysis.

Procedure

Extract 330 g of dried and powered plant material with 90% ethanol. Combine the extracts and reduce to aqueous phase. Partition this phase successively between *n*-heptane, ethyl acetate and *n*-butanol, and evaporate the extracts to dryness. The material extracted with *n*-heptane contains the furanocoumarins.

4.4.4 Lignans, Neolignans, Condensation Products and Norlignans

Lignans and related compounds are successively extracted from plant material with a polar and a non-polar solvent. These groups of compounds are formed by condensation of phenylpropane units. They occur either in free form or glycosidically linked to a wide variety of different carbohydrates.

Procedure I

Extract ground, pulverized or freeze-dried plant material with a non-polar organic solvent (hexane, pentane, petroleum ether, or dichloromethane). The hydrophilic compounds including lignans, can then subsequently be extracted from the residue with a polar solvent such as acetone, methanol, or ethanol. The addition of 5-10% of water to the solvent will facilitate solvent penetration and favours the extraction of more polar compounds such as lignan glycosides. The process can be accelerated using elevated temperatures and pressures for short time periods of extraction.

Furthermore, lipids and terpenoids can be removed easily from oils by solid phase extraction (SPE) on C_{18} cartridges.

Procedure II
Make a direct treatment of the plant source with a hot polar solvent, usually methanol or ethanol. A pre-treatment of the sample with water can be also made. Then, the methanolic extract can be directly analyzed or concentrated and diluted with water. Methanol-water solution is further fractionated with hexane to remove non-polar compounds including lipids, terpenes and chlorophylls and subsequently extracted with chloroform, dichloromethane or ethyl acetate to obtain a ligan enriched fraction. The methodology using 95% ethanol proved better for extracting the lignans from plant leaves.

5. FLAVONOID EXTRACTION

Flavonoids are virtually universal plant pigments. When they are not directly visible, they contribute to colour by acting as co-pigments. All flavonoids have a common biosynthetic origin and, therefore, possess the same basic structural elements, namely the 2-phenylchromane skeleton (3-phenylchromane skeleton for the isoflavonoids). The glycosidic form of flavonoids are water-soluble, accumulate in vacuoles and depending on the species, are either concentrated in the epidermis of the leaves or spread in both the epidermis and the mesophyll. In flowers, they are concentrated in epidermal cells. Whenever flavonoids are present in the leaf cuticle, they are always free aglycones, made even more lipophilic by the partial or total methylation of their hydroxyl groups (Harborne, 1967; Harborne *et al.*, 1975; Harborne and Mabry, 1982; Markham, 1982; Harborne, 1988; Harborne, 1994).

As a general rule, glycosides are water-soluble and soluble in alcohols, however, only a fair number are sparingly soluble (rutin, hesperidin). Aglycones are, for the most part, soluble in apolar organic solvents. When they have at least one free phenolic group, they dissolve in alkaline solutions.

5.1 Total Extraction of Flavonoids without Previous Tissue Hydrolysis

Procedure
Extract 4 g of fresh plant material or 0.5-1 g of dry material with 50 mL of 80% methanol and then re-extract the residue with equal volume of 50% methanol to ensure the extraction of glycosilated flavonoids. Bring together both extracts and evaporate the solvent to dryness under vacuum at 50°C. Redissolve the residue in a small volume of 100% methanol and analyze. If the extract contains higher concentration of flavonoids, make successive extractions of the residue with *n*-hexane, chloroform, ethyl acetate and water. Evaporate the solvents and analyze the different extracts.

5.2 Total Extraction of Flavonoids with Previous Tissue Hydrolysis

Procedure

Treat 1-2 g of fresh or dry tissue with 2 N HCl for 30-40 minutes at 100°C. Cool the extract, filter and extract the filtrate with ethyl acetate. If the solution is coloured, due to the presence of anthocyanins in the original tissue or to the formation of anthocyanidins, heat the aqueous extract to remove traces of ethyl acetate and re-extract the solution with amyl alcohol. Concentrate both extracts to dryness and analyze their composition.

5.3 Lipophilic Flavonoids

Procedure

Rinse air-dried (not ground) plant sample with acetone, dichloromethane or chloroform for several minutes to dissolve the surface flavonoids. Remove the solvent with a rotary vacuum evaporator and dissolve the residue in a small volume of 100% methanol. A step of purification through Sephadex LH-20 is advisable to separate the flavonoids from the terpenoids that are also present in the leaf surface (the flavonoids are eluted with methanol).

6. ANTHOCYANINS AND ANTHOCYANIDINS

Anthocyanins are a group of water-soluble pigments responsible for the red, pink, mauve, purple, blue, or violet colour of most flowers and fruits. These pigments occur as glycosides (anthocyanins) and their aglycones (anthocyanidins) are derived from the 2-phenylbenzopyrylium cation, commonly called the flavylium cation. This name emphasized the fact that these molecules belong to the group of flavonoids. Anthocyanins are soluble in water and alcohols, and insoluble in apolar organic solvents. Frequently, anthocyanins occur acylated with an organic acid such as malonic or with an aromatic acid such as p-cumaric acid. Acylation takes place on the sugar unit of the 3-position and both types of acylation may be present in the same molecule (Hattori, 1962; Proctor and Creasy, 1969; Harborne *et al.*, 1973; Harborne, 1984; Dick *et al.*, 1987; Macheix *et al.*, 1990).

Procedure I

Homogenize a small amount of fresh coloured petals with 1% HCl (v/v) in methanol for 20 minutes at room temperature, in darkness (anthocyanins are unstable in neutral or alkaline solution and the colour may slowly fade due to light exposure). Filter the mixture on a Buchner funnel and wash the remaining solids with 0.1% methanol until a clear solution is obtained. Dry the combined filtrates using a rotary evaporator at 30°C. Dissolve the concentrate in aqueous 0.01% HCl (v/v) and use the solution for further purification and analysis. Anthocyanin solutions are very unstable and they can only be kept under nitrogen, at low temperature and in darkness.

Procedure II

Grind frozen tissue (approximately 1 g of leaf, petiole and a small amount of fine stem material) in liquid nitrogen. Add 10 ml of 2 N HCl and heat at 55°C for 10 minutes, then cool and incubate overnight in darkness at 24°C to extract the anthocyanidins. Extraction under extremely harsh hydrolysis conditions (90°C, 15 minutes) is avoided, because it causes the red extract colour to turn brown. Clarify the extract by centrifugation and scan an aliquot from 240 to 600 nm in a spectrophotometer using 2 N HCl as a blank control. Partitionate anthocyanidin hydrolysates with water-saturated ethyl acetate, after which extract the red aqueous phase into isoamyl alcohol, dry and reconstitute in methanolic 1% HCl (Harborne, 1998). Control the presence of anthocyanidins by recording at the maximum wavelength (523 nm) and compare with a standard curve of commercial cyanide chloride.

7. TANNINS

Tannins are water-soluble phenolics of molecular weight between 500 and 3,000, which, in addition to displaying the classic reactions of phenols, can precipitate alkaloids and gelatine, and can react with proteins, forming stable water-insoluble co-polymers. Two main groups of tannins are generally distinguished, which differ by their structure, as well as their biogenetic origin: hydrolyzable tannins and condensed tannins. They are distributed unevenly throughout the plant kingdom.

The amount and type of tannins synthesized by plants varies considerably depending on plant species, cultivars, tissues, stage of development and environmental conditions. Due to the complexity of tannins, several methods have been developed for their extraction. None of them, however, is completely satisfactory. Condensed tannins or flavolanos can be considered as being formed biosynthetically by the condensation of single catechins (or gallocatechins) forming dimers and higher oligomers. The name proanthocyanidins is used to designate the condensed tannins because the monomer anthocyanidin is released after hydrolysis. Two classes of hydrolyzable tannins are distinguished: galloylglucose depside, in which a glucose core is surrounded by 5 or more galloyl ester groups and a second group that is present when the core molecule is a dimer of gallic acid attached to glucose. After hydrolysis they give ellagic acid.

7.1 Condensed Tannins

7.1.1 *Procedure:* Extraction from Heartwood Sawdust

Extract 100 g of heartwood sawdust (60 mesh) with 1,000 mL of acetone/water (7:3, v/v). After 24 hours of continuous shaking at room temperature, filtrate the extract and evaporate the acetone under reduced pressure at 30-40°C. Extract the residual solution three times with 30 mL of ethyl acetate. Mix the three organic layers and evaporate under reduced pressure. Solubilize the dried extract in MeOH/water (3:7, v/v).

7.1.2 *Procedure:* Extraction from Fresh Tissue

The optimal extraction yield is obtained from the fresh tissue, or from the frozen or lyophilized tissues, because in the dried drugs, part of the tannins is irreversibly combined to other polymers. Make the extraction with acetone/water (7:3, v/v). Eliminate the acetone as above, remove the pigments from the aqueous solution by a solvent extraction (e.g. with dichloromethane). Then, extract the aqueous solution with ethyl acetate that separates the dimeric proanthocyanidins and most gallotannins. The polymeric proanthocyanidins and high molecular weight gallotannins remain in the aqueous phase. Apply the appropriate chromatographic technique, most often one of gel filtration, to obtain pure compounds.

7.2 Hydrolyzable Tannins

Procedure

Use dried and ground samples from aerial plant parts. Make a Soxhlet extraction of the hydrolyzable tannins: place 5 g of plant sample in a Whatman 25 mm × 100 mm cellulose thimble. Carry out the extraction using standard Soxhlet method using 150 mL of solvent. Set the heating power to two (2) cycles per hour so that six (6) cycles of extraction are achieved within 3 hours of extraction time. Various organic and organic-aqueous solvents with different polarities can be used, one of the most used is acetone:water (70:30, v/v). Concentrate and dry the crude extract solution on a vacuum rotary evaporator at 60°C to remove the solvent. Avoid high temperatures to minimize the component degradation.

8. QUINONES

Quinones can be divided into four groups: benzoquinones, naphthoquinones, anthraquinones and isoprenoid quinones. Naphthoquinones and anthraquinones are among the most widely distributed natural quinones. The majority of them exist as coloured phenolic compounds, useful as dyes and pigments. Benzoquinones, naphthoquinones and anthraquinones are generally hydroxylated with phenolic properties, and can exist in nature as glycosidic derivatives or in a colourless, sometimes dimeric, quinol form. In such cases, acid hydrolysis is necessary to release the free quinone. Isoprenoid quinones are involved in important physiological functions such as cellular respiration (ubiquinones) and photosynthesis (plastoquinones).

8.1 Extraction of Free Quinone Pigments

Procedure

Suspend 40 g of plant material in 50 mM phosphate buffer (pH 7.0). Mix with 2.5 volumes of a chloroform-methanol mixture (2:1, v/v) sonicate for 1 minute on

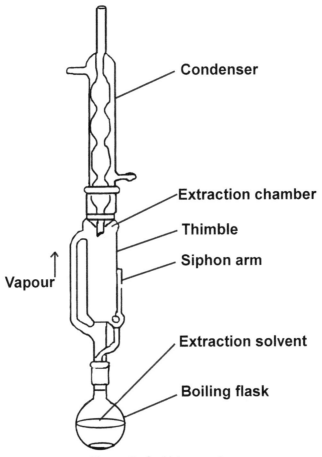

Figure 2 Soxhlet apparatus

ice (20 kHz; output power 100 W) and centrifuge at 5,000 × g for 10 minutes. Discard the resulting upper aqueous layer and collect the lipid layer with a pipette. Extract the residue with acetone and then twice (30 minutes each) with the chloroform-methanol mixture. Combine all extracts and evaporate the solvent under vacuum. Then, re-extract three times with *n*-hexane-1% saline (1:1, v/v). Concentrate the extract as above and analyze.

8.2 Extraction of Isoprenoid Quinones by Saponification

Procedure
Place 0.5 g of the sample in a screw-cap test tube (16 by 15 mm) with a Teflon-lined cap. Add 3 mL of 1% (w/v) pyrogallolin methanol and 0.2 mL of 50%

aqueous KOH. Reflux the mixture for 10 minutes at 100°C and immediately cool to room temperature by placing the tubes under running tap water. Add 1 mL of distilled water and 5 mL of *n*-hexane and shake the mixture vigorously for 5 minutes and then centrifuge at 1,200 × g for 10 minutes. Remove the hexane layer and re-extract again with 5 mL of *n*-hexane. Evaporate the combined hexane layers to dryness under N_2 and dry the resulting *n*-hexane layers. Then, take up the resulting residue in 0.1 mL of ethyl acetate or diethyl ether.

8.3 Direct Extraction of Isoprenoid Quinones

Procedure

Place 0.5 g of the sample in a screw-cap test tube (16 by 15 mm) with a Teflon-lined cap. Add 3 mL of methanol-hexane (6:4) and agitate the mixture for 3 minutes. Add 4 mL of *n*-hexane, shake the mixture vigorously for 5 minutes and centrifuge at 1,200 × g for 10 minutes. Remove the hexane layer and re-extract the aqueous phase twice with 4 mL of *n*-hexane. Combine the hexane layers and evaporate to dryness under N_2. Dissolve the residue in 0.1 mL of ethyl acetate or diethyl ether.

9. TERPENOIDS AND STEROIDS

Terpenoids and steroids are elaborated from common precursors. They constitute the largest known group of plant secondary metabolites. Majority of them are specific to the plant kingdom, but they also occur in animals, e.g. sesquiterpenoid insect pheromones and juvenile hormones, diterpenes of marine organisms (Coelenterate, Spongiae). Triterpenes are specific to the plant kingdom and like steroids, arise via squalene, from mevalonate. The mechanism of formation of these last compounds is slightly different from that of triterpenes, but their structure is perfectly specific to a plant group; this is true for cardenolides, steroidal alkamines, saponins, and phytosterols.

Terpenoids and steroids are formed by units of five carbons derived from 2-methylbutadiene ($CH_2C(CH_3)CHCH_2$, isoprene), although they can also contain other elements such as oxygen. Free isoprene is the only free widespread hemiterpene in plants. Generally, it only exists in extremely small amounts and its detection requires mass spectrometry, so its wide distribution has only recently come to light. Instead, the compound actually involved in the first steps of the biosynthetic pathway of terpenoids, and steroids is isopentenyl pyrophosphate ($CH_2C(CH_3)CH_2CH_2OPP$), which is formed in a biosynthetic route involving acetate and mevalonate. This product is present in living cells, in equilibrium with its isomer dimethylallylpyrophosphate ($(CH_3)_2CCHCH_2OPP$), an efficient alkylating agent. Both key intermediates exist only in trace amounts. However, hemiterpenes (C_5) are frequently found combined with non-terpenoid moieties as mixed terpenoids (some of them are purine derivatives which occur in free state, when they exert cytoquinin activity). Monoterpenes are biosynthetisized by condensation of a molecule of isopentenyl pyrophosphate and one of dimethylallyl pyrophosphate to give geranyl pyrophosphate (C_{10}), a key intermediate in monoterpene formation. Geranyl pyrophosphate and iso-pentenyl pyrophosphate are linked to give farnesyl pyrophosphate (C_{15}). Different combinations of C_5, C_{10} and C_{15} units and modifications of their basic structure are involved in the synthesis of higher terpenoids. The main classes of plant terpenoids are cited in Table 3.

Table 3   Main classes of plant terpenoids (Source: Harborne, 1998)

Number of isoprene units	Carbon number	Name of class	Main types and occurrence
1	C_5	Isoprene	Detected in *Hammamelis japonica* leaf
2	C_{10}	Monoterpenoids	Monoterpenes in plant essential oils (e.g. menthol from mint) Monoterpene lactones (e.g. nepeta-lactone) Tropolones (in gymnosperm woods)
3	C_{15}	Sesquiterpenoids	Sesquiterpenes in essential oils Sesquiterpene lactones (especially common in Compositae) Abscisins (e.g. abscisic acid)
4	C_{20}	Diterpenoids	Diterpene acids in plant resins Gibberellins (e.g. gibberellic acid)
6	C_{30}	Triterpenoids	Sterols (e.g. sitosterol) Triterpenes (e.g. β-amyrin) Saponins (e.g. yamogenin) Cardiac glycosides
8	C_{40}	Tetraterpenoids	Carotenoids* (e.g. β-carotene)
n	C_n	Polyisoprene	Rubber, e.g. in *Hevea brasiliensis*

*C50-based carotenoids are known in some bacteria.

Terpenoids and steroids are generally lipid-soluble and are located in the cytoplasm of the plant cell. Essential oils (EOs), the most volatile fraction of terpenoids, occur in special glandular cells on the leaf surface. Carotenoids, which are C_{40} tetraterpenoids (Britton, 1980; Goodwin, 1980; Straub, 1987) are specially associated with chloroplasts in the leaf and chromoplasts in petals. Terpenoids are normally extracted from plant tissue with organic solvents (hexane, ethyl ether or chloroform) and are separated by column chromatography on silica gel or alumina using the same solvents.

Monoterpenoids and sesquiterpenoids constitute a volatile steam-distillable fraction responsible for the odour, scent or smell found in many plants. They are important as bases of natural perfumes and also of spices and flavouring in the food industry. In plants, they are associated with other classes of chemical substances, i.e. hydrocarbons, alcohols, and ketones being the most common.

There may be 10 or 15 easily detectable components, and perhaps many other terpenoids which may be present in trace amounts. Monoterpenes, as well as sesquiterpenes, fall into chemical groups according to the basic carbon skeleton; the common ones are acyclic, monocyclic or bicyclic.

There are many methods of obtaining these EO compounds from plants (Muzika *et al.*, 1990; Laenger *et al.*, 1996; Anitescu *et al.*, 1997; Lis *et al.*, 1998; Ruberto *et al.*, 1999; Simandi *et al.*, 1999). The most commonly used methodology of extracting EOs from plants has been selected. The described techniques can be used to extract terpenoids and steroids. These substances are susceptible to chemical rearrangements, oxidations and reductions, substitutions and additions in structural modifications. Thus, careful storage at low temperature, neutral pH and absence of humidity is necessary.

9.1 Extraction by Steam Distillation (EOs)

Steam distillation is a method developed for separation of temperature sensitive materials. It enables the separation of mono and sesquiterpenes, which are essential oil (EO) constituents, according to their boiling points (140-180 and > 180°C).

Materials and equipments required
Plant sample dried or fresh; refrigerator; balloon of distillation; beakers; Erlenmeyer flasks; tubes.

Procedure A
Simple steam distillation consists in immersing the plant material to be treated (intact or crushed) directly into a flask filled with water, which is then boiled (Fig. 3). A steam current from another water boiling flask passes through the mixture. The heterogeneous vapours are condensed on a cold surface and the EOs are separated according to the differences in density and immiscibility.

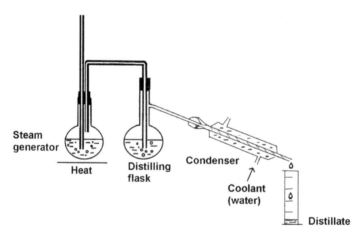

Figure 3 Laboratory set-up for steam distillation (Source: Moldoveanu and David, 2002).

Procedure B

In saturated steam distillation, the plant does not come in contact with the water; the steam is injected through the plant material placed on perforated trays (Fig. 4). To shorten the duration of treatment and limit the alteration of the constituents of the EO, it is possible to operate under moderate pressure (1-3 bar). The consequence of this pressurization is an increase in temperature, but the quality of the product may suffer.

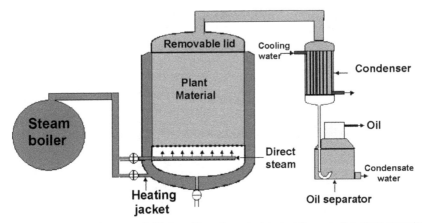

Figure 4 Saturated steam distillation. The steam is injected through the plant material placed on a perforated tray.

Procedure C

Hydro diffusion consists in sending pulses of steam under very low pressure (0.02-0.15 bar) through the plant material, from top to bottom.

Observations

Composition of resulting products is qualitatively different than products obtained by classic methods.

9.2 Simultaneous Distillation and Extraction (SDE)

It is a combined process of distillation and solvent extraction (SDE) and uses the steam distillation to remove the analytes from a sample with a complicated matrix. Then, the analytes are transferred from water solution to an organic solvent by a liquid/liquid extraction. A simplified scheme of the process is shown in Fig. 5 (Likens and Nickerson, 1964; Nickerson and Likens, 1966; Godefroot *et al.*, 1981). There are two versions of the apparatus used, one for solvents with densities higher than water (Fig. 6A) and the other for solvents with lower densities (Fig. 6B). Several solvents are used in SDE sample preparations, CH_2Cl_2 being the most common. Other solvents include ethyl acetate, pentane, hexane and diethyl ether. The decrease in the water solubility of various organic compounds is achieved by salting out the water solution.

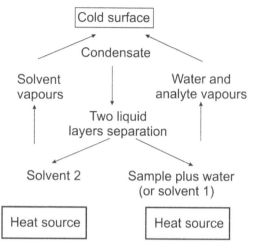

Figure 5 Simplified scheme of simultaneous distillation and extraction process (Source: Adapted from Moldoveanu and David, 2002).

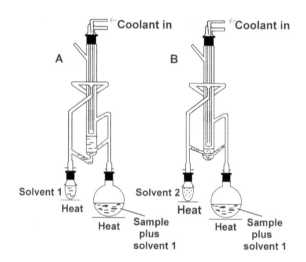

Figure 6 Versions of the apparatus used for simultaneous distillation and extraction procedures with solvents with (A) densities higher, and (B) densities lower than water (Source: Godefroot *et al.*, 1981).

This procedure increases the recovery efficiency (Bartsch and Hammerschmidt, 1993).

Materials and equipments required

The apparatus described in Figs. 6A and B; distilled water; the selected organic solvent, e.g. CH$_2$Cl$_2$ (Blanch *et al.*, 1993), or pentane-ethyl ether (1:1) (Wu and

Liou, 1992), or hexane (Janda and Marha, 1985); rotary vaccuum evaporator if necessary.

Procedure

The plant sample is introduced together with water (Solvent 1 in Figs. 6A and B) in a flask and the mixture is brought to boiling point. At the same time, a different solvent that is not miscible in water (Solvent 2 in Figs. 6A and B) is boiled in the second flask. The vapours of both flasks are condensed together and an intimate mixture is produced. The separation of the two layers allows each solvent to return to its own flask. The extraction process may take a number of hours to achieve efficient extraction. In some apparatus an external source of steam is used. A high cooling efficiency is necessary to avoid loss of more volatile analytes and of the solvent. The organic fraction is concentrated and analyzed by GC-MS.

Observations

SDE has some advantages and disadvantages in comparison to other extraction techniques such as supercritical fluid extraction (SFE) and Soxhlet extraction. As in other distillation procedures, some artefacts may be present in SDE samples. The sources of artefacts are the contamination with grease from glass joints, possible oxidation for compounds such as benzaldehyde and terpenes, and thermal reactions such as Maillard reactions, ester hydrolysis, and sugar degradations.

9.3 Extraction with Solvents

The components from dry or fresh plant tissue are soluble in solvents of low polarity (hexane, diethyl ether and chloroform).

Materials and equipments required

Chloroform; a fuming hood; beakers of different capacity; static or moving extractor; online extractor; rotary vacuum evaporator if necessary.

Procedure A

Suspend 10 g of fine powdered leaves from a fresh or an air-dried sample in 150 mL of chloroform and leave to extract for at least 2 days with occasional shaking. Filter through Whatman # 1 paper and evaporate the chloroform from the extract. Dissolve the preparation in a small volume of chloroform or ethanol. Examine the presence of terpenoids by TLC.

Procedure B

Dry test material is placed inside a "thimble" made from filter paper, which is loaded into the Soxhlet extractor. The extractor is attached to a flask containing a solvent (commonly diethyl ether or petroleum ether) and a condenser. The solvent is heated, causing it to evaporate. The hot solvent vapour travels up to the condenser, where it cools and drips down onto the test material. The chamber containing the test material slowly fills with warm solvent until, when it is almost full, is emptied by siphon action, back down to the flask. This cycle may be allowed to repeat many times. During each cycle, a portion of the material dissolves in the solvent. However, once the lipophilic substance reaches the solvent heating flask, it stays there. It does not participate in the extraction cycle any further. This is the key advantage of this type of extraction; only clean warm solvent is used to extract the solid in the thimble. This increases the efficiency of the extraction when compared with simply heating up the solid in a flask with the solvent. At the end of an extraction, the excess solvent may be removed using a rotary evaporator, leaving behind only the extracted material.

9.4 Extraction of Terpenoid Glycosides (Iridoids)

The extraction of these glycosides is particularly delicate due to their being highly unstable. This is the reason for plant darkening that takes place soon after plant collection in many iridoid-rich species.

Procedure
Extract the plant material with polar solvents (alcohols of various concentrations). Evaporate the solvent and dissolve the residue with water. Then, re-extract the water solution with solvents of increasing polarity.

9.5 Extraction of Sesquiterpene Lactones from Glandular Trichomes

To obtain the chemical profile of glandular trichomes, sesquiterpene lactones are frequently present in glandular trichomes of the leaf, flower or seeds of Compositae.

Procedure
Extract air-dried flower heads (30 g) with CH_2Cl_2 and evaporate the solvent. Remove insoluble parts and re-dissolve the residue in MeOH-water (1:1, v/v) and centrifuge. Analyze by HPLC.

9.6 Extraction of Abscisic Acid (Sesquiterpene)

Procedure
Extract fresh tissue with 80% MeOH, evaporate the MeOH from the filtered extract. Acidify and extract with ether. Crude abscisic acid is then removed from the ether extract into aqueous saturated $NaHCO_3$. Then recover with acidified ether and wash with ether and analyze.

9.7 Extraction of Diterpenoids (Gibberellins)

Principle
Diterpenes constitute a wide group of C_{20} compounds based on four isoprene units. The structure of diterpenes is highly variable and strictly dependent on their biogenesis. They can be acyclic, bicyclic, tri and tetracyclic diterpenes. The most frequently found diterpenes in the plant kingdom is the phytol, which is

present as the ester attachment in the molecule of chlorophyll, the resins and the gibberellins.

Procedure

Extract 30 g of dried leaf tissue in a Soxhlet with ether for 12 hours. Eliminate the solvent, dissolve the residue in benzene and analyze.

Procedure

Extract apical buds by homogenizing the tissue in a Varing blender, followed by extraction with 80% MeOH (v/v) for 24 hours at 20°C. Filter the extract and reduce to dryness in vacuo at 35°C on a rotary evaporator prior to paper chromatography.

9.8 Extraction of Triterpenoids

Most triterpenoids, which are made up of 30 carbons, are considered as a condensation of six isoprenes. Triterpenoids can be divided into at least four groups of compounds: true triterpenoids, steroids, saponins and cardiac glycosides. Saponins and cardiac glycosides are triterpenoids or steroids which occur mainly as glycosides. The structure of sterols and triterpenes is based on the cyclopentane perhydrophenantrene ring system. At first, sterols were mainly considered to be animal substances, but in recent years an increasing number of such compounds have been detected in plant tissue. Saponins constitute a vast group of glycosides which are ubiquitous in plants. They are characterized by their surface-active properties, i.e. they dissolve in water to form foamy solutions. Saponins are glycosides of triterpenes and sterols. The fourth group of triterpenoids to be considered is the cardiac glycosides or cardenolides. They constitute a well-defined and highly homogeneous group from a structural, as well as pharmacological, standpoint.

9.8.1 Extraction of Triterpenes

Procedure

Defat the dry plant material with ether. Once the ether is eliminated, extract the material with hot methanol. Concentrate the extract by vacuum evaporation and carry out the acid hydrolysis (0.5 M HCl for 2-6 hours at room temperature) to liberate the aglycones, if glycosides are present. Analyze the fractions.

9.8.2 Extraction of Saponins

Procedure

The extraction and separation of saponins are delicate processes. These compounds often occur in substantial quantities, but as complex mixtures; high molecular weight of their constituents often make it a lengthy and arduous task to obtain an intact and pure compound. Saponins are soluble in water, therefore, they can be extracted with this solvent, generally at the boiling point. However, an aqueous medium is favourable for the hydrolysis of some saponins and it is often better to use alcohols (methanol) or hydro alcoholic solutions after partial lipid removal.

Since polar solvents extract too many compounds, after the initial extraction, a partition between water and *n*-butanol is frequently carried out; *n*-butanol solubilizes the saponins which are then precipitated by adding a solvent such as diethyl ether.

9.8.3 Extraction of Sapogenins

Procedure

Sapogenins e.g. hecogenin, can be extracted from fresh agave leaves (Ikan, 1969). Extract 1 kg of agave with 11.95% EtOH at 100°C for 12 hours. Filter while hot, remove the solvent under reduced pressure and hydrolyze the residue by refluxing for 2 hours with 300 mL of 1 M HCl in ethanol. Filter the mixture and dilute the filtrate with 400 mL ether. Wash the solution with water, with 5% aqueous NaOH and again with water and evaporate the solvent. Hydrolyze the residue by refluxing with 10% alcoholic KOH for 30 minutes and after cooling, hecogenin is extracted into ether, which is then evaporated to dryness.

9.8.4 Extraction of Cardiac Glycosides

Procedure

Extract the pulverized drug with a mixture of 50% ethanol and lead acetate solution. After boiling, cooling and elimination of the residue by centrifugation, the cardiac glycosides present in the supernatant are extracted with chloroform. This chloroform solution undergoes the characterization reactions that can be due to the sugars or to the aglycones and chromatographic analyses.

9.9 Extraction of Carotenoids

9.9.1 From Maize Seeds

Procedure

Hydrate 3 g of dried ground maize seeds in a mortar in 10 mL of water for 30 minutes, at room temperature. Then, add 20 mL of acetone and leave for 15 minutes. Extract carotenoids by grinding the mixture in a mortar and pestle with 50 mL acetone. Vacuum filter the residue in a Buchner funnel equipped with filter paper Whatman # 2. Return the residue to the mortar and repeat the procedure until the residue is nearly colourless, usually once more. Transfer one-third of the filtrate to a separating funnel containing 20 mL of petroleum ether, to which 300 mL of distilled water is added. Discard the aqueous layer. Repeat the procedure for the remaining filtrate. Wash the organic phase 3 times with 200 mL of distilled water and pass through 15 g of anhydrous sodium sulfate into a round bottom flask. Concentrate the sample with a rotary evaporator and dry (Rodriguez-Amaya and Kimura, 2004).

9.9.2 From Alfalfa

Procedure

Mixed carotenoids are obtained by solvent extraction of alfalfa, removal of chlorophylls through saponification and subsequent purification of the carotenoids by solvent extraction. The main colouring principle consists of carotenoids, of which lutein accounts for the major part. Variable amounts of neoxanthin, violaxanthin and β-carotene are present.

Reagents

Acetonitrite, dichloromethane, ethyl acetate, acetone (all HPLC grade), butylated hydroxytoluene (BHT), potassium hydroxide, methanol, sodium sulfate (all analytical reagents of better grade). Add 1 g/L of BHT to all solvents.

Procedure

Grind dried alfalfa to pass through 1-mm mesh screen and thoroughly mix. Accurately weigh 1-2 g sample in a glass stoppered boiling tube. Add 30 mL dichloromethane:acetone (2:1) and shake. Let it stand overnight. Add 20 mL of dichloromethane, 2 mL of 40% (w/v) methanolic potassium hydroxide, shake and let it stand for 60 minutes. Add 30 mL of 10% (w/v) aqueous sodium sulfate, shake and let it stand for 60 minutes. Remove an aliquot of the lower layer and centrifuge for 10 minutes at 2,000 rpm. Remove a 3-mL aliquot of the lower yellow layer, mix with 3 mL acetonitrile and use this solution for analysis.

10. ORGANIC ACIDS, LIPIDS AND RELATED COMPOUNDS

10.1 Plant Acids

Aqueous extracts of plant tissues contain large amounts of organic acids. Citric, malic, tartaric, succinic and oxalic acids are usually the most abundant and the wide distribution of these particular acids is well recognized. These acids accumulate in plant cell vacuoles, sometimes in considerable amount, such as in Crassulaceae, where high quantities of malic, citric and isocitric acids are accumulated in the vacuoles during the day (Kluge and Ting, 1978). Plant organic acids can contain a variable number of carboxylic groups. Other chemical functions, such as hydroxyl and keto, can be present. Some widespread organic acids are present in small quantities in plant tissues, but have important plant functions. Oxalic, ascorbic, shikimic, quinic and monofluroacetic (inhibitor of the tricarboxylic acid cycle) acids are some of them. Among the dicarboxylic acids, malonic and succinic acids are worth mentioning. Plants also contain unsaturated and cyclic acids.

Determination of organic acids from plant tissues is difficult due to the diversity of acidic substances found in plants. Furthermore, the organic acids must be estimated in the presence of more or less mineral acids, a circumstance that places restrictions on the use of ordinary techniques.

Organic acids are water-soluble, colourless liquids with relatively low melting points. In general, they are non-volatile compounds. They are chemically stable, with the exception of α-ketoacids, which are easily descarboxylated. Their solutions in water are acidic and their presence is detected by the change in colour of acid-base indicators.

Procedure

Grind 2-5 g of fresh leaf lamina of a test plant with a little quantity of acid-washed sand in 5 mL of 0.1% formic acid in 80% ethanol. Make up the homogenate to 30 ml with further acidified 80% ethanol and occasionally stir during the next 30 minutes. Centrifuge at 5,000 g for 15 minutes and decant the supernatant and keep aside. Carry out all the operations up to this point at 4°C. Add 5 g of cation exchange resin (Dowex 50 W × 8 (H⁺) 20 × 50 mesh) to the residue of the centrifugation and extract the residue with resin with 50 mL of distilled water in a shaking water bath at 55°C for 4 hours. Filter the suspension through two layers of muslin cloth and combine with the ethanol extract. Reduce the volume to *ca.*20 mL under reduced pressure at 45°C. Shake the extract for 10 minutes and centrifuge for 20 minutes at 30,000 g. Remove any cation present in the supernatant by passing it through a 10 × 50 mm cation exchange column (Dowex 50 W × 8 (H⁺) 100 × 200 mesh). Allow the eluant to drip directly onto a 10 × 70 mm column of Dowex 1 × 8 (Formate) 200 to 400 mesh anion exchange resin. Elute the organic acids from the Dowex 1 with 25 mL of 8 M formic acid. Dry the eluant under reduced pressure at 45°C, re-dissolve in 10 mL of distilled water and dry to ensure the complete removal of formic acid.

10.2   Lipids and Fatty Methyl Esters Extraction

Principle

Fatty acids are present in plants in bound form, esterified to glycerol as fats or lipids. The fatty acid composition of lipid samples is usually determined by GLC [Gas liquid chromatography] of the corresponding methyl esters. Many methods are available to prepare the fatty acids methyl esters (FAMES) by transmethylation of lipids using either acid- or base-catalyzed reactions (Christie, 1989). Base-catalyzed transesterification is used to methylate lipids from tissues with high lipid and low water content (Barnes and Holaday, 1972; Conte *et al.*, 1989). However, these procedures are not fully quantitative, do not methylate the free fatty acids and give low yield with tissues containing water, because the hydrolysis competes with the transesterification reaction. Methanolic HCl or H_2SO_4 are efficient reagents to prepare methyl esters from free fatty acids and glycerolipids, the limiting factor is the solubility of the products during the transmethylation reaction. Free fatty acids, polar lipids and also the resulting methyl esters are readily solubilized in hot methanol, but non-polar lipids such as triacylglycerols (TAG) are less soluble and react very slowly, unless a further solvent like benzene, toluene, or tetra hydrofuran is added to improve the solubility. 2,2-Dimethoxypropane (DMP) improves the transmethylation of glycerolipids by converting the resultant glycerol to isopropylideneglycerol. Also, DMP reacts with the water excess and has been used in the combined digestion and FAMES preparation from fresh leaves (Browse *et al.*, 1986). However, even these simple methods need several manipulations, including the extraction of the tissue lipids, the resulting FAMES, or both.

Procedure I

Cut seeds, leaves or fruits of a test plant into small pieces. Treat samples of 50 mg (seeds and fruits) or 200 mg (leaves) with heptadecanoic acid as the internal standard and place the mixture in tubes with Teflon lined caps. The methylating mixture used contain methanol:benzene:DMP:H_2SO_4 (37:20:5:2, v/v); methanol:toluene:DMP:H_2SO_4 (39:20:5:2, v/v) or methanol:tetrahydrofuran: DMP:H_2SO_4 (31:20:5:2, v/v). Store the mixtures in dark coloured bottles at 4°C. Pipette an amount of one of the mixtures (3.2, 3.3 or 2.0 mL, respectively) and 5 mL (1.8, 1.7 and 2.5 mL, respectively) heptane. Flush nitrogen to the tubes and place them in a water bath at 80°C for 1 hour for isolated lipids and 2 hours for fresh tissues to ensure complete lipid extraction and methylation. Shake vigorously after 2-3 minutes of heating to mix all the components in a single phase. After heating, cool the tubes at room temperature and shake again. Two phases are formed, the upper phase contains the FAMES. Separate this phase and analyze by GLC.

Procedure II

Extract fresh leaf tissue of a test plant by maceration with 20 volumes of cold isopropanol to deactivate the hydrolytic enzymes. Re-extract the residue with a mixture of chloroform-methanol (2:1, v/v). Grind the seeds to a coarse powder and extract it with chloroform-methanol (2:1, v/v) or with light petroleum. If lipids are tightly bound to the seed tissue, use the chloroform-ethanol-water (20:9.5:0.5) mixture for their extraction. The extracted material must be stored at 4°C to avoid oxidation or in the presence of an antioxidant (e.g. 0.005% butylated hydroxy toluene, BHT).

10.3 Alkanes and Related Lipids

Alkanes are long-chain hydrocarbons distributed in the waxy coating film on the surface of leaves and fruits in plants . They are constituents of waxes. These with cutin, constitute the hydrophobic cuticle which limits water loss, controls gaseous exchange and participates in the protection against pathogenic agents. Waxes are a mixture of hydrocarbons, free and hydroxylated aliphatic acids, aliphatic alcohols, aliphatic aldehydes and aliphatic ketones, β-diketones and esters. They can also contain terpenoids and flavonoids.

Principle

All wax components are extractable by organic solvents (hexane and chloroform) and amenable to GC analysis.

Procedure I

Dip the unbroken leaves, stems or fruits of a test plant into ether or chloroform for short periods of time (e.g. 30 seconds) to remove the alkanes from the surface without attacking cytoplasmic constituents. Filter the preparation and analyze the filtrated components.

Procedure II

Extract the dry powdered leaves of a test plant with a Soxhlet for several hours using hexane as extractant. This extract needs additional purification, because it can be contaminated with cytoplasmic components.

Precautions

Use very clean glassware and redistilled solvents and avoid contact of the extraction solvent with the stopcock grease or plastic tubing. Be careful with the contact of the preparation with "Parafilm", a thermoplastic sealing material employed in the laboratories. This material is soluble in the mentioned solvents and gives a solution containing *n*-alkanes that can contaminate the extracted material. Otherwise, the solution of the dissolved "Parafilm" can be used as standard during the purification procedure of alkanes.

10.4 Alkanes, Aliphatic Alcohols and their Esters, Alkenes and Alkynes

These compounds are an unusual group of substances that are found in plants, fungi and algae (Bohlmann *et al.*, 1973). They have in their structure one or more acetylenic groups. Moreover, while acetylene is a highly reactive gas, the long-chain hydrocarbon derivatives can be manipulated during their isolation and purification. Alkynes (polyacetylenes) can be acyclic or cyclic and have different functional groups: alcohols, ketones, acids, esters, aromatics or furans. The extraction technique below can be applied to all of them.

Principle

The polyacetylenes are soluble in organic solvents.

Procedure

Extract 0.5 g fresh samples of a test plant by immersion in dichloromethane for 7 days at 5°C in darkness. Add a small aliquot of anhydrous $MgSO_4$ and mix. Filter the preparation through Whatman # 1 paper. Evaporate the solvent to dryness under nitrogen and resuspend the sample in HPLC-grade chloroform. Prepare samples for HPLC analysis by refiltering the extract through cellulose acetate HPLC syringe filters. Evaporate the chloroform solvent to dryness and resuspend the extract in 250 µl of HPLC-grade methanol (Flores *et al.*, 1988).

10.5 Sulfur Compounds

10.5.1 Glucosinolates

Glucosinolates are anionic glycosides responsible for the bitter taste and characteristic aroma of several Brassicaceae (i.e. mustard, radish, rutabaga and cabbage) and species belonging to the related families. When plant cell compartmentalization is lost (i.e. after being crushed by a phytophagous), stored glucosinolates are released from vacuoles and make contact with thioglucosidases (myrosinases). These enzymes hydrolyze the glucosinolates, releasing several compounds such as isothiocyanates, thiocyanates and nitriles, which are toxic for phytophagous and phytopathogenic organisms. Then, enzymes must be destroyed before glucosinolates isolation, often with boiling alcohol. Due to their ionic nature, glucosinolates are usually separated on ion-exchange resins. The sulfate ions or glucose may be quantitated following the action of thioglucosidases.

Procedure

Treat fresh tissue with boiling alcohol. In this step, it is important to avoid the enzymatic hydrolysis. Then, separate the components of the extract with anion exchange resins (Amberlite IR-400) or with acid washed "anionotropic" alumina (Kjaer, 1960).

10.5.2 Isothiocyanates

Extraction of isothiocyanate formed after glucosinolate hydrolysis aids in estimating the plant glucosinolate content.

Procedure

Extract volatile isothiocyanates by steam-distillation of triturated crucifer fresh tissue. They can also be obtained by enzymatic hydrolysis of glucosinolates. They posses the characteristic pungent smell and taste. Otherwise, extract non-volatile compounds with the technique similar to glucosinolates.

10.5.3 Thiophenes

These compounds are extracted and purified with similar techniques employed for polyacetylenes.

10.5.4 Sulfides

These compounds are present in the volatile fraction of *Allium* bulbs and leaves. The extraction and separation are made by GLC (Bernhard, 1970; Vernin and Metzger, 1991).

SUGGESTED READINGS

Anitescu, G., Doneanu, C. and Radulescu, V. (1997). Isolation of *Coriander* oil: comparison between steam distillation and supercritical CO_2 extraction. *Flavour and Fragrance Journal* **12**: 173-176.

Antolovich, M., Prenzler, P., Robards, K. and Ryan, D. (2000). Sample preparation in the determination of phenolic compounds in fruits. *Analyst* **125**: 989-993.

Barnes, P.C. and Holaday, C.E. (1972). Rapid preparation of fatty acid esters directly from ground peanuts. *Journal of Chromatographic Science* **10**: 181-183.

Bartsch, A. and Hammerschmidt, F.J. (1993). Separation of fragrance materials from perfumed consumer products. *Perfumer and Flavorist* **5**: 41-49.

Beckman, C.H. (1999). Phenolic-storing cells: Keys to programmed cell death and periderm formation in wilt disease resistance and in general defence responses in plants? *Physiological and Molecular Plant Pathology* **57**: 101-109.

Beevers, L. (1976). *Nitrogen Metabolism in Plants.* Edward Arnold, London.

Bengoechea, M.L., Sancho, A.I., Bartolomé, B., Estrella, I., Gómez-Cordovés, C. and Hernández, T. (1997). Phenolic composition of industrially manufactured purees and concentrates from peach and apple fruits. *Journal of Agricultural and Food Chemistry* **45**: 4071-4075.

Bernhard, R.A. (1970). Chemotaxonomy: Distribution studies of sulfur compounds in *Allium*. *Phytochemistry* **9**: 2019-2023.

Blanch, G.P., Herraiz, M., Reglero, G. and Tabera, J. (1993). Preconcentration of samples by steam distillation–solvent extraction at low temperature. *Journal of Chromatography* **655**: 141-149.

Bohlmann, F., Burkhardt, T. and Zdero, C. (1973). *Naturally Occurring Acetylenes.* Academic Press, London.

Britton, G. (1980). Biosynthesis of Carotenoids. In: *Plant Pigments.* (Ed., T.W. Goodwin). Academic Press, London.

Browse, J., McCourt, P.J. and Somerville, C.R. (1986). Fatty acid composition of leaf lipids determined after combined digestion and fatty acid methyl ester formation from fresh tissue. *Analytical Biochemistry* **152**: 141-145.

Christie, W.W. (1989). *Gas Chromatography and Lipids.* The Oily Press, Ayr, Scotland.

Conn, E.E. and Butler, G.W. (1969). Biosynthesis of cyanogenic glycosides and other simple nitrogen compounds. In: *Perspectives in Phytochemistry* (Eds., J.B. Harborne and T. Swain). Academic Press, London.

Conte, L.S., Leoni, O., Palmieri, S., Capella, P. and Lercker, G. (1989). Half-seed analysis: rapid chromatographic determination of the main fatty acids of sunflower seed. *Plant Breeding* **102**: 158-165.

Dey, P.M. and Harborne, J.B. (1966). *Plant Biochemistry.* Academic Press, London, UK.

Dey, P. M. (1990). *Methods in Plant Biochemistry. Vol. 2, Carbohydrates.* Academic Press, London.

Dick, A.J., Redden, P.R., DeMarco, A.C., Lidster, P.D. and Grindley, T.B. (1987). Flavonoid glycosides of spartan apple peel. *Journal of Agricultural and Food Chemistry* **35**: 529-531.

Dunnill, P.M. and Fowden, L. (1965). The amino acids of seeds of the Cucurbitaceae. *Phytochemistry* **4**: 933-944.

Flores, H.E., Pickard, J.J. and Hoy, M.W. (1988). Production of polyacetylenes and thiophenes in heterotrophic and photosynthetic root cultures of Asteraceae. In: *Chemistry and Biology of Naturally Occurring Acetylenes and Related Compounds.* (Eds., J. Lam, H. Breteler and T. Arnason). Elsevier, Amsterdam.

Godefroot, M., Sandra, P. and Verzele, M. (1981). New method for quantitative essential oil analysis. *Journal of Chromatography* **203**: 325-335.

Goodwin, T.W. and Britton, G. (1980). Distribution and Analysis of Carotenoids. In: *Plant Pigments* (Ed., T.W. Goodwin). Academic Press, London.

Harborne, J.B. (1967). *Comparative Biochemistry of the Flavonoids.* Academic Press, London.

Harborne, J.B. (1969). Occurrence of flavonol 5-methyl ethers in higher plants and their systematic significance. *Phytochemistry* **8**: 177-181.

Harborne, J.B. (1982). *Introduction to Ecological Biochemistry.* Academic Press, New York.

Harborne, J.B. (1984). *Phytochemical Methods.* Chapman & Hall, London.

Harborne, J.B. (1988). *The Flavonoids: Advances in Research since 1980.* Chapman & Hall, London.

Harborne, J.B. (1994). *The Flavonoids: Advances in Research since 1986*. Chapman & Hall, London.

Harborne, J.B. (1998). *Phytochemical Methods*. Chapman & Hall, New York.

Harborne, J.B. and Mabry, T.J. (1982). *The Flavonoids: Advances in Research*. Chapman & Hall, London.

Harborne, J.B. and Turner, B.L. (1984). *Plant Chemosystematics*. Academic Press. London.

Harborne, J.B., Mabry, T.J. and Mabry, H. (1973). *The Flavonoids*. Part 1. Academic Press, New York.

Harborne, J.B., Mabry, T.J. and Mabry, H. (1975). *The Flavonoids*. Chapman & Hall, London.

Hattori, S. (1962). Glycosides of flavones and flavonols. In: *The Chemistry of Flavonoid Compounds* (Ed., A. Geissman). The Macmillan Company, New York.

Hoyerová, K., Gaudinová, A., Malbeck, J., Dobrev, P., Kocábek, T., Solcová, B., Trávníčková, A. and Kamínek, M. (2006). Efficiency of different methods of extraction and purification of cytokinins. *Phytochemistry* **67**: 1151–1159.

Ikan, R. (1969). *Natural Products, A Laboratory Guide*. Academic Press, London.

Janda, V. and Marha, K. (1985). Recovery of s-triazines from water and their analysis by gas chromatography with photoionization detection. *Journal of Chromatography* **329**: 186-188.

Khanna, S.K., Viswanatham, P.N., Krishnan, P.S. and Sanwai, G.G. (1968). Extraction of total phenolics in the presence of reducing agents. *Phytochemistry* **7**: 1513-1518.

Kjaer, A. (1960). Naturally derived isothiocyanates (mustard oils) and their parent glucosides. *Fortschritte der Chemie Organischer Naturstoffe* **18**: 122-176.

Kluge, M. and Ting, I.P. (1978). *Crassulacean Acid Metabolism*. Springer-Verlag, Berlin.

Kofod, H. and Eyjolfsson, R. (1969). Cyanogenesis in species of the fern genera Cystopteris and Davallia 908. *Phytochemistry* **8**: 1509-1511.

Krygier, K., Sosulski, F. and Hogge, L. (1982). Free esterified and insoluble-bound phenolic acids: composition of phenolic acids in cereal and potato flours. *Journal of Agricultural and Food Chemistry* **3**: 337-340.

Laenger, R., Mechtler, C. and Jurenitsch, J. (1996). Composition of the essential oils of commercial samples of *Salvia officinalis* L. and *S. fruticosa* Miller: a comparison of oils obtained by extraction and steam distillation. *Phytochemical Analysis* **7**: 289-293.

Likens, S.T. and Nickerson, G.B. (1964). Detection of certain hop oil constituents in brewing products. *Proceedings of the American Society of Brewing Chemistry* **26**: 5-13.

Lis, B.M., Buchbauer, G., Ribisch, K. and Wenger, M.T. (1998). Comparative antibacterial effects of novel Pelargonium essential oils and solvent extracts. *Letters in Applied Microbiology* **27**: 135–141.

Loewus, F.A. and Tanner, W. (1982). *Plant Carbohydrates I. Intracellular Carbohydrates*. Springer-Verlag. Berlin.

Macheix, J.J., Fleuriet, A. and Billot, J. (1990). *Fruit Phenolics*. CRC Press, Florida, USA.

Maher, E.P. and Hughes, M.A. (1971). Isolation of linamarin-lotaustralin from *Trifolium repens*. *Phytochemistry* **10**: 3005-3010.

Markham, K.R. (1982). *Techniques of Flavonoid Identification*. Academic Press, London.

Miflin, B.J. (1981). *Biochemistry of Plants (Amino Acids and Derivatives)*. Academic Press, New York.

Moldoveanu, S.C. and David, R. (2002). *Sample Preparation in Chromatography*. Elsevier, Amsterdam.

Muzika, R.M., Campbell, C.L., Hanover, J.W. and Smith, A.L. (1990). Beta phellandrene kairomone for pine engraver ips-pini say Coleoptera Scolytidae. *Journal of Chemical Ecology* **16**: 2713-2722.

Naczk, M. and Shahidi, F. (1991). Critical evaluation of quantification methods of rapeseed tannins. *8th International Rapeseed Congress*, 9-11 July, Saskatoon, Canada.

Naczk, F., Shahidi, F. and Sullivan, A. (1992). Recovery of rapeseed tannins by various solvent systems. *Food Chemistry* **45**: 51-65.

Nardini, M., Cirillo, E., Natella, F., Mencarelli, D., Comisso, A. and Scaccini, C. (2002). Detection of bound phenolic acids: Prevention by ascorbic acid and ethylenediaminetetraacetic acid of degradation of phenolic acids during alkaline hydrolysis. *Food Chemistry* **79**: 119-124.

Nicholson, R. and Hammerschmidt, R. (1992). Phenolic compounds and their role in disease resistance. *Annual Review of Phytopathology* **30**: 369-389.

Nickerson, G.B. and Likens, S.T. (1966). Gas chromatographic evidence for the occurrence of hop oil compounds in beer. *Journal of Chromatography* **21**: 1-5.

Pridham, J.B. (1960). *Phenolics in Plants in Health and Disease*. Pergamon Press, New York.

Proctor, J.T. and Creasy, L.L. (1969). The anthocyanin of the mango fruit. *Phytochemistry* **8**: 2108.

Robbins, R. (2003). Phenolic acids in foods: an overview of analytical methodology. *Journal of Agricultural and Food Chemistry* **51**: 2866-2877.

Rodriguez-Amaya, D.B. and Kimura, M. (2004). *HarvestPlus Handbook for Carotenoid Analysis*. HarvestPlus, Washington, DC.

Ruberto, G., Biondi, D. and Renda, A. (1999). The composition of the volatile oil of *Ferulago nodosa* obtained by steam distillation and supercritical carbon dioxide extraction. *Phytochemical Analysis* **10**: 241-246.

Shahidi, F. (2000). Antioxidants in food and food antioxidants. *Nahrung* **44**: 158-163.

Shahidi, F. (2002). *Interface Friction and Energy Dissipation in Soft Solid Processing Applications*. ACS Symposium Series 807. American Chemical Society, Washington, DC.

Shahidi, M. and Naczk, M. (2004). *Phenolics in Food and Nutraceuticals: Sources, Applications and Health Effects*. CRC Press, Boca Raton, FL.

Simandi, B., Deak, A., Ronyai, E., Yanxiang, G., Veress, T., Lemberkovics, E., Then, M., Sasskiss, A. and Vamos-Falusi, Z. (1999). Supercritical carbon dioxide extraction and fractionation of fennel oil. *Journal of Agricultural and Food Chemistry* **47**: 1635-1640.

Straub, O. (1987). Index of the natural carotenoids. In: *Key to Carotenoids* (Ed., H. Pfander). Birkhäuser, Basel, Switzerland.

Tipson, R.S. and Horton, D. (1983). *Advances in Carbohydrate Chemistry and Biochemistry*. Academic Press, New York, USA.

Vernin, G. and Metzger, J. (1991). Essential oils and waxes. In: *Modern Methods of Plant Analysis: Essential Oils and Waxes* (Eds., H.F. Linskens and J.F. Jackson). Springer Verlag, Berlin, Germany.

Wanasundara, P.K., Amarowicz, R., Kara, M.T. and Shahidi, F. (1993). Removal of cyanogenic glycosides of flaxseed meal. *Food Chemistry* **48**: 263-266.

Wu, C.M. and Liou, S. E. (1992). Volatile components of water-boiled duck meat and cantonese style roasted duck. *Journal of Agricultural and Food Chemistry* **40**: 838-841.

Yan, Z., Jiao, C., Wang, Y. and Li, F. (2005). A method for the simultaneous determination of β-ODAP, α-ODAP and polyamines in *Lathyrus sativus* by liquid chromatography using a new extraction procedure. *Analytica Chimica Acta* **534**: 199-205.

SECTION

II

Isolation, Identification and Structural Elucidation

Colorimetric Reactions

**Diego A. Sampietro*, Melina A. Sgariglia, Jose R. Soberón,
Emma N. Quiroga and Marta A. Vattuone**

1. INTRODUCTION

Colorimetry, the measurement of the absorption of visible light, allows the first qualitative and quantitative estimation of the phytochemicals present in a plant, soil or water sample. Colorimetric methods allow the assessment of an unknown colour in reference to known colours (Thomas, 1996). In practice, they provide qualitative or semiquantitative estimations rather than quantitative determinations. Early colorimetric techniques consisted of the visual comparison between the colour in a sample and that of several permanent colour standards, which could be the same substance at known concentrations. The use of the human eye as a colour detector, however, includes a subjective perception reducing measurement reproducibility. The instrumental progress in visible spectroscopic techniques allows the development of reliable and sensitive instruments. Colorimeters and spectrophotometers, photoelectrically measure the amount of coloured light absorbed by a coloured sample with respect to a "blank" or colourless sample with reference. Modern spectrophotometers split the incident light into different wavelengths and measure the intensity of that radiation. Colorimetric measurements often involve the reaction of colourless substances under analysis with specific reagents (Hagerman, 2002). This allows the formation of coloured complexes with absorbance maxima at wavelengths comprised in the visible light. In this chapter, general reactions of colour

Authors' address: Instituto de Estudios Vegetales *"Dr. A. R. Sampietro"*, Facultad de Bioquímica, Química y Farmacia, Universidad Nacional de Tucumán, España 2903, 4000, San Miguel de Tucumán, Tucumán, Argentina.
Corresponding author: E-mail: dasampietro2006@yahoo.com.ar

development are provided for the qualitative and quantitative determination of natural products in a sample.

2. GENERAL COLORIMETRIC PRINCIPLES

2.1 Beer–Lambert Law

Electromagnetic radiation can be characterized as a wave with frequency (v) and wavelength (λ). This wave has energy (E) proportional to its frequency and is given by the equation:

$$E = hv = hq/\lambda$$

where, h is Plank's constant and q is light velocity. When a wave of radiation finds a substance and the energy is absorbed, a molecule of the substance is promoted to an excited state (Thomas, 1996). The absorption of visible light generally excites electrons of a molecule from a ground electronic state to an excited one.

The amount of light absorbed by a substance in a solution is quantitatively related to its concentration. This relationship is expressed by the Beer–Lambert Law (Fig. 1):

$$\text{Log}_{10}\,(I_0/I) = -\log_{10} T = abc = A \quad \text{or} \quad I_0/I = 1/T = 10^{abc}$$

where, I_0 is the intensity of incident light, I is the intensity of transmitted light, T is the transmittance, a is the absorptivity or extinction coefficient, b is the path length through an absorbing solution, c is the concentration of the absorbing substance, and A is the absorbance (or optical density in old literature). Absorptivity is constant for a given compound and is a function of the wavelength. When c is expressed in moles per L and b is expressed in centimeters, the constant is known as the molar absorptivity (ε) and is expressed in L per mole per centimeter (L mol^{-1} cm^{-1}).

Beer–Lambert Law expresses a linear relationship between the absorbance and the concentration of a given solution, if the length of the path and the

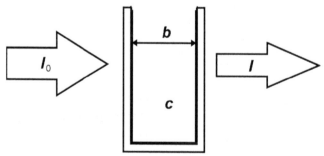

Figure 1 Absorption of radiation by a sample. I_0: Intensity of incident light; I, Intensity of transmitted light; b, path length of a sample; c, concentration of a sample.

wavelength of light are kept constant. This enables determination of the concentration of a substance through the measure of the transmittance or absorbance.

2.2 Deviations from the Beer–Lambert Law

Deviations from the linear relationship between concentration and absorbance often occur and can lead to errors in the application of Beer–Lambert Law (Fig. 2). High concentrations of substances in the solution, the use of radiant energy that is not monochromatic or conditions within the solution causing shifts in the chemical equilibrium, are common deviation sources (Brown, 1997). Linear relationship can be evaluated by building a calibration curve, where the absorbance of standard solutions is plotted as a function of their concentrations. If the absorbance of an unknown sample is then measured, the concentration of the absorbing component can be determined from this graph. In general, transmittance should be in the range of 10-90% (absorbance 1.0-0.1) to avoid high instrumental errors.

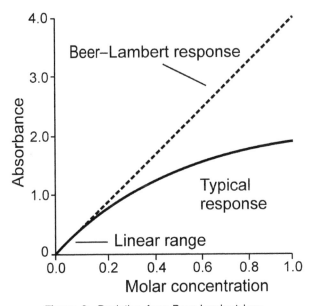

Figure 2 Deviation from Beer–Lambert Law

2.3 Details of Spectrophotometer Operation

Colorimetric analysis is often performed using a spectrophotometer (Brown, 1997). The cheapest spectrophotometers are single beam instruments (Fig. 3A). They have a broadband lamp as the light source. A wavelength selector narrows the wavelength range to a typical band width of 20 nm. This light then passes through one compartment with a typical path length of 1 cm and the transmittance is measured by a photodetector. The instrument is calibrated to zero when a shutter is lowered to 0% transmittance (or infinite absorbance). Then, a blank (a reference solution) is placed in the cell and the instrument is adjusted to 0 absorbance (100% transmittance). A double beam instrument operates similarly, but the beam is split to pass through a reference cell containing the blank and the sample cell, simultaneously (Fig. 3B). A blank is kept in the reference cell for all standardization and sample measurements, as the instrument reports the difference in transmittance or absorbance measured from the two cells.

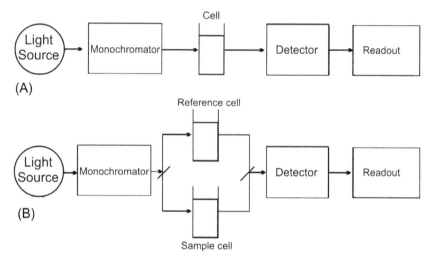

Figure 3 Typical designs of (A) single, and (B) dual beam spectrophotometers (Source: Adapted from Brown, 1997).

Most modern instruments allow computer interfaces for data-logging or automated operation. Also, flow-through cells may be used for continuous measurements. Fibre optic units with modern instruments are computerized and have sophisticated optical charged coupled device (CCD) arrays to monitor the emission intensity at many wavelengths simultaneously. Statistical correlations are used to generate highly selective methods and to reduce interference levels.

CCD detectors are becoming quite popular, as complete UV-Vis spectra can be obtained in seconds, *in situ* sensing is available and computer interfacing is enhanced by digital electronics.

2.4 Use of Chromogenic Reagents

Some natural products are coloured substances and can be directly analyzed through colorimetry. Natural colourless or weakly coloured compounds, however, must be converted to more coloured derivatives. The reaction for colour formation may include enzymatic conversion, chemical modification of a substance to produce coloured products or formation of complexes between a substance and a colour-forming reagent. Several colour forming reagents are commercially available for analysis of many substances.

2.5 General Clues for Colorimetric Reactions

I. Chromogenic reagents usually react with a mixture of compounds belonging to the (correspond to the) same chemical class or many related chemical classes. Colour reactions of these compounds are controlled by the different kinetics producing variations in colour development (Fig. 4) and reducing the accuracy of the colorimetric method for quantitative analysis. Standard compounds should be selected according to past knowledge of the main representative molecules that are known components in a sample. Construction of several calibration curves with different standard compounds help to validate the use of a colorimetric method for quantitative analysis.

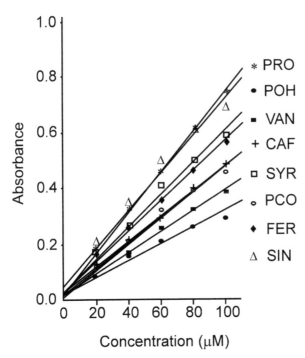

Figure 4   Calibration curves for protocatechuic (PRO), p-hydroxybenzoic (POH), vanillic (VAN), caffeic (CAF), syringic (SYR), p-coumaric (PCO), ferulic (FER), and sinapic (SIN) acids determined after reaction with Folin Ciocalteu reagent. Absorbance was measured at 750 nm (Source: Blum *et al.*, 1991).

II. A sample composition should be evaluated using several colorimetric methods (different chemical, reactions). Although water is the solvent of many colorimetric reactions, it interferes with some of them, hence is replaced by organic solvents. For this reason, the solvent used in the samples must be the same as that used to dissolve the standard compounds.

III. Colorimetric measures of a sample must be in the range of absorbances comprised in the calibration curve. Extreme absorbance values (too high or low) must be avoided. This ensures that the colorimetric measurements will be as per norms of Beer–Lambert Law. Little changes in reagents used in a colorimetric method can modify the slope of the calibration curve. Hence, calibration curves should be built each time a colorimetric method is applied for sample analysis.

IV. Sometimes components of the sample solution absorb at the wavelength used for colorimetric measurements or react with the chromogenic reagent increasing colour development. Partial isolation of the compounds of interest before colorimetric measurements aid in overcoming or reducing this kind of interference. Absorbance can also be read in the sample solution before and after colour development at the same wavelength or at the wavelength of maximum absorbance of interfering compounds. These values are further used to correct the absorbance lectures of the colorimetric method. General UV-Vis analysis can help to check if major interferences from other compounds are detected in the sample.

3. COLORIMETRIC METHODS

3.1 Phenolic Compounds

Experiment 1: Total Phenolic Compounds (Folin Ciocalteu Reagent)

The Folin Ciocalteu reagent is used to quantify concentrations of easily oxidizable compounds, such as phenols, by colour changes accompanying a redox reaction. The hydroxylated compounds reduce Cu^{2+} to Cu^+ in an alkaline medium, which in turn reduces the phosphotungsten-phosphomolybdic acid of the Folin Ciocalteu reagent to form an intensely blue complex (Singleton *et al.*, 1999). This complex is often quantified spectrophotometrically at 750 nm. A lithium sulfate salt is also a component of the reagent added to reduce the amounts of precipitate that can appear when high concentrations of reagent are used to increase the assay's reactivity. Folin Ciocalteu reagent has been widely used in phytochemistry and chemical ecology for preliminary analysis of phenolic compounds (Blum *et al.*, 1991).

Materials and equipments required

Assay tubes, Erlenmeyer flasks and pipettes; analytical balance with 0.01 mg sensitivity; spectrophotometer or filter photometer capable of measurement at 750 nm, equipped with a 1 cm path length microcuvette; vortex mixer.

Reagents

0.5 N NaOH (dissolve 20 g of NaOH in 500 mL of distilled water and make up to 1 L); 1% $CuSO_4$ (dissolve 1 g of $CuSO_4$ in 50 mL of distilled water and make up to 100 mL); 2% Na_2CO_3 (dissolve 2 g of Na_2CO_3 in 50 mL of distilled water and make up to 100 mL); 2.7% sodium and potassium tartrate (dissolve 2.7 g of sodium and potassium tartrate in 50 mL of distilled water and make up to 100 mL); solution A (mix 10 mL of 2% Na_2CO_3 with 0.1 mL of $CuSO_4$ and 0.1 mL of sodium and potassium tartrate in an Erlenmeyer flask, prepare this solution immediately before use); Folin Ciocalteu reagent (dilute the Folin Ciocalteu reagent with distilled water in the ratio 1:1 v/v, just before use); 1 mM ferulic acid standard (dissolve 1 mg of ferulic acid in 5.15 mL of distilled water, if ferulic acid is not completely dissolved, warm the tube in water at less than 50°C); aqueous extract from a plant material.

Procedure

(i) Set up a calibration curve in five tubes: dilute 1 mM ferulic acid with distilled water to prepare solutions 0.25, 0.50 and 0.75 mM. Label tubes from 1 to 5. Then, add 0.2 mL of 0, 0.25, 0.50, 0.75 and 1 mM ferulic acid to tubes 1, 2, 3, 4 and 5, respectively.

(ii) Set up your experimental assay tubes: prepare tubes 6 and 7 adding 0.1 and 0.2 mL of a plant extract, respectively. Complete volume of tube 6 to 0.2 mL with distilled water.

(iii) Add 2 mL of solution A in each tube and vortex. Then, add 0.4 mL of 0.5 N NaOH. Stand for 10 minutes at room temperature.

(iv) Add 0.2 mL of the Folin Ciocalteu diluted in water (1:1) and vortex. Leave to stand for 30 minutes at room temperature.

(v) Read absorbance at 750 nm using a 1 cm path length (1 mL) microcuvette.

Calculations

Step 1. Graph the calibration curve relating to absorbance at 750 nm to ferulic acid concentration. Fit a line function to the measured data points applying regression analysis ($r^2 > 0.80$). A calibration curve as shown in Fig. 5 should be obtained.

Step 2. Calculate concentration of total phenolic compounds (C) in the plant extract, expressed in equivalent concentrations of ferulic acid. For example, if absorbance at 750 nm (A_{750}) = 0.5 and extinction coefficient (ε) obtained from linear regression = 0.89, concentration (C) for tube 6 should be:

$$C \text{ (mM)} = (A_{750} \times \varepsilon) \times 2 \quad \text{hence, } C = (0.5 \times 0.89) \times 2$$
$$C = 0.72 \text{ mM}$$

In the example above, if calculations were on tube 7, concentration (C) should be:

$$C \text{ (mM)} = A_{750} \times \varepsilon \quad \text{hence, } C = 0.5 \times 0.89$$
$$C = 0.36 \text{ mM}$$

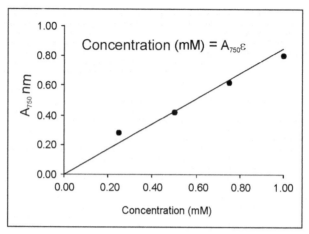

Figure 5 Absorbance at 750 nm vs ferulic acid concentration (Source: Adapted from Singleton *et al.*, 1999). ε = extinction coefficient. A_{750} nm = Absorbance at 750 nm.

Observations

(i) Make at least three independent experiments (N = 3).

(ii) The Folin Ciocalteu reagent can react with many oxidizable compounds. Although phenols are often the most abundant reactive compounds in a sample, samples may also contain other reactive compounds such as tyrosine, tryptophan, and ascorbic acid. To solve this problem, extraction with solutions of acetone-trichloroacetic acid (TCA) can be used to precipitate proteins eliminating tyrosine, tryptophan and other non-phenolic reactive compounds (Hatch *et al.*, 1993). Remotion of non-protein oxidizable compounds is also achieved with polyvinylpolypyrrolidone (PVPP): phenolic compounds bind to non-soluble PVPP under acidic conditions and are then removed by filtration or centrifugation. Reaction with Folin Ciocalteu reagent on sample aliquots occurs before and after precipitation with PVPP. Then, absorbance of phenolic compounds (A_{750}) is corrected as under:

A_{750} (corrected) = A_{750} (before PVPP precipitation) – A_{750} (after PVPP precipitation)

Concentration is determined with the corrected absorbance value as previously indicated in *Calculations*.

Experiment 2: Total Flavonols and Flavones

Carbonyl and hydroxyl groups of flavonols and flavones form complexes with Al^{3+} which absorb at 415 nm (Fig. 6).

Materials and equipments required

Assay tubes, Erlenmeyer flasks, pipettes; analytical balance with 0.01 mg sensitivity; spectrophotometer or filter photometer capable of measurement at 415 nm, equipped with a 1 cm path length microcuvette; vortex mixer.

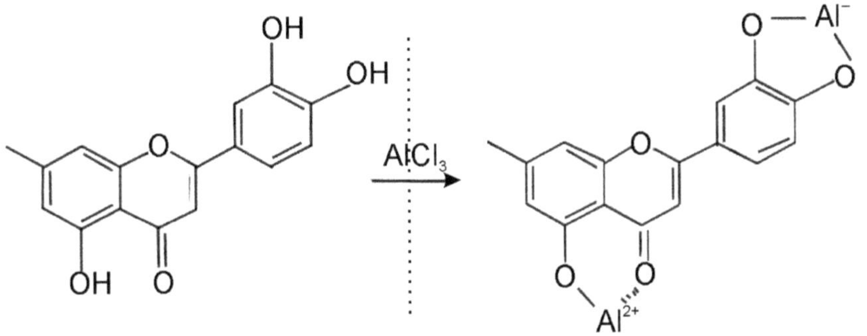

Figure 6 Reaction of Al^{3+} with a flavonoid molecule (Source: Adapted from Woisky and Salatino, 1998).

Reagents

10% $AlCl_3$ (dissolve 10 g of $AlCl_3$ in 50 mL of 98% ethanol and make up to 100 mL); 1 M potassium acetate (dissolve 19.63 g of potassium acetate in 80 mL of distilled water and adjust volume to 100 mL); 1 mg/mL stock quercetin solution (dissolve 100 mg of quercetin in 80 mL of 98% ethanol and make up to 100 mL; this solution can be kept in a freezer for at least one month); 0.1 mg/mL standard quercetin solution (dilute 1 mL of quercetin stock solution in 9 mL of 98% ethanol; this solution can be kept in the refrigerator for at least one week); an ethanolic plant extract.

Procedure

(i) Set up a calibration curve in five tubes: dilute 0.1 mg/mL quercetin standard with 98% ethanol to obtain solutions with concentrations of 12, 24, 36 and 48 µM. Label tubes from 1 to 5. Then, add 0.5 mL of 0, 12, 24, 36 and 48 µM quercetin to tubes 1, 2, 3, 4 and 5, respectively.

(ii) Set up your experimental assay tubes: Prepare tubes 6 and 7 adding 0.25 and 0.5 mL of an ethanolic plant extract, respectively. Complete volume of tube 6 to 0.5 mL with 98% ethanol.

(iii) Add 1.5 mL of 98% ethanol and swirl.

(iv) Add 0.1 mL of 10% $AlCl_3$ and 0.1 mL of 1 M potassium acetate and swirl.

(v) Add 2.8 mL of distilled water. Let stand for 30 minutes at room temperature.

(vi) Read absorbance at 415 nm using a 1 cm path length (1 mL) microcuvette.

Calculations

Step 1. Graph the calibration curve relating to absorbance at 415 nm to quercetin concentration. Fit a line function to the measured data points applying regression analysis with $r^2 > 0.80$ (refer *Calculations* in Experiment 1).

Step 2. Calculate total concentration of flavonol and flavones (C) in the plant extract expressed in equivalent concentrations of quercetin:

$$C\ (mM) = (A_{415} \times \varepsilon) \times 2\ \text{(tube 6)}\quad \text{or}\quad C\ (mM) = A_{415} \times \varepsilon\ \text{(tube 7)}$$

where, A_{415} is absorbance at 415 nm and ε is extinction coefficient obtained from linear regression.

Observations

(i) Make at least three independent experiments (N = 3).

(ii) 10% $AlCl_3$ solution should be used within a week after preparation. The solution must be stored at room temperature in a dark coloured bottle. If precipitates are observed, solution must be decanted.

(iii) This method was originally developed to measure concentrations of quercetin, the most common flavonol in plants (Woisky and Salatino, 1998). It eliminates the interference of several phenolic compounds in the spectrophotometric analysis. However, accuracy of the method is high only when flavonols are the dominant flavonoids in a sample. Data will be less representative in samples with high levels of flavones because Al^{3+}-flavones complexes have maximum absorbance at wavelengths less than 415 nm.

Experiment 3: Condensed Tannins (Proanthocyanidins)

Oxidative cleavage of proanthocyanidins leads to formation of coloured anthocyanidins that are detected spectrophotometrically at 550 nm (Fig. 7).

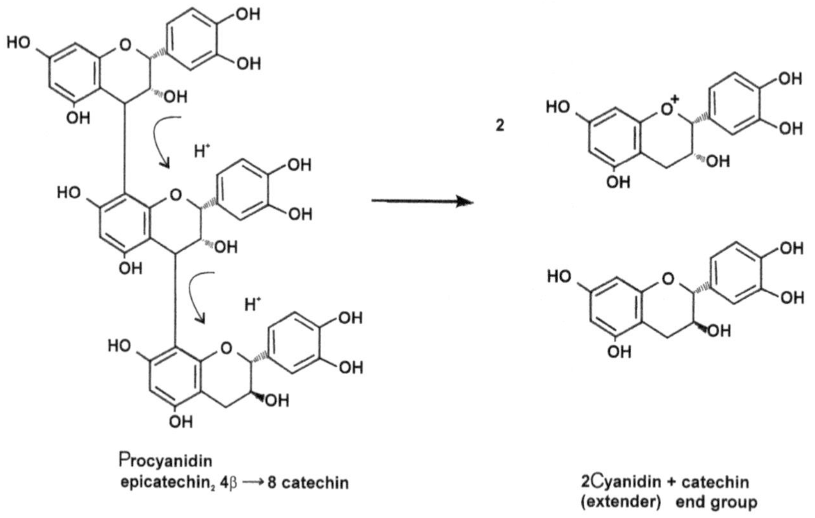

Procyanidin
epicatechin, 4β → 8 catechin

2Cyanidin + catechin
(extender) end group

Figure 7 Conversion of proanthocyanidins to coloured anthocyanidins in the butanol acid assay (Source: Hagerman, 2002).

Materials and equipments required

Assay tubes, Erlenmeyer flasks, marbles, pipettes; analytical balance with 0.01 mg sensitivity; water bath; spectrophotometer or filter photometer capable of measurement at 550 nm, equipped with a 1 cm path length microcuvette; vortex mixer.

Reagents

Butanol-HCl reagent, 95:5 v/v (add 950 mL n-butanol to 50 mL concentrated HCl); 2% ferric reagent (make up 16.6 mL of concentrated HCl to 100 mL with distilled water and add 2.0 g of ferric ammonium sulfate); 1 mg/mL stock catechin solution (weigh 100 mg catechin and dissolve in 80 mL of 98% ethanol. Then, make up the volume to 100 mL); 0.1 mg/mL standard catechin solution (dilute 1 mL of catechin stock solution in 9 mL of 98% ethanol); an ethanolic plant extract.

Procedure

(i) Set up a calibration curve in five tubes: dilute 0.1 mg/mL catechin solution with 98% ethanol to prepare solutions 17, 35, 50 and 70 μM. Label tubes from 1 to 5. Then, add 1 mL of 0, 17, 35, 50 and 70 μM catechin to tubes 1, 2, 3, 4 and 5, respectively.

(ii) Set up your experimental assay tubes: prepare tubes 6 and 7 adding 0.5 and 1 mL of an ethanolic plant extract, respectively. Complete volume of tube 6 to 1 mL with 98% ethanol.

(iii) Add 6 mL of the Butanol-HCl reagent to each tube.

(iv) Add 0.2 mL of ferric reagent and vortex. Cover the mouth of each tube with a glass marble and put the tubes in a boiling water bath for 50 minutes.

(v) Let the tubes cool at room temperature. Read absorbance at 550 nm using a 1 cm path length (1 mL) microcuvette.

Calculations

Step 1. Graph the calibration curve relating absorbance at 550 nm to catechin concentration. Fit a line function to the measured data points applying regression analysis with $r^2 > 0.80$ (refer *Calculations* in Experiment 1).

Step 2. Calculate concentration of total condensed tannins (C), considering absorbance at 550 nm (A_{550}) of tubes 6 and 7 and extinction coefficient (ε) obtained from linear regression:

$$C\,(\mu M) = (A_{550} \times \varepsilon) \times 2 \text{ (tube 6)} \quad \text{or} \quad C\,(\mu M) = A_{550} \times \varepsilon \text{ (tube 7)}$$

Observations

(i) Make at least three independent experiments (N = 3).

(ii) Ferric reagent should be stored in a dark coloured bottle.

(iii) Store catechin solution in a freezer for not more than a month after preparation.

(iv) Sometimes chemical characteristics of certain plant tannins, such as position of the interflavan bond and oxygenation pattern, significantly affect the colour development, underestimating tannin content of the sample.

(v) The presence of pigments, particularly chlorophyll, may interfere in this method. An alternative is to read absorbance before heating and subtract this value from the absorbance determined after heating. Chlorophyll may also be eliminated by precipitation with Polyvinylpyrrolidone (PVP) or by chlorophyll extraction of the dry sample with petroleum ether.

Experiment 4: Gallotannins

Hydrolyzable tannins, such as gallotannins, release gallic acid after hydrolysis. Gallic acid reacts with rhodanine dye producing coloured complexes that are detected spectrophotometrically at 520 nm. Plant samples often contain gallic acid in free forms. Hence, not only total hydrolizable gallic acid (free and bound) is measured, but also the free gallic acid found in the sample before hydrolysis (Makkar, 2000).

Materials and equipments required

Assay tubes, Erlenmeyer flasks, pipettes, marbles; analytical balance with 0.01 mg sensitivity; spectrophotometer or filter photometer capable of measurement at 520 nm, equipped with a 1 cm path length microcuvette; vortex mixer; rotary evaporator; water bath; vacuum pump.

Reagents

Rhodanine solution 0.667% (weigh 667 mg rhodanine and dissolve it in 100 mL methanol; this solution is stable for 2 weeks when stored in a refrigerator); 0.5 N KOH (dissolve 2.8 g potassium hydroxide [KOH] in 100 mL distilled water); H_2SO_4 solutions (prepare 0.2 and 22 N solutions diluting concentrated H_2SO_4); 1 mg/mL gallic acid stock solution (weigh 100 mg gallic acid and dissolve in approximately 80 mL of 0.2 N H_2SO_4 and then make up the volume to 100 mL with 0.2 N H_2SO_4; it can be kept frozen for at least one month); 0.1 mg/mL standard gallic acid solution (dilute 1 mL of gallic acid stock solution with 9 mL of 0.2 N H_2SO_4; it can be stored in a refrigerator for at least two weeks); tannin extract (place 200 g of finely ground plant material in a 25 mL glass beaker; add 10 mL of 70% aqueous acetone and shake for 20 minutes at room temperature; transfer the liquid to centrifuge tubes and centrifugate for 10 minutes at $3,000 \times g$ and 4°C; collect the supernatant and keep it on ice for tannin analysis).

Procedure

Calibration Curve

(i) Set up a calibration curve in five tubes: dilute 0.1 mg/mL gallic acid solution with 0.2 N H_2SO_4 to prepare solutions 0.12, 0.25, 0.38 and 0.50 mM. Label tubes from 1 to 5. Then, add 0.2 mL of 0, 0.12, 0.25, 0.38 and 0.50 mM gallic acid to tubes 1, 2, 3, 4 and 5, respectively.

(ii) Add 0.3 mL of the rhodanine solution to each tube.

(iii) After 5 minutes, add 0.2 mL of 0.5 N KOH solution to each tube. Wait for 2.5 minutes and then add 4.3 mL of distilled water.

(iv) After 15 minutes, read absorbance at 520 nm using a 1cm path length (1 mL) microcuvette.

Determination of Free Gallic Acid

(i) Set up four experimental assay tubes and add 0.2 mL of tannin extract to each tube. Eliminate acetone drying in a rotary evaporator. Then, add 0.2 mL of 0.2 N H_2SO_4.

(ii) Proceed as previously indicated for calibration curve.

Determination of Gallic Acid Present in Free and in Gallotannin Forms

(i) Pipette 3.34 mL of tannin extract in a tube. Remove acetone by drying in a rotary evaporator. Make up to 1 mL. Then add 0.1 mL of 22 N H_2SO_4, so that the final H_2SO_4 concentration is up to 2 N. Freeze the contents and remove air from these culture tubes by using a vacuum pump.

(ii) Cap each tube with a glass marble and put the tubes in a boiling water bath for 4 hours to hydrolyze gallotannins to gallic acid. After hydrolysis, make up the volume to 11 mL by adding 9.9 mL distilled water. H_2SO_4 concentration in this solution is 0.2 N. Prepare four tubes. Add 0.2 mL of this solution in three of them. Add 0.2 ml of 0.2 N H_2SO_4 to the fourth tube. This last tube will be used as the control.

(iii) Add 0.3 mL of the rhodanine solution to each tube.

(iv) Wait for 5 minutes and then add 0.2 mL of 0.5 N KOH solution to each tube. Again wait for 2.5 minutes and add 4.3 mL distilled water. After 10 minutes, measure absorbance at 520 nm.

Calculations

Step 1. Graph the calibration curve relating absorbance at 520 nm to gallic acid concentration. Fit a line function to the measured data points applying regression analysis with $r^2 > 0.80$ (refer *Calculations* in Experiment 1).

Step 2. Calculate concentration of free gallic acid (C_{free}) in the tannin extract as under:

$$C_{free} \text{ (mM)} = A_{520} \times \varepsilon$$

where, A_{520} is absorbance at 520 nm and ε is extinction coefficient obtained from linear regression.

Step 3. Calculate concentration of total gallic acid in free and gallotannin forms in 0.2 ml of the sample (Cs):

$$Cs \text{ (mM)} = A_{520} \times \varepsilon$$

However, original volume of tannin extract was 3.34 mL. Hence, total concentration (C_T) should be:

$$C_T \text{ (mM)} = C_{total} \times 0.2/3.34$$

Step 4. Calculate concentration of gallic acid in gallotannin forms ($C_{g\ form}$), subtracting free gallic acid found in tannin extract:

$$C_{g\ form} \text{ (mM)} = C_T - C_{free}$$

Precautions

H_2SO_4 is highly corrosive. Rhodanine is toxic by inhalation, contact or swallowing. Avoid exposure. Obtain special instructions before use.

Observations

(i) Ascorbic acid (generally added to prevent oxidation of phenols) interferes in this assay and, therefore, should not be added to the solvent used for tannin extraction.

(ii) Rhodanine solution is stable for two weeks when stored in a refrigerator, stock solution of gallic acid diluted in H_2SO_4 can be kept in the refrigerator for one week and gallic acid dissolved in H_2SO_4 can be kept frozen for at least one month.

Experiment 5: Lignin Content (Refer Chapter 14, Experiment 6)

3.2 Terpenoids

Experiment 6: Total Azadirachtin-like Limonoids

Triterpenoids form conjugated diene groupings with H_2SO_4. Addition of vanillin transforms this complex to a blue chromophore that is detected at 577 nm (Dai *et al.*, 1999).

Materials and equipments required

Assay tubes, pipettes, Erlenmeyer flasks; spectrophotometer or filter photometer capable of measurement at 577 nm, equipped with a 1 cm path length microcuvette; analytical balance with 0.01 mg sensitivity; neem seeds or other plant material containing limonoids.

Reagents

Dichloromethane; petroleum ether; methanol; concentrated H_2SO_4 (98%); 0.02 mg/mL vanillin reagent (dissolve 100 mg of vanillin in 50 mL methanol; take 0.5 mL of this solution and make up to 50 mL with methanol); 0.1 mg/mL standard azadirachtin solution (dissolve 10 mg of azadirachtin in 100 mL of dichloromethane); neem seed extract (stir blended 2 g of neem seeds overnight in 60 mL of petroleum ether at room temperature; filter and recuperate the residues; stir the residue in 50 mL methanol at room temperature for 10 minutes; then, filter and evaporate the filtrate solution in a rotary evaporator; an orange amorphous solid should be obtained; then, dissolve this material in dichloromethane).

Procedure

(i) Set up a calibration curve in five tubes: dilute 0.1 mg/mL azadirachtin solution with dichloromethane to prepare solutions 1.0, 2.6, 5.1 and 7.2 mM. Label tubes from 1 to 5. Then add 0.7 mL of 0, 1.0, 2.6, 5.1 and 7.2 mM azadirachtin to tubes 1, 2, 3, 4 and 5, respectively.

(ii) Set up your experimental assay tubes and prepare tubes 6 and 7 adding 0.35 and 0.70 mL of the dichloromethane plant extract, respectively. Complete volume of tube 6 to 0.7 mL with dichloromethane.

(iii) Add 0.2 mL of the vanillin solution to each tube and swirl. Stand for 2 minutes at room temperature.

(iv) Add 0.3 mL of concentrated H_2SO_4 to each tube in three portions (0.1 mL each) and mix carefully.

(v) Add 0.7 mL of methanol. Stand for 5 minutes at room temperature.

(vi) Read absorbance at 577 nm using a glass (1 mL) microcuvette.

Calculations

Step 1. Graph the calibration curve relating absorbance at 577 nm to azadirachtin concentration. Fit a line function to the measured data points applying regression analysis with $r^2 > 0.80$ (refer *Calculations* in Experiment 1).

Step 2. Calculate content of azadirachtin like limonoids in equivalent azadirachtin concentration (C), considering absorbance at 577 nm (A_{577}) and extinction coefficient (ε) obtained from linear regression:

$$C \text{ (mM)} = (A_{557} \times \varepsilon) \times 2 \text{ (tube 6)} \quad \text{or} \quad C \text{ (mM)} = A_{557} \times \varepsilon \text{ (tube 7)}$$

Observations

Although this method was specifically described for azadirachtin, vanillin–H_2SO_4 also reacts with triterpenoid sapogenins, sterols and steroidal sapogenins producing stable red-purple colours with λ^{max} in the region 515-545 nm.

Precautions

H_2SO_4 is highly corrosive. Methanol, petroleum ether and dichloromethane are toxic by inhalation, contact or swallowing. Avoid exposure. Obtain special instructions before use.

3.3 Alkaloids

Experiment 7: Precipitable Alkaloids (Dragendorff Reagent)

Dragendorff´s reagent form red precipitates with alkaloids in solution. The precipitates are solubilized in NaI, being possible to read absorbance of the coloured solution at 467 nm (Stumpf, 1984).

Materials and equipments required

Assay tubes, Erlenmeyer flasks, Pasteur pipettes, pipettes; analytical balance with 0.01 mg sensitivity; spectrophotometer or filter photometer capable of measurement at 595 nm, equipped with a 1 cm path length glass microcuvette; centrifuge capable to reach a speed of $7,000 \times g$.

Reagents

Dragendorff reagent (Solution A: Prepare 0.35 M bismuth nitrate in 20% acetic acid (v/v); Solution B: Prepare 2.45 M NaI in distilled water; prepare Dragendorff reagent by mixing solutions A and B in a 1:1 (v/v) relation)); 0.1 mg/mL standard papaverine hydrochloride solution (dissolve 1 mg of papaverine hydrochloride in 100 mL distilled water); a plant extract.

Procedure

(i) Set up a calibration curve in four tubes: dilute 0.1 mg/mL papaverine solution with distilled water to prepare solutions 0.1, 0.2 and 0.3 mM. Label tubes from 1 to 4 Then, add 0.2 mL of 0, 0.1, 0.2 and 0.3 mM papaverine to tubes 1, 2, 3 and 4, respectively.

(ii) Set up your experimental assay tubes. Prepare tubes 5 and 6 adding 0.15 and 0.3 mL of a plant extract, respectively. Complete volume of tube 6 to 0.3 mL with distilled water.

(iii) Add 0.1 mL of Dragendorff reagent to each tube.

(iv) Centrifugate the tubes for 1 minute at $7,000 \times g$. Carefully remove the supernatant with a Pasteur pipette.

(v) Dissolve each pellet in 1mL of 2.45 M NaI. Add 10 µL aliquots of each tube to 1 mL of 0.49 M NaI and swirl.

(vi) Read absorbance at 467 nm using the glass microcuvette.

Calculations

Step 1. Graph the calibration curve relating absorbance at 467 nm to papaverine concentration. Fit a line function to the measured data points applying regression analysis with $r^2 > 0.80$ (refer *Calculations* in Experiment 1).

Step 2. Calculate content of total alkaloids in equivalent papaverine concentration (C), considering absorbance at 467 nm (A_{467}) and extinction coefficient (ε) obtained from linear regression:

$$C \text{ (mM)} = (A_{467} \times \varepsilon) \times 2 \text{ (tube 5)} \quad \text{or} \quad C \text{ (mM)} = A_{467} \times \varepsilon \text{ (tube 6)}$$

Observations

After centrifugation, the supernatant must be completely removed or the remaining bismuth nitrate will cause an increased background. Proteins can precipitate in the presence of Dragendorff reagent. It is recommended to remove the proteins from the samples before addition of the reagent.

Experiment 8: Pyrrolizidine Alkaloids (Methyl Orange Method)

Protonation of pyrrolizidine alkaloids in chloroform solution with acetic acid and their stoichiometric reaction with aqueous methyl orange lead to the formation of yellow complex that can be detected at 525 nm (Birecka *et al.*, 1981).

Materials and equipments required

Assay tubes, Erlenmeyer flasks; analytical balance with 0.01 mg sensitivity, centrifuge capable to reach a speed of $400 \times g$, Whatman # 1 paper, spectrophotometer or filter photometer capable of measurement at 525 nm, equipped with a 1 cm path length glass microcuvette.

Reagents

Methyl orange reagent (dissolve 500 mg powdered methyl orange in 100 mL distilled water at 40°C for 20 minutes; let cool at room temperature and then filter with Whatman # 1 paper); 1.25% acetic acid in distilled water; 2% H_2SO_4 in ethanol (dissolve 2 mL H_2SO_4 in 100 mL ethanol); 0.25 M H_2SO_4 (dissolve 1.40 mL H_2SO_4 in 100 mL ethanol); Plant Extract (Soak 1 kg of dried milled plant material in methanol for 24 hours; filter and evaporate the solution to dryness; dissolve the residue in 0.25 M H_2SO_4; extract chlorophyll with petroleum ether; add Zn dust to reduce N oxide forms of alkaloids to tertiary alkaloids; filter and add NH_3 to reach pH > 10. Add chloroform and vortex. Aspirate the chloroform phase with a Pasteur pipette and use as the plant extract.); standard monocrotaline solution, 0.1 mg/mL (dissolve 10 mg of monocrotaline in 100 mL of chloroform).

Procedure

(i) Set up a calibration curve in five tubes: dilute 0.1 mg/mL monocrotaline solution with chloroform to prepare solutions 6, 12, 25 and 50 µM. Label tubes from 1 to 5. Then, add 5 mL of 0, 6, 12, 25 and 50 µM monocrotaline to tube 1, 2, 3, 4 and 5, respectively.

(ii) Set up your experimental assay tubes: prepare tubes 6 and 7 adding 2.5 and 5 mL of the chloroform plant extract, respectively. Complete volume of tube 6 to 5 mL with chloroform.

(iii) Add 10 µl of 1.25% acetic acid and vortex.

(iv) Add 25 µl of methyl orange solution and vortex.

(v) Stand for 5 minutes. Transfer the chloroform phase to glass centrifuge tubes.

(vi) Centrifugate at 400 × g for 2 minutes.

(vii) Transfer 1.5 mL of each supernatant to a tube.

(viii) Add 0.1 mL of 2% ethanolic H_2SO_4 to each tube and vortex.

(ix) Read absorbance at 525 nm using the glass microcuvette.

Calculations

Step 1. Graph the calibration curve relating to absorbance at 525 nm to monocrotaline concentration. Fit a line function to the measured data points applying regression analysis with $r^2 > 0.80$ (refer *Calculations* in Experiment 1).

Step 2. Calculate total pyrrolizidine alkaloids in equivalent monocrotaline concentration (C), considering absorbance at 525 nm (A_{525}) and extinction coefficient (ε) obtained from linear regression:

$$C \text{ (µM)} = (A_{525} \times \varepsilon) \times 2 \text{ (tube 6)} \quad \text{or} \quad C \text{ (µM)} = A_{525} \times \varepsilon \text{ (tube 7)}$$

Precautions

Chloroform and methyl orange are toxic by inhalation, contact or swallowing. Avoid exposure. Obtain special instructions before use.

3.4 Lipids

Experiment 9: Total Sterol Content (Lieberman–Buchard Method)

Free sterols and sterol esters react with acetic anhydride and concentrated H_2SO_4, forming a blue-green complex which absorbance can be read at 550 nm.

Materials and equipments required

Assay tubes, pipettes, Erlenmeyer flasks; analytical balance with 0.01 mg sensitivity, water bath; Vortex mixer, spectrophotometer or filter photometer capable of measurement at 550 nm, equipped with a 1 cm path length glass microcuvette.

Reagents

Chloroform–methanol mixture: Mix chloroform with methanol in 2:1 ratio; Lieberman–Buchard reagent [add slowly, while stirring continuously, 10 mL of concentrated H_2SO_4 to 60 mL of acetic anhydride in an Erlenmeyer flask. As the reaction is strongly exothermic, cooling the Erlenmeyer flask in an ice bath is recommendable. Then, add 30 mL of acetic acid and 0.6 g of anhydrous sodium sulfate]; Standard cholesterol solution (0.4 mg/mL): Dissolve 40 mg of cholesterol in 100 mL of chloroform:methanol (2:1); walnut extract: Grind 1 g of walnut thoroughly in a mortar with 5 mL of chloroform:methanol mixture. Transfer to a test tube washing the mortar with 5 mL of the same solvent into the same tube; cap the tube and place it in a water bath at 50°C for 10 minutes; filter using Whatman #1 paper. Measure the volume of filtrate and add enough $CaCl_2$ to prepare a 0.02% solution; vortex; leave the tube until there are two separate phases; remove the upper phase carefully with a Pasteur pipette; double the total volume of the lower phase by the addition of chloroform:methanol:water (2:50:50); vortex and leave until two phases separate; remove the upper phase with a Pasteur pipette and measure the volume of the lower phase, containing the lipids; measure 0.5 mL of the obtained extract and transfer to another tube for sterol determination; the remaining lipid extract is placed in a water bath at 80°C and evaporated to dryness; weigh the dry lipid residue.

Procedure

(i) Set up a calibration curve in five tubes: dilute 0.4 mg/mL cholesterol solution with chloroform:methanol (2:1) to prepare solutions 0.2, 0.4, 0.6 and 0.8 µM. Label tubes from 1 to 5. Then, add 0.5 mL of 0, 0.2, 0.4, 0.6 and 0.8 µM cholesterol to tubes 1, 2, 3, 4 and 5, respectively.

(ii) Set up your experimental assay tubes and prepare tubes 6 and 7 adding 0.25 and 0.50 mL of a plant extract, respectively. Complete volume of tube 6 to 0.5 mL with chloroform:methanol (2:1).

(iii) Add 5 mL of Lieberman–Buchard reagent to each tube.

(iv) Cap the test tubes, swirl and place in a water bath at 35°C for 10 minutes.

(v) Measure the absorbance at 550 nm.

Calculations

Step 1. Graph the calibration curve relating absorbance at 550 nm to cholesterol concentration. Fit a line function to the measured data points applying regression analysis with $r^2 > 0.80$ (refer *Calculations* in Experiment 1).

Step 2. Calculate concentration of total sterols (C) in the plant extract expressed as cholesterol equivalent concentration. Consider absorbance at 550 nm (A_{550}) from experimental tubes and extinction coefficient (ε) obtained from linear regression as under:

$$C\ (\mu M) = (A_{550} \times \varepsilon) \times 2 \ (\text{tube 6}) \quad \text{or} \quad C\ (\mu M) = A_{550} \times \varepsilon \ (\text{tube 7})$$

Observations

The Lieberman–Buchard test can be done at room temperature. However, placing the test tubes in a water bath at 35°C speed up the reaction.

3.5 Carbohydrates

Experiment 10: Total Neutral Carbohydrates (Phenol Sulfuric Method)

H_2SO_4 in hot conditions dehydrates the carbohydrates to form either furfural (from pentoses, Fig. 8A) or 5-hydroxymethyl furfural (from hexoses, Fig. 8B). These products of carbohydrate dehydration can react with phenol to form a coloured complex that can be detected spectrophotometrically at 490 nm (Dubois *et al.*, 1956).

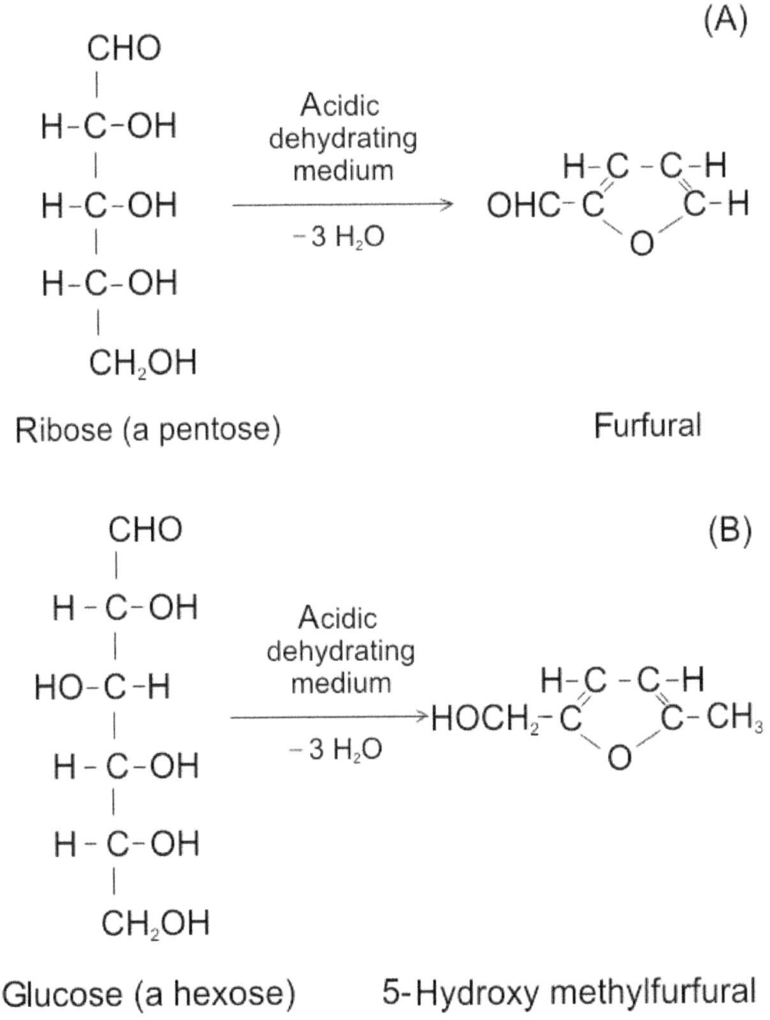

Figure 8 Hot acid dehydration of (A) pentoses, and (B) hexoses.

Materials and equipments required

Assay tubes, marbles, Erlenmeyer flasks, pipettes; analytical balance with 0.01 mg sensitivity, water bath, spectrophotometer or filter photometer capable of measurement at 490 nm, equipped with a 1 cm path length microcuvette.

Reagents

H_2SO_4 concentrated; 80% phenol, w/v (Dissolve 0.8 g phenol in a boiling water bath. Then cool and dilute with distilled water to reach 1 L. This solution should be prepared with caution because phenol is hydroscopic and toxic.); 1 mg/mL stock glucose solution (Weigh 90 mg glucose and dissolve in 25 mL of distilled water; make up to 50 mL. It can be kept frozen for at least six months.); 0.18 mg/mL standard glucose solution (dilute 1 mL the glucose stock solution in 9 mL of distilled water); an aqueous plant extract.

Procedure

(i) Set up a calibration curve in five tubes: dilute 0.18 mg/mL glucose solution with distilled water to prepare solutions 60, 120, 190 and 250 μM. Label tubes from 1 to 5. Then, add 0.8 mL of 0, 60, 120, 190 and 250 μM glucose to tubes 1, 2, 3, 4 and 5, respectively.

(ii) Set up your experimental assay tubes and prepare tubes 6 and 7 adding 0.4 and 0.8 mL of the plant extract, respectively. Complete volume of tube 6 to 0.8 mL with distilled water.

(iii) Add 0.04 mL of 80% phenol to each tube and swirl. Add 2 mL of H_2SO_4 concentrated. Cover the mouth of each tube with a glass marble and put the tubes in a boiling water bath for 20 minutes. Let the tubes cool to room temperature.

(iv) Read absorbance at 490 nm using a 1 cm path length (1 mL) microcuvette.

Calculations

Step 1. Graph the calibration curve relating absorbance at 490 nm to glucose concentration. Fit a line function to the measured data points applying regression analysis with $r^2 > 0.80$ (refer *Calculations* in Experiment 1).

Step 2. Calculate concentration of total sugars (C) in the plant extract expressed as glucose equivalent concentration. Consider absorbance at 490 nm (A_{490}) from experimental tubes and extinction coefficient (ε) obtained from linear regression:

$$C \, (\mu M) = (A_{490} \times \varepsilon) \times 2 \text{ (tube 6)} \quad \text{or} \quad C \, (\mu M) = A_{490} \times \varepsilon \text{ (tube 7)}$$

Precautions

H_2SO_4 is extremely corrosive and can cause immediate and serious skin or eye damage. It can also cause damage to clothing and equipment. Avoid exposure. Obtain special instructions before use.

Experiment 11: Total Reducing Carbohydrates (Somogyi–Nelson Method)

Carbohydrates with free aldehyde or keto groups, in alkaline conditions, reduce Cu^{2+} to Cu^+ which in turn reduces the arseno-molybdate complex of Nelson reagent to molybdenum blue which absorbs at 520 nm (Fig. 9).

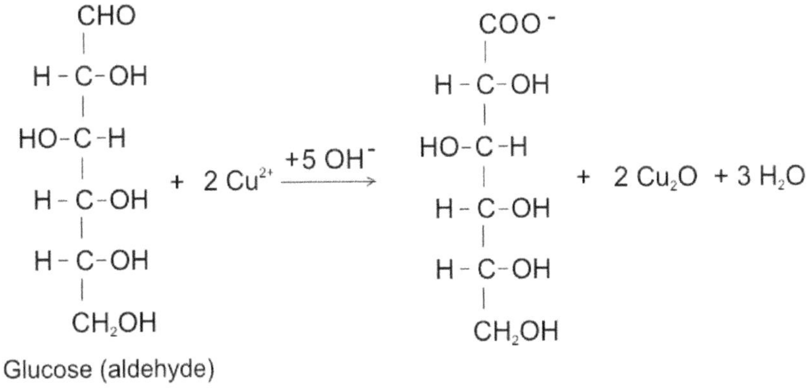

Glucose (aldehyde)

Figure 9 Somogyi–Nelson reaction (Source: Adapted from Somogyi, 1952).

Materials and equipments required

Assay tubes, marbles, pipettes, Erlenmeyer flasks; analytical balance with 0.01 mg sensitivity; water bath; spectrophotometer or filter photometer capable of measurement at 520 nm, equipped with a 1 cm path length microcuvette.

Reagents

Somogyi reagent [Solution A: 10% $CuSO_4$; Solution B: Dissolve 28 g Na_2HPO_4 (anhydrous) in 700 mL of distilled water. Add 40 g of tetra hydrated sodium and potassium tartrate; then, add 100 mL 1 N NaOH and 120 g Na_2SO_4 (anhydrous); reach to 900 mL with distilled water, stand 48 hours at room temperature and filter through Whatman # 1 paper; mix solution A with solution B in the ratio 1:9 just before use.]; Nelson reagent (dissolve 25 g of $(NH_4)_6Mo_7O_{24} \cdot 4 \, H_2O$ in 450 mL of distilled water; add 21 mL of concentrated H_2SO_4; after mixing, add 3 g $Na_2HAsO_4 \cdot 7 \, H_2O$ previously dissolved in 25 mL distilled water; keep at 37°C for 2 days and store in a brown bottle at room temperature); 1 mg/mL stock glucose solution (weigh 100 mg glucose and dissolve in 80 mL of distilled water; then, make up the volume to 100 mL. It can be kept frozen for at least six months); 0.1 mg/mL standard glucose solution (dilute 1 mL of glucose stock solution in 10 mL of distilled water); an aqueous plant extract.

Procedure

(i) Set up a calibration curve in five tubes: dilute 0.18 mg/mL glucose solution with distilled water to prepare solutions 0, 0.1, 0.2, 0.3 and 0.4 mM. Label tubes from 1 to 5. Then, add 0.5 mL of 0, 0.1, 0.2, 0.3 and 0.4 mM glucose to tube 1, 2, 3, 4 and 5, respectively.

(ii) Set up your experimental assay tubes and prepare tubes 6 and 7 adding 0.25 and 0.50 mL of a plant extract, respectively. Complete volume of tube 6 to 0.5 mL with distilled water.

(iii) Add 0.5 mL of Somogyi reagent to each tube. Cover the mouth of each tube with a glass marble and put the tubes in a boiling water bath for

15 minutes. Let cool to room temperature. Add 0.5 mL of Nelson reagent to each tube and swirl.

(iv) Add 1 mL of distilled water to each tube and vortex.

(v) Read absorbance at 520 nm using a 1 cm path length (1 mL) microcuvette.

Calculations

Step 1. Graph the calibration curve relating absorbance at 520 nm to glucose concentration. Fit a line function to the measured data points applying regression analysis with $r^2 > 0.80$ (refer *Calculations* in Experiment 1).

Step 2. Calculate concentration of total reducing sugars (C) in the plant extract expressed as glucose equivalent concentration. Consider absorbance at 520 nm (A_{520}) from experimental tubes and extinction coefficient (ε) obtained from linear regression:

$$C \text{ (mM)} = (A_{520} \times \varepsilon) \times 2 \text{ (tube 6)} \quad \text{or} \quad C \text{ (mM)} = A_{520} \times \varepsilon \text{ (tube 7)}$$

Observations

Compare total carbohydrate and total reducing carbohydrate contents in the extract and calculate the relative participation of reducing carbohydrates referred to total carbohydrate content.

Experiment 12: Starch

Addition of an adequate quantity of KI to a starch solution (weakly acid) and dilute hydrogen peroxide leads to the formation of a blue starch-iodide complex. This complex can be detected at 620 nm (Chinoy, 1939).

Materials and equipments required

Assay tubes, Erlenmeyer flasks, pipettes; analytical balance with 0.01 mg sensitivity; water bath; spectrophotometer or filter photometer capable of measurement at 620 nm, equipped with a 1 cm path length microcuvette.

Reagents

Chloroform; H_2O_2 of 20 volumes; 0.1 N KI (Dissolve 8.3 g in 300 mL of distilled water. Make up to a final volume of 500 mL); 2.5 mg/mL standard starch solution (dissolve 1 g of soluble starch in 400 mL of distilled water by heating for 15 minutes in a water bath; cool to room temperature and make up to 500 mL in a flask).

Procedure

(i) Set up a calibration curve in five tubes: dilute 2.5 mg/mL starch solution with distilled water to prepare solutions 0.3, 0.6, 0.9 and 1.3 mg/mL. Label tubes from 1 to 5. Then, add 2 mL of 0, 0.3, 0.6, 0.9 and 1.3 mg/mL starch to tubes 1, 2, 3, 4 and 5, respectively.

(ii) Set up your experimental assay tubes and prepare tubes 6 and 7 adding 1 and 2 mL of a plant extract, respectively. Complete volume of tube 6 to 2 mL with distilled water.

(iii) Add 2 mL of 0.1 N KI to each tube and swirl.

(iv) Add 2 mL of H_2O_2 to each tube and swirl.

(v) Stand tubes for 1 hour at room temperature.
(vi) Add 5 mL of chloroform to each tube and swirl.
(vii) Remove the chloroform phase of each tube and discard the chloroform phase.
(viii) Repeat steps 6 and 7 two times to eliminate the iodine excess from the aqueous phase.
(ix) Read absorbance of the aqueous phase at 620 nm.

Calculations

Step 1. Graph the calibration curve relating absorbance at 620 nm to starch concentration. Fit a line function to the measured data points applying regression analysis with $r^2 > 0.80$ (refer *Calculations* in Experiment 1).

Step 2. Calculate concentration of total starch (C) in the plant extract expressed as starch equivalent concentration. Consider absorbance at 620 nm (A_{620}) from experimental tubes and extinction coefficient (ε) obtained from linear regression:

$$C \text{ (mM)} = (A_{620} \times \varepsilon) \times 2 \text{ (tube 6)} \quad \text{or} \quad C \text{ (mM)} = A_{620} \times \varepsilon \text{ (tube 7)}$$

Observations

After chloroform addition, it is advisable to shake the tubes gently avoiding emulsion of the organic solvent with the aqueous solution. Turning tubes from end to end ensures efficient shaking and removal of iodine from the aqueous solution. Three or four changes of chloroform are adequate to remove all the iodine from the solution. This is easily noticed because the last volume of chloroform remains colourless.

3.6 Proteins and Amino Acids

Experiment 13: Protein Content (Bradford Method)

The absorbance maximum of an acidic solution of the dye Coomassie® Brilliant Blue G-250 shifts from 465 nm to 595 nm when binding to protein occurs. The dye has been assumed to bind via electrostatic attraction of the dye's sulfonic groups to the arginine residues of the proteins. Formation of the coloured complexes is measured spectrophotometrically at 595 nm (Bonjoch and Tamayo, 2001).

Materials and equipments required

Assay tubes, Erlenmeyer flasks, pipettes; analytical balance with 0.01 mg sensitivity, a spectrophotometer or filter photometer capable of measurement at 595 nm, equipped with a 1 cm path length microcuvette.

Reagents

Bradford reagent: dissolve 100 mg Coomassie Brilliant Blue G-250 in 50 mL 95% ethanol. Add 100 mL 85% (w/v) phosphoric acid; dilute to 1 L with distilled water when the dye has completely dissolved and filter through Whatman #1 paper just before use; stock solution of commercial bovine serum albumin, BSA (dissolve

100 mg of BSA in 100 mL of the same buffer used to prepare the plant extract); 0.1 mg/mL standard BSA solution (dilute 1 mL of BSA stock solution in 9 mL of buffer); this solution can be stored at −20°C and used whenever needed); plant extract (proteins are often extracted from plant tissues in buffer solutions; literature can be selected according to the tissue and protein to be extracted).

Procedure
(i) Set up a calibration curve in five tubes: dilute 0.1 mg/mL BSA solution with buffer to prepare solutions 25, 50 and 75 µg/mL. Label tubes from 1 to 5. Then, add 2 mL of 0, 25, 50, 75 and 100 µg/mL BSA to tubes 1, 2, 3, 4 and 5, respectively.
(ii) Set up your experimental assay tubes. Prepare tubes 6 and 7 adding 0.10 and 0.20 mL of a plant extract, respectively. Complete volume of tube 6 to 0.20 mL with buffer solution.
(iii) Add 1 mL Coomassie Brilliant Blue solution and swirl. Allow to stand 5 minutes at room temperature.
(iv) Read absorbance at 595 nm using a 1 cm path length (1 mL) microcuvette.

Calculations
Step 1. Graph the calibration curve relating absorbance at 595 nm to BSA concentration. Fit a line function to the measured data points applying regression analysis with $r^2 > 0.80$ (refer *Calculations* in Experiment 1).

Step 2. Calculate total protein concentration (C) in the plant extract expressed as BSA equivalent concentration. Consider absorbance at 595 nm (A_{595}) from experimental tubes and extinction coefficient (ε) obtained from linear regression:

$$C\ (\mu g/mL) = (A_{595} \times \varepsilon) \times 2 \text{ (tube 6)} \quad \text{or} \quad C\ (\mu g/mL) = A_{595} \times \varepsilon \text{ (tube 7)}$$

Observations
If the unknown protein concentration is too high, dilute the protein, assay a smaller aliquot of the unknown, or generate another calibration curve in a higher concentration range (e.g. 10 to 100 µg).

Experiment 14: Amino Acid Content (Acid-Ninhydrin Method)
Ninhydrin produces the oxidative deamination of the amino groups of amino acids releasing ammonium, the corresponding aldehyde and ninhydrin in reduced form. Released ammonium reacts with an additional mol of ninhydrin and with reduced ninhydrin, thus producing a complex (Ruhemann's purple) spectrophotometrically detected at 570 nm (Fig. 10).

Materials and equipments required
Assay tubes, marbles, Pasteur pipettes, pipettes; analytical balance with 0.01 mg sensitivity; water bath; spectrophotometer or filter photometer capable of measurement at 520 nm, equipped with a 1 cm path length glass microcuvette.

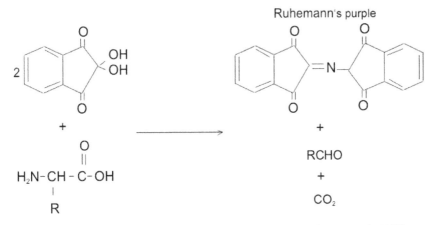

Figure 10 Reaction with Ninhydrin (Source: Adapted from Bates *et al.*, 1973).

Reagents

Sulfosalicylic acid 3% (w/v); phosphoric acid 6 M; glacial acetic acid; acid-ninhydrin solution (Warm 1.25 g ninhydrin in 30 mL glacial acetic acid and add 20 mL of 6 M phosphoric acid. Acid ninhydrin will keep stable only for 24 hours, at 4°C); toluene; 1 mg/mL stock proline solution (weigh 100 mg of proline and dissolve in 80 mL of 3% sulfosalicylic acid; then, make up the volume to 100 mL with 3% sulfosalicylic acid); 0.1 mg/mL standard proline solution (dilute 1 mL of proline stock solution in 9 mL 3% sulfosalicylic acid); plant extract (0.5 g of plant material was homogeinzed in 10 mL of 3% aqueous sulfosalicylic acid and the homogenate filtered through Whatman # 2 filter paper).

Procedure

(i) Set up a calibration curve in five tubes: dilute 0.1 mg/mL proline solution with 3% sulfosalicylic acid to prepare solutions 10, 20, 40 and 60 µg/mL. Label tubes from 1 to 5. Then, add 0.50 mL of 0, 10, 20, 40 and 60 µg/mL proline to tubes 1, 2, 3, 4 and 5, respectively.

(ii) Set up your experimental assay tubes and prepare tubes 6 and 7 adding 0.25 and 0.50 mL of a test plant extract, respectively. Complete volume of tube 6 to 0.50 mL with buffer solution.

(iii) Add 0.5 mL acid ninhydrin to each tube.

(iv) Add 0.5 mL glacial acetic acid to each tube.

(v) Cover the mouth of each tube with a glass marble and put the tubes in a boiling water bath for 1 hour. Let cool in an ice bath.

(vi) Add 1 mL toluene to each tube and vortex. Stand the tubes for 1 minute.

(vii) Aspirate the toluene phase containing the chromophore with a glass Pasteur pipette.

(viii) Read absorbance of the toluene phase at 520 nm using the glass microcuvette.

Calculations

Step 1. Graph the calibration curve relating absorbance at 520 nm to proline concentration. Fit a line function to the measured data points applying regression analysis with $r^2 > 0.80$ (refer *Calculations* in Experiment 1).

Step 2. Calculate total protein concentration (C) in the plant extract expressed as proline equivalent concentration. Consider absorbance at 520 nm (A_{520}) from experimental tubes and extinction coefficient (ε) obtained from linear regression:

$$C\ (\mu g/mL) = (A_{520} \times \varepsilon) \times 2 \ \text{(tube 6)} \quad \text{or} \quad C\ (\mu g/mL) = A_{520} \times \varepsilon \ \text{(tube 7)}$$

Observations

It is advisable to express free proline concentration relative to dry weight of plant material.

3.7 Glycosides

Experiment 15: Cyanide Production (Konig Reaction)

Total cyanide reaction is usually used to determine the release of cyanide from plant cyanogenic glycosides (Goldstein and Spencer, 1985). Cyanide subjected to an oxidant is transformed into cyanogen halide, CN^+ (Fig. 11). The cyanogen halide reacts with pyridine to produce an intermediate which hydrolizes to a conjugated dialdehyde, the glutaconic aldehyde. This compound is then coupled with a primary amine or a compound containing reactive methylene hydrogens to form coloured compounds that can be read at 580 nm.

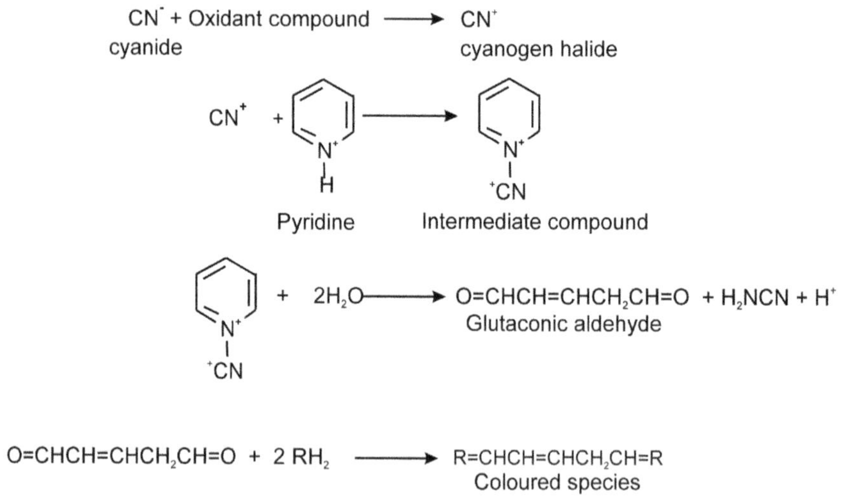

Figure 11   Reactions involved in detection of cyanide production.

Materials and equipments required

Assay tubes, marbles, pipettes; analytical balance with 0.01 mg sensitivity; spectrophotometer or filter photometer capable of measurement at 580 nm, equipped with a 1 cm path length glass microcuvette.

Reagents

1 N NaOH (Dissolve 40 g of NaOH in 400 mL of distilled water. Make up to a final volume of 1 L); acetic acid 2 N (dilute 57.2 mL of acetic acid to a final volume of 1 L); oxidant reagent (dissolve 2.5 g of succinimide in 300 mL of distilled water; add 0.25 g N-chlorosuccinimide and stir to dissolve; dilute the solution to 1 L); coupling compound reagent (dissolve 12 g of barbituric acid in 120 mL of pyridine; make up to a final volume of 400 mL with distilled water); 0.5 mg/mL stock NaCN solution (dissolve 50 mg of NaCN in 100 mL of 1 N NaOH); standard NaCN solution (dilute 1 mL of stock NaCN solution in 49 mL of 1 N NaOH); cassava extract (collect root peels of cassava; grind 1 g of root peel in a mortar; stand for 2 hours. Add 7 mL of 1 N NaOH; grind again in the mortar; filter and use for the assay).

Procedure

(i) Set up a calibration curve in five tubes: dilute standard NaCN solution with 1 N NaOH to prepare solutions 20, 40, 60 and 80 µM. Label tubes from 1 to 5. Then, add 1 mL of 0, 20, 40, 60 and 80 µM NaCN to tubes 1, 2, 3, 4 and 5, respectively.

(ii) Set up your experimental assay tubes and prepare tubes 6 and 7 adding 0.50 and 1 mL of cassava extract, respectively. Complete volume of tube 6 to 1 mL with 1 N NaOH.

(iii) Add 0.5 mL of acetic acid 2 N and 5 mL of the oxidant reagent to each tube.

(iv) Add 1 mL of the coupling compound reagent and swirl. Stand tubes for 10 minutes.

(v) Read absorbance at 580 nm.

Precautions

NaCN is extremely poisonous, eating or inhaling it is harmful, or even fatal.

Calculations

Step 1. Graph the calibration curve relating absorbance at 580 nm to NaCN concentration. Fit a line function to the measured data points applying regression analysis with $r^2 > 0.80$ (refer *Calculations* in Experiment 1).

Step 2. Calculate cyanide concentration (C) produced from cassava extract expressed as NaCN equivalent concentration. Consider absorbance at 580 nm (A_{580}) from experimental tubes and extinction coefficient (ε) obtained from linear regression:

$$C \ (\mu M) = (A_{580} \times \varepsilon) \times 2 \ (\text{tube } 6) \quad \text{or} \quad C \ (\mu M) = A_{580} \times \varepsilon \ (\text{tube } 7)$$

Experiment 16: Total Cardenolide Content (Cardiac Glycosides)

In alkaline conditions, cardenolides are formed with 2,2',4,4'-tetranitrobiphenyl (TNBP) reagent from a stable coloured complex, whose absorbance is measured at 620 nm (Genkina and Eídler, 1975).

Materials and equipments required

Assay tubes, marbles, Erlenmeyer flasks, pipettes; analytical balance with 0.01 mg sensitivity, water bath, spectrophotometer or filter photometer capable of measurement at 620 nm, equipped with a 1 cm path length glass microcuvette.

Reagents

2,2',4,4'-tetranitrobiphenyl (TNBP) reagent (dissolve 150 mg of TNBP in 100 mL of absolute ethanol); stock digitoxin solution (dissolve 100 mg of digitoxin in 10 mL ethanol); 50 μg/mL standard digitoxin solution (dilute 0.5 mL of the stock digitoxin solution to 100 mL with absolute ethanol); *Asclepias* leaf extract (add 5 g of fresh young leaves of *Asclepias glaucescens* [or other *Asclepias* spp] to 50 mL of 95% ethanol; boil for 5 minutes; after cooling, filter the extract).

Procedure

(i) Set up a calibration curve in five tubes: dilute 50 μg/mL digitoxin solution with ethanol to prepare solutions 1.8, 3.5, 5.3 and 7 μM. Label tubes from 1 to 5. Then, add 3.7 mL of 0, 1.8, 3.5, 5.3 and 7 μM digitoxin to tubes 1, 2, 3, 4 and 5, respectively.

(ii) Set up your experimental assay tubes and prepare tubes 6 and 7 adding 1.85 and 3.7 mL of *Asclepias* leaf extract, respectively. Complete volume of tube 6 to 3.7 mL with ethanol.

(iii) Add 1 mL of TNBP solution and 1 mL of 0.1 N NaOH to each tube.

(iv) Stand tubes for 40 minutes.

(v) Read the absorbance at 620 nm.

Calculations

Step 1. Graph the calibration curve relating absorbance at 620 nm to digitoxin concentration. Fit a line function to the measured data points applying regression analysis with $r^2 > 0.80$ (refer *Calculations* in Experiment 1).

Step 2. Calculate total cardenolide concentration (C) in plant extract expressed as digitoxin equivalent concentration. Consider absorbance at 620 nm (A_{620}) from experimental tubes and extinction coefficient (ε) obtained from linear regression:

$$C \ (\mu M) = (A_{620} \times \varepsilon) \times 2 \ \text{(tube 6)} \quad \text{or} \quad C \ (\mu M) = A_{620} \times \varepsilon \ \text{(tube 7)}$$

Observations

The epigeal plant parts contain large amounts of substances (i.e. pigments, chlorophyll and resins) that are soluble in ethanol, which can interfere with this colorimetric method. Chromatographic methods can overcome this kind of interference.

SUGGESTED READINGS

Bates, L.S., Waldren, R.P. and Teare, I.D. (1973). Rapid determination of free proline for water stress studies. *Plant and Soil* **39**: 205-207.

Birecka, H., Catalfamo, J.L. and Eisen, R.N. (1981). A sensitive method for detection and quantitative determination of pyrrolizidine alkaloids. *Phytochemistry* **20**: 343-344.

Blum, U., Wentworth, T.R., Klein, K., Worsham, A.D., King, L.D., Gerig, T.M. and Lyu, S.W. (1991). Phenolic acid content of soils from wheat-no till, wheat-conventional till and fallow-conventional till soybean cropping systems. *Journal of Chemical Ecology* **17**: 1045-1068.

Bonjoch, N. and Tamayo, P. (2001). Protein content quantification by Bradford method. In: *Handbook of Plant Ecophysiology Techniques* (Ed., M.J. Reigosa). Kluwer Academic Publications, New York.

Brown, C.W. (1997). Ultraviolet, Visible, and Near-Infrared Spectrophotometers. In: *Analytical Instrumentation Handbook* (Ed., G.W. Ewing). Marcel Dekker, New York.

Chinoy, J.J. (1939). A new colorimetric method for the determination of starch applied to soluble starch, natural starches, and flour. *Microchimica Acta* **26**: 132-142

Dai, J., Yaylayan, V.A., Raghavan, G.S.V. and Pare, J.R. (1999). Extraction and colorimetric determination of azadirachtin-related limonoids in neem seed kernel. *Journal of Agricultural and Food Chemistry* **47**: 3738-3742.

Dubois, M., Gilles, K.A., Hamilton, J.K., Rebers P.A. and Smith, P. (1956). Colorimetric method for determination of sugars and related substances. *Analytical Chemistry* **28**: 350-356.

Genkina, G.L. and Eídler, Y.I. (1975). Spectrophotometric method of analyzing cardenolides of the Sthophanthidin group with 2,2,4,4-tetranitrobiphenyl. *Chemistry of Natural Compounds* **10**: 29-30.

Goldstein, W.S. and Spencer, K.C. (1985). Inhibition of cyanogenesis by tannins. *Journal of Chemical Ecology* **11**: 47-858.

Hagerman, A.E. (2002). *The Tannin Handbook*. (http://www.users.muohio.edu/hagermae/tannin.pdf).

Hatch, W., Tanner, C.E., Butler, N.M. and O'Brien, E.P. (1993). A micro-Folin-Denis method for the rapid quantification of phenolic compounds in marine plants and animals. International Society of Chemical Ecology. (*Abstract*). Tampa, Florida.

Makkar, H.P.S. (2000). *Quantification of Tannins in Tree Foliage — A Laboratory Manual*. FAO/IAEA Working Document, Vienna, Austria.

Singleton, V.L., Orthofer, R. and Lamuela-Raventos, R.M. (1999). Analysis of total phenols and other oxidation substrates and antioxidants by means of Folin Ciocalteu reagent. *Methods in Enzymology* **299**: 152-178.

Somogyi, M. (1952). Notes on sugar determination. *The Journal of Biological Chemistry* **195**: 19-23.

Stumpf, D.K. (1984). Quantitation and purification of quaternary ammonium compounds from halophyte tissue. *Plant Physiology* **75**: 273-274.

Woisky, R.G. and Salatino, A. (1998). Analysis of propolis: some parameters and procedures for chemical quality control. *Journal of Apicultural Research* **37**: 99-105.

5

TLC and PC

Silvia L. Debenedetti

1. INTRODUCTION

Chromatography is a process used for separating mixtures into their components for analyzing, identifying purifying, and/or quantifying compounds. It is a very useful technique applied to complex biological extracts. Analysis, separation and purification of organic compounds from complex mixtures are necessary and important steps in the chemical industry or in phytochemical research. The methodology used for separating and isolating substances is of great importance in chemistry. In fact, *Chemistry* is a term derived from the Dutch and originally means '*Art of separation*'. The word *Chromatography* (from Greek *Chroma* = colour) was used for the first time by the Russian botanist Tswett in 1903 to describe the separation of the pigments in green leaves by passing a leaf extract through a column packed with powdered calcium carbonate (Ettre and Sakodynskii, 1993). Although the original term has been retained, modern chromatographic techniques are far more advanced. In separation processes, differences in the chemistry of the analyte and the other components are exploited. The components to be separated are distributed between two phases: a *stationary phase* bed and a *mobile phase* which percolates through the stationary phase. In chromatography, molecules from the sample have different interactions with the stationary support and migrate on it, leading to their separation. Analytes, which have more affinity for the mobile phase, move more rapidly, whereas, those with higher affinity for the stationary phase move more slowly. In this way, different types of molecules can be separated from each other as they move through the stationary phase.

Author's address: Department of Biological Sciences, Faculty of Sciences, National University of La Plata, Calle 47 y 115, 1900 La Plata, Argentina.
E-mail: sdebenedetti@biol.unlp.edu.ar

1.1 Chromatography Applications

(i) **Analysis**: To examine a mixture, its components and relations among them.
(ii) **Identification**: To determine the identity of a mixture or components based on known components.
(iii) **Purification**: To separate components for isolating the one of interest for further study.
(iv) **Quantification**: To determine the amount of one or more components found in the sample.

1.2 Types of Chromatography

All chromatographic techniques use the same basic principle, the separation of components in a sample by their distribution between a **mobile phase** and a **stationary phase**. These two phases are common to all chromatographic techniques.

The chromatography technique could be classified according to the shape of the chromatographic bed, physical state of the mobile phase and mechanism of separation.

1.2.1 Shape of the Chromatographic Bed

Column Chromatography: A separation technique in which the stationary bed is within a tube.

(i) *Packed Column*: The particles of the solid stationary phase or the support coated with a liquid stationary phase may fill the whole inside volume of the tube.
(ii) *Open-Tubular Column*: The particles of the stationary phase may be concentrated on or along the inside tube wall leaving an open, unrestricted path for the mobile phase in the middle part of the tube.

Planar Chromatography: A separation technique in which the stationary phase is present as or on a plane.

(i) *Paper Chromatography* (PC): The plane can be a paper, serving as such or impregnated by a substance as the stationary bed.
(ii) *Thin Layer Chromatography* (TLC): The plane can be a layer of solid particles spread on a support, e.g. a glass plate.

1.2.2 Physical State of the Mobile Phase

Chromatographic techniques are often classified by specifying the physical state of *both* phases used (Table 1).

Gas Chromatography (GC): A separation technique in which the mobile phase is a gas. Gas chromatography is always carried out in a column.

(i) Gas-liquid chromatography (GLC)
(ii) Gas-solid chromatography (GSC)

Table 1 Classification of chromatographic methods

Type	Stationary phase	Mobile phase
Liquid Chromatography LC and HPLC	Particles in a column or coated tubing	Liquid
Thin layer or Paper Chromatography (TLC or PC)	Coated plates or paper	Liquid
Gas Chromatography (GC)	Particles in a tube or coated tubing	Gas

Liquid Chromatography **(LC):** A separation technique in which the mobile phase is a liquid. Liquid chromatography can be carried out either in a column or on a plane.

(i) Liquid-liquid chromatography (LLC)
(ii) Liquid-solid chromatography (LSC)

High-Performance (or *High-Pressure) Liquid Chromatography* (HPLC) is a term used to characterize a kind of column liquid chromatography utilizing very small particles and a relatively high inlet pressure.

1.2.3 Mechanism of Separation

Adsorption Chromatography: Separation in this chromatographic system involves competing interactions between adsorption at the solid sorbent surface and dissolution in a liquid mobile phase. Substances are separated according to polarity of their groups, when they migrate on and interact with the sorbent surface. Using silica gel or aluminium oxide layers as stationary phase and dichloromethane as mobile phase, for instance, ethers and esters lie in the upper part of the chromatogram, while ketones and aldeyhydes are about in its middle, alcohols below and acids at the start.

Partition Chromatography: This chromatography is a liquid-liquid phase system and is based on the NERNST's law. Separation is based mainly on differential solubilization of sample components in mobile and stationary phases. In TLC, the stationary phase is achieved by impregnating the solid support of the TLC plate with a liquid (e.g. ethylene glycol, water or acetic acid).

Ion-Exchange Chromatography: Separation is based mainly on differences in the ion exchange affinities of the sample components.

Exclusion Chromatography: Separation is based mainly on exclusion effects, such as differences in molecular size and/or shape, or in charge. The term *Size-Exclusion Chromatography* may also be used when separation is based on molecular size. The terms *Gel Filtration* and *Gel Permeation Chromatography* (GPC) were used earlier to describe this process when the stationary phase is a swollen gel. The term *Ion-Exclusion Chromatography* is specifically used for the separation of ions in an aqueous phase.

Affinity Chromatography: This expression characterizes a particular variant of chromatography in which the unique biological specificity of the analyte and ligand interaction is utilized for the separation.

2. PLANAR CHROMATOGRAPHY

The planar chromatography is a separation technique in which the stationary phase is present as or on a plane. The plane can be a paper (PC) or a layer of solid particles spread on a support (TLC). Sometimes planar chromatography is also termed Open-bed chromatography.

In PC and TLC, the mobile phase is a solvent. The stationary phase in PC is the strip or piece of paper that is placed in the solvent. In TLC the stationary phase is a plate coated with a layer of solid particles. In both PC and TLC, the solvent rises up the stationary phase by capillary action (Fig. 1).

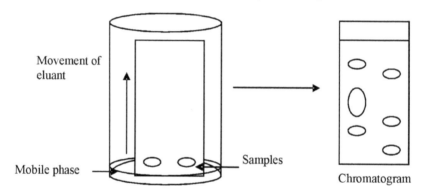

Figure 1 Planar chromatography.

2.1 Retention Factor

The retention factor (retardation factor), R_f, is a quantitative indication of how far a particular compound travels in a particular solvent on a given stationary phase. If the R_f value of an unknown compound is close to, or the same as the R_f value of a known compound, then the two compounds are most likely similar or identical. The R_f is defined as the ratio between the distance travelled by the centre of the spot (a sample component) and the distance simultaneously travelled by the mobile phase (Fig. 2):

$$R_f = b/a$$

where, a is the distance travelled by the mobile phase travelling along the medium from the starting (application) front or the line to the mobile phase front (or solvent front); b is the distance between the starting (application) point of the solute sample and the centre of the solute spot. If the solute spot is not circular, an imaginary circle is used whose diameter is the smallest axis of the spot; by definition, the R_f values are always less than unity. They are usually given to two

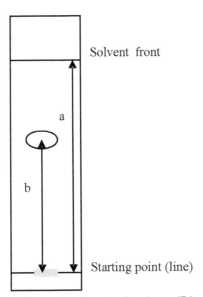

Solvent front

a

b

Starting point (line)

Figure 2 The retention factor (R_f).

decimal places. The *Starting Point* or *Line* is the point or line on a chromatographic paper or layer, where, the substance to be chromatographed is applied.

2.2 Preparing the Chamber

The developing chamber should be sealed well and should be large enough to hold the paper or TLC sheet that will be developed. It should also be clean and dry before use. The developing solvent is placed in the chamber (Fig. 3). The

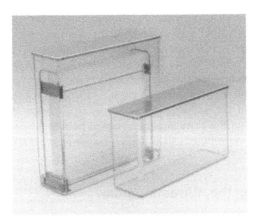

Figure 3 Chambers for TLC and PC.

solvent should completely cover the bottom of the chamber to a depth of approximately 0.5 cm. Then, the chamber is closed and shaken. The chamber is kept covered so that evaporation does not change the composition of the developing solvent mixture. After 15 minutes, the chamber will be saturated with the solvent vapour. A saturated atmosphere allows a more effective development of the chromatograms. Larger chambers need longer saturation times.

2.3   Preparing the Stationary Phase

A square piece of high-quality filter paper or TLC plate is cut to fit into the development chamber. With a pencil, a straight line about 1 cm above the bottom edge of the paper or the plate is drawn. Each strip is labelled with its corresponding solution

2.4   Capillary Spotters

A melting point glass capillary tube is placed in the dark blue zone of a Bunsen burner flame. The capillary is held in the flame until it softens and starts to sag. Quickly the capillary is removed from the flame and is pulled from both ends to about two to three times its original length. If the capillary is pulled inside the flame, a "piece of art" is obtained, but not a good spotter. The capillary is left to cool down and then is broken in the middle.

2.5   Spotting the Samples

Each sample should be dissolved in an appropriate solvent. The dissolved samples may be spotted on the paper or the TLC. A line is drawn with a pencil approximately 1.5 cm from the bottom of the plate. The coating of the plate should not be scraped. The sample is touched with the tip of a clean drawn-out capillary tube. If too much solution is drawn into the capillary, the tip is touched with a piece of clean filter paper. To spot the sample, the chromatogram is touched with the capillary in the selected starting point. Make sure the spot does not exceed 0.3-0.5 cm in diameter. The solvent is allowed to evaporate and the solution may be re-spotted in the starting point, if the amount of spotted sample is not enough for chromatography. To do it, leave the spot to dry and repeat capillary spotting in the same point, avoiding an increase in spot diameter. However, all spots on the chromatogram should be 1 to 1.5 cm away from the edges of the paper and different spots should be 0.5-1 cm (TLC) or 1-1.5 cm (PC) away from each other.

2.6   Developing the Chromatogram

After preparing the chamber and spotting the samples, the paper strip or the TLC plate is ready for development. The chamber is covered and the paper or

TLC plate is placed in the chamber. Be careful to handle the paper or TLC plate only by its edges and try to leave the development chamber uncovered for as little time as possible. The bottom edge of the chromatogram is lowered into the developing solution, making sure the spots themselves do not touch the liquid, and then the chamber is closed.

An example of a PC developing system is shown in Fig. 4. As the solvent rises up mainly by capillary action through the paper or TLC plate, the components of the mixture are partitioned between the mobile phase (solvent) and the stationary phase (silica gel) due to their different adsorption and solubility strength. As the stronger a given component is adsorbed to the stationary phase, the less easily is it carried out by the mobile phase. On the other hand, the running distance depends on the solubility of the compound in the solvent (i.e. low polar compounds will be best dissolved in a non-polar solvent and will be carried out to long running distances than polar compounds).

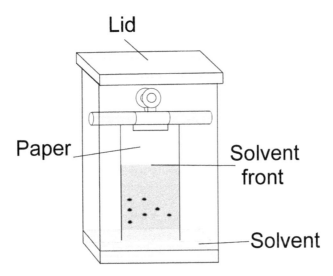

Figure 4   An example of PC development.

2.7   Identifying the Spots

After development, the spots corresponding to different compounds may be located by direct observations, under the ultraviolet (UV) light, or by treatment (spraying) with different reagents (Fig. 5). If the spots can be seen by direct observation, they are outlined with a pencil. If the spots are not obvious, the most common visualization technique is to hold the paper under a UV lamp (Caution: Do not look directly into the UV light). Many organic compounds can be seen using this technique. The spots are outlined with a pencil.

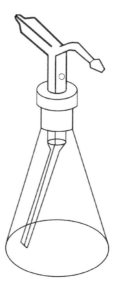

Figure 5 TLC flask for spray reagents.

2.8   Interpretation of the Results

The R_f value of each spot should be calculated and is characteristic for each compound. Hence, known R_f values could be compared to those of unknown substances to aid in compound identification. As R_f values depend on several factors, the most effective way to identify a compound is to spot the known substances in a point close to that of the spot sample, in the same starting line. If a sample develops as only one spot, it may or may not be pure because the sample may contain another compound which did not separate under the selected chromatographic conditions. Purity of samples should be determined in conjunction with other techniques, such as measuring a sample's melting point or recording its nuclear magnetic resonance spectrum.

2.9   Multi-dimensional Chromatography

In planar chromatography, two-dimensional chromatography refers to the chromatographic process in which the chromatogram is first developed in one direction and, after rotating the paper or the TLC plate 90°, is subsequently developed in a direction that is the right angle of the first one; the two elutions are carried out with different eluents. This is useful for separating complex mixtures of similar compounds, for example, amino acids, peptides, carbohydrates, steroids and many other organic compounds. This technique provides an easy way to separate the components of a mixture. Procedure for multi-dimensional chromatography is as follows:

(i) A drop of mixture is placed in the right corner of the chromatogram.

(ii) The paper or TLC plate is immersed in the solvent as shown in Fig. 6A.

(iii) The solvent migrates up the sheet or plate.

(iv) As it does so, the substances in the drop are carried along at different rates (Fig. 6B).

(v) Each compound migrates at a rate that reflects the affinities of each component for the stationary phase and its solubility in the solvent.

(vi) A second run is achieved using a different solvent, previously turning the paper sheet or the TLC plate 90° clockwise as shown in Fig. 6C. The various substances will be spread out at distinct spots across the paper sheet or the TLC plate.

(vii) The identity of each spot can be determined by comparing its position with the position occupied by known substances under the same conditions.

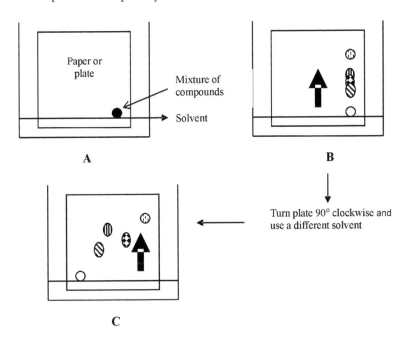

Figure 6 Two-dimensional chromatography.

3. THIN LAYER CHROMATOGRAPHY (TLC)

TLC is a very common, simple, quick and inexpensive procedure that gives the chemist a quick answer of how many components are present in a mixture. It is used for identifying compounds, determining their purity and following the progress of a reaction in synthetic chemistry. It also allows the optimization of the solvent system for a given separation problem. TLC is also used to support the identity of a compound in a mixture, when the R_f of a compound is compared with the R_f of a known compound (preferably both run on the same TLC plate). In comparison with column chromatography, it only requires small quantities of the compound (~ng) and is much faster as well. Among many chromatographic methods presently available, TLC is widely used for rapid analysis of herbs and herbal products. There are several reasons for this:

- The time required for visualization of most of the characteristic constituents of a sample by TLC is very short.
- In addition to quality detection, TLC also provides semiquantitative information of the major sample or product constituents.
- TLC provides a chromatographic herb fingerprint. It is, therefore, suitable for monitoring the identity and purity of herbs, and for detecting the presence of different herbs in the sample.
- With the aid of appropriate separation procedures, TLC can be used to analyze phytochemical profiles of a plant or plant mixtures.

3.1 Stationary Phase

The plate is a layer of solid particles (stationary phase) spread on a support such as a glass, aluminium or plastic sheet, as a thin layer (~0.25 mm). In addition, a binder such as gypsum, is mixed into the stationary phase to make it stick better to the slide. In many cases, a fluorescent powder is mixed into the stationary phase to simplify further visualization (e.g. bright green when the stationary phase is exposed to 254 nm UV light). There are different types of stationary phases:

- Unmodified silica gel
- Nano-TLC or HPTLC
- Modified silica gel: RP-18, RP-18 chiral modified, amino, cyano
- Aluminium oxide
- Cellulose (fibres, microcrystalline)
- Polyamide

I. **Silica gel:** This is the sorbent most frequently used in both adsorption and partition chromatography. It is available with pore sizes of 40 Å and 60 Å.

II. **Nano-TLC (or HPTLC, High Performance TLC):** Particles size of the sorbent material (e.g. silica gel 60) is smaller than that of TLC. Size variations are also narrower than TLC. HPTLC layers have improved surface homogeneity and are thinner. The advantages are improved resolution, shorter analysis time, higher detection sensibility and *in-situ* quantitation. It is sufficient to apply nL or ng of sample.

III. **RP-18 Reversed phase precoated plates:** Chemical reaction of the polar silanol groups (=Si–OH) of silica gel with alkylchlorosilanes, differing in chain lengths, yields reversed phases, e.g. RP-2, RP-8, RP-18. With the exception of RP-2 (dimethyl), the numbers 8 and 18 stand for the number of carbon atoms of the alkyl chain. The hydrophobic character increases with increasing chain length. Chromatography is preferably carried out using a polar aqueous mobile phase.

IV. **Chiral precoated plates (CHIR):** Precoated plates are used for the separation of enantiomers. Layers used are produced from a special RP-18 modified silica gel impregnated with a copper (II) salt and an optically active pure enantiomerical hydroxyproline.

V. **Amino phases:** Silica gel containing an NH_2-function is chemically bonded by means of a spacer. Short-chain n-alkyl groups, e.g. n-propyl, can be used as spacers. Aminopropyl silica gel is a weakly basic anion exchanger. This layer is completely wettable by water.

VI. **Cyano phases:** Are made of silica gel with a CN function chemically bonded to it, as a polar phase, by means of a hydrophobic spacer. The polarity of the CN layer is intermediate between that of silica gel and reversed phase silica gels. The layer is completely wettable by water.

VII. **Aluminium oxide:** It is a naturally basic reacting polar sorbent that is prepared by thermal dehydration of water containing oxide hydrates (hydrargillite, *-alumina) at 400-700°C. It is commercially available in various pore sizes (e.g. 60 or 150 Å, i.e. 6 or 15 nm).

VIII. **Cellulose (native, microcrystalline):** It is an organic sorbent suitable for the separation of hydrophilic substances such as amino acids, carbohydrates, nucleic acid derivatives and phosphates by partition chromatography. Native, fibrous and chemically treated microcrystalline (rod-shaped) celluloses are commercially available. The separation properties of the two types of celluloses are different.

IX. **Polyamide:** Two types of polycondensation products are used: polyundecanamide (PA 11) and polycaprolactam (PA 6). They are suitable for partition chromatography of nitrophenols, phenolcarboxylic acids, tannins and flavones. They have their own eluotropic series: water is the least polar and dimethylformamide the most polar mobile phase component.

The solvent is allowed to travel up the plate with the sample spotted on the sorbent just above the solvent. Depending on the sorbent, the separation can be either partition or adsorption chromatography (cellulose, silica gel and alumina are commonly used). The technique came to prominence during the late 1930s; however, it did not become popular until Merck and Desaga developed commercial plates that provided reproducible separations. The major advantage of TLC is the disposable nature of the plates. Samples do not have to undergo the extensive clean-up steps required for HPLC. The other major advantage is the ability to detect a wide range of compounds cheaply, using very reactive reagents (iodine vapours and sulfuric acid) or indicators. Non-destructive detection (fluorescent indicators in the plates and examination under a UV lamp) also means that purified samples can be scraped off from the plate and can be analyzed by other techniques. There are special plates for such preparative separations.

3.2 Mobile Phase

The mobile phase in TLC is a solvent or mixture of solvents moving forward through the porous stationary phase as a result of capillary forces. The term *'eluent'* is also used for the mobile phase. Solvents suitable for chromatography can be classified according to their eluting power (polarity). To a first approximation, the dielectric constant (DC) is a measure of polarity. Parameters such as surface tension, viscosity and vapour pressure also serve to characterize the solvent. When the solvent has reached the top of the plate, the plate is removed from the developing chamber, dried, and the separated components of the mixture are visualized.

3.3 Sample Application

(i) After clean up, the sample is applied at the starting point of the TLC plate either as a specific point or as a band. Position of the application should be exact and — particularly for quantitative analyses — the applied sample volume should be precise and accurate.

(ii) The chromatographic layer should not be damaged. The methods applied to transfer a sample onto the plate are contact spotting and spray-on application.

(iii) The spotting zone must be as small as possible to maximize the separation power of the chromatographic system. (i.e. 2 to 4 mm diameter should not be exceeded on conventional layers).

(iv) Spotting sample volumes should be between 5 to 100 μL but when using HPTLC (coated with particles of 5-7 μm in diameter), only a maximum loading of about 100-200 nL can be tolerated on a circular area not greater than 1 mm in diameter.

(v) Concentration of large sample volumes: the sample is often applied onto the plate with a micropipette or a capillary tube. Microsyringes can also be employed. With care and a little local heating, the sample can be concentrated. The contents of the syringe are slowly but continuously discharged onto the plate and at the same time, the solvent is progressively evaporated.

(vi) Application positions should not be marked on the layer, because marks can damage the plate surface interfering with the sample transfer onto the layer. Pencil marks can also affect evaluation by scanning densitometry. Samples should be applied at least 2 mm/4 mm from the side edges of the plate, because they may have irregularities in thickness. When development starts, the application position must be clearly above the level of the developing solvent (i.e. if the solvent level is maintained at 5 mm (HPTLC)/ 10 mm (TLC) a suitable spotting position above the lower edge of the plate would be 8 mm/15 mm). The centres of spots are usually spaced 5 mm apart on HPTLC plates/10 mm on TLC plates.

3.4   Detection

If the compounds are coloured, visualization is straightforward. Usually the compounds are not coloured, so a UV lamp is used to visualize the plates. Colourless components are often visualized, on TLC plates containing a fluorescent indicator, under short wavelength UV light (254 nm). Colourless substances that absorb at short-wavelength UV light (254 nm), appear as dark chromatogram zones on a green or pale blue fluorescent background. The fluorescent indicators commonly used in TLC are:

- $F254$ = manganese-activated zinc-silicate, which is stimulated to green fluorescence emission by 254 nm.
- $F254s$ = acid-stable alkaline earth tungstate that is stimulated to pale blue fluorescence emission by 254 nm.
- $F366$ = optical brightener, that is stimulated to intense blue fluorescence emission by 366 nm.

Spray reagents can also be used to visualize the sample spots on a TLC plate. These reagents are mixtures of chemicals, which when sprayed on the plate, react with the colourless compounds producing coloured derivatives. Before spraying, the plates should be well dried in the hood ensuring solvent elimination. All these reagents should always be sprayed onto plates in a well ventilated hood while wearing safety glasses. Moderate amounts should be applied to the plate so that it always appears dull and flat (if it looks wet, too much has been sprayed). One can always overspray to enhance the detection.

3.5   Chromatogram

After chromatographic development, record the TLC plate as photocopy, photograph, sketch, absorption or fluorescence scans.

3.6   Application of TLC to Different Phytochemical Groups (Wagner and Bladt, 1996)

3.6.1   *Coumarins and Phenolic Carboxylic Acids*

I. **Sample solution (methanolic extract):** 1 g of powdered sample is extracted with 10 mL methanol for 30 minutes under reflux. The filtrate is used for TLC analysis.

II. **Reference solutions:** Coumarins are prepared as 0.1% methanolic solutions. Phenolic carboxylic acids are prepared as 0.05% methanolic solutions.

III. **Chromatographic system:**

(i) **Adsorbent:** Silica gel 60 F_{254}

(ii) **Mobile phase:**

- Ethyl acetate-formic acid-glacial acetic acid-water (100:11.11:26) for

coumarin glycosides and phenolic carboxylic derivatives (e.g. chlorogenic acid).
- Toluene-diethyl ether (1:1, saturated with 10% acetic acid) for coumarin aglycones and phenolic carboxylic acids (e.g. caffeic acid). Fifty mL of toluene and 50 mL of diethyl ether are shaken for 5 minutes with 50 mL of 10% acetic acid in a separating funnel. The lower phase is discarded and the saturated toluene-ether mixture is used for TLC.

(iii) **Sampling:** 20 µL of *sample solution* and 5-10 µL of *standard solutions*.

(iv) **Detection:**

UV 254 nm: Distinct fluorescence quenching of all coumarins and phenolic carboxylic acids.

UV 365 nm: Intense blue or green fluorescence (simple coumarins and phenolic carboxylic acids) yellow, brown, blue or blue-green fluorescence (furano and pyrano coumarins).

(v) **Visualization Reagents:**
- **Potassium hydroxide:** The fluorescence of the coumarins and phenolic carboxylic acids (with phenolic hydroxyl group) are intensified.
- **Ammonia vapour (concentrated):** The fluorescence of the coumarins and phenolic carboxylic acids (with phenolic hydroxyl group) are intensified.
- **Natural products-polyethylene glycol reagent (NP/PEG):** This reagent intensifies and stabilizes the existing fluorescence of the coumarins. Phenolic carboxylic acids fluoresce blue or blue-green.

3.6.2 Anthracene Derivatives

I. **Sample solution (methanolic extract):** Powered sample (0.5 g) is extracted with methanol (5 mL) for 10 minutes heating on a water bath. The filtrate is evaporated to about 1 mL and is used for TLC.

II. **Reference solutions:** Glycosides: aloin, frangulin A/B, rhein, aloe-emodin; aglycon 1,8-dihydroxyanthraquinone are prepared as 0.1% methanolic solutions.

III. **Chromatographic system:**

(i) **Adsorbent:** Silica gel 60 F_{254}

(ii) **Mobile phase:**
- Ethyl acetate-methanol-water (100:13.5:10) for glycosides.
- Light petroleum-ethyl acetate-formic acid (75:25:1) or toluene-ethyl formiate-formic acid (50:40:10) for aglycones.

(iii) **Sampling:** 20 µL of *sample solution* and 10 µL of *standard solutions*.

(iv) **Detection:**

UV-254 nm: All anthracene derivatives quench fluorescence.

UV 365 nm: All anthracene derivatives give yellow, orange or red fluorescence.

(v) **Visualization Reagents:**
- **Potassium hydroxide:** After spraying with 5-10% ethanolic KOH, anthraquinones appear orange or red in the visible light and show an

intensified orange (C-glycosides) or red (aglycones and O-glycosides) fluorescence in UV 365 nm.

- **Anthrones and anthranols:** Yellow (visible light), bright yellow fluorescence (UV 365 nm).
- **Ammonia vapour (concentrated):** The fluorescence of the anthraquinones are intensified showing similar colours, rather than with 5-10% ethanolic KOH.
- **Natural products-polyethylene glycol reagent (NP/PEG):** This reagent intensifies and stabilizes the fluorescence of the anthraquinones showing orange (C-glycosides) or red (aglycones and O-glycosides) fluorescence in UV 365 nm.

3.6.3 *Flavonoids*

I. **Sample solution (methanolic extract):** Powered sample (1 g) is extracted with methanol (10 mL) for 5 minutes on a water bath at about 60 °C and then filtered. This method extracts both lipophilic and hydrophilic flavonoids.

II. **Standard solutions:** Standard compounds are prepared as 0.05% solutions in methanol. For a general description of the flavonoid pattern of a herb, rutin, hyperoside, and isoquercitrin, are used as standard.

III. **Chromatographic system:**

  (i) **Adsorbent:** Silica gel 60 F_{254}

  (ii) **Mobile phase:**
- Ethyl acetate-formic acid-glacial acetic acid-water (100:11.11:26), ethyl acetate-formic acid-glacial acetic acid-ethylmethyl ketone-water (50:7:3:30:10) or ethyl acetate-formic acid-water (88:6:6) for flavonoid glycosides.
- Chloroform-acetone-formic acid (75:16.5:8.5) for flavolignans, biflavonoids.
- Toluene-ethylformiate-formic acid (50:40:10) or toluene-dioxan-glacial acetic acid (90:25:4) for flavonoid aglycones.

  (iii) **Sampling:** 20-30 µL of *sample solution* and 10 µL of *standard solutions*.

  (iv) **Detection:** The solvent (acids) must be thoroughly removed from the silica gel layer before detection.

    UV-254 nm: All flavonoid derivatives have quench fluorescence.

    UV 365 nm: Depending on the structure, flavonoids show dark yellow, orange, brown, green or blue fluorescence. Colours are intensified and change by the use of ammonium vapour.

  (iv) **Visualization Reagents:**
- **Natural products-polyethylene glycol reagent (NP/PEG):** Intense fluorescence at 365 nm is produced immediately after spraying. Addition of polyethylene glycol solution lowers the detection limit and intensifies the fluorescence, which depends on the oxidation pattern and position of the hydroxyl groups in the flavonoid nucleus. Flavonoids are visualized as follows:

- **Flavonols:** Quercetin, myricetin and their glycosides → orange-yellow colours

  Kaempferol, isorhamnetin and their glycosides → yellow-green

- **Flavones:** Luteolin and their glycosides → orange

  Apigenin and their glycosides → yellow-green

3.6.4 Essential Oils

I. **Sample solutions:**

Dichloromethane extract (DCM): 1 g of powdered sample is extracted by shaking for 15 minutes with 10 mL dichloromethane. The suspension is filtered and the clear filtrate evaporated to dryness. The residue is dissolved in 1 mL toluene and used for TLC analysis.

Essential oil: 1 mL essential oil is diluted with 9 mL of toluene.

II. **Standard solutions:** Solutions of commercially available compounds are prepared in toluene (1:30).

III. **Chromatographic system:**

(i) **Adsorbent:** Silica gel 60 F_{254}.

(ii) **Mobile phase:**
- **Toluene-ethyl acetate (93:7).** This system is suitable for the analysis and comparison of all important essential oils.
- **Other solvents:** Chloroform, dichloromethane, toluene-ethyl acetate (95:5), toluene-ethyl acetate (90:10) or chloroform-toluene (75:25).

(iii) **Sampling:** 30-100 μL of DCM *sample solution* or 5 μL of distillated essential oil diluted in toluene (1:9) or 3 μL (100 μg) of *standard solutions* (thymol and anethol are detectable in quantities of 10 μg and less).

(iv) **Detection:**

UV 254 nm: Compounds containing at least two conjugated double bonds quench fluorescence and appear as dark zones against the light green fluorescent background of the TLC plate (all phenylpropane derivatives present in the essential oil, e.g. anethole, apiole and eugenol).

UV 365 nm: No characteristic fluorescence of terpenoids and propylphenols is noticed.

(v) **Visualization Reagents:**
- **Anisaldehyde-sulfuric acid reagent (AS):** Evaluation in visible light, essential oil compounds show strong blue, green, red and brown colours. Most of them develop fluorescence under UV 365 nm.
- **Vainillin-sulfuric acid reagent (VS).** Evaluation in visible light, colours very similar to those obtained with the AS reagent.

3.6.5 Sesquiterpene Lactones

I. **Sample solution:** 2.5 g of powdered sample is extracted with 30 mL of hot water. After 5 minutes, the mixture is filtered through a wet filter, washing the filter with 10 mL of water; 15 mL of $CHCl_3$ is added to the water extract and shaken carefully several times. The $CHCl_3$ phase is separated and reduced to dryness. The residue is dissolved in 0.5 mL of $CHCl_3$.

II. **Standard solutions**: Sesquiterpene lactones (e.g. helenalin, dihydrohelenalin and partenolide) are prepared as 0.1% methanolic solutions.

III. **Chromatographic system**:

  (i) **Adsorbent:** Silica gel 60 F_{254}

  (ii) **Mobile phase:**
- n-pentane-ether (25:75)
- Cyclohexane-ethyl acetate (1:1)

  (iii) **Sampling:** 10-30 µL of *sample solution* or 10 µL of *standard solutions*

  (iv) **Detection:**
UV 254 nm: Compounds containing at least two conjugated double bonds quench fluorescence and appear as dark zones against the light green fluorescent background of the TLC plate.
UV 365 nm: No characteristic fluorescence of sesquiterpene lactones is noticed.

  (v) **Visualization Reagents:**
Zimmermann reagent violet, grey zones

3.6.6 Saponins

I. **Sample solution (methanolic extract):** 2 g of powdered herb is extracted with 10 mL 70-90% ethanol for 15 minutes heating on a water bath. The filtrate is evaporated to about 5 mL and is used for TLC analysis.

II. **Standard solutions:** The commercially available standard saponins, such as aescin, hederin and asiaticoside, are prepared as 0.1% methanolic solutions.

III. **Chromatographic system:**

  (i) **Adsorbent:** Silica gel 60 F_{254}

  (ii) **Mobile phase:** Chloroform-glacial acetic acid-methanol-water (64:32:12:8), chloroform-methanol-water (65:50:10) or (70:30:4), or ethyl acetate-formic acid-glacial acetic acid-water (100:11:11:26)

  (iii) **Sampling:** 20-40 µL of *sample solution* and 10 µL of *standard solutions.*

  (iv) **Detection:** With a few exceptions, saponins are not detectable by exposure to UV 254 or UV 365 nm.

  (v) **Visualization Reagents:**
- **Vainillin-sulfuric acid reagent (VS):** Evaluation in visible light, saponins give mainly blue, blue-violet, red zones.
- **Anisaldehyde-sulfuric acid reagent (AS):** Evaluation in visible light, saponins give similar colours to those with VS reagent; inspection under UV 365 nm light results in blue, violet and green fluorescent zones.

3.6.7 Alkaloids

I. **Sample solution:** The samples applied to the TLC plate should contain between 50 and 100 µg total alkaloids, which have to be calculated according to the average alkaloid content of the sample. Powered herbs or sample (1 g) with a total alkaloid content of 0.3% extracted with one of the following methods will yield 3 mg in 5 mL methanolic solution.

Alkaloid sample with medium to high alkaloid contents (>1%): 1 g of powered sample is mixed thoroughly with 1 mL of 10% ammonia solution or 10% Na_2CO_3 solution and then extracted for 10 minutes with methanol under reflux. The filtrate is concentrated according to the total alkaloids of the specific drug, so that 100 μL contains 50-100 μg total alkaloids.

Alkaloid sample with low total alkaloid contents (<1%): 2 g of powered sample is ground in a mortar for about 1 minute with 2 mL 10% ammonia solution and then thoroughly mixed with 7 g basic aluminium oxide (active grade I). This mixture is packed loosely into a glass column (diameter 1.5 cm; length 20 cm) and 10 mL $CHCl_3$, is added. Alkaloid bases are eluted with about 5 mL $CHCl_3$, and the eluate is collected and evaporated to 1 mL and used for TLC.

II. **Standard solutions:** Commercially available compounds are usually prepared in 1% alcoholic solution, e.g. atropine, brucine, codeine, berberine and strychnine. Alkaloid references can also be obtained from pharmaceutical products by a simple methanol extraction. The sample solution contains between 50 and 100 μg of alkaloid.

III. **Chromatographic system:**

   (i) **Adsorbent:** Silica gel 60 $F_{254.}$

   (ii) **Mobile phase:** Toluene-ethyl acetate-diethylamine (70:20:10) or acetone-light petroleum-diethylamine (20:70:10), cyclohexane-ethanol-diethylamine (80:10:10) or ethyl acetate-methanol-water (100:13.5:10).

   (iii) **Sampling:** 20-30 μL of *sample solution* and 10 μL of *standard solutions.*

   (iv) **Detection:**

   UV-254 nm: Pronounced quenching of some alkaloid types such as indoles, quinolines, isoquinolines and purines; weak quenching of e.g. tropine alkaloids.

   UV 365 nm: Blue, blue-green or violet fluorescence of alkaloids.

   (v) **Visualization Reagents:**

   - **Dragendorff reagent:** Inspect under visible light, alkaloids appear as orange zones.

   - **Marquis reagent:** Opium alkaloids morphine and codeine are immediately stained typically violet, papaverine weak violet and noscapine weak yellow-brown colour.

   - **Iodine/$CHCl_3$ reagent:** Ipecacuanha alkaloids appear as light blue zones.

   - **Iodoplatinate reagent (IP):** Detection of nitrogen-containing compounds (blue-violet). Detection of Cinchona alkaloids: the plate is first sprayed with 10% ethanolic H_2SO_4 and then with IP reagent.

3.7 Preparation of Spray Reagents

I. p-Anisaldehyde – sulfuric acid

(i) *Phenols, sugars, steroids and terpenes:* Spray with a solution of freshly prepared 0.5 mL p-anisaldehyde in 50 mL glacial acetic acid and 1 mL 97% sulfuric acid.

Another formulation of this reagent: 0.5 mL p-anisaldehyde is mixed with 10 mL glacial acetic acid, followed by 85 mL methanol and 5 mL concentrated sulfuric acid, in that order. This reagent is colourless and should be stored in the refrigerator. If a colour develops, the reagent must be discarded.

The TLC plate must be sprayed with about 10 mL, heated at 100°C for 5-10 minutes and then evaluated in visible light or long wavelength UV light (365 nm).

Results: Lichen constituents, phenols, terpenes, sugars, and steroids turn violet, blue, red, grey or green.

(ii) *Sugars:* Spray with a solution of freshly prepared 1 mL p-anisaldehyde, 1 mL 97% sulfuric acid in 18 mL ethanol and heat at 110°C.

Results: Sugar phenylhydrazones produce green-yellow spots in 3 minutes. Sugars will produce blue, green, violet spots in 10 minutes. Also detects *Digitalis* glycosides.

II. Dichloroquinonechlorimide reagent (DQC):

Polyphenols

Dissolve 1 g dichloroquinonechlorimide in 100 mL methanol.

Dissolve 10 g anhydrous sodium carbonate in 100 mL water.

Spray first with DQC and dry. Then, spray with sodium carbonate solution and dry. Inspect under visible light.

III. Dragendorff reagent

Nitrogen compounds, alkaloids

Solution 1: 1.7 g basic bismuth nitrate and 20 g tartaric acid in 80 mL water

Solution 2: 16 g potassium iodide in 40 mL water

Stock solution (stable for several weeks in a refrigerator): Mix equal volumes of solutions 1 and 2.

Procedure: Spray with a solution of 10 g tartaric acid, 50 mL water and 5 mL stock solution. The alkaloids appear as a brown or orange brown zone.

IV. Iodine-chloroform reagent ($I/CHCl_3$)

Ipecacuanha alkaloids

0.5% Iodine in chloroform.

The sprayed plate is warmed at 60°C for about 5 minutes. After 20 minutes the chromatogram is evaluated in visible light or in long wavelength UV light (365 nm).

V. Iodoplatinate reagent (IP)

Organic nitrogen compounds, alkaloids, e.g. cocaine metabolites

Spray with a freshly prepared mixture of 0.3 g hydrogen hexachloroplatinate (IV) dissolved in 100 mL water and mixed with 100 mL 6% aqueous potassium iodide solution.

The plate is sprayed with 10 mL and evaluated in visible light.

VI. **Natural products reagent**: 2-Aminoethyl diphenylborate or Ethanolamine diphenylborate (flavone reagent according to Neu).

Flavonoids and phenols

Spray with a 1% solution of ethanolamine diphenylborate in methanol.

Spray with a 2-5% ethanolic solution of polyethylene glycol 400 for fluorescence stabilization.

Irradiate 2 minutes with intense long wavelength UV light (365 nm).

View under the same UV light.

VII. **Potassium hydroxide reagent (KOH)**

Polyphenols

5-10% ethanolic KOH. The plate is sprayed with 10 mL and evaluated in visible or in long wavelength UV light (365 nm). Intensifies the fluorescence of polyphenol compounds under the same UV light.

VIII. **Vanillin/sulfuric acid**

Terpenoids (essential oils, triterpenes and steroids). Note, this reagent can only be used with G (gypsum) binder plates since it will char the polymer binders in the harder layer plates.

Spray with a 1% solution of vanillin in concentrated sulfuric acid. Another formulations of this reagent:

0.5 g vanillin in 100 mL sulfuric acid/ethanol (40:10) or

1% ethanolic vanillin (solution I) and 10% ethanolic sulfuric acid (solution II). The plate is sprayed with 10 mL solution I, followed immediately by 10 mL solution II.

For all reagents after heating at 110°C for 5-10 minutes under observation, the plate is evaluated in visible light.

IX. **Zimmermann reagent**

Sesquiterpenes and sesquiterpenelactones

(a) 10 g dinitro benzene + 90 mL of toluene

(b) 6 g NaOH + 25 mL water + 45 mL methanol

The TLC is first sprayed with (a), followed by (b)

Results: In visible light, sesquiterpenes are visualized as violet, grey zones.

3.8 Other Detection Reagents

I. **Aluminium chloride**

Flavonoids

Spray plate with a 1% ethanolic solution of aluminium chloride.

Results: Yellow fluorescence in long wavelength UV light (365 nm).

II. Aniline phthalate

Reducing sugars

Dry the developed chromatogram.

Spray with 0.93 g aniline and 1.66 g o-phthalic acid dissolved in 100 mL n-butanol saturated with water.

Briefly dry with hot air, and then heat to 105°C for 10 minutes.

Results: Substance spots show different colours on an almost colourless background. Some spots give fluorescence at 365 nm.

III. p-Anisidine Hydrochloride

Carbohydrates/sugars

Mix a solution of 3% p-anisidine hydrochloride in n-butanol.

Spray and heat at 100°C for 2-10 minutes.

Results: Aldohexoses are seen as green-brown spots, ketohexoses as yellow spots, aldopentoses as green spots, and uronic acids as red spots.

IV. Anisidine phthalate

Carbohydrates and reducing sugars

Spray with a solution of 1.23 g p-anisidine and 1.66 g phthalic acid in 100 mL 95% ethanol.

Results: Hexoses, green; pentoses, red-violet — sensitivity 0.5 µg; methylpentoses, yellow-green; uronic acids, brown — sensitivity 0.1-0.2 µg.

V. Antimony (III) chloride

Flavonoids

Spray with a 10% solution of antimony (III) chloride in chloroform.

Results: Fluorescing spots in long wavelength UV light (365 nm).

VI. Antimony (III) chloride

Vitamins A and D, carotenoids, steroids, sapogenins, steroid glycosides, terpenes

Spray with a solution of 25 g antimony (III) chloride in 75 mL chloroform (generally a saturated solution of antiomony (III) chloride in chloroform or carbon tetrachloride is used).

Heat for 10 minutes at 100°C, view under long wavelength UV light (365 nm).

VII. Fast Blue B reagent

Cannabinoids, phenols, tanning agents, amines which can be coupled

Spray with a solution of 0.5 g Fast Blue B (tetraazotized di-o-anisidine) in acetone-water (9:1, v/v), always prepared fresh. Then overspray with 0.1 M sodium hydroxide solution.

Results: Cannabinoids turn dark red/purple in colour.

VIII. Iodine vapour

Relatively unspecific universal reagent for many organic compounds

Charge chamber with some crystals of iodine. Place developed, dried chromatogram in iodine vapour.

Results: Spots turn tan-brown in colour.

3.9   Phytochemical Analysis Using TLC Systems

3.9.1   *TLC Analysis of the Polyphenol Content of Bearberry Leaf (Arctostaphylos uva-ursi)*

I. **Sample:** In a flask 0.5 g of powdered drug is heated under reflux with mL of a methanol-water mixture 1:1 for 10 minutes. The hot mixture is filtered. The flask and filter are washed with methanol-water 1:1 and the filtrate is diluted to a final volume of 5 ml. This is the test solution.

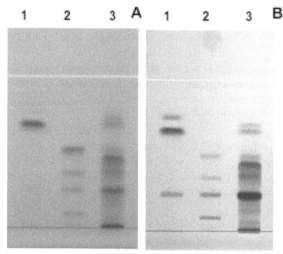

Figure 7   TLC chromatogram of polyphenols from *Arctostaphylos uva-ursi* visualized with (A) short-wavelength UV light (254 nm), and (B) DQC reagent (see colours below).
Track assignment (in each track, in order of increasing R_fs):
1. arbutin (cyan), gallic acid (brown), hydroquinone (brown)
2. rutin (brown), hyperoside (dark), quercetin-3-O-arabinopyranoside (pale brown), quercetin-3-O-arabinofuranoside (pale brown).
3. *Acrtostaphylos uva-ursi*
Source: Camag herbal application notes. http://www.camag.ch/l/herbal/index.htm

II. **Standard (optional):** 1 mg each of quercetin-3-O-arabinofuranoside, quercetin-3-O-arabinopyranoside, hyperoside, and rutin are dissolved in 1 ml methanol. This is standard mixture A. 1 mg each of arbutin, hydroquinone, and gallic acid are dissolved in 1 ml methanol. This is standard mixture B.

III. **Dichloroquinonechlorimide reagent (DQC):** Dissolve 1 g dichloroquinonechlorimide in 100 ml methanol. Dissolve 10 g anhydrous sodium carbonate in 100 mL water.

IV. **Chromatographic conditions:**

V. **Stationary phase:** HPTLC plates 10 × 10 cm silica gel 60 F254.

  (i) **Mobile phase:** Ethyl acetate-formic acid-water (88:6:6).

(ii) **Sample application:** 2 ml of test solution and standard as 10 mm bands, space 6 mm, 8 mm from lower edge.

(iii) **Development:** 10 × 10 cm chamber, saturated for 10 minutes (filter paper), developing distance 5.5 cm.

(iv) **Detection:** a) UV 254 nm and b) DQC reagent: spray first with DQC, dry, then spray with sodium carbonate solution, dry, inspect under white light.

3.9.2 TLC Analysis of the Polyphenol Content of Different Samples of *Equisetum arvense*

I. **Sample:** In a water bath at 80°C reflux 500 mg of the powdered drug in a conical flask with 5 mL of methanol-water 8:2 for 15 minutes. Filter through filter paper. Wash flask twice with 1 mL of methanol-water. Evaporate combined filtrates to dryness at 40°C using the rotary evaporator. Reconstitute residue with 1 mL methanol while sonicating for 1 minute. This is the test solution.

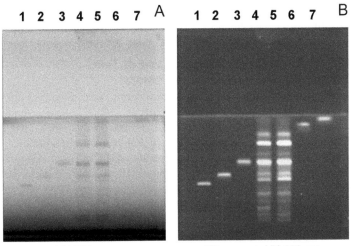

UV 254 nm   NP-reagent, UV 366 nm

Figure 8 TLC chromatogram of polyphenols from *Equisetum arvense* visualized with (A) short-wavelength UV light (254 nm) and (B) NP under long-wavelength UV light (see spot colours below).
Track assignment:
1. Rutin (orange)
2. Chlorogenic acid (cyan)
3. Isoquercitrin (orange)
4. *Equisetum arvense*
5. *Equisetum arvense* (East Asian chemotype)
6. Caffeic acid (cyan)
7. Kaempferol (green)
Source: Camag herbal application notes, http://www.camag.ch/l/herbal/index.html

II. **Standard (optional):** 1 mg each of chlorogenic acid, isoquercitrin, rutin and kaempferol and 0.5 mg caffeic acid are individually dissolved in 10.0 mL methanol each.

III. **Reagents:**

IV. **Natural products reagent:** 1.0 g of diphenylborinic acid aminoethylester is dissolved in 100 mL methanol.

PEG-reagent: 5% solution of polyethylene glycol 400 in ethanol.

Chromatographic conditions:

(i) **Stationary phase:** HPTLC plates 10 × 10 cm silica gel 60 F_{254}.

(ii) **Mobile phase:** Ethyl acetate-formic acid-acetic acid-water (100:11:11:26).

(iii) **Sample application:** 2 μL volumes of test solution and standards, applied as 8 mm bands, 4 mm apart, 8 mm from lower edge of plate.

(iv) **Development:** 10 × 10 cm chamber, saturated for 10 minutes (filter paper), 5 mL developing solvent per trough, developing distance 6.0 cm from lower edge of plate. Dry plate with warm air (hair dryer) until there is no odour of the acids.

(v) **Detection:** (a) UV 254 nm, and (b) NP reagent: heat plate to 110°C for 5 minutes. Spray the warm plate with NP reagent then immediately with PEG reagent. Examine plate under 366 nm.

3.9.3 TLC Analysis for the Identification of Two Crataegus Species (Leaves & Flowers)

I. **Sample:** In a test tube heat 1.0 g of powdered plant material with 10 ml methanol for 5 minutes in a water bath at 65°C. Filter after cooling to room temperature. Dilute the filtrate to 10 mL with methanol. 1 mL of this solution is transferred to a small sample vial labelled test solution.

II. **Standard (optional):** Dissolve 1 mg of caffeic acid, 2 mg of chlorogenic acid, 5 mg of hyperoside, 5 mg of rutin, 5 mg of vitexin, and 5 mg of vitexin-2"-O-rhamnoside in 20 mL of methanol.

III. **Natural products reagent:** 1.0 g of diphenylborinic acid aminoethylester (Natural Product Reagent) is dissolved in 100 mL of methanol.

IV. **Chromatographic conditions:**

(i) **Stationary phase:** HPTLC plates 10 × 10 cm silica gel 60 F_{254}.

(ii) **Mobile phase:** Ethyl acetate-methanol-water-formic acid (50:2:3:6).

(iii) **Sample application:** 2 ml volumes of test solution and standard are applied each as an 8 mm band, 4 mm apart and 8 mm from lower edge of plate.

(iv) **Development:** 10 × 10 cm chamber, saturated for 10 minutes, 5 ml developing solvent per trough, developing distance 70 mm from lower edge of plate. Dry plate in an oven at 110°C for 5 minutes.

(v) **Detection:** (a) UV 254 nm and (b) NP reagent, immerse plate in reagent for 1 second, dry in stream of cold air, heat to 110°C for 5 minutes. Examine plate under UV 366.

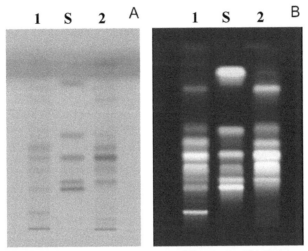

UV 254 nm   NP-reagent, UV 366 nm

Figure 9 TLC chromatogram of constituents from *Crataegus* species visualized with (A) short-wavelength UV light (254 nm) and (B) NP under long-wavelength UV light (see spot colours below).
Track assignment (in each track, in order of increasing R_fs):
1. *Crataegus laevigata*
2. *Crataegus monogyna*
S. Standards: Rutin (orange), Vitexin-2"-O-rhamnoside (green), Chlorogenic acid (blue), Hyperoside (orange), Vitexin (green), Caffeic acid (cyan).
Source: Camag herbal application notes. http://www.camag.ch/l/herbal/index.html

3.9.4 *TLC Analysis for the Identification of Valerian (Valeriana officinalis)*

I. **Sample:** Shake 0.2 g of freshly powdered valerian with 5 ml of CH_2Cl_2 for 1 minute, let stand for 5 minutes and then filter. Wash filter with 2 ml of CH_2Cl_2. Evaporate filtrate and washings to dryness. Dissolve residue in 0.2 mL of CH_2Cl_2.

II. **Standard (optional):** 1 mg of valerenic acid in 0.5 ml of CH_2Cl_2.
Anisaldehyde reagent: In an ice bath mix 9 mL 98% H_2SO_4, 85 mL methanol, 10 mL acetic acid and 0.5 mL anisaldehyde.
HCl-Acetic acid reagent: Mix 20 mL of acetic acid with 80 mL HCl.

III. **Chromatographic conditions:**

   (i) **Stationary phase:** HPTLC plates 10 × 10 cm Silica gel 60 F_{254} (Merck or equivalent).

   (ii) **Mobile phase:** hexane:Ethyl acetate:acetic acid (65:35:0.5).

   (iii) **Sample application:** 3 mL of test solution and standard as 10 mm bands, space 6 mm, 8 mm from lower edge.

   (iv) **Development:** 10 × 10 cm chamber, saturated for 10 minutes (filter paper), developing distance 5.5 cm.

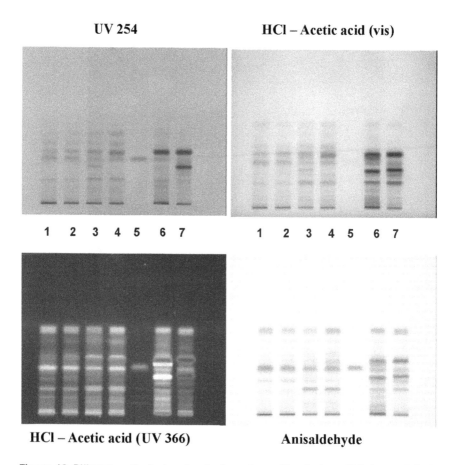

UV 254

HCl – Acetic acid (vis)

HCl – Acetic acid (UV 366)

Anisaldehyde

Figure 10 Different methods for visualization of constituents from *Valeriana officinalis* (Source: Camag herbal application notes. http://www.camag.ch/l/herbal/index.html).

(v) **Detection:** (a) UV 254 nm, (b) HCl-acetic acid: spray, dry in cold air, heat to 110°C for 5 minutes, inspect in visible light and UV 366 nm and (c) anisaldehyde: spray, dry in cold air, heat to 120°C for 2 minutes.

4. PAPER CHROMATOGRAPHY (PC)

PC was the first chromatographic technique developed in organic chemistry. This technique enables separation of chemical substances, carried out by a liquid mobile phase, according to their different rates of migration across sheets of paper. PC has been largely replaced by TLC. However, it is still used because it is an inexpensive but powerful analytical tool that requires very small quantities of material.

4.1 Sample Application

Paper is prepared as previously indicated. The sample is applied at the starting point of the paper chromatogram and methods for sample application are the same as those described above for TLC. For *ascending* paper chromatography (Fig. 4), the piece of paper is ruled near the bottom edge and, if possible, known standards are spotted at points along the starting line.

4.2 Stationary and Mobile Phases

The stationary phase is usually a piece of high quality paper Whatman 1 MM or 3 MM. Hence, the cellulose of the paper is the inert support, and the water adsorbed (hydrogen bonded) from air onto the hydroxyl groups of the cellulose becomes the real stationary phase.

The mobile phase is a developing solution that travels up the stationary phase, by capillary action, carrying the sample components with it. The principle of PC separation is partition chromatography, where components of the sample are differently solubilized between the water tightly bound to the paper and the mobile liquid eluent. Then, retention of a sample molecule depends on its tendency to escape into the mobile phase *versus* its solubility in the stationary phase. This is quantitatively expressed by the Partition Coefficient, K_D:

$$K_D = \frac{\text{(solute in mobile phase)}}{\text{(solute in stationary phase)}}$$

Different mobile phases are used for PC analysis of flavonoid and polyphenol compounds, depending on polarity:

(i) **Phenyl propanoids (e.g. chlorogenic acid, isochlorogenic acid and caffeic acid):** 0.1 N HCl
(ii) **Flavonoid glycosides and phenylpropanoid derivatives:** HOAc 15%
(iii) **Flavonoid aglycones:**
   - HOAc 60% (60 mL of glacial acetic acid mixed with 40 mL of water).
   - HOAc 40% (6-OH flavonoids) (60 mL glacial acetic acid mixed with 40 mL of water).
   - Butanol-acetic acid-water (BAW) (4:1:5) upper phase.
   - BAW (4:1:1).
   - Tertiary butanol-glacial acetic acid-water (TeBA) (3:1:1).
(iv) **Sugars:** EtOAc-pyridine-water (12:5:4).

4.3 Developing the Chromatogram

Initially, the chromatogram should be suspended in the chamber without touching the solvent. To do it, the chromatogram is held to a paper support in the chamber. If needed, paper support could be a glass rod adjusted to the chamber walls with cork rings. The paper should be hung for at least half an hour, high enough to avoid direct contact with the solvent at the bottom of the chamber. This allows paper saturation with the mobile phase.

4.3.1 Immersing the Paper

The chamber is opened and the paper is lowered into the developing solution (Bates and Schaefer, 1971). For ascending PC, the piece of paper is suspended in the chamber with the bottom edge immersed in the solvent (Fig. 11A). Then, the chamber is immediately closed. Ensure that the spots at the starting line are not immersed in the solvent at the bottom of the chamber. The eluent moves up through the paper and when the solvent front nears the top of the paper, the paper is removed from the chamber and the edge of the solvent front is marked. In a variation called *descending* paper chromatography, a solvent tray is placed at the top of the chamber and the solvent is allowed to move down the strip of paper. The flow of solvent is much more rapid in this case, as it is aided by gravity. The paper should be raised out of the tray slightly before descending, to avoid too rapid siphoning action, as shown in Fig. 11B. Depending on the PC system, chromatogram development can take from several minutes to several hours.

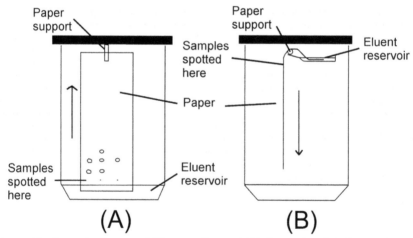

Figure 11 General schemes of (A) ascending and (B) descending PC (Source: Bates and Schaefer, 1971).

4.3.2 Removing the Chromatogram

After the solvent front is within 1.5 cm of the top of the paper, the chromatogram is removed. First, the lid of the development chamber is opened. Then, the chromatogram is lifted out of the chamber. It should be ensured that the chromatogram is handled only from the sides. The solvent front is carefully marked with a pencil and the chromatogram is allowed to dry in a well-ventilated room.

4.4   Detection

After development, the spots corresponding to different compounds may be located by their colour, by long wavelength UV light (365 nm) alone, with ammonium vapour (for polyphenol compounds), or by treatment with spray reagents (for more details, refer to above-mentioned spray reagents for TLC). The final chromatogram can be compared with known chromatograms to identify components of the sample using the R_f value. As in other chromatographic processes, PC uses R_f values and standard compounds that aid in identifying sample compounds. For detection of sugars, the dried chromatogram is sprayed with 1 g of a solution of p-anisidine hydrochloride and 0.1 g sodium hydrosulfite in 10 ml of methanol diluted to 100 mL with n-butanol. The sprayed chromatogram is heated at 130°C in an oven for 10 minutes. Sugar spots turn brown. In phenolic compounds, spectral UV-Vis properties are often obtained by cutting out the section of the dried chromatogram containing a component, soaking this piece of paper in a proper solvent and then recording the spectrum of the component in the solvent.

4.5   Chromatogram

In some cases, compounds are not completely separated in PC; this occurs when two substances have similar R_f values in a particular solvent. In these cases, two-way chromatography (two dimensional PC) is used to separate the multiple-spots. The chromatogram is turned by 90° and placed in a different solvent; some spots separate into multiple spots, showing the presence of more than one compound.

5.   APPLICATION OF TWO-DIMENSIONAL PC IN ANALYSIS OF FLAVONOIDS

Two-dimensional PC is an old technique but is still used for the separation of flavonoids in complex mixtures (Mabry *et al.*, 1970).

5.1   Stationary Phase

Whatman 3 MM chromatographic paper (46 × 57) has proved to be satisfactory for qualitative analysis of complex mixtures of flavonoids.

5.2   Mobile Phase

(a) **For first dimension:** TeBA.
(b) **For second dimension:** HOAc 15% (15 mL of glacial acetic acid mixed with 85 mL of water).
    The TeBA and HOAc solvent systems are appropriated for the two-

dimensional PC analysis of all kinds of flavonoids. The TeBA solvent is unstable when stored for long periods. It is recommended that a fresh solvent is prepared each month and stored in the dark.

5.3 Sample Application

About 0.1 g of the sample is dissolved in methanol. Methanolic solution is spotted on the lower right-hand corner of a sheet of Whatman 3 MM. A hair dryer is used for solvent evaporation between repeated applications of the solution to the paper. The final spot is about 4 mm in diameter and 4 cm from each edge of the paper. The chromatogram is folded for descending PC as shown in Fig. 12.

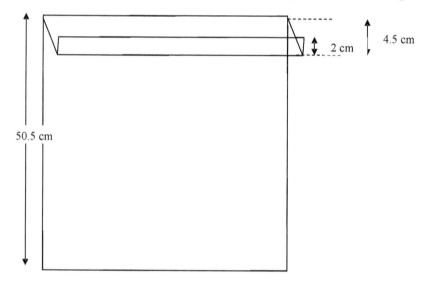

Figure 12   Stationary phase for descending PC.

The chromatogram is developed in the chromatographic chamber using TeBA as the solvent. When the solvent front reaches about 3 cm of the lower paper edge (22-26 hours), the chromatogram is removed from the cabinet and allowed to dry in a fume hood.

The dry chromatogram is folded along the edge adjacent to the band containing the flavonoids and then is developed as indicated above in the second direction with HOAc solvent (4 hours).

5.4 Detection

The dried chromatogram is observed in the visible light. If the spots can be seen, they are outlined with a pencil. Occasionally the colour fades over a period of time, hence the pencil outline is important.

If the spots are not obvious, the most common visualization technique is to hold the paper under an ultraviolet lamp in UV at 365 nm alone and with ammonium vapour. The position of the spots in the chromatogram could give an idea of the type of flavonoid present in the sample (Fig. 13).

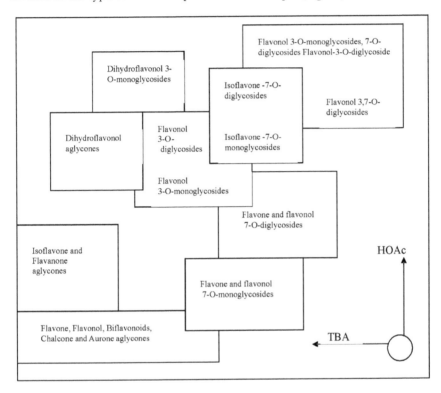

Figure 13 Distribution of flavonoids on a TeBA/HOAc, two-dimensional paper chromatogram (Source: Mabry *et al.*, 1970).

SUGGESTED READINGS

Bates, R.B. and Schaefer, J.P. (1971). *Research Techniques in Organic Chemistry*. Prentice Hall Inc., Englewood Cliffs, NJ.

Camag herbal application notes. Available from: http://www.camag.ch/l/herbal/index.html; *20 October 2006.*

Christian, G.D. and O'Relly, J.E. (1986). *Instrumental Analysis*. 2nd edition. Allyn and Bacon, Boston, Country.

Ettre, L.S. and Sakodynskii (1993). M. S. Tswett and the Discovery of Chromatography I: Early Work (1899-1903). *Chromatographia* **35**: 223-231.

Fried, B. and Sherma, J. (1996). *Handbook of Thin-Layer Chromatography*. 2nd edition. Marcel Dekker, New York.

Fried, B. and Sherma, J. (1996). *Practical Thin-Layer Chromatography — A Multidisciplinary Approach*. CRC Press, FL, USA.

Fried, B. and Sherma, J. (1999). *Thin-Layer Chromatography*. 4th edition, revised and expanded. Marcel Dekker, New York.

Geiss, F. (1987). *Fundamentals of Thin-Layer Chromatography*. Huethig, Heidelberg, Germany.

Hahn-Deinstrop, E. (2000). *Applied Thin-Layer Chromatography*. Wiley-VCH Verlag, Weinheim, Germany.

Harborne, J.B. (1984). *Phytochemical Methods: A Guide to Modern Techniques of Plant Analysis*. 2nd edition. Chapman & Hall, London.

Inczedy, J., Lengyel, T., and Ure, A.M. (1998). *Compendium of Analytical Nomenclature. The Orange Book — 3rd edition*. Blackwell Science, New York.

Jork, H., Funk, W., Fischer, W. and Wimmer, H. (1990). *Thin Layer Chromatography, Volume 1a, Physical and Chemical Detection Methods*. VCH Verlagsgesellschaft mbH, Weinheim, Germany.

Jork, H., Funk, W., Fischer, W., and Wimmer, H. (1994). *Thin-Layer Chromatography, Volume 1b, Reagents and Detection Methods*. VCH Verlagsgesellschaft mbH, Weinheim, Germany.

Mabry, T.J., Markham, K.R. and Thomas, M.B. (1970). *The Systematic Identification of Flavonoids*. Springer-Verlag, New York.

Nyiredy, S. (2001). *Planar Chromatography — A Retrospective View for the Third Millennium*. Springer Scientific Publisher, Budapest, Hungary.

Ravindranath, B. (1989). *Principles and Practice of Chromatography*. Halsted Press, New York.

Scott, R.P.W. *Principles and Practice of Chromatography*. Ed. Series. (*http://www.chromatography-online.org/1/contents.html*).

Stahl, E. (1969). *Thin Layer Chromatography*. Springer-Verlag, New York,

Wagner, H. and Bladt, S. (1996). *Plant Drug Analysis: A Thin Layer Chromatography Atlas*. 2nd edition. Springer-Verlag, New York.

HPLC and LC-MS

Erica G. Wilson

1. INTRODUCTION

High performance liquid chromatography, better known as HPLC, is the most widely used instrumental technique for analytical purposes in natural product studies and also in pharmaceutical, cosmetic and food industries, surpassing the use of Gas Chromatography (GC). The reason for this is, basically, its versatility and the relatively simple sample treatment required for analysis. While GC requires all analytes to be volatile, HPLC requires only the analyte to be soluble in the mobile phase (Snyder *et al.*, 1997). While GC requires the analytes of interest to be stable at the high temperatures required to reach volatility, HPLC is carried out most frequently at room temperature or at temperatures no higher than 45°C, if necessary for improved efficiency, as is discussed in this chapter.

These characteristics are particularly suited for the analysis of natural samples, with two major exceptions, the study of volatile fractions such as essential oils and lipids. In the first case, these low weight non-polar compounds are practically impossible to detect with the detectors available in the market. In the second, given the nature of the lipid fraction — mainly triglycerides or other esters of fatty acids — the efficiency required to separate the components of their mixtures can only be provided by GC. Another field where HPLC has severe limitations is in the detection of insecticides or herbicides — organophosphorous or chlorinated — which cannot be detected either as a consequence of the low tolerated levels and/or their low response to most detectors.

Allelochemicals include compounds such as benzoic acid derivatives, flavonoids

Author's address: Department of Pharmacognosy, School of Pharmacy and Biochemistry, University of Buenos Aires, Junin 956, Ciudad Autónoma de Buenos Aires, Argentina. E-mail: erica.g.wilson@gmail.com

and cynammic acid derivatives that are ideal for HPLC analysis, since they have high UV absorptivity at relatively high UV wavelengths and are easily and reproducibly separated using bonded phase chromatography (BPC). For this reason, special attention has been paid to this type of chromatography in this chapter, though an outline of other methods is also provided.

2. HPLC SYSTEMS

HPLC can be most easily understood if it is considered as a sophisticated instrumentalization of open column chromatography (Miller, 2000). The latter, which together with paper chromatography were the initiators of chromatography, is still used nowadays for the preparative or semipreparative separation of fractions of plant extracts. Anyone who uses this type of method regularly is aware of its advantages and disadvantages! Long hours of elution, which though reduced by techniques such as flash chromatography, are still relatively high, yield only semipure compounds and nobody would try to use these for analytical means without suffering the frustration derived from the enormous band-spreading caused by the low efficiency of stationary phases. The downscaling of particle size in the search for higher efficiency brought with it the unwelcome consequence of the high operating pressures required to force the mobile phase through these highly compacted beds of stationary phase. Of course, the high operating pressure (125 atm or 2,000 psi are normal working pressures) required the design of special devices for injecting the sample and all kinds of leak free connections. The efficiency achieved by the reduction of particle size could not be lost by slow fraction collection for later UV detection. For example, flow-through low volume cell detectors were attached to the output of the column, simplifying the monitoring of column effluents. The result of all this was the liquid chromatograph (Fig. 1).

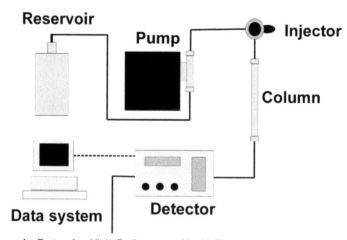

Figure 1 Parts of a High Performance Liquid Chromatograph (HPLC) system.

2.1 Basics of HPLC Separation

An instrumental chromatographic separation is typically visualized by the plotting of a chromatogram (Fig. 2) which could be done by a simple pen-recorder, but is commonly nowadays by some data processing system (Lee, 2000).

In this graph, the detector response is plotted against time rather than volume of elution, and the eluting compounds will appear as peaks, the apex of which is considered to be the retention time (t_R) that identifies the substance in the same manner that retention factor (R_f) is used in planar chromatography.

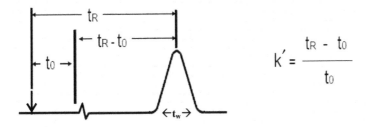

Figure 2 Chromatogram indicating detector response vs time.

Retention time (t_R): The time between injection and the appearance of the peak maximum. The adjusted retention time t_R' adjusts for the column void volume ($t_R - t_0$).

Void volume (t_0): Void volume, time taken for an unretained molecule to travel from the injector to the detector.

2.2 Resolution of a Chromatographic System

The reliability of a chromatographic separation depends on the ability of the method to fully separate the analytes of interest from any other components of the mixture as the ultimate goal, or, to be more precise, at least to be able to provide sufficient guarantee for the purity of the peak of interest for the detection system that is being used. This point is made because with the exception of universal detectors and especially a MS detector — and even then, according to the concentration – no system can give absolute certainty of peak purity. This is related to the **Resolution** which can be defined as a characteristic of the separation of two adjacent peaks. It can be calculated from a chromatogram using the equation described below. In Fig. 3, two other factors have been described, k' and α, that stand for capacity and selectivity, respectively.

Capacity (k') is defined as the degree by which a system retains a certain compound and is calculated using the following equation:

$$k' = t_R' - t_0/t_0 = t_R - t_0/t_0$$

Selectivity (α), as can be seen in Fig. 3, is a ratio between the k′ values for the two adjacent peaks.

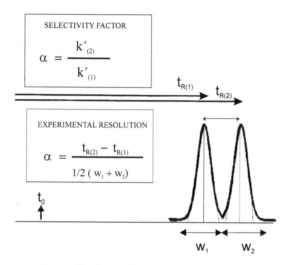

Figure 3   Resolution of a chromatographic system.

Resolution is a comprehensive concept that takes into account the width of the peaks and their separation. The width of the peak is a measure of the band-spreading which normally occurs in any chromatographic system; this is quantified by the calculation of the number of plates and reflects the efficiency of the system.

Efficiency (N) of a column is normally expressed as the number of plates per unit of length. At the same time, the number of plates (N) can be calculated using several equations and the final number will depend on the height of the peak at which its width is measured. Therefore:

$$N = 16 \ (t_R)^2/t_{wb} \quad \text{or} \quad 5{,}54 \ (t_R)^2/t_{wh}$$

where, t_{wb} is the width of the base of the peak expressed as time, while t_{wh} is the width measured at half-height also expressed as time.

2.3   Reverse and Normal Phase Chromatography

The first chromatographic separations were carried out using polar adsorbents such as calcium carbonate and later paper or silica. This type of separation, in which the analytes to be separated are less polar than the stationary phase, is known as "normal phase" chromatography. Naturally, the compounds that are separated using these systems will elute in the increasing order of polarity. Accordingly, their retention (k′) will be decreased, if the polarity of the mobile phase is increased and vice versa.

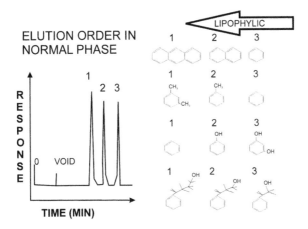

Figure 4 Elution in normal phase chromatography. Hydrophilic compounds are more retained on polar stationary phases such as silica gel.

Later on, stable non-polar stationary phases were achieved by bonding non-polar radicals to the active silanol groups on silica gel surface. These were called "reverse phases" as the resulting stationary phase subsequently proved to be less polar than the samples that could be analyzed on it. When these systems are used, the elution order is exactly the opposite to that expected with normal phase chromatography and lipophylic substances are retained while hydrophilic, polar compounds elute beforehand as illustrated in Fig. 5. It is very important to determine whether a system is behaving as a reverse or normal phase chromatography, since it enables us to: (i) predict the order of elution and (ii) be able to make the correct decision when optimizing the method.

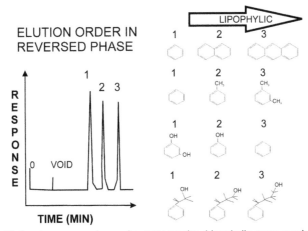

Figure 5 Elution in reverse phase chromatography. Lipophylic compounds are retained on non-polar stationary phase and elute later than the polar, hydrophilic compounds.

3. INSTRUMENTATION

3.1 Pumps

The main part of a HPLC instrument is the pump, which has the difficult task of pumping the mobile phase through the column and system against back pressures that may vary between approximately 600 psi and 3,000 psi (40 and 200 atm) at a constant, accurate, reproducible and pulseless flow rate. It is important to note that the range of flow rates which an analytical pump must produce is between 0.50 and 2.00 mL/min.

The main requirements of a pump are:

(i) It must be chemically inert
(ii) Resist high pressure (5,000 psi)
(iii) It must provide flow rates between 0.5 and 10 mL/min
(iv) Provide a pulse free or damped flow
(v) Flow reprodnucibility > 1.0%
(vi) Gradient, rapid solvent change

There are several designs of HPLC pumps that fulfil these requirements in different degrees and at different costs. The most popular by far are any of the two versions of dual-headed reciprocating piston pumps. In one case, the two pump heads are placed in parallel and operate 180° out of phase, allowing one pump head to deliver high pressure mobile phase while the other pump head refills with mobile phase.

The other design consists of two pump heads assembled in series, the first piston being about twice the size of the second. This allows the first piston to deliver the solvent to the column at high pressure while simultaneously refilling the second piston. When the first piston chamber is empty, the second piston assumes the function of pumping the solvent to the column, while the first piston refills its chamber.

In both cases the resulting flow rate is practically pulse-free.

Vital and vulnerable parts of this design are the valves that are placed at the entrance and exit of the solvent chamber, known as "check valves" (Fig. 6).

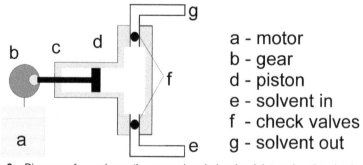

a - motor
b - gear
d - piston
e - solvent in
f - check valves
g - solvent out

Figure 6   Diagram of a reciprocating pump-head showing inlet and outlet check valves

The function of a check valve is to prevent the solvent from back flowing into the solvent reservoir, when the piston is pushing it towards the column (outlet check valve), or to prevent it from being sucked into the chamber, when the piston withdraws to allow the solvent to come into the chamber from the reservoir (inlet check valve). This is achieved by the presence of a small ball bearing made of highly resistant materials, such as ruby. The main and relatively frequent problem, is that these bearings can get "stuck" in a certain position, usually due to the presence of salt crystals or the use of a solvent that reacts with them in some way. As a result of this, no solvent can be delivered by that pump head, resulting in a definite pulse and lower flow rates. Some hints are provided in the troubleshooting section to solve this very usual problem.

3.2 Isocratic and Gradient Elution Systems

HPLC elution can be carried out in two ways: by the delivery of a mobile phase with a constant composition — isocratic elution — or with a mobile phase of two or more components, the composition of which is varied throughout the chromatographic run in a gradual or step-like manner (gradient elution). This type of elution is similar to the temperature gradient in GC; in this case increasing the percentage of the stronger solvent in the mobile phase increases the strength of the mobile phase. A brief description is given below of the available instrumental set-ups. Special attention has been given to this, because the use of gradient elution is inevitable when working with natural samples, owing to the diversity of polarity and chemical properties exhibited by the components of the sample.

There are two instrumental solutions to achieve a gradient system. In Fig. 7, the gradient is formed at high pressure with a system consisting of two pumps,

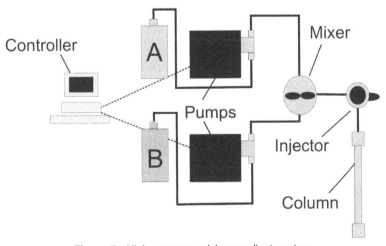

Figure 7 High pressure mixing gradient system.

the flow of which is synchronized by the system controller to contribute to the programmed gradient in the right proportion. The solvent from each pump is mixed in the mixing chamber and delivered to the column. This system is more precise and leads to less bubble formation when the solvents are mixed. However, it is limited to binary gradients — if two pumps are used — and is more expensive.

The other system (Fig. 8) consists of the use of one pump, which takes up the solvent from a mixing chamber which has a proportioning valve that regulates the intake of solvent from each solvent reservoir according to the gradient programme. The obvious advantages lie in the fact that only one pump is necessary and that binary, tertiary or up to quaternary gradients are possible. However, the gradient tends to be less reproducible.

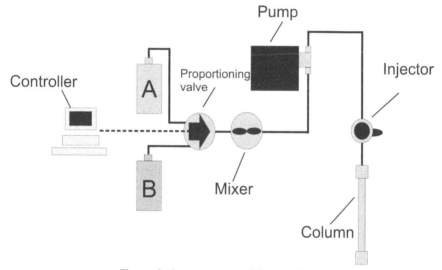

Figure 8 Low pressure mixing gradient.

In both cases, the variation of the composition of the mobile phase is achieved by the mixture of different components of the mobile phase contained in separate solvent reservoirs.

Whichever the system used, the most important issue to consider when using and/or purchasing these systems, is the size of the mixing chamber, since a very large chamber will not allow for very accurate gradients when using low flows, while a very small chamber will not ensure the correct proportions at high flows.

3.3 Injectors

There are basically two types of injectors: automatic and manual. HPLC injectors have very special designs since it is necessary to introduce the sample into a system which is operating at very high pressure. It is, therefore, necessary to isolate the flow stream mechanically from the sample. To accomplish this, multiport switching valves are used.

Both automatic and manual systems follow the same approach. The valve consists of two positions: load and inject (Fig. 9). In the load position, the flow is diverted from the pump to the column, thus leaving the loop at room pressure. The sample is loaded into this loop with the aid of a manual syringe or through the use of an automated loading device (autosampler). The valve is then rotated into the inject position, redirecting the flow from the pump, through the loop and into the column.

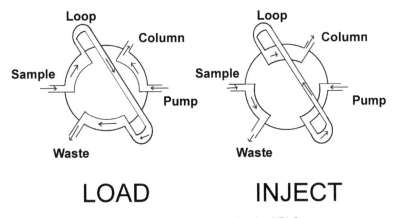

LOAD   **INJECT**

Figure 9   Manual sampling valve for HPLC.

For manual injectors, the highest precision (reproducibility) is achieved by fully loading the loop, the limitation being that the injection volume is thus inflexible. However, it is generally accepted that a volume of 20 μL is standard for analytical samples, while semipreparative work requires the injection of much higher volumes in the range of 100 μL. However, it is very important to note that the injection of high volumes of samples can severely distort the resolution of a system, especially if the injection solvent is different from the mobile phase.

It is very difficult to establish clear advantages of one system over the other. The results of routine analyses are more favourable if an autosampler is used, as is also the case of LC-MS systems that usually run for long durations and can be programmed to work during the night for example, increasing the working time of a very expensive system. On the other hand, method development is very often easier to handle with a manual injector, the ideal situation being thus, to have an instrument which is equipped with both types of sampling devices.

Some autosamplers have interesting features, such as the possibility of loading biphasic samples and programming the injector on the depth from which the sample must be drawn, avoiding the separation of the phases for example. Vital characteristics of autosamplers are the dead volume that they may introduce in the system and the rinsing and needle wash systems which often produce distortion or cross contamination of samples. The price and/or possible recycling of vials can be another factor that should be looked into if one must rely solely on this type of injector.

3.4 Detectors

The second most important hardware item in a HPLC system is the detector. When using detectors and/or deciding which type will best fit one's needs, it is important to bear in mind that common HPLC detectors are not unique to HPLC but rather, are adaptations of instruments used in analytical chemistry. These adaptations are designed to meet the requirements and conditions for HPLC, the most important of which are a low-volume flow through cell and a fast response time, also known as time constant. HPLC detectors most commonly used are listed in Table 1.

Table 1 HPLC detectors

- Absorbance detectors
Ultraviolet, fixed
Ultraviolet, spectrophotometric (UV-Vis)
Photodiode array (known as PDA or DAD)
- Fluorescence
- Refractive Index
- Electrochemical
- Electrochemical detector (ECD)
- Conductivity detector
- Light scattering detector (available as LALLS, ELSD)
- Mass spectrometer (MS)

3.4.1 Characteristics of a Detector

When selecting a appropriate detector, it is important to consider certain characteristics which have to be analyzed in order to determine its suitability to solve one's needs. These are:

(i) Noise (actual noise/signal ratio)
(ii) Time constant
(iii) Signal:
   Sensitivity
   Detectability
   Linearity
   Universal or selective response

Characteristics such as the noise and time constant refer to the quality of the design of the detector or to the type of detector, and their analyses exceed the scope of this book. However, most state-of-the-art instruments have adequately solved these issues or at least, have little difference between them.

The signal refers to the type of response which can be expected from each detector and is chosen according to the chemical properties of the samples and the approximate concentration range used for working. Sensitivity and linearity are once more intrinsically related to instrument design, whereas, the selectivity of the response is a very important issue to consider. In general, universal detectors measure a property which all compounds have, albeit in varying degrees. A typical example is the Refractive Index (RI) detector which detects changes in the refractive index of the column eluate produced when compounds are eluted. On the other hand, selective detectors, such as the UV-Vis detector, can only be applied to compounds that have a relatively large UV or visible light absorptivity in the range of 210 to 750 nm. The choice of the best detector is based on the type and concentration of analytes under analysis. For example, for pharmaceutical, cosmetics or natural products, a spectrophotometric detector (UV-Vis) would be the best choice, whereas if samples are mainly simple sugars or polysaccharides, the choice would be limited to an RI, ELS or other universal type detector. Main features of HPLC detectors are shown in Table 2. A brief description of each type of detector is provided below.

Table 2 Summary of the main features of different HPLC detectors

Detector	Response	Sensitivity	Applications
UV-Vis	Selective	1×10^{-10} g	Most organic compounds, including most natural products
RI/ELS	Universal	1×10^{-9} g	Sugars, amino acids, preparative chromatography. Polymers, proteins (macromolecules)
Fluorescence	Selective	1×10^{-13} g	Aflatoxins, derivatized amino acids
Electrochemical conductivity	Selective	1×10^{-13} g	Amino acids, ionic species, catecholamines, peptides
Mass spectrometer	Universal	1×10^{-10} g	All types of compounds

3.4.2 Ultraviolet Absorbance Detectors

There are two basic types of detectors, those with the older technology based on a typical design by which HPLC eluates are irradiated with an energy of a certain wavelength known to be absorbed by the analytes of interest, and the other relatively new technology introduced in the early 1990s — the photodiode array detectors (DAD).

I. **Fixed wavelength UV detector.** This was the first type of detector that appeared in the market and is basically a spectrometer that employs a UV energy source — typically a mercury lamp that emits a narrow range of wavelengths, the strongest being at approximately 254 nm. Other wavelengths can be achieved by the use of filters. The main advantage of this detector is its cost and sensitivity, in the case of compounds that have a maximum absorbance in this range of wavelengths. It has been displaced by the spectrophotometric absorbance detectors owing to its limitations.

II. **Variable-wavelength UV detector.** This detector has a design similar to a spectrophotometer, that is, it has an energy source that can emit a broad range of wavelengths in the UV zone — typically a deuterium lamp — between 190 and 350 nm, and a lamp for the visible region. This light is directed onto a grating which can select the specific wavelength required for the best detection of the compound of interest. This is a relatively cheap detector, which is very useful for routine analysis since it requires very simple software and usually has some scanning capabilities for additional qualitative information. A simplified diagram of this device is shown in Fig. 10.

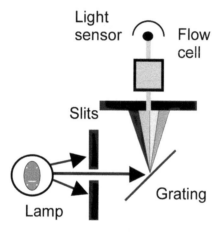

Figure 10 Simple diagram of a spectrophotometric UV-Vis detector.

III. **Photodiode Array Detector (PDA or DAD).** The chief characteristic of DAD is the fact that while in the other cases the sample is irradiated with a certain wavelength energy and the amount of absorbed energy is measured, in this case the light provided by a continuum source is focussed on the sample cell and the transmitted light impacts an array of photodiodes which can, therefore, translate the information of the energy transmitted into absorption information (Fig. 11). The process enables recovery of information every 0.1 second, resulting in very accurate data of what is eluting through the cell. This information can thus be processed in order to gain very reliable evaluations of peak purity and identity if compared with the spectra of a standard or compound libraries. Naturally, this detector requires a powerful computer to process and store this information and is by far, the most expensive UV detector in the market.

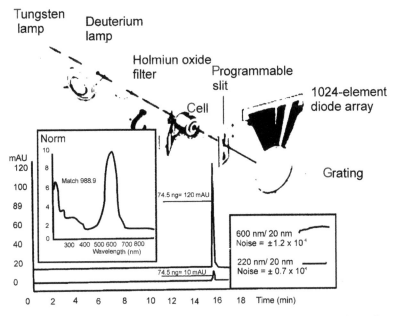

Figure 11 Diagram of a PDA detector showing its application for peak identity and purity determination.

Unfortunately, not all compounds will prove to be detectable using a UV-Vis detector, either because it has no UV absorptivity above 205 nm or, more often, because it is very low ($\varepsilon < 100$). Considering that the molecules of a sample have to be detected among an overwhelming majority of mobile phase molecules, their detection is impossible. A list of the absorption wavelength of some of the most frequent chromophores is given below (Table 3). Clearly there are many compounds with structures that cannot be distinctly detected with a UV detector, for example, sugars, most amino acids, low MW principally aliphatic acids and bases, among other common natural products.

Table 3 Absorbance wavelength of some common chromophores

Chromophore	λ *max* (nm)
Aldehyde	210
Amine	195
Benzene	202,255
Carboxyl	200-210
Dibenzene	220,275
Ester	205
Ethylene	190
Ketone	195
Nitro	310

Naturally other detectors must be used for those compounds that absorb below 205 nm.

3.4.3 Refractive Index (RI) Detector

This detector is based on the fact that the refraction of a beam of light in a liquid media will vary according to its qualitative and quantitative composition. Therefore, it consists in a light source and a system of filters and grids that allows two equal beams to focus on a cell with pure mobile phase (the reference cell) and another cell (the sampling cell) through which the eluate flows. The baseline response corresponds to the refractive index of the pure mobile phase in the reference cell and the sample cell is monitored in relation to this. Therefore, any change in the composition of the eluate is recorded as a peak, which will be negative or positive according to the higher or lower refraction of the eluate. Unfortunately, this detector has two major drawbacks. One is that refraction depends greatly on temperature, and the other is that the detector has a universal response, being not very sensitive. These characteristics imply that the system must be thermally stable for quantitative response and that it can never be used for the analysis of low concentration of analytes.

3.4.4 Evaporative Light-Scattering Detector (ELSD)

A diagram of this detector is shown in Fig. 12. The effluent of the column must be nebulized, thus entering the light scattering cell as a fine spray. A beam of light is made to pass through the cell and the particles are detected by their scattering properties. Naturally this detector, in the same manner as the RI detector, is universal but is much easier to handle, as it is not sensitive to temperature changes. Sensitivity is also an issue, since it is not very sensitive (about two orders of magnitude lower than UV detection).

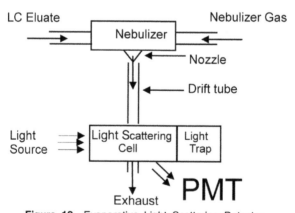

Figure 12 Evaporative Light Scattering Detector.

3.4.5 Fluorescence Detector

This detector is similar to a spectrofluorometer, and shares its advantages and of course, disadvantages. On one hand, it is extremely sensitive but on the other it is extremely selective, only detecting compounds that naturally fluoresce. It is also used when compounds that are difficult to detect by other means, such as amino acids, are derivatized with fluorescent reagents to give fluorescent substances. It has a source of light, typically a deuterium lamp that can emit light in the UV range for excitation of the compounds and a system which allows the selection of the emission wavelength detection.

3.4.6 Electrochemical Detector (ECD)

This is another sensitive and selective detector which is used for substances that can easily oxidize or reduce. It is typically equipped with three electrodes (working, counter and reference electrode) that can be adjusted to the necessary potential according to the compound. The ensuing redox reaction produces a flow of electrons which is translated as a peak. It is a difficult detector to work with, requiring long stabilization periods. However, compounds such as catecholamines or non-UV absorbing antioxidants such as ascorbic acid, can be successfully analyzed using this detector.

3.5 Mass Spectrometer (MS)

This is the only detector that provides positive identification of analytes. It is the most difficult detector to use when coupled to a LC system because it requires volatilization of the analytes, a condition very difficult to achieve considering that they are dissolved in the liquid mobile phase in an extremely low ratio.

A MS detector consists of three distinct features: (i) the source, (ii) the analyzer, and (iii) the detector. The difference in the results obtained with the various types available of each of these parts distinguishes the types of MS techniques that might be useful for HPLC applications. As illustrated in Fig. 13, all MS techniques require previous analyte ionization in the source. The resulting ions, which have

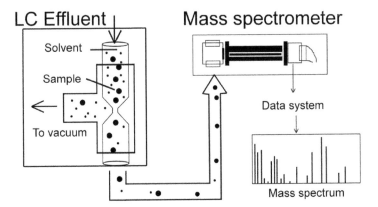

Figure 13 LC-MS interface.

discrete mass/charge ratios (m/z) are separated and focussed in the mass analyzer. The final focussed beam impinges on a detector that determines the intensity of the beam.

Ion sources. There are several possible ionization methods, such as electron ionization (EI), chemical ionization (CI), electrospray ionization (ESI), thermospray (TSP) and fast-atom bombardment (FAB). However, the most popular are undoubtedly the atmospheric-pressure ionization interface (API), electrospray (ESI), ion spray (ISP) and thermospray (TSP) sources. These give different products according to the amount of energy that impacts the analyte. Matrix-assisted laser desorption ionization (MALDI) is used especially for the analysis of proteins or large peptides. Once the ions are obtained, these will flow into the mass analyzer. There are several types, including magnetic and electrostatic sectors, quadrupole, ion trap (ITP) and time of flight (TOF). The selection of the MS should be related to the kind of sample which is to be analyzed and the kind of required information. The MS ion source must be compatible with the expected molecular weights of the analytes. The ideal situation is to have a high-resolution separation coupled with a high-resolution MS through an interface or ion source that will provide both individually charged and multiply charged ions derived from every analyte species. It is important to remember that two types of information are provided by this spectrometry: (i) the MW — symbolized as M^+ or MH^+, or (ii) fragmentation patterns if using an ion trap (MS/MS/MS for example), or triple quadrupole system (MS/MS), thus obtaining much more qualitative information on the analytes.

In spite of increasing technological innovations and improvements in the state-of-the-art of LC-MS systems as they are known, there are still severe limitations regarding the composition and the conditions of the mobile phase. In the first case, methods must be adapted to lower flow rates, typically around 0.5 mL/minute. Secondly, no salts should be used in the mobile phase and the choice of solvents should be restricted to organic solvents and water using volatile substances, such as ethanoic or methanoid acid for acid pH, and amines for alkaline conditions. This imposes quite severe limitations to the possible resolution of the system, that is of course, counterbalanced by the fact that the detector itself can pick up the real purity of the peaks because of the richness of the qualitative information provided by it.

3.6 Columns

HPLC columns come in a great variety of sizes and materials. A summary of the most usual dimensions and their applications can be seen in Table 4. Columns are usually made of high grade stainless steel, as in the rest of the chromatograph, with a certain thickness which allows them to resist the high back pressures that build up inside them. Throughout the history of HPLC, a number of variations were attempted, such as radial compressed plastic cartridges, fused silica within special metal casings, and PEEK [polyetheretherketone] columns, with the objective of reducing their cost, which is greatly due to the stainless steel. However, as time went by, none of those solutions proved to be acceptable for the market and to date, there is a strong prevalence of metal columns. Another characteristic of a column, apart from the type of stationary phase is discussed in Section 3.7 of this chapter, is the size of the particles of this solid. As this is generally associated with the dimensions of the column, it is also included in Table 4.

Table 4 Column configuration

Type	Length (mm)	Inner diameter (mm)	Particle size (m)
Analytical	30-250; usually 125/150 or 250	3-4.60; most typical are in the 4-4.60 range	3-10
Semipreparative	100-250	8-10	7-20
Preparative	100-250	10-50	7-20

3.7   Column Troubleshooting

There are two main sources of troubleshooting when working with a HPLC system — one of them is the pump, and the other is the column. Those related to the pump have been partially dealt with above. Those related to the column are briefly addressed below. HPLC columns are quite expensive items which have a limited "lifetime". This lifetime can, however, be extended if some care is given to certain aspects. The most common causes for column life decrease are:

(i)   Partially blocked or plugged inlet frit or column bed.
(ii) Adsorbed or strongly retained sample components.
(iii) Chemical attack on the support or stationary phase.

The symptoms which will appear as a result of these events are:

(i)   High back pressure.
(ii) Tailing bands.
(iii) Loss of resolution, evidenced by decrease of selectivity or retention.
(iv) Band-widening (loss in plate number).

3.7.1   Blocked or Plugged Column Inlet Frit

This is certainly one of the most common problems encountered by HPLC users. The reason is often related to a guard column if used or, more frequently, the injection of samples containing particulates. These could be avoided by filtering the sample through a 0.25 μm filter membrane. It is important to note that the injection of unfiltered samples will also damage the seals and rotors of the injector valve, causing leaks. Particles from worn pump seals that are released into the mobile phase flow are a factor that might plug the inlet filters. Changing the pump and rotor valve seals regularly will also help to avoid frit plugging.

A possible solution is to carefully replace the column inlet frits by unscrewing the column inlet and installing a new one with great care, avoiding disturbance to the stationary phase. If this were not available, a good solution is to invert the inlet and outlet fittings and flush the column with mobile phase, having disconnected it from the detector. It is important to avoid inverting the whole column unless every other strategy has failed.

A third cause for pressure build-up might be that the sample or some component of the sample is precipitating due to solubility problems with the mobile phase. It is important to emphasize, once more, that the sample and all its components have to be soluble in the mobile phase. Moreover, the best peak shape is generally achieved when the injection solvent is the mobile phase itself. However, if sample precipitation is suspected, the best way to confirm it is to inject a pure strong solvent such as THF, which will dissolve the precipitate.

3.7.2 Adsorbed or Strongly Retained Sample Components

When samples with complex matrixes such as plant extracts or biological fluids are injected, it is possible that some of the components of the sample might be strongly retained on the stationary phase. If the inlet fitting of a column is unscrewed, it is quite usual to find that the visible packing is dark green due to the retention of the lipophylic chlorophylls, for example. As a result of this, the resolution of the column will decrease and additionally, when using a gradient, ghost peaks might appear in the last part eluting with the stronger mobile phase composition. The solution for this type of problem is to flush the column with a sequence of solvents, starting from methanol:water (50:50); methanol 100%; methanol:dichloromethane (50:50); dichloromethane 100% and then work back to the first step in the inverse sequence. Approximately 10 column volumes should be pumped through the column. If this does not work, the column can be inverted but this should the last resort, since pumping solvent through a column at high pressure might disturb the column bed packing and produce a severe loss in efficiency.

The long-term solution is to avoid this type of problem by doing three things on a routine basis:

(i) Flushing the column when finished with a stronger solvent than the mobile phase and leaving it in acetonitrile:water. (50:50)
(ii) Cleaning-up the sample, as explained in Section 3.11 at the end of this chapter.
(iii) Dissolving the sample in mobile phase.

3.7.3 Chemical Attack

The evidence of this type of deterioration is band-widening, loss of resolution or compound retention. As is explained further on, certain combinations of stationary/mobile phases should be avoided since they will lead to the elimination of the stationary phase material. Tailing will appear if stationary phases have been stripped in C18 or C8 columns due to the use of mobile phases with a pH below 2 or above 8.5. High concentrations of acids such as TFA will also produce irreversible damage and working at temperatures above 50°C will decrease packing stability. Lastly, the use of certain mobile phase modifiers, such as ion-pair reagents will produce changes in the packings by adsorption on the silanol groups, which will not necessarily damage it but will change its behaviour.

3.7.4 Stationary Phases

There are three types of materials that are used as stationary phases or support-materials in HPLC: (i) silica gel (and chemically modified silica gel), (ii) hydrophilic gels (Sephadex) and (iii) resins (polysterene cross-linked with divinyl benzene). Lately there have been some interesting developments in the preparation of hybrid columns made of silica and resins which seem to combine the advantages of these materials.

In general, all HPLC columns are packed with the stationary phase which is usually a powder, but once again, there are a few variations, such as the appearance of monolithic columns that achieve good separations in a shorter time at lower back pressure.

Particles used in stationary phases can have different formats and thereby, different applications. They can be porous irregular, spherical or pellicular, consisting in an impervious glass bead which is coated with stationary phase. The characteristics are summarized in Table 5.

Table 5 Type of particles used as stationary phases or support material

Macroporous	Pelicular	Microporous
Preparative separations	Guard columns	Preparative and analytical separations
High capacity	Low capacity	High capacity
Easily packed	Easily packed	Less easily packed
Low cost	High cost	Very expensive
Low efficiency	High efficiency	Highly efficient
		High speed

   I. **Silica gel:** The characteristics of this hydrated silicic acid have been explained in Chapter 5, Section 3.1. There are two features of silica that acquire more importance in its HPLC application: firstly, the fact that because of its acidic nature it is especially vulnerable to a high pH, quickly dissolving above 9 approximately. The other characteristic is its water adsorbent properties, which must be considered to avoid variation in the retention times of sample components. In all other aspects, silica gel constitutes a very rugged and efficient stationary phase or support.

  II. **Chemically bonded silica:** Owing greatly to the fact that the analysis of water-soluble compounds was especially restricted by the strong hydrogen bonding ability of silica, column packing material made by covalently bonding an organosilane or a polymeric organic layer to silica appeared 30 years ago. As a result, HPLC became an extremely versatile tool for analyzing all types of samples with relatively simple pre-treatment. The chemical binding to the active acid silanol groups of the silicic acid resulted in the formation of quite stable Si–O–Si esters known as siloxanes or substituted silanols. A relatively large array of chemically bonded silica phases is available at present, ranging from non-polar/hydrophobic phases to very polar and hydrophilic bonded silanes, as shown in Table 5. The most popular columns by far are C18 and C8 columns, which are used to analyze 95% of natural product samples. The remaining samples can be solved usually using an aminopropyl bonded phase.

Table 6 Useful bonded phases for HPLC

Stationary phase	Characteristics
C18 (octadecylsilane)	Hydrophobic; widely available; reverse-phase.
C8 (octylsilane)	Also hydrophobic; reverse phase: adequate for basic compounds. Both can be used for ion-pair chromatography.
Phenyl, phenyethyl	Moderately hydrophobic
CN (cyano)	Moderately lipophylic, more polar than C18 and C8, used for both reverse and normal phase
NH_2 (aminopropyl)	Hydrophylic, can replace silica applications but is less adsorbent. Can be used for reverse and normal phase. Not very stable. Can be used as a weak anion exchanger. Used for sugar analysis/saponins.
OH (diol)	Highly hydrophylic and polar; not very stable. Used for normal phase separations.

III. **Stability of bonded phase columns:** Column life depends greatly on two factors: (i) the conditions in which a sample is injected (sample pre-treatment), and (ii) the attention paid to the mobile phases which are used. Good stationary phase stability minimizes the need for adjustments of separation parameters or eventually the replacement of the expensive column. This is related to:

(i) Type of bonded phase: C18 and C8 are more stable than shorter alkyl group columns.

(ii) Low mobile phase pH: Loss of silane bonded phases results from the hydrolysis of the Si–O–Si bond that binds the silane to the support. This occurs at a pH < 2.5 and typically results in tailing due to the exposure of Si–OH groups that will retain the compounds in a very different manner to the bonded phase.

(iii) High mobile phase pH: Rapid dissolution of the silica support occurs at a pH > 8; that is the siloxane bond resists this pH, but the whole support dissolves and the bonded silane is undermined and falls from the surface. End-capping greatly solves this problem and should always be used when dealing with samples that require intermediate to high pH values.

(iv) The use of certain acids, like trifluoracetic acid (TFA), and hydrochloric acid (HCl) should be restricted to low percentages or avoided altogether if possible. TFA can be replaced by ethanoic or methanoic acid, and phosphoric acid can easily replace HCl unless precipitation of phosphates is feared.

(v) Buffers: the use of buffers is very popular; however, a column should be thoroughly rinsed with water and then left in acetonitrile/water if the flow is to be stopped.

(vi) Certain organic modifiers, e.g. ion-pair reagents like tetrabutylammonium or alkylsulfonate salts, may cause irreversible changes in a bonded phase column due to their adsorption on the phase. It is convenient to keep columns exclusively for this use.

IV. **Porous polymers.** This type of stationary phase has been used traditionally for molecular filtration techniques, specifically for the size-exclusion separation of non water-soluble polymers or proteins. Attaching strongly acid or basic groups to their surface resulted in the typical resins for ion-exchange separations.

In the last few years, new stationary phases have appeared using these resins as a support for bonded phases such as C18 and C8. These columns are rapidly acquiring popularity since they can be used in all the pH ranges, avoiding the use of ion-pairing reagents or even replacing ion exchange based methods. The disadvantage of these packings is a somewhat lower efficiency and slower separations. However, the possibility of working at a high pH constitutes a definite edge when injecting biological materials that can be heavily contaminated with water-soluble compounds that would otherwise bond tightly to the silica, shortening its life dramatically. It is important to note that new columns, hybrids containing silica and resin particles (Gemini columns) have appeared in the market allegedly having the advantages of both these phases and few of the disadvantages.

3.8 Mobile Phases

Liquid chromatography owes its versatility to the fact that most solvents can be used as a mobile phase, so long as it is compatible with the detector and the stationary phase stability (Table 7). There are also a few solvents that are corrosive to the stainless steel that makes up most of the HPLC system. Additionally, special care should be taken if PEEK tubing or other polymers are used as tubing, since some organic solvents might dissolve them.

Table 7 Physical and chemical characteristics of the main solvents used in liquid chromatography

Solvent	MW	BP [°C]	RI	UV cut-off [nm]	Viscosity [cP]	μ (dipole moment) [Debye]
Acetonitrile	41	82	1.341	195	0.358	3.37
Dioxane	88	101	1.421	215	1.260	0.45
Ethanol	46	78	1.359	205	1.190	1.68
Methanol	32	65	1.326	205	0.584	1.66
Isopropanol	60	82	1.375	205	2.390	1.68
Tetrahydrofuran	72	66	1.404	215	2.200	1.70
Water	18	100	1.333	185	1.000	1.84

MW: Molecular weight, BP: Boiling Point, RI: Refractive Index, UV cut-off: wavelength at which the solvent absorbance in a 1 cm path length cell is equal to 1 AU (absorbance unit) using water in the reference cell.

There are a few basic requirements that a mobile phase must meet:

(i) Compatible with detector.
(ii) Free of particles: must be filtered through a 45 μm pore diameter filter.
(iii) Degassed by a suitable method (sonication, boiling, vacuum degas, Helium sparging or online vacuum degasser).
(iv) Low viscosity.
(v) Boiling point: compatible with working temperature.
(vi) Non-reactive: with column packing and sample.

Besides these general considerations, there is a basic principle in liquid chromatography that should never be ignored: the sample MUST always be soluble in the mobile phase. Basic rules for mobile phases to avoid column damage are:

(i) pH compatible with the stationary phase: silica based BPC columns resist solutions with a pH between 2 and 8.5. Since mobile phases are mixtures of water with organic solvents such as methanol or acetonitrile, it should be taken into account that each percentage of methanol means that the real acidity is actually 1 unit below that read on a pH meter.

(ii) When using phosphate buffers, these must be thoroughly flushed out from the HPLC system with water, but the column must be stored in acetonitrile:water (50:50).

(iii) Use of HCl solutions should be avoided.

3.9 Types of Chromatography

There are four basic types of chromatography according to the separation mechanism that occurs in each case. The type of chromatography to be used for a specific separation depends on the chemical characteristics of the sample and/

or its matrix, and on the expected results. A general classification of LC methods is shown Fig. 14.

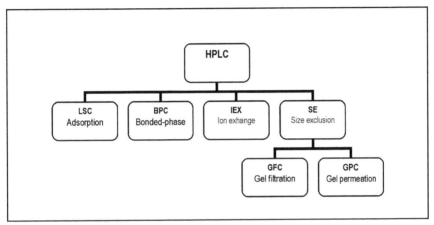

Figure 14 Classification of LC methods.

3.9.1 Which Type of Chromatography will Provide the Best Separation for a Specific Sample?

The first major issue is the molecular weight (MW) of the analyte. If analytes have a MW above 2,000 Da, it is most likely that the polymer or macromolecule has to be submitted to a size exclusion separation (general considerations on this method have been given in Section 3.9.8 of this chapter). If not, any of the other three methods will have to be used. The choice will depend on two characteristics of the sample: (i) its solubility in water and (ii) whether it is ionic/ionizable or not. Ionic compounds that are charged even at a pH below 2.5 or above 9 or inorganic ions will most likely be analyzed using an ion-exchange method. All other compounds can be separated using adsorption for non-polar compounds or bonded phases (normal or reverse phase) according to their polarity. A brief description of each of these methods is given below.

3.9.2 Adsorption Chromatography

Liquid-solid chromatography was the first type of chromatography to be used in HPLC, adapted directly from column chromatography. As in that case, the stationary phases used were similar to those applied in TLC, namely silica gel and aluminium oxide also known as alumina. Alumina use decreased owing to its high reactivity and extremely basic characteristics, but silica was used as a stationary phase until well into the 1970's. High efficiency, ruggedness, large loading capacity, and simple method optimization were clear advantages of this phase. However, its use in HPLC was severely limited due to its hygroscopic properties, which made retention time reproducibility very difficult to control. When bonded phases appeared, silica gel was dropped and is only used for isomer separations that cannot be handled with any other phase.

As silica has been described in TLC (Chapter 5, Section 3), only its characteristics and applications in HPLC are summarized below:

(i) Amorphous, totally porous, polar and acidic.
(ii) **Particle types:** Macroporous, pellicular, and microporous.
(iii) **Solvents:** Hexane, methylene chloride, methanol. (*Mobile phases are saturated with water to improve Rt reproducibility.*)
(iv) **Modifiers:** Ethanoic/formic acid for acid ionization suppression; diethylamine, triethylamine and ammonium hydroxide for basic compounds.
(v) TLC is not frequently used nowadays, except for isomer separation. It has been replaced by BPC normal phases.

Optimization of adsorption chromatography. Retention times will be managed by varying the polarity of the mobile phase. Increasing polarity will reduce t_R, while decreasing the polarity will result in increased t_R. Selectivity can be increased changing the stronger, polar component of the mobile phase as shown in Fig. 15.

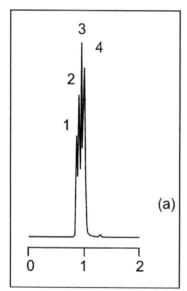 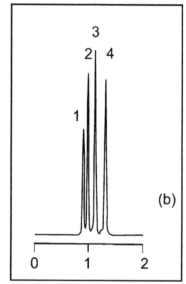

Figure 15 Separation of biphenyls performed on a silica column. Conditions: Chromolith® Performance Si, 100 - 4.6 mm) column with two different mobile phases (a) n-Heptane/ethyl acetate (97/3, v/v); (b) n-Heptane/dioxane (97/3); flow rate: 2 mL/min; detection: UV 254 nm; Temperature:ambient:injection volume:10 μL. 1. Biphenyl; 2. m-Terphenyl; 3. m-Quaterphenyl; 4. m-Quinquephenyl.

In the example given above, a good separation is carried out in 3 minutes. The non-polar compounds are separated using a mobile phase composed of a non-polar solvent (n-heptane) in a large proportion and two much more polar solvents, (a) ethyl acetate, and (b) dioxane, in a low proportion. In order to increase retention times, the proportion of dioxane should be further reduced. Conversely, if lower retention times are required, the proportion of ethyl acetate should be increased. The difference in selectivity of the solvents is illustrated by the fact that in chromatogram (b) full baseline separation is achieved in the same time as with mobile phase in (a). The saturation of the non-polar water immiscible solvents with water improves the reproducibility of retention times drastically and, while not mentioned specifically, is almost a routine practice. It is important to note that this separation is carried out on relatively newly released silica based columns in which the stationary phase is monolithic, allowing very high flows (2 mL/min) because of a reduced back pressure.

3.9.3 Bonded Phase Chromatography (BPC)

This type of chromatography is, by far, the most popular, owing to its enormous versatility, reproducibility and predictability. These phases were developed originally to replace liquid non-polar stationary phases that were highly unstable. C18 and C8 phases appeared as an excellent opportunity for RPC separations, since they always behave as reverse chromatography phases, given their low polarity and highly hydrophobic behaviour. Conversely, when the groups that are bonded to the silanols are hydrophilic, such as aminopropyl, cyanopropyl or diols, the resulting phase can act as normal phases. As most natural products can be analyzed using one of these stationary phases, maximum attention has been paid to development of method strategies for optimization of these separations.

3.9.4 Reverse Phase Chromatography (RPC)

I. **Retention mechanism:** RPC separations are typically carried out using non-polar stationary phases and mixtures of water with a polar organic solvent as mobile phases. The explanations given for the real nature of the retention mechanism of these systems are not totally clear. However, from a practical standpoint, it is possible to consider it as being comparable to a liquid-liquid extraction of different compounds from an organic solvent such as decane into water, where the most hydrophobic compounds will preferentially remain in the organic phase, while those that are more hydrophilic will be extracted more readily. Thus, for a given mobile phase composition, there will be a differential retention of the sample components according to their hydrophobicity in a manner such that the more hydrophilic (polar) compounds will be less retained in the stationary phase and elute first, while the more hydrophobic (non-polar) compounds will be more retained and elute last. If this basic principle is understood it is possible to develop, optimize or customize methods predictably.

II. **Optimization of RP separations:** A RP separation depends on the right choice of the stationary phase, the mobile phase and the temperature at which the separation is carried out. As explained above, with very few exceptions, C18 and C8 will constitute the best choice. It is very likely that the best separation of basic samples will be achieved with C8 columns.

III. **Mobile phase:** While the selection of the right column plays an important role in the success of a separation, the major role is undoubtedly played by the mobile phase. A proof of this is the relatively small range of stationary phases that exist and are used in HPLC.

The following aspects must be considered here:

(i) Water/organic solvent ratio
(ii) Type of organic solvent
(iii) Mobile phase pH
(iv) Use of salts or complexing agents

IV. Water/organic solvent ratio: As mentioned earlier, mobile phases for RPC are typically made of water and either acetonitrile (ACN) or/and methanol (Fig. 16). These solvents are used because they are miscible with water, have relatively low viscosity and low UV absorption. Additionally, they exhibit different solvating properties, methanol interacting predominantly due to its hydrogen-bonding capacity, while ACN has a high dipole moment. This will provide the right polarity and different selectivity for better separations.

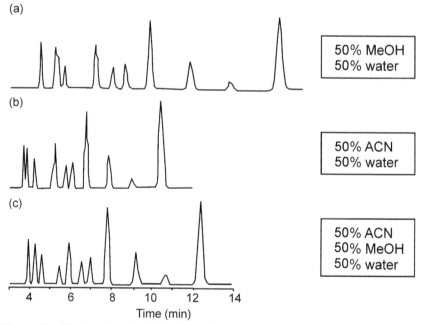

Figure 16 Effect of changing the mobile phase composition. (a) adequate resolution of late eluters with MeOH; (b) lower analysis time but poor resolution of first peaks with stronger ACN; (c) adequate resolution for all peaks in lower analysis time with methanol: ACN:water mixture.

The proportion of water in a mobile phase will be directly proportional to the retention of the compounds in the system. Thus, increasing the amount of water will produce higher T_R while reducing it, will decrease the retention times. This is logical, considering this separation as an extraction, as discussed above. The compounds that have affinity with the non-polar hydrophobic C18 or C8 will be more successfully extracted into the mobile phase and thus elute sooner if it has a lower percentage of water and a higher percentage of methanol and/or ACN. Conversely, increasing the water will force the compounds to remain in the stationary phase, thus increasing their retention (Fig. 17).

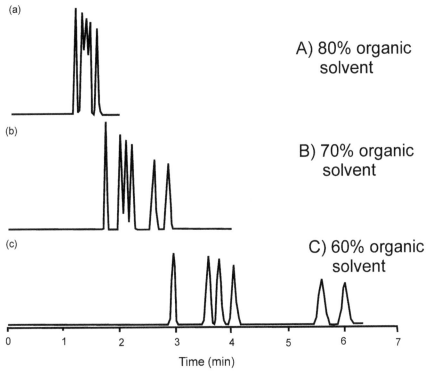

Figure 17 Solvent strength. Effect of the change in percentage of organic solvent on a reverse phase chromatography separation.

   V. **Type of organic solvent:** The strength of organic solvents increases as their polarity increases. Thus, the eluting power of ACN in RPC separations is higher than that of methanol (Fig. 16). The use of less polar solvents such as 2-propanol will produce even lower retention, as in the case of tetrahydrofuran, which is one of the strongest solvents that can be mixed with water.
  VI. **Mobile phase pH:** The effect of the pH of the mobile phase on sample separation will depend on the ionic nature of the analytes of interest (Fig. 18). Thus, if samples are ionic or ionizable, suppressing their ionization will increase their t_R. Typically, ethanoic, methanoic or phosphoric acid is used to achieve an acid pH, whereas in the case of alkyl amines, preferably

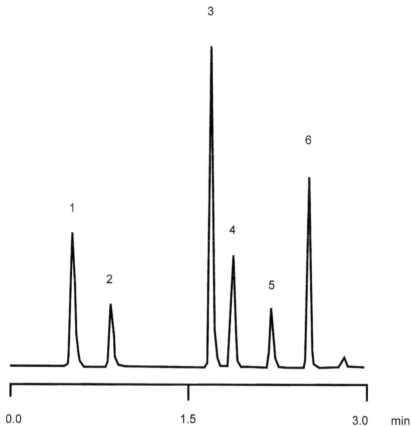

Figure 18 Separation of basic substances. Conditions: Chromolith ® Performance RP-18 endcapped, 100-4.6 mm column, A: Acetonitrile, B: 20 mM phosphate buffer pH 7.7; gradient 0.0 min 90% B; 2.5 min 40% B; 3.0 min 40% B; 7.0 mL/min flow; Detection: 254 nm. Sample: 1. Caffeine; 2. Aniline; 3. N-Methyl aniline; 4. 2-Ethyl aniline; 5. 4-Nitroanisole; 6. N,N-Dimethyl aniline.

triethyl amine is added for a basic pH. If maintaining mobile phase pH is critical, buffers might be used, taking into account that the increase of ionic strength of the aqueous mobile phase may also affect retention, as is discussed below. Mono and dibasic potassium or sodium phosphate, ammonium phosphate or citrate buffers are the most common buffers. The use of citrate buffers is controversial since these might produce some irreversible changes in the stationary phases or might form complexes with some basic compounds such as alkaloids. It is important to note that when basic compounds such as alkaloids are analyzed, C8 is the best choice and the inclusion of an alkyl amine in the mobile phase will greatly improve the peak shape as it blocks the free acid silanol groups which can bind to the basic analytes, causing different degrees of tailing.

VII. **Effect of saline concentration:** Sample retention is handled using the above-mentioned strategies, i.e. adjustment of water content of the mobile phase, polarity of the organic solvent and pH. However, when even high concentrations of water are unable to produce an acceptable retention on the column, the addition of a salt to the mobile phase might enhance sample retention due to a salting-out effect. The increase of ionic strength and the reduction of solvation molecules of soluble solutes decrease their solubility in water, thus increasing their retention in the stationary phase. This is also an interesting way to attempt to solve selectivity problems. The most commonly used salts are potassium or sodium phosphates (mono or dibasic). When adding these salts, it is important to consider the effect on the mobile phase pH. In Fig. 19, four organic acids are separated on a C18 column using a mixture of a phosphate buffer at acid pH and methanol. Considering the hydrophilic nature of these acids, the use of a saline buffer instead of an acid such as o-phosphoric acid, increases their retention and resolution.

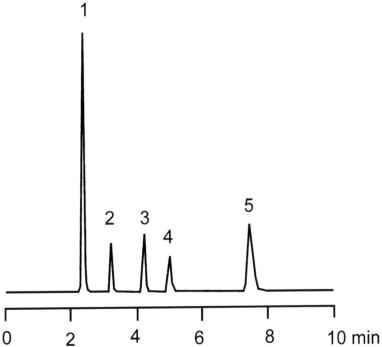

Figure 19 Separation of organic acids. 1. Oxalic acid; 2. Formic acid; 3. Malic acid; 4. Lactic acid; 5.Maleic acid. Conditions: Prevail C18, 5 µ, 150 × 4.6 mm column; 25 mM KH_2PO_4, pH 2.5 mobile phase; 0.7 mL/min, detection 210 nm.

VIII. **Gradient elution:** When analyzing plant extracts or natural products, it is necessary to use gradient elution in order to get satisfactory separations.

This is because the chemical diversity of compounds is very great and their extraction is often increased by the presence of compounds such as saponins, for example, that help to dissolve insoluble compounds due to their tensioactive properties. In RPC, gradient elution is carried out using mobile phase composition programmes in which the water proportion is reduced as the run progresses, while the organic solvent is increased. An example is shown in Fig. 20. In this case, a mixture of chlorogenic acids, flavonols and tannins is separated in one analysis with a gradient of acidified water and methanol. It is important to note that when an acid aqueous solution is used in gradient elution, the same acid is added to the organic solvent in order to avoid a gradient of acidity throughout the run. Naturally the polar acids, gallic and chlorogenic acids, elute firstly (having a large MW) followed by flavonoid glycosides such as rutin, and lastly the flavonol aglycones, quercetin and kaempferol (having the smallest MW).

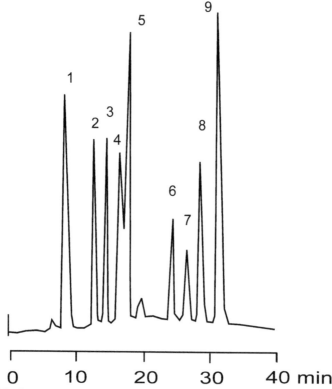

Figure 20 Separation of polyphenols. Conditions: Econosil C18, 10 µm, 250 × 4.6 mm column; gradient elution A: 0.5% H_3PO_4 in water B: 0.5% H_3PO_4 in methanol 60 to 100% B in 30 minutes; 1.0 mL/min; UV at 310 nm. 1. Gallic acid, 2. Protocatechuic acid, 3. Catechin, 4. Chlorogenic acid, 5. Epicatechin, 6. Rutin, 7. Fisetine, 8. Quercitin, 9. Kaempferol.

3.9.5 Normal-Phase Chromatography (NPC)

This type of chromatography has replaced silica based adsorption chromatography in most cases due to the fact that it is simpler and more reproducible. Additionally, the use of high proportions of water in the mobile phases allows the inclusion of salts and other compounds, greatly increasing versatility. The stationary phases that can behave as normal phases are: CN (cyano), NH_2 (amino) and Diol. Of these three phases, the first two can act as normal or reverse phases according to the sample or analyte, whereas the Diol phase always acts as a normal phase such as silica. The importance of properly assessing whether a system behaves as a normal or reverse phase system, lies in the strategy required to optimize it. In Fig. 21, if the decrease of k' for raffinose is attempted, the water content should be increased or ACN proportion may be decreased.

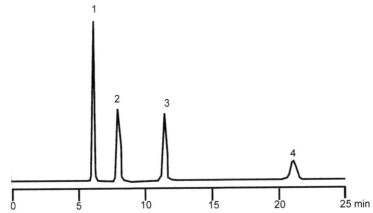

Figure 21   Separation of disaccharides. 1. Fructose; 2. Glucose; 3. Sucrose; 4. Raffinose. Conditions: Astec Amino, 5 μm, 250 × 4.6 mm column; ACN:Water:Methanol (70:20:10); 1.0 mL/min; detection: ELSD.

Figure 22 depicts a chromatogram obtained for the separation of triterpene saponins from ginseng by NPC. This separation is very difficult to achieve by

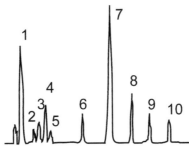

Figure 22   Separation of triterpene saponins from P.ginseng. 1. Rh1, 2. Rg2, 3. Rg3, 4. Rg1, 5. Rf, 6. Re, 7. Rd, 8. Rc, 9. Rb2, 10. Rb1. Conditions: Nucleosil NH2, 5 μm, 250 × 4.6 mm column; ACN: H_2O:IPA (80:16:4); 2 mL/min; detection: ELSD.

RPC because of the structural and chemical similarity, and the number of compounds generally present in extracts of this plant.

Gradient elution. As in the case of RPC, the complete resolution of some samples cannot be carried out with isocratic elution. Gradient elution can be achieved by following a pattern of increasing solvent strength (Fig. 23). In this case, the most polar component of the solvent will be increased, as can be seen in the example provided below. Comparing this separation with the previous isocratic separation on a similar stationary phase, it is clear that the same or even better resolution has been achieved in half the time. The evidence of normal phase behaviour lies in the fact that the gradient is performed with a gradual increase of water, the more polar of the two mobile phase components.

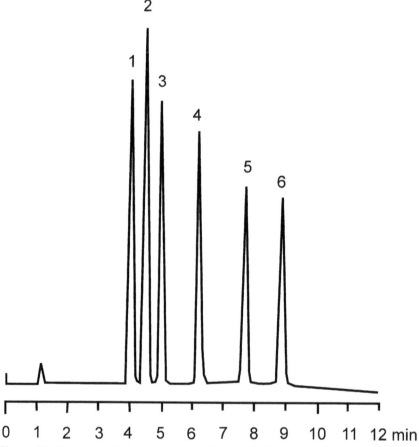

Figure 23 Separation of simple sugars. 1. Fructose, 2. Mannitol, 3. Glucose, 4. Sucrose, 5. Raffinose, 6. Stachyose Prevail Carbohydrate ES 300 Å, 5 µm, 100 × 7.0 mm column; Gradient: A: ACN/B:Water (Time, %B), (0, 20), (10, 50); 2.0 mL/min; detection: ELSD.

3.9.6 Ion-Pair Chromatography

This type of chromatography has not been mentioned among the possible types, basically because it uses BPC reverse phase stationary phases such as C18 or C8, and includes special ion-pairing reagents in the mobile phase that can form lipophylic ionic pairs with ionic or ionized samples. This type of method is applied to hydrophilic ionic compounds in which ion suppression would require a pH which would be incompatible with the column stability. A clear example of this is its use with organic acids, amines, water-soluble vitamins, etc.

I. **Ion-pair reagents:** Basic samples, such as vitamins or amines, are typically separated using mobile phases containing alkylsulfonated salts, in which the alkyl chain can have 5 to 8 carbons. In this case, retention will increase as the C chain is increased, due to increased lipophylicity. Therefore, pentanesulphonate is the least retaining ion-pairing reagent while octanesulphonate produces the highest retention.

Conversely, acid samples can be complexed by adding alkylammonium salts, generally tetrabutylammonium (TBA). Mobile phases used in ion-pair chromatography (IPC) therefore contain an aqueous buffer or pH controlled aqueous solution, the ion-pairing reagent (usually in a 25 mM concentration) and an organic solvent. pH regulation is fundamental to ensure the ionization of the sample and the ion-pairing reagent, while the organic solvent will contribute to adjust the sample k'. Sample retention can also be controlled by the type of ion-pairing reagent, as explained above.

II. **Sample preparation:** Best results are obtained by preparing the sample in the mobile phase to contribute to the pairing of all analyte molecules.

3.9.7 Ion Exchange Chromatography (IEC)

This type of chromatography can be applied to separate any mixture of charged molecules, from amino acids to peptides and proteins. Protein analysis often comprises a size exclusion fractionation followed by ion exchange separation. Amino acids are also separated using IEC because of their amphoteric behaviour, which means that they will be ionized at most pHs with the exception of their individual isoelectric points. For all other molecules, its use is restricted to those applications in which BPC or IPC are not successful, basically due to the difficulties that can be encountered in the day to day routine. Apart from the above-mentioned uses, it is applied for separation of small highly charged amines, such as catecholamines or compounds with similar structures. Naturally, it is also very useful to separate inorganic cations and anions.

Stationary phases consist of a silica or resin support, to which the following four types of molecules are chemically bonded:

(i) **Weak anion exchange (WAX):** $N(R)_3^+$; di- or tri-substituted amines; the NH_2 (amino) column can also act as a weak anion exchanger if used at an acid pH.

(ii) **Strong anion exchange (SAX):** $N(R)_4^+$, quaternary ammonium groups.

(iii) **Weak cation exchange (WCX)** : COO^-; carboxylate or carboxymethyl groups behave as weak cation retainers.

(iv) **Strong cation exchange (SCX):** SO_3^-; typically charged sulphonate groups act as strong cation exchangers.

Mobile phases are as follows:

(i) Aqueous buffers at a pH such that the stationary phase and the sample are ionized.

(ii) Polar water-miscible organic solvents are added as modifiers to achieve selectivity.

The pH and ionic strength are vital for successful separations.

INSIDE A PORE IN THE STATIONARY PHASE

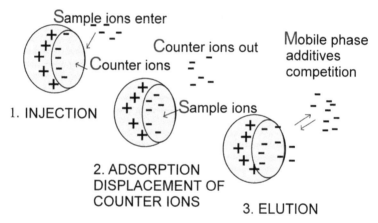

Figure 24 Separation mechanism in IEX. Consists of competitive equilibrium between sample and solvent ions for active stationary phase sites.

I. **Optimizing an IEC method:** Retention and selectivity in ion exchange chromatography depend exclusively on the charge density (charge/radius) of the hydrated ion. In organic acids and bases the elution order is determined by their pKa or pKb (strength of acid or base). Their successful separation can be controlled by the pH and ionic strength of the aqueous mobile phase. Given below is a brief outline of how these factors affect the resolution parameter.

(i) **Increasing pH:** It increases the ionization of acids, therefore increases retention. Decreases ionization of bases, hence, decreases retention.

(ii) **Increasing buffer strength:** Decreases sample retention due to increased competition for exchange site.

(iii) **Increasing temperature:** Retention generally decreases as K_{eq} is shifted.

II. **Applications:** In theory, any compound that can be ionized between pH 1-14, such as inorganic ions, carboxylic acids, sulfonic acids, phosphonic acids, organic amines, amino acids, peptides and proteins. However, in practice, it is used especially for those compounds that are charged at pH < 2 and > 8.

3.9.8 Size Exclusion Chromatography (SEC)

As was explained in the first part of this chapter (Section 3.9.1), samples containing analytes of interest with a molecular weight larger than 2,000 Da will be analyzed by SEC. According to the solubility, the system will be a Gel Permeation Chromatography (GPC) for non water-soluble polymers such as plastics, or Gel Filtration Chromatography (GFC) for hydrophilic polymers such as dextrins, starches, etc.

A brief outline of both applications is given below.

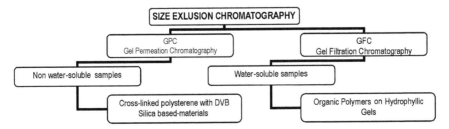

Figure 25 Types and applications of Size Exclusion Chromatography.

I. **Mechanism of separation:** SEC is an instrumental application of molecular filtration, where the chromatographic interactions typical of BPC or adsorption systems are minimized in order to achieve a separation based on the MW of the analytes. It is important to note that these separations are not based on the absolute size of the polymers, but on their "hydrodynamic volume" or effective volume in solution.

In order to separate a sample of a resin or carbohydrate (which is typically composed of a range of molecular weights), it has to be dissolved in a solvent and injected into the chromatograph. The mobile phase flows through the stationary phase which is made up of highly porous, rigid particles with pore sizes that are controlled and available in a range of sizes. The stationary phase pore size must be related to the sample MW since it will be effective in a certain range only.

Sample molecules will elute according to their MW. The largest will be the first to elute and the smallest, typically the sample solvent molecules will be the last, as they can spend more time in the stationary phase pores. A simple illustration of the mechanism is given below.

II. **Stationary phases:** Silica or resin based polymers are used for GPC while hydrophilic gels are used for GFC separations.

III. **Mobile phases:** The polymer must be dissolved, which is quite a challenge. Generally GFC solvents include water, THF or saline solutions; GPC separation can usually be carried out with toluene, m-cresol, pyridine,

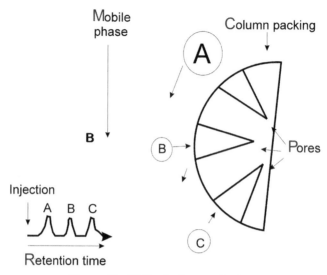

Figure 26 SEC retention mechanism.

dichloromethane or decalin. Mobile phases must be compatible with both the stationary phase and the detector, and have a low viscosity since the polymers themselves are often viscous.

IV. **Optimizing a SEC separation:** There are very few parameters than can be adjusted in order to improve a separation. The first and foremost is the

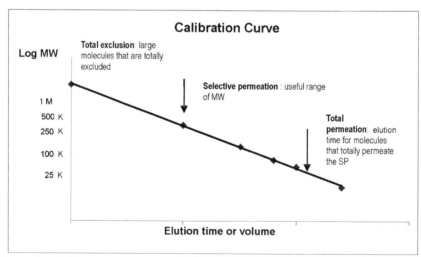

Figure 27 Calibration curve of SEC column. Arrows show the range in which the column will separate molecules selectively; i.e. this column will be useful to separate polymers with MWs within the range of approximately 5×10^5 to 5×10^3.

choice of the right pore size. This must be within the range of selective permeation/filtration of the column, as is illustrated in the figure below. Samples with a larger MW will be totally excluded and, therefore, will not be retained, while those that are smaller will fully permeate and thus not be separated from other similar sized molecules.

The other factor than can be modified is temperature. Increasing temperature will decrease the viscosity and thus the back pressure allowing higher flow rates and shorter run times. It will also generally increase sample solubility, though great care should be taken when dealing with proteins or some plastics.

V. **Applications:** SEC can thus provide very useful information about polymers, namely *number average molecular weight, weight average molecular weight, Z weight average molecular weight,* and the most fundamental characteristic of a polymer is its *molecular weight distribution.*

3.10 Column Selection

Different types of chromatography have been described, the bonded phase in more detail, while IEX and SEC have only been outlined, the reason for this being that most allelochemicals can be analyzed using BPC techniques. In order to review these options and assist in the column choice, Fig. 28 provides a guideline that may help to select the best type of chromatography according to characteristics of the sample.

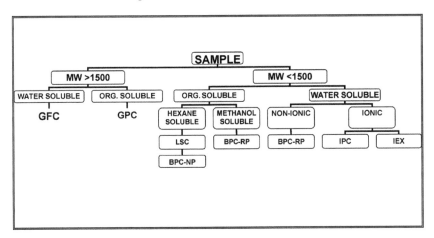

Figure 28 Column selection guideline.

3.11 Sample Clean-up

Special care has to be taken before injecting a sample into a HPLC system:

(i) Samples must be soluble in the mobile phase. Failure to comply with this requirement will inevitably produce precipitation in the column.

(ii) Samples have to be filtered through a 0.20 μm membrane filter to avoid injector blockage.

(iii) It is convenient to clean up the sample, eliminating compounds that might be highly retained on the stationary phase. This can be achieved by liquid-liquid extraction of water-soluble samples with a non-polar solvent such as toluene or dichloromethane when this will be injected onto a RP column.

Another convenient method for sample clean-up can be the use of Solid Phase Extraction (SPE). This consists in the elution of the sample through cartridges that contain a solid with very similar characteristics to the stationary phase but are less retentive due to their larger particle size.

SPE can be carried out on all types of phases: C18, C8, IEX, etc. according to the impurities that must be removed. It is, therefore, possible to load the sample on the cartridge and elute the impurities that have a lower affinity for this stationary phase, or rather to choose a SPE phase in which the impurities are more retained than the sample and can be retained, while the sample is eluted.

4. LC-MS

This chromatography is a hyphenated technique, in which the versatility and efficiency of a HPLC separation is combined with compound identification by mass spectrometry (He *et al.*, 1997). The mass spectrometer detector is certainly the best available for positive peak identification and purity determination. The major challenge in connecting a LC to a mass spectrometer is the interface that must be able to eliminate the solvent and generate gas phase ions needed for compounds detection and analysis.

HPLC operates using flow rates of 1 to 1.5 mL/min, high pressure and liquid phase at room temperature, while MS requires low flow rates, high vacuum and gas-phase operation at high temperatures. The base of MS is the ionization of molecules to produce ions that must then be separated according to their mass/charge ratio (m/z). A LC-MS interface must accomplish nebulization and vaporization of the eluate, ionization of the analyte, removal of the excess solvent vapour and extraction of the ions into the mass detector. Several interfaces can be used for this, but the types most commonly used in natural products are electrospray (ESI) and thermospray (TSP). When a high potential is applied between a narrow bore capillary filled with a flowing liquid and a nearby plate, a mist of charged droplets is obtained. This is known as electric atomization or electrospray (ESP). The ions thus formed are typically protonated molecules. The major drawback of this method is that it requires the use of very low flow rates.

The thermospray interface is based on the formation of a mist of vapour and small droplets from a heated vaporizer tube. With thermally labile ionic compounds, electrospray ionization (ESI) shows better performance than TSP.

Many instruments now use atmospheric pressure ionization (API) technique in which solvent elimination and ionization steps are combined in the source and take place at atmospheric pressure. When electron impact ionization (EI) is the choice, the solvent elimination and ionization steps are separate. The interface is a particle beam (PB) type, which separates the sample from the solvent, and allows the introduction of the sample in the form of dry particles into the high vacuum region.

A diagram of a LC-MS with the different possible parts, as explained in the Detectors section (Sec. 3.4 in this chapter) is shown in Fig. 29.

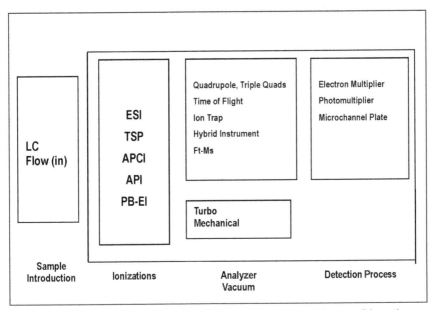

Figure 29 LC-MS. Diagram showing the four main parts and the possible options.

Applications of LC-MS. The application field of LC-MS is extremely large and is covered by a wide range of instruments and techniques. According to this, different ionization techniques and analyzers will be chosen. Some uses of LC-MS are:

(i) To obtain the mass information (MW or fragments); the quantitative aspect might be of no or little importance. This might be used to monitor or confirm an organic chemistry synthesis or to confirm the presence or absence of certain molecules in a mixture.

(ii) To get a very selective and sensitive detection of certain targeted molecules that are already identified, so that the quantitative aspect is important, but the mass information (fragmentation pattern) is of secondary importance.

(iii) Confirmation of identity and quantitation of certain targeted molecules. The MW and the presence of a few specific fragments which comply with the expected abundance are as important as the sensitivity and selectivity.

SUGGESTED READINGS

He, X.G., Lian, L.Z., Lin, M.W., and Bernart, J. (1997). High-performance liquid chromatography–electrospray mass spectrometry in phytochemical analysis of sour orange (*Citrus aurantium* L.). *Journal of Chromatography* A **791**: 127.

Lee, N.P. (2000). Liquid Chromatography: basic overview. In: *Analytical Chemistry in a GMP Environment*. (Eds., J.M. Miller and B. Crowther). Wiley Interscience, New York.

Miller, J.M. (2000). Chromatographic principles. In: *Analytical Chemistry in a GMP Environment*. (Eds., J.M. Miller and B. Crowther). Wiley Interscience, New York.

Snyder L.R., Kirkland, J.J. and Glajch, J.L. (1997). *Practical HPLC Method Development*. 2nd edition. Wiley Interscience, New York.

Chromatography and Spectroscopy of Alkaloids

Silvia R. Leicach*, Hugo D. Chludil and Margarita A. Yaber Grass

1. INTRODUCTION

Plants can neither move to avoid herbivory nor defend themselves through an immune system against the attack of bacteria, fungi or viruses. Since herbivores and pathogenic microorganisms were already present before angiosperms evolution, plants developed defence mechanisms to survive them (Wink, 2003). Tissue toughness, lignification, suberization, deposition of silica and calcium carbonate around vascular boundless or throughout tissues and trichomes, are physical defenses, even though some of them involve a chemical contribution. Defensive chemicals are mainly secondary metabolites that include phenolic derivatives, terpenoids, hydrocarbons, long chain alcohols, aldehydes, acids and esters and alkaloids.

Alkaloids are a very large and heterogeneous subgroup of nitrogenous compounds. They are basic compounds containing one or more nitrogen atoms in their chemical structures. Based on their biosynthetic pathway and chemical structures, alkaloids have been classified in three main groups:

(i) **True alkaloids:** These are derived from amino acids and have a heterocyclic ring, which includes a nitrogen atom. *Lupinus* alkaloids derived from lysine, via cadaverine, and *Senecio* alkaloids derived from ornithine, are examples of true alkaloids containing one or more nitrogen atoms from the first step of their biosynthesis.

Authors' address: *Cátedra de Química de Biomoléculas, Facultad de Agronomía, Universidad de Buenos Aires, Avda, San Martín 4453, Buenos Aires (1417), Argentina. **Corresponding author:* E-mail: leicach@agro.uba.ar

(ii) **Proto-alkaloids:** These are also derived from amino acids but the nitrogen atom is not contained in a heterocyclic ring.

(iii) **Pseudo-alkaloids:** These are not derived from amino acids but from terpenoids (solanine) or purines (caffeine) and have a heterocyclic ring with a nitrogen atom in their chemical structures.

1.2 Distribution, Localization and Concentration

Alkaloids are rarely found in gymnosperms, ferns, mosses and lower plants. They are more often found in genera of angiosperm plant families including Asteraceae (*Senecio*), Leguminosae (*Lupinus, Crotalaria, Swainsona, Astragalus, Oxytropis*), Loganiaceae (*Strychnos*), Boraginaceae (*Heliotroprium*), Solanaceae (*Solanum, Atropa, Nicotiana*), Lilaceae (*Colchicum*) and Papaveraceae (*Papaver*) (Harborne, 1998).

Alkaloid producing species have developed different storage methods, to prevent autotoxicity. Cell compartmentalization is very common. Depending on the chemical structure, alkaloids can be found in different subcellular compartments such as cytosol, vacuole, tonoplast membrane, endoplasmic reticulum, chloroplast, stroma and thylakoid membranes. Biosynthetic or transport vesicles have the same purpose (Facchini, 2001). Other plants avoid autotoxicity through glycosilation, esterification or salt formation with organic acids such as benzoic, tropic or tiglic acids.

Alkaloid concentration in plant tissues depends on environmental factors such as soil and climate conditions, and on plant developmental stage and physiological condition. Alkaloids are present in plants in a dynamic state, with fluctuations in their total concentrations and rate of turnover (Wink and Witte, 1984; Sharam and Turkington, 2005). Alkaloids half life depends on the particular alkaloid and on the species that produce it (Levitt and Lovett, 1984).

1.3 Biological Activity

Alkaloids are highly reactive substances, exhibiting biological activity in very low doses. Besides their defensive role, some alkaloids were recognized as growth regulators and insect repellents or attractants. Many alkaloids such as strychnine and coniine are poisonous, whereas others are used in medicine as pain relievers, anaesthetics or antitusives, like morphine and codeine. Most alkaloids are bitter and can be easily detected in fresh fruits or leaves by tasting them (Wink, 1983).

1.4 Physical and Chemical Properties

Alkaloids exist in plants in free state, as salts, or as esters, glycosides and *N*-oxides. They are usually colourless optically active crystalline substances. Very few, like nicotine, are in liquid form at room temperature. Some alkaloids have low polarity structures with different levels of complexity, from the monocyclic coniine to the polycyclic strychnine; they are readily soluble in organic solvents like hexane, chloroform, methylene chloride or ethyl acetate, making liquid-liquid extraction possible. Others, bearing several hydroxyl groups or glycosidic linkages, are very soluble in aqueous or hydro alcoholic solutions.

1.5 True Alkaloids

I. **Quinolizidine alkaloids:** They are naturally occurring bitter compounds, contained in several species within *Lupinus* genus, which make them resistant to herbivores (Fig. 1).

Lupanine   Tigloyloxylupanine

Figure 1   Chemical structures of lupanine and tigloyloxylupanine

II. **Pyrrolizidine alkaloids:** These are esters of 1-hydroxymethyl dehydropyrrolizidine (Fig. 2). Representative genera containing them are *Senecio* (Asteraceae), *Crotalaria* (Leguminaceae), *Heliotroprium*, *Trichodesma*, and *Symphytum* (Boraginaceae). Plants containing these hepatotoxic compounds also have significant carcinogenic and toxic effects on animals that feed on them.

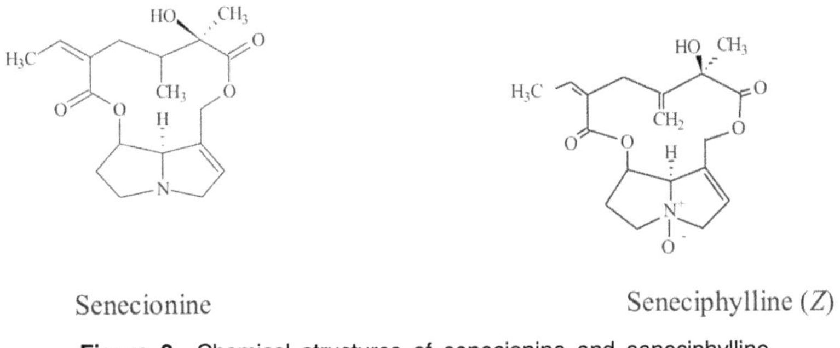

Senecionine   Seneciphylline (*Z*)

Figure 2   Chemical structures of senecionine and seneciphylline

III. **Tropane alkaloids:** These are derivatives from 8-azabicyclo[3.2.1]octane-3-ol (Fig. 3). Atropine, the racemic mixture of (–) and (+)-hyoscyamine, is biosynthesized by several members of Solanaceae family. It can be extracted from *Atropa belladonna* L., *Datura innoxia* Mill., *D. metel* L., *D. stramoniun* L., and can also be found in *Brugmansia*, *Erythroxylum*, and *Hyoscyamus* species. More than 200 tropane alkaloids had been isolated from Erythroxylaceae, Convolvulaceae, Solanaceae, Proteaceae, Rhizophoraceae, Brassicaceae, and

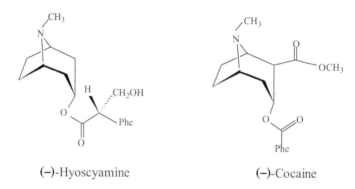

(−)-Hyoscyamine   (−)-Cocaine

Figure 3   Chemical structures of (−)-hyoscyamine and (−)-cocaine

Euphorbiaceae families, comprising mono-, di- and triesters, carboxylated and benzoylated tropanes.

IV. **Polyhydroxy alkaloids:** Monocyclic or bicyclic alkaloids of the pyrrolidine, piperidine, pyrrolizidine, indolizidine, and tropane classes, bearing two or more hydroxyl groups are representative structures of polyhydroxy alkaloids. 1-deoxynojirimycin, swainsonine, castanospermine, and calystegine C_1 are a few such examples, all highly hydrophilic compounds (Fig. 4). Most of them lack of chromophore group, disabling the possibility of detecting them by UV absorption.

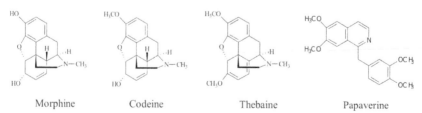

1-Deoxynojirimycin   Swainsonine   Castanospermine   Calystegine C_1

Figure 4   Chemical structures of 1-deoxynojirimycin, swainsonine, castanospermine and calystegine C_1

V. **Morphinan alkaloids** (opiates): These are found in *Papaver* species. Morphine, codeine, thebaine, noscapine, and papaverine, are known as opium alkaloids (Fig. 5). Morphine, codeine, and thebaine are used in the drug industry and are under international control (Pothier and Galand, 2005).

Morphine   Codeine   Thebaine   Papaverine
Figure 5   Chemical structures of morphine, codeine, thebaine and papaverine

1.6 Proto-Alkaloids

Proto-alkaloids are derived from amino acids but they do not contain a nitrogenated heterocyclic ring. Hordenine present in *Hordeum* genus, ephedrine from *Ephedra* species and colchicine from *Colchinum autumnale* L., are a few such examples, where the nitrogen atom is placed in side chain of a ring (Fig. 6).

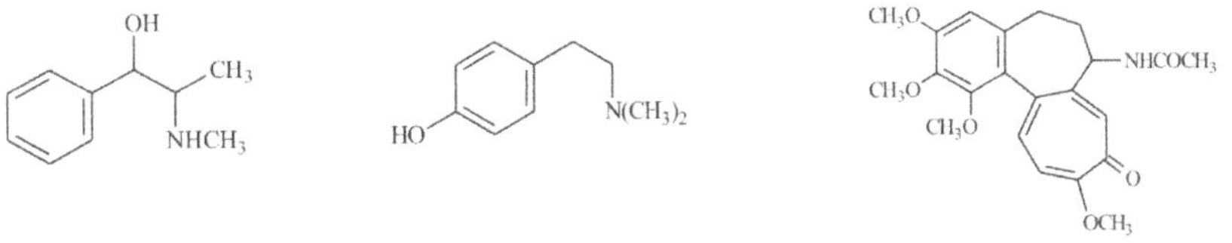

Ephedrine   Hordenine   Colchicine

Figure 6 Chemical structures of ephedrine, hordenine and colchicine

1.7 Pseudo-Alkaloids

Alkaloids with a terpenoid structure like solanine can be considered as modified terpenoids since they do not contain nitrogen initially; instead it is incorporated much later in their biosynthetic pathway (Fig. 7). Pseudo-alkaloids are derived from terpenoids (solanine) or purines (caffeine) and have a heterocyclic ring including a nitrogen atom. *Solanum* alkaloids can be found in nearly all organs of certain species within the Solanaceae family, such as *Solanum* (potato) and *Lycopersicon* (tomato) species; however, they also occur in other families, like Asclepiadaceae. Solanidine and tomatidine are the aglycones of these natural glycoalkaloids (Fig. 8).

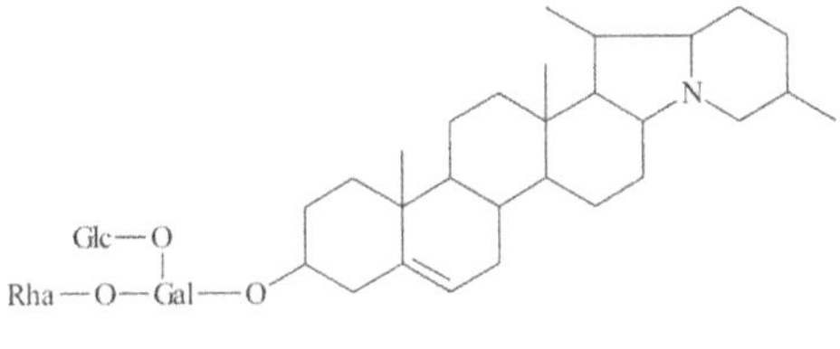

α-Solanine

Figure 7 Chemical structure of α-solanine

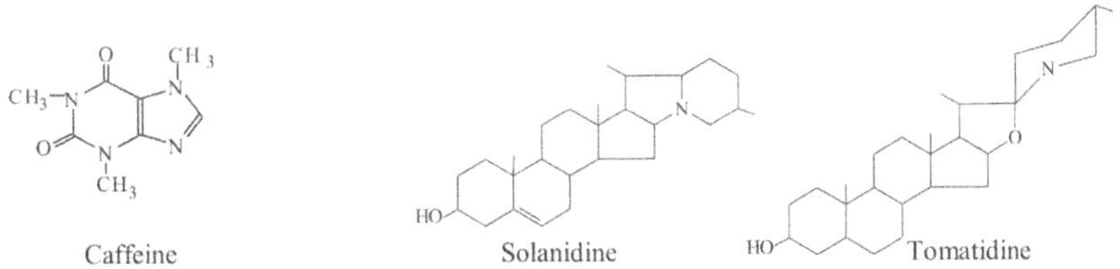

Caffeine   Solanidine   Tomatidine

Figure 8 Chemical structures of caffeine, solanidine and tomatidine

1.8 Extraction

The procedure to be used for separation of alkaloids depends on their structural characteristics as indicated below.

1.8.1 Less Polar Alkaloids

For less polar alkaloids, separation methods based on the different solubility of the salt form compared with the free base are performed by liquid-liquid extraction (LLE). Generally the acidic aqueous solution containing alkaloid salts is basified, and alkaloids as free bases are readily extracted with hydrophobic organic solvents. In this step, ester alkaloids such as tropane and pyrrolizidine alkaloids may be labile, therefore, moderate alkalinization with sodium carbonate or ammonia is recommended.

Less polar alkaloids can be extracted directly from dried plant material into a weak acid alcoholic solvent. The obtained solution is defatted with a non-polar solvent, and after addition of ammonia, the alkaline solution is extracted again with an adequate organic solvent. Further purification procedures are usually needed. Other procedures start with the separation of lipidic constituents.

1.8.2 Hydrophilic Alkaloids

For hydrophilic alkaloids, extraction methods are quite different from those just described. The higher solubility in water enables their selective extraction from leaf material with aqueous acids. Alkaloids bearing several hydroxyl groups and glycoalkaloids are insoluble in non-hydroxylic solvents. Most polyhydroxy alkaloids are extracted from milled plant material using hydro-alcoholic solvents (methanol or ethanol with 25-50% water) or just with water. This procedure is performed in neutral or acidic conditions, suppressing alkalinity, which may decompose some alkaloids such as pyrrolidines (Fellows and Fleet, 1989). The slight solubility in chloroform of alkaloids bearing only two hydroxyl groups may help to separate them from the more hydroxylated ones after this first extraction.

In general, sugars and amino acids are co-extracted with more hydrophilic alkaloids. Commonly employed resins to remove undesired compounds, in both anionic and cationic forms, are either Dowex 50 or Amberlite CG120 in their NH_4^+ or H^+ ion forms. For plant tissues with a high proportion of saccharides, methanol is a better option than water as an extracting solvent, avoiding the interference of carbohydrates with subsequent analysis.

1.9 Sample Preparation

Even after selective extractions, interfering compounds can still be present, and sometimes alkaloids need to be concentrated before chromatography. Sample preparation can be accomplished by different methods, depending on the chemical and physical characteristics of the particular alkaloid class. Liquid-liquid extraction

(LLE) with immiscible solvents and solid-phase extraction (SPE) are commonly used for this purpose.

1.9.1   Liquid-Liquid Extraction (LLE)

When LLE is performed, potential losses by alkaline hydrolysis if ester linkages are present, insufficient extraction due to good water solubility of the free bases, and emulsion formation, should be considered as disadvantages. Saponins, antipathic compounds abundantly present in plants, generate emulsions when shaking immiscible liquids and the phases can be difficult to separate even after centrifugation.

1.9.2   Solid Phase Extraction (SPE)

In SPE this problem does not arise, and smaller amounts of solvent are required, exhibiting high versatility, since different matrices can be used depending on the chemical structure of the alkaloids being separated. This technique can be less expensive if refillable laboratory-made cartridges are used. Many phases are available in cartridges for solid phase extraction (SPE). When reverse-phases like RP C_{18} are used, the alkalinized aqueous extract is applied to the column, where most lipophilic compounds are retained, and alkaloids can be eluted with acidic methanol. Tropane alkaloids had been purified following this procedure (Papadoyannis *et al.*, 1993). Other authors use diatomaceous earth, kieselguhr (Extrelut). The alkaline aqueous extract is applied to the dry column, and the adsorbed free alkaloids exposed on kieselguhr particles surface, can be eluted with organic solvents such as chloroform (Namera *et al.*, 2002). For some small alkaloids with one hydroxyl groups, the octanol : water partition coefficient is low and they can only be partially extracted by the organic solvent in a LLE procedure, even as free bases. SPE offers a better alternative to clean and concentrate these samples. In these cases plant extracts are made slightly basic (pH 10-11) with carbonate or ammonia, and applied to the Extrelut column but elution volumes need to be larger, starting with chloroform, and then raising solvent polarity by the addition of methanol (Portsteffen *et al.*, 1994).

SPE, a fast method to concentrate alkaloids, can also be used to purify samples prior to HPLC analysis. It is much more advantageous to use cation-exchange SPE phases. Figure 9 shows efficiency of different commercial solid phase extraction sorbents to purify samples of glycoalkaloids before HPLC analysis (Väänänen *et al.*, 2000). The SCX (Strong Cation Exchange) was the most selective and efficient matrix to remove undesired components from *Solanum brevidens* phil. leaf extracts previously obtained by extraction with 5% acetic acid.

Cation-exchange SPE and HPLC with a diode array detection system were successfully used for the screening of 100 basic substances in urine, atropine, cocaine, etc. (Logan *et al.*, 1990).

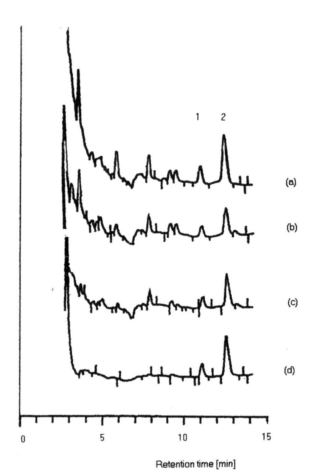

Figure 9 RPLC chromatograms of *S. brevidens* extract (a) Before the clean-up and after clean-up by SPE with (b) C18, (c) Certify and (d) SCX sorbents. Chromatographic conditions: column 250 × 4.6 mm I.D., Zorbax Rx-C18, 5 μm; column temperature 40°C; flow rate 1 ml/minute; stepwise acetonitrile: 25 mM TEAP buffer (pH 3.0) gradient 20, 25, 35, 45 and 65% acetonitrile at times 0, 12, 15, 17, 25 minutes, respectively; UV absorbance detection at 205 nm. 1: dehydrotomatine, 2: α-tomatine.

2. CHROMATOGRAPHY

2.1 Thin Layer Chromatography (TLC)

Plant species commonly produce complex mixtures of structurally related alkaloids and only those in higher proportions can be easily detected by TLC. Multiple details about stationary phases, solvent systems, and detection reagents used in TLC for different families of alkaloids can be found in the literature (Baerheim-

Svendsen and Verpoorte, 1983; Popl *et al.*, 1990). Hydrophilic stationary phases, such as silica gel or alumina, are used with non-polar solvent systems to separate alkaloids, but reverse phases can also be employed. Alkaloids can be detected in the crude extract by means of selective chromogenic reactions based on the presence of the nitrogen atom. Most alkaloids give precipitates with heavy metal iodides, likewise alkaloids are precipitated from neutral or slightly acidic solution by Mayer reagent (potassium and mercuric iodides solution) giving a cream colored precipitate. Dragendorff reagent, a solution of potassium and bismuth iodides, can also be used to detect alkaloids, giving orange colored precipitates. Different colors, depending on the particular alkaloid, can be obtained with iodoplatinate reagent. Pothier and Galand (2005) reported deep blue for morphine, pink violet for codeine, brown violet for thebaine, light pink for papaverine, and pink brown for noscapine.

The chromatographic behaviour of a given compound in a TLC is described by the R_f **value,** which is defined as the ratio between distances **a** and **b**, where **a** is the distance from the point where the sample was first applied and the center of the corresponding spot after elution, and **b** is the distance from that point to the solvent front (Fig. 10). The shape and colour of each spot in the sample and its R_f value can be compared with those of authentic standards allowing a tentative preliminary identification.

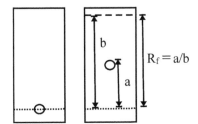

$$R_f = a/b$$

Figure 10 Example of a TLC system and calculation of a R_f value.

Densitometry: Approximate quantification by TLC can only be achieved by means of instrumental densitometric evaluation of TLC plates, but extensive and repeated calibration procedures are needed to obtain reliable results (Monforte *et al.*, 1992). High performance TLC plates (HPTLC) prepared with more homogeneous and smaller size particles, exhibit higher selectivity, and are preferred when subsequent densitometric quantification is planned.

As a preliminary analytical method, TLC is a very simple technique. Nevertheless, it only provides semiquantitative analysis of secondary metabolites (Dräger, 2002).

2.2 Paper Chromatography (PC)

PC is not often used and may offer some advantages only in extremely polar alkaloids. The bands can be detected by chromogenic reactions and the individual alkaloids can be extracted from the paper with large volumes of water (Molyneux *et al.*, 2002). Preparative PC was used 20 years ago to separate polyhydroxypipecolic acid alkaloids but has not been applied to other classes (Bleeker and Romeo, 1983). Pyrrolizidine alkaloids can be separated by PC on Whatman No. 1 filter paper using *n*-butanol saturated with 5% aqueous acetic acid.

2.3 Counter-current Chromatography (CCC)

CCC is a liquid-liquid partition chromatographic procedure. It is well-suited for separation and purification of natural products on a preparative scale (Conway, 1990). Two immiscible liquids become uniformly segmented in the coils of a helical column when the latter is rotated. The liquid playing the role of the stationary phase is retained by centrifugal force, while the mobile phase is pumped through the rotating column. Some advantages of this technique are: total sample recovery, good resolution (350-1,000 plates), and high reproducibility. Cooper *et al.* (1996) applied high-speed CCC to separate and purify pyrrolizidine alkaloids from several species of the Boraginaceae family and from *Senecio douglasii* var. *longilobus* (Asteraceae).

2.4 Column Chromatography (CC)

Although some alkaloids have been separated by CC, high performance procedures are far more selective and sensitive, certainly avoiding all the tedious and time consuming work related to classical column chromatography. Less polar alkaloids had been separated by CC with silica gel (Ambrozin *et al.*, 2005). Pyrrolizidine alkaloids can be separated by CC using neutral alumina and benzene:chloroform:methanol mixtures as eluent. Silica gel columns can also be used to separate them eluting with benzene-ethyl acetate-diethylamine.

2.5 Gas Chromatography (GC)

GC proved to be faster and more precise than other quantization methods previously used for alkaloid analysis in plant tissues. Packed columns, mostly self-made, with a wide inner diameter (3 to 6 mm) were used initially, with SE30 (silicon elastomer) as the coating of the stationary phase. Extracted alkaloids were converted into salts, which decomposed in the hot injector and migrated as free bases. Atropine was one of the first alkaloids to be analyzed by GC (Lloyd *et al.*, 1960).

Packed columns have been mostly abandoned for plant alkaloids analysis, and had been replaced by capillary columns to perform GC. Columns more frequently used for alkaloids have lengths from 15 to 60 m, with I.D.: 0.32 or 0.25 mm and are coated with: 100% dimethyl polysiloxane (e.g. DB1), 5% phenyl polysiloxane, 95% dimethyl poly-siloxane (DB5), 14% cyanopropyl phenyl polysiloxane, 86% dimethyl polysiloxane (DB1701), 50% phenyl polysiloxane, and

50% dimethyl polysiloxane (DB17). Film thickness varies from 0.1 to 0.5 µm. Split injection of 0.5 to 2 µl is generally used with split ratios 1:10 to 1:50. For very dilute samples, splitless injection for 10 to 30 seconds, and then high split ratio (1:50 to 1:70) for a few minutes to prevent a broad solvent signal, is preferred. In such situations the injection should be done on-column, or by means of a programmed temperature vaporizer (PTV), for compounds that tend to decompose or be retained in the split injector. Nitrogen (except with nitrogen-phosphorus detection) or helium are commonly used, with a gas flow of 15-30 cm/s. The temperature programme is variable, depending on the class of alkaloids to be separated, the column characteristics and the kind of injection. The time per analysis, which depends on the particular sample, can be from 15 to 60 minutes. Relative abundances (in percentage) are recorded as a function of retention times (in minutes).

2.6 High-Performance Liquid Chromatography (HPLC)

HPLC analysis has been improved with the development of novel techniques for sample clean-up, a wide range of stationary phases, and different detection systems. HPLC is still one of the most useful techniques in the purification of particular alkaloid families (Fig. 11). Nevertheless, conventional HPLC with UV

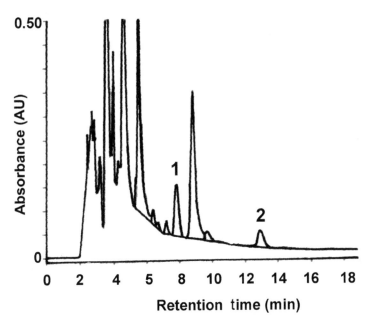

Figure 11 HPLC of glycoalkaloids from *Solanum*. Peak 1 correspond to α-chaconine and peak 2 to α-solanine (Source: Phillips *et al.*, 1996).

detection is not reliable for compounds lacking suitable chromophore groups, such as polyhydroxy alkaloids. The pre- or post-column derivatization of the hydroxyl groups with appropriate chromophores or fluorophores overcame this disadvantage, but has added complexity to the analytical procedure. Pulsed amperometric detection represents another possible solution.

In general terms, alkaloids separation can be performed by HPLC using columns of 10 to 25 cm length, 4 to 5 mm I.D. for regular samples, or microbore, 5 to 15 cm length and 1 to 2 mm I.D. for smaller samples, with mostly spherical (from 3 to 10 µm) porous or nonporous silica based particles. The injection volumes vary from 10–20 µl for normal columns, to <1 µl for microbore columns. Solvent systems depend on the selected column. For RP columns ion-pair reagents in an acetonitrile:water mixture are widely applied to counteract the alkaline properties of alkaloids.

Detection limit depends on the selected detection method, e.g. 1 µg/mL of hyoscyamine and scopolamine for UV 220 nm, 8.5 ng/mL hyoscyamine (200 µl injection) for UV 215 nm, and 145 ng/mL hyoscyamine for ECD (electron capture detection).

2.7 Capillary Electrophoresis (CE)

CE was developed in the last 15 years and successfully applied to the analysis of plant extracts due to its high resolution, minimum sample volume, short analysis time, and high separation efficiencies. CE appears to be specially suited for alkaloid analysis because these compounds are natural cations in the appropriate acidic buffer pH. Migration occurs in the cationic mode, as the sample is introduced at the anode and the detection occurs at the cathode, where the outlet is placed, and its displacement is caused by the charged nitrogen atom. Since the sample volume is very low in CE, a few nanolitres, very sensitive detection methods are required.

Alkaloids were determined in herbal plants by different CE modes, such as capillary zone electrophoresis (CZE), non-aqueous capillary electrophoresis (NACE), and micellar electrokinetic chromatography (MEKC). CE is an excellent method for detection and quantization of toxic pyrrolizidine alkaloids in plants (Lebada *et al.*, 2000). Detection in CE is usually achieved by UV, but alkaloids lacking of a chromophore group can be detected by a diode array detection system (DAD).

3. STRUCTURAL ELUCIDATION

3.1 Mass Spectrometry (MS)

MS has been improved in the last two decades, becoming a widely used research tool in the analysis of alkaloids. When instruments employed in separation techniques were coupled directly to those used to elucidate chemical structures, the term 'hyphenated technique' became popular (Wilkins, 1983). It has been applied mostly to the coupling of mass spectrometry (MS) to chromatographic techniques such as gas chromatography (GC), high performance liquid chromatography (LC) or capillary electrophoresis (CE), and to the combination of LC with nuclear magnetic resonance spectroscopy (NMR).

Fast methods were developed in order to evaluate the potential health hazard posed by quinolizidine alkaloids concentration in foods related to *Lupinus* species (Reinhard *et al.*, 2006). Figure 12 shows the alkaloid pattern found in *L. albus* L.

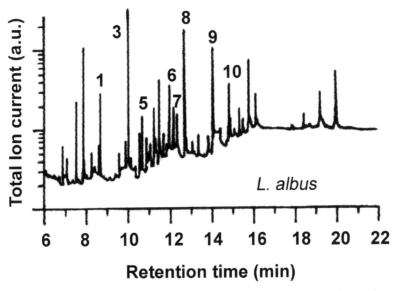

Figure 12 GC-MS chromatograms for underivatized seeds of *L. albus* (the ordinate is in log scale in order to visualize small peaks). 1. C_{16}, 3. caffeine (IS), 5. ammodendrine, 6. angustifoline, 7. α-isolupanine, 8. lupanine, 9. 13-α-hydroxylupanine, 10. C_{26}.

3.2   GC-MS

Coupled to gas chromatography, mass spectrometry has been employed for analysis of polyhydroxy alkaloids. The derivatization with MSTFA [N-methyl-N-trimethylsilyltrifluoracetamide] at 60°C for 1 hour is suitable for most polyhydroxy alkaloids, and since the capillary column retention times correlate closely with the degree of hydroxylation, it is generally possible to estimate the number of hydroxyl groups in a novel alkaloid from its retention time (Molyneux *et al.*, 2002). Swainsonine and calystegines B_2 and C_1 co-occur in *Ipomoea calobra* W. Hill & E. Muell. and *I. polpha* R.W. Johnson from Australia (Molyneux *et al.*, 1995), *I. carnea* Jacq. from Mozambique (de Balogh *et al.*, 1999), and *I. asarifolia* (Desr.) Roem. & Schult. from Brazil. The GC trace corresponding to their determination by GC-MS is shown in Fig. 13.

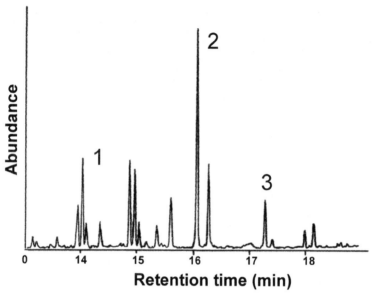

Figure 13 GC-MS identification of polyhydroxy alkaloids in *Ipomoea calobra*. 1. swainsonine, 2. calystegine B_2, 3. calystegine C_1.

3.3 HPLC-MS

The potential of LC-MS in phytochemical research is related to the development of adequate interfaces to transfer the molecules eluting in solution from a liquid chromatograph, into gas phase ions suitable for mass analysis. Thermospray, particle beam and continuous-flow fast atom bombardment are the most common interfaces, which have made LC-MS a sensitive, robust and more widely used technique. Even when these interfaces had been superseded in the last two decades by electrospray (ES) and atmospheric pressure chemical ionization (APCI) (Kite *et al.*, 2003), they are still employed for particular alkaloid mixtures (Chen *et al.*, 1990; Egan *et al.*, 1999).

3.4 NMR Spectroscopy

Nuclear magnetic resonance (NMR) is a very useful tool for structural elucidation of alkaloids. Development of two-dimensional ^{1}H- and ^{13}C-NMR techniques has contributed to the unambiguous assignments of proton and carbon chemical shifts, in particular in complex molecules. The frequently used techniques include direct correlations through homonuclear (COSY [correlation spectroscopy], TOCSY [total correlated spectroscopy], ROESY [rotating-frame overhauser enhancement spectroscopy], and NOESY [nuclear overhauser enhancement spectroscopy]) and heteronuclear (HMQC [heteronuclear multiple-quantum coherence experiment],

HMBC [heteronuclear multiple bond correlation]) couplings. HETCOR (heteronuclear correlation spectroscopy) a 2D ^{13}C, ^{1}H correlation experiment, which gives information about the connectivity of protons with ^{13}C nuclei over a single bond can also useful in structural elucidation of alkaloids (Nakanishi, 1990).

3.5 X-ray Crystallography

This technique has been used for examining alkaloid structures to obtain information on their spatial distribution. It accurately determines the relative position of pyrrolizidine alkaloid structural components, the pyrrolizidine nucleus and the macrocyclic diester. X-ray crystallography has helped in visualizing that the two 5-membered rings in 1,2-unsaturated necines determine a 115-130° angle, and that the saturated ring is puckered *endo* in some PA alkaloids such as heliotridine, and *exo* in others such as retronecine.

EXPERIMENTAL

Experiment 1: Detection of Opium Alkaloids by TLC

Papaver somniferum L. is cultivated in different countries for medicinal purposes. It produces a narcotic compound that induces very deep sleep, rendering the person insensible. Opium is the air-dried milky exudation obtained from excised unripe fruits, and is extensively used for obtaining morphine, codeine, narcotine, laudenine, papaverine, and other alkaloids for pharmacological purposes. All organs of the poppy plant contain alkaloids and can be used for extraction of morphine. Even when classical TLC of opium alkaloids can be performed, complex eluents with strong alkaline substances like diethylamine are used to obtain a clean separation (Baerheim-Svendsen and Verpoorte, 1983; Wagner and Bladt, 1996). The presence of diethylamine, not easily eliminated from the chromatographic plate after elution, can generate difficulties at the visualization step, and ammonia has been proposed as substitute. (Kalász *et al.*, 1995; Večerkova, 1996)

Materials and equipments required

Papaver somniferum L. leaves; lyophilizer; reflux equipment; heating mantle; TLC silica gel F_{254} plate; capillary glass tube; brown glass bottles; analytical balance; rotavapor; TLC cylindrical developing tank; TLC chromatography reagent sprayer.

Reagents

Chloroform, concentrated ammonium (28%). Authentic standards: morphine, papaverine, and codeine.

Chromatographic grade methanol, toluene, acetone, ethanol and ammonia, iodoplatinate reagent.

Iodoplatinate reagent, modified for alkaloids (spray solution): Mix 3 mL 10% platinum chloride solution with 97 mL water to which 100 mL of a 6% aqueous solution of KI were added (keep for some time in brown glass bottles).

10% platinum chloride: dissolve 10 mg platinum chloride in 100 mL water.

6% aqueous KI: dissolve 12 mg KI in 200 mL water.

Standard solutions: dissolve 10 mg of the authentic standard in 1 mL of methanol.

Procedure

(i) Lyophilize 10 g *Papaver somniferum* L. leaves for 48 hours to eliminate water. Extract morphinan alkaloids under reflux with a mixture of 147 mL chloroform and 3 mL concentrated ammonium (49/1, v/v) for 2 hours. Evaporate to dryness at 40°C under reduced pressure.

(ii) Dissolve the opium extract residue in 2 mL methanol. Apply the solution with a capillary glass tube as three different spots on a TLC silica gel F_{254} plate (layer thickness 0.2 mm), 5 drops in the first spot, 10 drops in the second, and 15 drops in the third spot, to elute three different concentrations. The solution should be applied drop by drop, carefully drying between drops. Apply the three authentic standard solutions (5 drops) in three spots, respectively. Dry the plate at room temperature.

(iii) Elute with a mixture of (22.5 mL) toluene: (22.5 mL) acetone: (3.5 mL) ethanol: (0.5 mL) 28% ammonia (45:45:7:1). After drying the chromatogram, spray with iodoplatinate reagent.

Observations

With iodoplatinate reagent each alkaloid show a specific colour: deep blue for morphine, pink violet for codeine, brown violet for thebaine, light pink for papaverine, and pink brown for noscapine. Compare spots in sample chromatogram with those of authentic samples for a preliminary identification of individual alkaloids.

Calculations

Calculate R_f values and compare them with those of authentic standards. Compare the spot intensities obtained with the different concentrations.

Precautions

Methanol, toluene, acetone and ethanol are inflammable, keep away from sources of ignition. All organic solvents, reagents, and concentrated ammonium (28%) are toxic or allergenic by inhalation, contact or swallowing.

Experiment 2: Quantification of Glycoalkaloids in Potatoes by HPLC (LiChrospher-5 NH$_2$ Column)

Potato is included among the main horticultural crops in South America. Due to its broad availability and nutritional characteristics, it has been considered as one of the most important components of human diet. Nevertheless, potato tubers (*Solanum tuberosum* L.) contain two classes of naturally occurring toxins: (i) α-solanine, and α-chaconine, (trisaccharide glycoalkaloids), and (ii) the polyhydroxy alkaloids calystegines A_3 and B_2. Glycoalkaloids toxicity affects commercial acceptability of potatoes and related food products. TLC, HPLC, GC-MS, and UV spectroscopy, electrophoresis-MS had been applied to the study of potato glycoalkaloids, α-chaconine and α-solanine, their hydrolysis products (metabolites) with two, one, and zero monosaccharide groups; and potato water-soluble *nor*-tropane alkaloids calystegine A_3 and B_2. HPLC has many advantages, such as the analysis at room temperature and simultaneous analysis of individual structurally related glycoalkaloids and their hydrolysis products without derivatization (Friedman, 2004).

Materials and equipments required

Potato tuber; Celite; glass-fibre filter (1.2 µm); domestic food processor; tubes; pipettes; analytical balance; Sep-pak C_{18} chromatography cartridge; HPLC equipment with UV detector; LiChrospher-5 NH_2 column (25 cm × 4.6 mm, 5 µm); LiChrospher-5 NH_2 pre-column (1 cm × 4.6 mm, 5 µm).

Reagents

α-solanine and α-chaconine (authentic standards), 5% (v/v) acetic acid solution, methanol, 20 mM KH_2PO_4 solution, acetonitrile (HPLC grade).

Standard solutions: Dissolve 10 mg of each alkaloid in 100 mL of methanol.

5% (v/v) acetic acid solution: Add 5 mL of acetic acid to100 mL of distilled water.

20 mM KH_2PO_4 solution: Dissolve 272 mg of KH_2PO_4 in 100 mL of distilled water.

Chromatographic mobile phase: Add 23 mL of 20 mM KH_2PO_4 solution to 77 mL acetonitrile.

Procedure

(i) Cut as finely as possible potato tops or tubers using a domestic food processor.

(ii) Mix 2 g sample with 40 mL 5% acetic acid and 2 g Celite, homogenize for 3 minutes with a homogenizer or a stirrer. Filter through a 1.2 µm glass-fibre filter. Wash the residue in 40 mL 5% acetic acid and filter again.

(iii) Increase filtrate volume to 80 mL with 5% acetic acid solution. Purify the filtrate (20 mL) using a Sep-pak C_{18} chromatography cartridge and elute with 5 mL methanol. Evaporate to dryness at 40°C under reduced pressure and dissolve in methanol (1 mL).

(iv) Filter sample and standards through a 0.45 disposable syringe filter.

(v) Perform HPLC at room temperature with UV detection at 205 nm, using a LiChrospher-5 NH_2 column protected by a LiChrospher-5 NH_2 pre-column. Injection volume: 50 µL. Mobile phase: acetonitrile: 20 mM KH_2PO_4 solution (77:23, v/v) with a flow rate of 0.5-1 mL/minute.

Observations

(i) Filter mobile phase (acetonitrile: 20 mM KH_2PO_4 solution (77:23, v/v)) through a filter membrane (Nylon 66®) using an all-glass vacuum mobile phase filtration system. Sample and standards solutions should be filtered through a 0.45 disposable syringe filter.

(ii) Identification of α-solanine and α-chaconine is achieved comparing their peaks retention times (R_t) obtained from standard solutions, with the corresponding peaks in the sample.

(iii) R_t α-chaconine 7.4 minutes, R_t α-solanine 12.3 minutes (Phillips *et al.*, 1996).

Calculations

Glycoalkaloid quantification by external standard plot method: Plot α-solanine and α-chaconine standard curves in a concentration range of 0.5 to 60 µg mL^{-1}. Compare peak areas with those corresponding to standard solutions and obtain their concentration value in the sample from the standard curves using the corresponding area value.

Precautions

Methanol and acetonitrile are inflammable, keep away from sources of ignition. Methanol, acetonitrile and acetic acid are toxic by inhalation, contact or swallowing.

Experiment 3: Quantification of Glycoalkaloids in Potatoes by HPLC (ODS-Hypersil Column)

Materials and equipments required

Potato tuber; lyophilizer; centrifuge; centrifuge tubes; pipettes; analytical balance; pH meter (or pH indicator sticks); Sep-Pak C$_{18}$ chromatographic cartridges; MilliQ-Plus purification system; Membrane filter (0.45 µm) (Nylon 66®); HPLC equipment with photodiode array detector; (25 cm × 4.6 mm, 5 µm) ODS-Hypersil column.

Reagents

Acetic acid, α-solanine and α-chaconine (standards), methanol, sodium bisulfite, potassium phosphate buffer (pH: 7.6).

Solvents: Water, acetonitrile and ethanolamine (HPLC grade).

α-solanine and α-chaconine standard solutions: Dissolve 10 mg of each alkaloid in 1 mL of 0.05 M phosphate buffer (KH_2PO_4).

0.05 M KH_2PO_4 solution: Dissolve 680 mg of KH_2PO_4 in 100 mL distilled water.

0.022 M KH_2PO_4 solution: Dissolve 300 mg of KH_2PO_4 in 100 mL distilled water.

10% phosphoric acid: Add 10 mL phosphoric acid in 90 mL distilled water.

Extraction solution: Dilute 20 mL acetic acid in 100 mL distilled water; add 5 g sodium bisulfate, stir until it is completely dissolved and add distilled water to reach a 1,000 mL volume.

Water:acetonitrile (85:15): Add 15 mL of acetonitrile to 85 mL distilled water chromatographic grade.

SPE Eluent: Acetonitrile: 0.022 M potassium phosphate buffer, pH = 7.6, (60:40) (v/v).

Mobile phase for HPLC analysis (65% water, 35% acetonitrile and 0.05% ethanolamine, pH 4.5-4.6 adjusted with phosphoric acid (10%). Mix 70 mL acetonitrile with 130 mL distilled water and add 0.1 mL of ethanolamine. Adjust to pH 4.5-4.6 with drops of phosphoric acid.

Procedure

(i) Wash the tuber under running water and allow to dry at room temperature (25°C).

(ii) Lyophilize sliced tuber (1 g), mince sample and keep at −15°C until analysis.

(iii) Shake freeze-dried sample (0.5 g) with 10 mL of extraction solution for 15 minutes and centrifuge at 4,000 rpm for 20 minutes. Separate supernatant.

(iv) Sample cleaning by SPE: Condition cartridge (100 mg Sep-Pak C-18 cartridges) with 5 mL acetonitrile, followed by 5 mL extraction solution. Apply 5 mL supernatant to preconditioned cartridge and wash with 4 mL water:acetonitrile 85:15 (v/v). Elute glycoalkaloids with 4 mL acetonitrile: 0.022 M potassium phosphate buffer pH 7.6, 60:40 (v/v); adjust eluate volume to 5 mL with the same solution.

(v) Filter sample and standars through a 0.45 disposable syringe filter. Perform HPLC analysis with a ODS-Hypersil (25 cm × 4.6 mm, 5 μm) column and a photodiode array detector. Mobile phase: 65% water, 35% acetonitrile, 0.05% ethanolamine, phosphoric acid (10%), pH 4.5-4.6. Injection volume: 20 μl, flow rate 1.0 mL minute^{-1}.

Observations

(i) Solvents: HPLC grade acetonitrile, water previously purified in a MilliQ-Plus purification system or HPLC grade. Ethanolamine and phosphoric acid (HPLC grade).

(ii) Identification of α-solanine and α-chaconine is possible by comparison of its peak retention time (Rt) in the sample with that obtained with α-solanine and α-chaconine authentic samples at the same conditions.

Calculations

Glycoalkaloid quantification by external standard plot method: Plot α-solanine and α-chaconine standard curves in a concentration range of 0.5 to 60 μg mL^{-1}. Compare peak areas with those from authentic α-solanine and α-chaconine and obtain their concentration value in the sample from the standard curves using the corresponding area value.

Precautions

Methanol and acetonitrile are inflammable, keep away from sources of ignition. All organic reagents are toxic or allergenic by inhalation, contact or swallowing, phosphoric acid is caustic.

Experiment 4: Determination of Quinolizidine Alkaloids from *Lupinus albus* L. Leaves

Quinolizidine alkaloids, naturally present in *Lupinus* species, are bitter substances related to its defense mechanism. Plant breeders developed sweet varieties in order to enhance their palatability, with lower levels of quinolizidine alkaloids.

Quinolizidine alkaloid levels can be modulated by biotic and abiotic environmental factors. CG and CG-MS are particularly useful for detecting any changes in relative composition of quinolizidine alkaloids in *Lupinus* species.

Materials and equipments required

Milled dry *Lupinus* leaves; Soxhlet apparatus; separatory funnel; rotavapor; vacuum desiccator; capillary glass tube; brown glass bottles; analytical balance; rotary evaporator; TLC cylindrical developing tank; TLC chromatography reagent sprayer silica gel GF-254 plates; GC and GC-MS equipment.

Reagents

Methanol, dichloromethane, chloroform, ethyl acetate, 0.5 M hydrochloric acid, ammonium hydroxide 4M, 28% ammonium hydroxide (see exp. 6), hexadecane (chromatographic grade), anhydrous sodium sulfate, Silica gel GF-254 plates, TLC cylindrical developing tank, TLC chromatography reagent sprayer, Dragendorff reagent, even number hydrocarbons from C_{10} to C_{28}.

Dragendorff reagent:

Stock solution: Boil for a few minutes 2.6 g bismuth carbonate and 7.0 g sodium iodide with 25 mL glacial acetic acid. Keep for 12 hours at room temperature. Filtrate precipitated sodium acetate using a sintered glass funnel. Mix 20 mL of the clear, red-brown filtrate with 8 mL ethyl acetate and store in a brown coloured bottle.

Spray solution: add 25 mL glacial acetic acid and 60 mL ethyl acetate to 10 mL stock solution. After spraying and drying the plate, alkaloids appear as orange or yellow spots.

Procedure

 (i) Extract dried plant material (20 g) with methanol (200 mL) in a Soxhlet apparatus, for 3 hours. Evaporate the solution to dryness under reduced pressure. Add 0.5 M hydrochloric acid (80 mL) to the dry extract.
 (ii) Defat the acid solution by partitioning with dichloromethane (3×70 mL) using a separatory funnel.
 (iii) Alkalinize the aqueous acid solution with 4 M ammonium hydroxide to pH 10.
 (iv) Extract the alkaline aqueous solution with dichloromethane (3×70 mL).
 (v) Dry the organic extract with anhydrous sodium sulfate, filtrate and evaporate to dryness under reduced pressure.
 (vi). Keep in a vacuum desiccator for 24 hours; weigh the alkaloid extract (AE).
 (vii) Qualitative analysis of alkaloids in AE by thin layer chromatography (TLC).
 (viii) Silica gel GF-254 (250 μm layers) plates, developing tank mobile phase:chloroform:methanol:28% ammonium hydroxide (85:15:1). Visualize major alkaloids with Dragendorff reagent.
 (ix) Quantitative analysis of alkaloids in AE by GC: Dissolve a weighed sample of alkaloid extract (5 mg) in chloroform (1 mL). Perform GC with a SPB-1 (30 m, I.D. 0.25 mm) capillary column and Helium as gas carrier under programmed temperature from 100 to 270°C. Use hexadecane as internal standard. Relative abundances (in percentage) are recorded in relation to retention times (in minutes), to obtain the levels of different alkaloids.

(x) GC-MS: Mass spectrometry (MS) coupled to GC provides a preliminary identification of alkaloids, and can be performed on Shimadsu GCMS-QP 5000 equipment, with the same column and under the same temperature program. Data can be processed using the corresponding software, the CLASS 5000 program.

Calculations

Calculate alkaloid percentage in plant dry matter (DM) as:

$$\% \text{ Alkaloids in DM} = \frac{\text{AE(g)} \times \%A}{\text{DM(g)}}$$

where, AE is the alkaloid extract, %A is the percentage of alkaloids in the alkaloid extract, and DM is the weight of the dry matter subjected to methanol extraction.

Calculate retention indices (RI) with respect to a set of co-injected even number hydrocarbons (C_{10}–C_{28}). Compare RI and MS spectra with published data in order to confirm the identity of each alkaloid.

Observations

Identification of major quinolizidine alkaloids by GC-MS.

RI, M^+ and EI-MS five significant fragments and their relative abundance (Wink *et al.*, 1995).

Lupanine: RI 2165, M^+ 248, EI-MS: 136(100), 149(60), 248(40), 150(34), 219(8).

Multiflorine: RI 2310, M^+ 246, EI-MS: 134(100), 246(65), 148(20), 110(15), 217(5).

13α-Hydroxylupanine: RI 2400, M^+ 264. EI-MS: 152(100), 246(40), 165(40), 264(40), 134(30).

13α–Tigloyloxylupanine : RI 2753, M^+ 346. EI-MS: 246(100), 134(30), 148(15), 112(12), 55(10).

Angustifoline: RI 2083, M^+ 234, EI-MS: 193(100), 112(85), 150(15), 94(11), 55(20).

Precautions

Methanol is inflammable, keep away from sources of ignition. All organic solvents and reagents are toxic or allergenic by inhalation, contact or swallowing. Hydrochloric acid and concentrated ammonium are toxic.

Experiment 5: Determination of Tropane Alkaloids by GC-MS

There are several known cases of human fatalities caused due to poisoning by accidental ingestion of herbal decoctions or food contaminants prepared from *Datura* and *Scopolia* species (which contain tropane alkaloids). *Datura stramonium* L. (jimson weed) seeds have been found to contaminate grain crops and animal feed.

Materials and equipments required

Datura seeds; hood; rotavapor; sonic bath; glass column (10 mm I.D. × 150 mm); centrifuge; GC and GC-MS equipment; 30 m × 0.25 mm (I.D.) HP-5MS capillary column, film thickness 0.25 mm.

Reagents

Diethyl ether, atropine, dichloromethane, ammonium hydroxide, methanol, 100 mM borate buffer (pH 9), Extrelut NT20 granules, 1% ammonium hydroxide in methanol, nitrogen gas. *N,O*-Bis (trimethylsilyl)-trifluoroacetamide (BSTFA)–trimethylchlorosilane (TMCS) (99:1, v/v). Authentic samples of *L*-Hyoscyamine, *L*-scopolamine.

Procedure

Sample I: Urine or serum (0.5 mL)

(i) Wash 2.0 g Extrelut granules with two volumes of diethyl ether, dry for 1 hour at room temperature and heat for 1 hour at 40°C; pack into a glass column (10 mm I.D. × 150 mm).

(ii) Pass a mixture of sample I with 5 mL atropine (0.1 mg/mL) as internal standard and 1.0 mL 100 mM borate buffer (pH 9), through the Extrelut column. Elute adsorbed tropane alkaloids with 10 mL dichloromethane. Evaporate in vacuo to dryness at 40°C.

Sample II: Plant material (seeds)

(i) Macerate in a sonic bath, a mixture of 100 mg crushed *Datura* seeds with 0.4 mL 1% ammonium hydroxide in methanol for 10 minutes, to extract tropane alkaloids without heating.

(ii) Centrifuge (3,000 rpm, 3 minutes) and evaporate to dryness 0.1 mL of the supernatant under nitrogen flow.

GC-MS analysis:

(i) Derivatization: Dissolve the residue in 20 mL of BSTFA-TMCS. Heat the mixture at 80°C for 15 minutes. Add 100 mL of dichloromethane.

(ii) Perform GC-MS using a 30 m × 0.25 mm (I.D.) fused silica semipolar capillary column HP-5MS, film thickness 0.25 mm with splitless injection at 250°C, detector temperature 280°C, carrier gas: Helium, 0.8 mL/minute. Temperature programme: 50°C for 1 minute, then up to 300°C at a 20°C/minute rate and kept at 300°C for 5 minutes. MS-EI (70 eV), scan range from *m/z* 50 to 550.

Observations

Retention times and identification of major tropane alkaloids by GC-MS.

RT, M$^+$, EI-MS (significant fragments and their relative abundance) (Namera *et al.*, 2002).

Hyoscyamine: 13.14 minutes, M$^+$ 289, EI-MS: 124(100), 272(22), 140(50), 94(75), 82(80), 83(80).

L-scopolamine: 13.71 minutes, M$^+$ 303, EI-MS: 94(100), 108(90), 120(60), 138(95), 154(80), 273(20).

Precautions

Diethyl ether is highly inflammable, work in a hood away from ignition and heat sources. Methanol is inflammable, keep away from ignition sources. All organic reagents are toxic or allergenic by inhalation, contact or swallowing. Hydrochloric acid is caustic and toxic by inhalation, contact or swallowing, concentrated ammonium is toxic by inhalation or swallowing.

Experiment 6: Determination of Pyrrolizidine Alkaloids from *Senecio grisebachii* Baker Leaves

Pyrrolizidine alkaloids (PA) represent a potential hazard to human health. World Health Organization (WHO) published a report on chemical, botanical, biochemical and toxicological characteristics of pyrrolizidine alkaloids, including a complete review on most common analytical procedures to analyze them [Australian New Zealand Food Authority (2001)]. Several methods, such as TLC, GC, HPLC, and CE have been used for their separation, and quantification in plant sources. GC-MS and HPLC-MS have also been used. Sensitivity, specificity, and reliability of the different methods are continuously being improved. Nowadays capillary columns are commonly used in association with MS.

Materials and equipments required

Milled dry *Senecio* leaves; Soxhlet apparatus; separatory funnel; rotary evaporator; vacuum desiccator; silica gel GF-254 plates; hexadecane (chromatographic grade); GC and GC-MS equipment.

Reagents

Methanol, dichloromethane, 0.5 M hydrochloric acid, 4 M ammonium hydroxide, 28% ammonium hydroxide, anhydrous sodium sulfate, Dragendorff reagent, **0.5 M hydrochloric acid:** Add 44.25 mL concentrated hydrochloric acid to 955.75 mL water, **4 M ammonium hydroxide:** Add 280 mL ammonium hydroxide 28% to 720 mL water, **10% acetic anhydride:** add 10 mL acetic anhydride to in 90 mL benzene, **Ehrlich Reagent:** 0.5 g of *p*-dimethylaminobenzaldehyde (*p*-DMAB) in a mixture containing 50 mL of ethanol and 50 mL concentrated hydrochloric acid.

Procedure
Sample I

(i) Extract for 3 hours 20 g of dried plant material with methanol (200 mL) in a Soxhlet apparatus. Evaporate the solution to dryness under reduced pressure.

(ii) Add 0.5 M hydrochloric acid (80 mL) to the dry extract and pour the solution into a separatory funnel. Remove chlorophyll and lipids by partitioning with dichloromethane (3 × 70 mL).

(iii) Add Zn dust to the acid solution, keep for 4 hours under continuous stirring maintaining pH=1 and an excess of Zn in order to reduce *N*-oxides. Filter the acidic solution.

(iv) Alkalinize the acidic aqueous solution to pH=11 with 4M ammonium hydroxide. Extract the alkaline aqueous solution with dichloromethane (3 × 70 mL). Dry the organic extract with anhydrous sodium sulfate, filter and evaporate to dryness under reduced pressure. Keep in vacuum desiccator for 24 hours, weigh the alkaloid extract (AE).

Sample II

Repeat the procedure performed for sample I, without the reduction step, in the following sequence: 1, 2, and 4. Obtain alkaloid extract (AE1).

Dissolve AE (5 mg) and AE1 (5 mg), in 1 mL chloroform each.

TLC analysis: Apply each chloroform solution in three spots (5, 10 and 15 drops) on a silica gel F254 plate. Prepare two plates. Dry both at room temperature.

Run the two plates using chloroform:methanol:concentrated ammonia (85:14:1 v/v) as solvent system. Dry the plates and analyze the chromatogram under UV light at 254 nm, alkaloids are recognized as dark, light blue or blue absorbing spots.

Spray one chromatogram with Dragendorff reagent and heat at 100°C for 1 minute.

Spray the other chromatogram with a dual reagent spray system consisting of 10% acetic anhydride in benzene, followed by Ehrlich reagent (only *N*-oxides produce an intense purple color on heating the plate). Compare both chromatograms.

GC: Inject AE chloroform solution. Perform GC with a HP-5MS (30 m, DI 0.25 mm) capillary column and Helium as gas carrier under programmed temperature from 100°C to 270°C, with hexadecane as internal standard.

GC-MS analysis: GC system with a fused capillary column (30 m × 0.25 mm) × film thickness 0.25 µm, HP-5MS (Crossbond 5%-phenyl–95%-dimethyl-polysiloxane) directly coupled to a Hewlett Packard 5973 selective mass detector. Conditions: injection temperature 250°C; temperature program 70–300°C, 4°C/minute; splitless during 1.50 minutes, carrier gas He: 1 mL/minute, constant flow; sample volume 0.2 or 1 µl. Data can be processed using the corresponding software. Calculate retention indices (RI) with respect to a set of co-injected even number hydrocarbons (C_{10}–C_{28}). Compare RI and MS spectra with those of known standards, or with published data in order to confirm the identity of each alkaloid.

Observations

Identification by GC-MS, of major pyrrolizidine alkaloids.

RI, M^+ and EI-MS (significant fragments and their relative abundance) (Witte *et al.*, 1993).

Senecionine: RI 2290, M^+ 335, EI-MS: 136(100), 93(64), 106(13), 120(90), 220(40), 246(16), 335(9).

Seneciphylline: RI 2305, M^+ 333, EI-MS: 120(100), 94(60), 108(15), 119(70), 136(71), 218(3), 246(8), 335(2).

Integerrimine: RI 2335, M^+ 335, EI-MS: 120(100), 93(84), 109(19), 119(88), 136(98), 220(23), 248(12), 335(4).

Retrorsine: RI 2510, M^+ 351, EI-MS: 120(100), 93(70), 119(84), 136(98), 220(20), 246(8), 351(5).

Calculations

Calculate alkaloid percentage in plant dry matter (DM) as:

$$\% \text{ Alkaloids in DM} = \frac{\text{AE(g)} \times \%\text{A}}{\text{DM(g)}}$$

where, AE is the alkaloid extract, %A is the percentage of alkaloids in the organic extract, and DM is the weight of the dry matter subjected to methanol extraction.

Precautions

Methanol is inflammable, keep away from sources of ignition. All organic reagents and concentrated ammonium are toxic by inhalation, contact or swallowing.

SUGGESTED READINGS

Ambrozin, A.R. P., Mafezoli, J., Vieira, P.C., Fernández, J.B., da Silva, M.F., Das, G.F., Ellena, J.A. and de Albuquerque, S. (2005). New pyrone and quinoline alkaloid from *Almeidea rubra* and their trypanocidal activity. *Journal of the Brazilian Chemical Society* **16**: 434-439.

Australia New Zealand Food Authority (2001). Pyrrolizidine Alkaloids in Food: A Toxicological Review and Risk Assessment. Technical Report Series No. 2. Canberra, Australia: Australia New Zealand Food Authority. (http://www.foodstandards .gov.au/_srcfiles/TR2.pdf).

Baerheim-Svendsen, A. and Verpoorte, R. (1983). *Chromatography of Alkaloids, Part A: TLC.* Elsevier, Amsterdam.

Bleeker, A.B. and Romeo, J.T. (1983). 2,4-cis-4,5-cis-4,5-Dihyroxypipecolic acid—a naturally occurring amino acid from *Calliandra pittieri*. *Phytochemistry* **22**: 1025-1026.

Chen, T.M., George, R.C., Weir, J.L. and Leapheart, T. (1990). Thermospray liquid chromatographic-mass spectrometric analysis of castanospermine-related alkaloids in *Castanospermum australe*. *Journal of Natural Products* **53**: 359-365.

Conway, W.D. (1990). *Countercurrent Chromatography: Apparatus, Theory and Applications.* VCH, New York.

Cooper, R.A., Bowers, R.J., Beckham, C.J. and Huxtable, R.J. (1996). Preparative separation of pyrrolizidine alkaloids by high-speed counter-current chromatography. *Journal of Chromatography A* **732**: 43-50.

de Balogh, K.K.I.M., Dimande, A.P., van der Lugt, J.J., Molyneux, R.J., Naude, T.W. and Welmans, W.G. (1999). A lysosomal storage disease induced by *Ipomoea carnea* in goats in Mozambique. *Journal of Veterinary Diagnostic Investigation* **11**: 266-273.

Dräger, B. (2002). Analysis of tropane and related alkaloids. *Journal of Chromatography A* **978**: 1-34.

Egan, M.J., Porter, E.A., Kite, G.C., Simmonds, M.S.J., Barker, J. and Howells, S. (1999). High performance liquid chromatography quadrupole ion trap and gas chromatography/mass spectrometry studies of polyhydroxyalkaloids in bluebells. *Rapid Communications in Mass Spectrometry* **13**: 195-200.

Facchini, P.J. (2001). Alkaloid biosynthesis in plants: biochemistry, cell biology, molecular regulation, and metabolic engineering applications. *Annual Review of Plant Physiology and Plant Molecular Biology* **52**: 29-66.

Fellows, L.E. and Fleet, G.W.J. (1989). Alkaloidal Gylcosidase Inhibitors from Plants. In: *Natural Products Isolation. Separation Methods for Antimicrobials, Antivirals and Enzyme Inhibitors*, (Eds., G.H. Wagman and R. Cooper). Elsevier, Amsterdam.

Friedman, M. (2004). Analysis of biologically active compounds in potatoes (*Solanum tuberosum*), tomatoes (*Lycopersicon esculentum*), and jimson weed (*Datura stramonium*) seeds. *Journal of Chromatography A* **1054**: 143-155.

Harborne, J.B. (1998). Nitrogen compounds. In: *Phytochemical Methods. A Guide to Modern Techniques of Plant Analysis*. Chapman and Hall, London.

Kite, G.C., Veitch, N.C., Grayer, R.J. and Simmonds, M.S.J. (2003). The use of hyphenated techniques in comparative phytochemical studies of legumes. *Biochemical Systematics and Ecology* **31**: 813-843.

Lebada, R., Schreier, A., Scherz, S., Resch, C., Krenn L. and L. Kopp, B. (2000). Quantitative analysis of the pyrrolizidine alkaloids senkirkine and senecionine in *Tussilago farfara* L. by capillary electrophoresis. *Phytochemical Analysis* **11**: 366-369.

Levitt, J. and Lovett, J.V. (1984). *Datura stramonium* L.: alkaloids and allelopathy. *Australian Weeds* **3**: 108-112.

Lloyd, H.A., Fales, H.M., Highet, P.F., VandenHeuvel, W.J.A. and Wildman, W.C. (1960). Separation of alkaloids by gas chromatography. *Journal of the American Chemical Society* **82**: 3791.

Logan, B.K., Stafford, D.T., Tebbett, I.R. and Moore, C.M. (1990). Rapid screening for 100 basic drugs and metabolites in urine using cation exchange solid-phase extraction and high-performance liquid chromatography with diode array detection. *Journal of Analytical Toxicology* **14**: 154-159.

Molyneux, R.J., McKenzie, R.A., O'Sullivan, B.M. and Elbein, A.D. (1995). Identification of the glycosidase inhibitors swainsonine and calystegine B, in weir vine (*Ipomoea* sp. (Q6 EAFF. Calobra)) and correlation with toxicity. *Journal of Natural Products* **58**: 878-886.

Molyneux, R.J., Gardner, D.R., James, L.F. and Colegate, S.M. (2002). Polyhydroxy alkaloids: chromatographic analysis. *Journal of Chromatography A* **967**: 57-74.

Monforte, G.M., Ayora, T.T., Maldonado, M. and Loyhola, I.V. (1992). Quantitative analysis of serpentine and ajmalicine in plant tissues of *Catharanthus roseus* and hyoscyamine and scopolamine in root tissues of *Datura stramonium*. *Phytochemical Analysis* **3**: 117- 121.

Nakanishi, K. (1990). *One-Dimensional and Two-dimensional NMR Spectra by Modern Pulse Techniques*. Co-published by Kodansha Ltd., Tokyo and University Science Books, California.

Namera, A., Yashiki, M., Hirose, Y., Yamaji, S., Tani, T. and Kojima, T. (2002). Quantitative analysis of tropane alkaloids in biological materials by gas chromatography–mass spectrometry. *Forensic Science International* **130**: 34-43.

Papadoyannis, I.N., Samanidou, V.F., Theodoridis, G.A., Vasilikiotis, G.S., Van-Kempen, G.J. and Beelen, G.M. (1993). A simple and quick solid phase extraction and reversed phase HPLC analysis of some tropane alkaloids in feedstuffs and biological samples. *Journal of Liquid Chromatography* **16**: 975-998.

Phillips, B.J., Hughes, J.A., Phillips, J.C., Walters, D.G., Anderson, D. and Tahourdin, C.S.M. (1996). A study of the toxic hazard that might be associated with the consumption of green potato tops. *Food and Chemical Toxicology* **34**: 439-448.

Popl, M., Faehnrich, J. and Tatar, V. (1990). *Chromatographic Analysis of Alkaloids*. Marcel Dekker, New York.

Portsteffen, A., Dräger, B. and Nahrstedt, A., (1994). The reduction of tropinone in *Datura stramonium* root cultures by two specific reductases. *Phytochemistry* **37**: 391-400.

Pothier, J. and Galand, N. (2005). Automated multiple development thin-layer chromatography for separation of opiate alkaloids and derivatives. *Journal of Chromatography A* **1080**: 186-191.

Reinhard, H., Rupp, H., Sager, F., Streule, M. and Zoller, O. (2006). Quinolizidine alkaloids and phomopsins in lupin seeds and lupin containing food. *Journal of Chromatography A* **1112**: 353-60.

Sharam, G. and Turkington, R. (2005). Diurnal cycle of sparteine production in *Lupinus arcticus*. *Canadian Journal of Botany* **83**: 1345-1348.

Väänänen, T., Kuronen, P. and Pehu, E. (2000). Comparison of commercial solid-phase extraction sorbents for the sample preparation of potato glycoalkaloids. *Journal of Chromatography A* **869**: 301-305.

Wagner, H. and Bladt, S. (1996). *Plant Drug Analysis*. 2nd edition. Springer, Berlin.

Wilkins, C.L. (1983). Hyphenated techniques for analysis of complex organic mixtures. *Science* **222**: 291-296.

Wink, M. (1983). Wounding-induced increase of quinolizidine alkaloid accumulation in lupin leaves. *Zeitschrift für Naturforschung* **38C**: 905-909.

Wink, M. (2003). Evolution of secondary metabolites from an ecological and molecular phylogenetic perspective. *Phytochemistry* **64**: 3-19.

Wink, M. and Witte, L. (1984). Turnover and transport of quinolizidine alkaloids: Diurnal variation of lupanine in the phloem sap, leaves and fruits of *Lupinus albus* L. *Planta* **161**: 519–524.

Wink, M., Meibner, C. and Witte, L. (1995). Patterns of quinolizidine alkaloids in 56 species of the genus *Lupinus*. *Phytochemistry* **38**: 139.

Witte, L., Rubiolo, P., Bicchi, C. and Hartmann, T. (1993). Comparative analysis of pyrrolizidine alkaloids from natural sources by gas chromatography–mass spectrometry. *Phytochemistry* **32**: 187-196.

GC and GC-MS of Terpenoids

María L. López, José S. Dambolena, María P. Zunino and
Julio A. Zygadlo*

1. INTRODUCTION

Allelopathy is defined as the effect of one plant on growth and distribution of other plants (including microorganisms) through the release of chemical compounds into the environment. This definition includes not only positive and negative effects, but also direct and indirect effects of plant allelochemicals (Inderjit and Nilsen, 2003). Plant secondary compounds involved in are known as allelopathic agents. These compounds have been characterized according to their modes of release, phytotoxic action, bioactive concentration and persistence and fate in the environment (Inderjit and Duke, 2003). Terpenes are components of essential oils and accumulate in aromatic plants. They have multiple ecological functions in plants, such as protection against herbivores (Banchio *et al.*, 2005) and microbial diseases, attraction of pollinators and allelopathy (Fisher, 1986; 1991; Weidenhamer *et al.*, 1993). An allelopathic phenomenon involving terpene compounds comprises the following components (Fig. 1):

I. **Terpene production**: Allelochemicals in aromatic plants are usually produced in secretory structures such as glandular hairs, secretory cavities and ducts. Several environmental and physiological factors affect biosynthesis and accumulation of essential oils (Sangwan *et al.*, 2001). Essential oil composition also changes with geographical location, altitude, intraspecific variation,

Authors' address: *Instituto Multidisciplinario de Biología Vegetal (IMBIV-CONICET), Facultad de Ciencias Exactas Físicas y Naturales, Cátedra de Química Orgánica y Productos Naturales, Universidad Nacional de Córdoba, Av. Vélez Sarsfield 1611. Cp X 5016 GCA Córdoba, Argentina.
Corresponding author: E-mail: jzygadlo@efn.uncor.edu

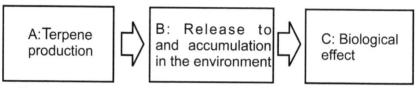

Figure 1   Main components of an allelopathic phenomenon in aromatic plants.

soil type, nutrient status and phenology (Graven *et al.*, 1991; Gil *et al.*, 2000; Azevedo *et al.*, 2001; Hudaib *et al.*, 2002).

II.   **Release and accumulation:** Terpene release and accumulation in the environment occur almost sequenced. Release can occur by evaporation (volatilization), leaching, root exudation and decomposition of plant residues in the soil (An *et al.*, 2001; Walker *et al.*, 2001; Dudareva *et al.*, 2004). Accumulation is the outcome of several factors such as edaphic properties, climatic conditions and periodic replenishment of terpenes from donor plants.

III.   **Biological effects:** Allelopathic effect must be corroborated in field conditions. One of the best ways to ensure that a test for allelopathy reflects natural (field) conditions is to perform a bioassay *in situ*. However, these assays often provide limited information because allelopathy is only one of the complex interacting processes that simultaneously occur in a natural system (Inderjit and Nilsen, 2003). Laboratory bioassays usually allow a better elucidation of allelopathic mechanisms because the experimental system can be modelled and factors can be kept constant. Studies carried out in the laboratory have provided valuable information about action mechanisms of terpenes on plant physiology (Zunino and Zygadlo, 2005; Zunino *et al.*, 2005).

2. GAS CHROMATOGRAPHY

GC has been the preferred technique for terpenes separation since 1950. Uses of this technique include characterization of new essential oils (EO), biotechnological research and detection of terpenes in different environments. Analysis of complex samples, such as essential oils, is difficult because a GC chromatographic column usually does not allow a simultaneous separation of all constituents. In these cases, combining the results of two chromatograms carried out on two GC columns with different polarities allows a better interpretation of the sample composition. First, a non-polar column with high content of polymethylsiloxane is used and compounds are separated according to their differences in boiling points. In a second step, a polar column (i.e. packed with polyethyleneglycol) separates compounds according to differences of solubility into the sorbent phase. In this way, compounds with similar boiling points (e.g. limonene and cineole) can be separated by differences in solubility. Similarly, those terpenes which can not be separated because of similar solubility (e.g. caryophyllene and terpinen-4-ol) will be separated by differences in boiling points. The GC technique can be used in qualitative analysis (characterization), as well as in quantitative analysis of EO and other terpene samples.

2.1 Qualitative Analysis

Comparison of retention index (RI) is the strategy most frequently used for terpene detection and identification. RI is retention time of a known compound and retention time of a homologous series of n-alkanes, n-alcohols or fatty acids. The procedure for calculation of the index is summarized in Section 2.2 of this chapter.

Identification of a compound based on the RI as the sole criterion can generate ambiguous results. In this case, is recommendable to combine the RI information with some other analytical technique. The most frequently used technique is GC coupled to a mass detector (GC-MS). There are abundant and complete references to mass spectra of numerous organic compounds. However, isomer compounds have similar structures leading to undistinguishable RI and fragmentation patterns. In these cases, alternative detection methods must be combined, such as GC coupled to Fourier transformed infrared (GC-FTIR). The RI and the mass spectra are always compared with those corresponding to pure compounds obtained in the same conditions. Furthermore, they should be compared with others available in the literature and in mass spectral libraries (NIST, WILEY and ADAMS).

2.2 Retention Indices (RI)

Kovats index must be defined for proper understanding and calculation of retention indices. This index allows the characterization of components from a mixture with respect to a homologous series of n-alkanes. Adjusted retention time (R_t') must be determined before calculation of Kovats index:

$$R_t' = R_t - R_{t\,A}$$

where, R_t is retention time of the compound and $R_{t\,A}$ is retention time of a compound not trapped in the column (i.e. methane). Kovats index of the compound (I) is calculated as follows:

$$I = 100\,N + 100\,n\ \frac{\log R_t'\,(x) - \log R_t'\,(N)}{\log R_t'\,(N+n) - \log R_t'\,(N)}$$

where, N is carbons number from the n-alkane eluting immediately before the interest compound, N+n is carbons number of the n-alkane eluting immediately after the interest compound, $R_t'\,(x)$ is adjusted retention time of the compound of interest, $R_t'\,(N)$ is the adjusted retention time from the n-alkane of N carbon atoms, and $R_t'\,(N+n)$ is the adjusted retention time from the n-alkane of N+n carbon atoms.

$$R_{t\,(N)}' \leq R_{t\,(x)}' \leq R_{t\,(N+n)}'$$

Kovats retention index is very useful because peak identification. However, this index can only be applied at constant temperature and high quality chromatographic separations of complex mixtures are difficult to achieve in isothermal conditions. For this reason, a new index, named Retention Index (RI), was developed to work under temperature intervals. The RI relates the adjusted chromatographic parameters and the carbon numbers to the Kovats index according to the following formula:

$$RI = 100\,N + 100\,n\,\frac{T_R(x) - T_R(N)}{T_R(N+n) - T_R(N)}$$

where, RI is the retention index of the interest compound, N is carbons number from the n-alkane eluting immediately before the interest compound, N+n is carbons number from n-alkane eluting immediately after the interest compound, $T_R(x)$ is the retention or elution temperature of the interest compound, $T_R(N)$ is the retention and/or elution temperature of the n-alkane of N carbon atoms, and $T_R(N + n)$ is the retention or elution temperature of the n-alkane of N+n carbon atoms.

$$T_{R(N)} \leq T_{R(x)} \leq T_{R(N+n)}$$

Retention or elution temperature can be replaced by the retention or elution time.

2.3 Quantitative Analysis

GC with flame ionization detector (GC-FID) is the most frequently used method for detection and quantification of terpenoids in complex samples. Peak areas in the chromatogram are transformed in relative amounts by internal normalization and in concentrations by internal and/or external standard methods.

Internal normalization: relates area of each peak with total peak area in the chromatogram.

$$\%A = \frac{\text{Peak Area}}{\text{Total Area}} \times 100$$

FID detector gives different responses for different compounds. Hence the peak area is not lineally proportional to the concentration. Therefore, a correction factor must be calculated (Orio *et al.*, 1986).

2.3.1 Internal Standard

It is a common quantification method. A proper choice for an internal standard requires several conditions:

(a) To give a satisfactory sign.
(b) To elute near the middle of the chromatogram.
(c) Not to be a natural compound found in the analyzed mixture.
(d) Its chemical structure must be similar to that of the compound to be analyzed.

The internal standard method is carried out in two steps.

Step 1: A solution of known concentration of the internal standard and the component of interest is prepared. This solution is chromatographied and the ratio between the peak areas obtained is plotted against the weights ratio (Fig. 2).

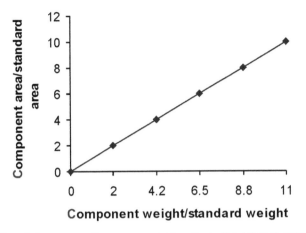

Figure 2   Plot of the areas-ratio *versus* weights-ratio in the internal standard method used for quantitative analysis of terpenoids by Gas Chromatography.

Step 2: The sample and the internal standard added in a known concentration are chromatographied. The weight of the component in the sample is then calculated as:

$$W_i = \frac{f_i\, A_i\, W_0}{A_0}$$

where, W_0 is the weight of the internal standard, A_0 is the area of the internal standard, W_i is the weight of the component i, A_i is the area of the component i, and f_i is the slope of the linear function. Terpenes commonly used as internal standards are 2-carene, 4-methyl-2-pentanol, tetradecane, octadecane, isobutylbenzene, n-octane and nonyl-acetate.

2.3.2   External Standard
Pure standard compounds are chromatographied at different concentrations and area vs weight of the standard compound is plotted. Compound weight in the sample is calculated by interpolation.

3. TEMPERATURE PROGRAMMES

Several temperature programmes for monoterpene analysis are found in literature. Ranges of temperatures comprised in these programmes are between 50 and 290°C. Table 1 summarizes temperature programmes, column properties, and types of terpenes studied from different biological samples.

Table 1 Temperature programmes and column properties

Matrix	Temperature programme (°C)	Gas carrier and flow	GC stationary phase	Column (I.D. × length × film)	Detector used	Terpenes identified	Reference identified
1 Essential oil (hydrodisti-lation)	70(0)/5/250(0) 70(0)/5/190(0)	N_2 1 mL/min	Rtx-5 HP-innowax	0.25 × 30 × 0.25 0.2 × 50 × 0.2	MS (EI)	Limonene, β-pinene, α-pinene, myristicin, terpinen-4-ol	Velasco-Negueruela et al., 2003
2 Essential oil (hydrodisti-lation)	60(0)/2/230(35) 60(0)/2/230(35)	He 1 mL/min	Rtx-1 Rtx-wax	0.22 × 60 × 0.25 0.22 × 60 × 0.25	MS (EI, CI)	33 sesquiterpenes (43.3%) and 12 mono-terpenes (33.5%). ger-macrene (28.5%), -phe landrene (19.0%) and para-cymene (5.2%).	Paolini et al., 2005
3 VOCs (SPME)	45(5)/4/100(3)/ 2/125/6/180(5) 40(5)/5/120/3/ 165/5/180(2)	He 1 mL/min	Permethylated cyclodextrin 2,3-di-acetoxy-6-O butyl dimethylsilyl cyclodextrin	0.25 × 25 × 0.25 0.25 × 30 × 0.25	MS	(+)-α -thujene, α-pinene, (+)-sabinene, (+)-α-phe-landrene n.d., (+)-limonene, β-pinene, (-)-menthyl ace-tate, (-)-menthone, neo-menthol, (-)-menthol, (+)-isomenthol, (+)-α-terpineol, (-)-trans-β-caryophyllene	Ruiz del Castillo et al., 2004
4 Air indoor (solid adsorbents)	40(8)/5/90(0.10)/ 10/280(12.3)	He 170 kpa (injector and column)	Rtx-5	250 × 60 × 0.25	MS	3-camphene, camphor, Δ-carene, 1,8-cineol, limonene, linalool, α-pinene, β-pinene, α-terpinene,	Hollender et al., 2002
5 Volatile (SPME)	40(1)/9/130/2/230 40(1)/9/130/2/230	He 10 psi	DB-5 CPSIL8CB	0.25 × 30 × 0.25	MS	γ-terpinene, fenchyl alcohol. Limonene, β-myrcene, α-pinene, β-caryophyllene, isoincensole acetate	Hamm et al., 2003

#	Sample	Temperature program	Carrier gas	Column	Dimensions	Detector	Compounds	Reference
6	Volatile ambient air (canister and solid adsorbent)	30(5)/3/55/5/ 150/10/200(5)	He 10 mL/min	Permabond OV-624-DF	0.22 × 50 1.40	MS	α-pinene, β-pinene	Tolnai et al., 2000
7	VOCs ambient air (Tenax tubes)	40(1)/5/210/ 20/250(8)		HP-5	0.2 × 50 × 0.5	MS	Sabinene, 3-carene, limonene, α-pinene, β-felandrene, β-pinene myrcene, α-terpinene, γ-terpinene	Manninen et al., 2002
8	Essential oil (hydrodistillation)	80/2, 4, 6/290	He 1 mL/min	HP-5MS DB-35MS	0.25 × 30 × 0.25	MS	Camphor, borneol, limonene, linalool oxide, eucalyptol, p-menth-1-en-8-ol, β-phelandrene	Zhao et al., 2005
9	Volatiles of aromatics plants (gas syringe)	30(2)/5/150/ 25/250	He	HP-5	0.25 × 30 × 0.25	MS	α-pinene, 1,8-cineole, camphor	Dunkel and Sears, 1998
10	Terpenes standard (gas syringe)	Isothermally to 50, 75 or 100	He 1 mL/min	DB-1	0.25 × 15 × 0.25	FID	Menthol, cineole, limonene, pinene, camphor, pinene	Vaughn and Spencer, 1991
11	Seed parts (extracted with methyl-tert-buthyl ether)	70(2)/4/200	He	HP-5	0.25 × 30 × 0.25	MS	Citral, nerol, neral, geraniol, geranial, neric acid, geranic acid, citronellal, citronellol, citronellic acid, vanillin, vanillic acid, vanillyl alcohol, pulegone, isopulegone, iso-menthone, menthone, carvacrol, thymoquinone, thymohydroquinone.	Dudai et al., 2001
12	Penicillium spores (Tenax Tubes)	50/5/150/10/200	He 1 mL/min	DB-5 FSOT	0.32 × 50 × 0.2	MS	Nerol, neral, geranial	Demyttenaere and De Pooter 1998)

#	Sample	Temp program	Carrier	Column	Film	Detector	Compounds	Reference
13	Floral organs volatile (SPME)	50(1)/30/220/10/290(5)	He 4 mL/min	HP-1	0.32 × 25 × 0.52	MS	α-pinene, β-pinene, myrcene, limonene, hotrienol, β-ionone, caryophyllene, humulene, bicyclogermacrene	Mac Tavish et al., 2000
14	Plant volatile (SPME)	30(5)/10/280	N_2 2 mL/min	BP-X5	0.2 × 25 × 0.25	FID	α-pinene, camphene, β-pinene, myrcene, limonene, hotrienol, β-ionone, caryophyllene, humulene, bicyclogermacrene, cis-β-ocimene	Mac Tavish et al., 2000
15	Soil volatile (hydrodistillation)	60(10)/5/220 60(10)/5/220 60/3/240	N_2 2 mL/min He 1 mL/min	HP-5 HP-wax DB-5	0.25 × 30 × 0.25 0.25 × 30 × 0.25 0.25 × 30 × 0.25	FID FID MS	α-terpineol, α-pinene, β-caryophyllene, germacrene D, δ-cadinene	Angellini et al., 2003
16	Essential oil (hydro-distillation)	45(10)/2.5/180 50(8)/4/180 (10)/5/220	He 1 mL/min	Rtx-5MS Stabilwax	0.25 × 30 × 0.25 0.25 × 30 × 0.25	MS	Thymol, linalool, carvacrol, α-terpinene, p-cymene	Hudaib et al., 2002
17	Plant volatile (SPME)	40(1)/4/220	He 50 mL/min	CP Sil 5 CP Sil 52CB DB wax	0.25 × 30 × 0.25 0.32 × 30 × 0.25 0.32 × 60 × 0.25	MS	α-pinene, β-pinene, limonene, linalool, camphene, farnesene	Rohloff and Bones, 2005
18	Volatile (SPME)	40(1)/10/28(17) 40(1)/5/210/10/280	nd	Omegawax ZB-5M	0.32 × 30 × 0.25 0.32 × 30 × 0.25	FID MS β	α-phellandrene, α-terpinene, p-cimene, farnesene, limonene, naphthalene, β-trans-ocimene, β-cis-ocimene, curcumene, sabinene, linalool, 4-terpineol, terpineol, cedrol, nerol, trans-nerolidol, bornil formate, geraniol	Peña-Alvarez et al., 2004
19	Terpenes from plant tissues (solvent extraction)	50(10)/5/200(5)	He	SP2331	0.25 × 60	FID MS	α-pinene, camphene, β-pinene, sabinene, myrcene, limonene, geranyl acetate, caritol	Bowman et al., 1997

No.	Sample	Temperature programme	Carrier gas	Column	Dimensions	Detector	Compounds	Reference
20	Grape volatile HS-SPME	70(5)/1/95(10)/2/190(40)	He 1 mL/min	BP-21	0.32 × 50 × 0.32	MS	Benzyl alcohol, linalool, α-terpineol, citronellol, nerol, geraniol	Sanchez-Palomo et al., 2005
21	Volatile flavour Penicillium (Elmer tubes)	35(1)/4/175/25/250(10)	He	DB-1701	0.25 × 30 × 1.0	MS	Limonene, β-caryophylene, unidentified sesquiterpenes, and two unidentified diterpenes	Larsen, 1998
22	Eighteen essential oils (commercial)	70/1/80/10/190/30/250	He 30 cm/seg	CBP	0.25 × 25	MS	Borneol, camphor, carvacrol, citral, 1,8-cineol, linalool, limonenne, thymol, thujone, geraniol	Lee et al., 2001

(a) Capillary column parameters are given in inner diameter (I.D.) in mm^3, length in m and film thickness in μm, numbers in parentheses are isothermic times in min, numbers between two "/" are oven programme rates.

(b) Numbers are temperatures in ºC.

nd = no data

Programes are clustered into three groups:

(i) Programmes with constant temperature ramps
(ii) Isothermal programmes
(iii) Programmes with one or more temperature ramps with or without isothermal stages

In the first group, retention indices for monoterpenes do not vary too much, being temperature ramps of 2, 4 and 6°C/minute (Zhao *et al.*, 2005). Sometimes, overlapped compounds or a low ratio signal-noise are obtained. In these cases, changes in temperature ramps can improve the separation.

In the second group, the chromatography is carried out at a constant temperature. However, isothermal programes require longer times for analysis than programes with temperature ramps, and are inappropriate for terpene analysis as a consequence of poor chromatographic separation.

Programes from the third group are specially used for complex samples and allow separations with good quality. Examples of these programes are provided by Paolini *et al.* (2005), who used 60 to 230°C at 2°C/minute and then held isothermally at 230°C (35 minutes). Hollender *et al.* (2002) used 40°C with 5°C/minute up to 90°C (isothermally 0.1 minute) and then 10°C/minute up to 280 °C for 12.3 minutes for terpene analysis ranging from 10 to 60% of the total volatiles from an air sample.

4. INJECTOR TEMPERATURE AND GAS CARRIER ON GC ANALYSIS

Carrier gas used for terpene analysis can be helium, hydrogen or nitrogen. The flow programes usually are in the 0.5 to 2 mL/minute range (Table 1). Helium is the carrier gas most frequently used at a flow rate of 1 mL/minute. The column resolution depends on the amount of sample injected. For example, a slow injection of a large sample can give a wide peak in the chromatogram. The injector temperature must be 50°C over the boiling point of the least volatile compound of the sample. In terpene analysis, the injector is usually set at 250°C. "Split injection" is the most commonly used method for the study of terpenes. Only 0.1 to 1% of the sample enters into the column, and the rest is vented. On the other hand, in the "splitless" mode the entire quantity of sample injected enters into the column. The "on-column injection" method is used for samples that decompose above their boiling point, and is not a very frequently used technique for terpene analysis.

5. COLUMNS

One of the most important parameters for terpene analysis is the column. There is no specific column for analysis of terpenes because their molecular diversity is huge. Non-polar columns are temperature and water resistant. These columns separate compounds by differences in their boiling points. On the other hand, the polar columns are sensitive to high temperature, so they only can be used for molecules of low boiling point. Besides, they are oxygen and water sensitive. The columns most frequently used are 25 to 60 m long and the film thickness is in the range of 0.2 to 10 µm. The stationary phase material is dimethylpolysiloxane for non-polar columns and polyethyleneglycol for polar columns. Table 2 summarizes columns most commonly used. Specific columns can be used for particular systems or compounds. For example, columns with greater affinity for aromatic compounds can be used for terpenes as thymol and carvacrol. Table 3 summarizes the main properties of column sorbents and its potential use.

Table 2   Columns most commonly used for terpene analysis

Polarity	Stationary phase	Similar phase
Low	100% Methyl polysiloxane	VF-1ms, DB-1, DB-1MS, Rtx-1, Rtx-1MS, HP-1, HP-1MS, Ultra-1, SPB-1, CP-SIL 1CB
Low	Methyl 5% Phenyl polysiloxane	VF-5ms, DB-5, DB-5 MS, Rtx-5, Rtx-5 MS, HP-5, HP-5 MS, Ultra-2, SPB-5, CP-SIL 8CB
High	Polyethyleneglycol	DB-wax, Carbowax, STABIL WAX, Supelco-wax, CP-Wax 52CB

Table 3   Characteristic of the columns and potential use

Polarity	Stationary phase	Applications	Temperature
Non- polar	100% Methyl polysiloxane	General use, wide range of applications	50-325°C Excellent thermal stability
Non- polar	5% Phenyl, 95% Methyl polysiloxane	General use, wide range of applications	50-325°C Excellent thermal stability
Medium	35% Phenyl, 65% dimethyl-polysiloxane	Sterols, pesticides, herbi-cides, aromatic compounds	40-325°C
Medium	50% Trifluoro-propyl 50% Methyl polysi-loxane	Ketones, aldehydes, nitro or chloro compounds, un-saturated compounds	40-300°C Good thermal stability
Polar	Cyanopropyl polysiloxane	FAME, and other separa-tions requiring unique selectivity and polarity	40-280°C
Polar	Polyethylene-glycol	General use, selectivity hydrogen bonding-type molecules. Particularly useful for the analysis of complex oxygenated samples	20-260°C Susceptible to oxygen degradation. Not recommended for the analysis of mixtures containing silylating reagents

In 1990s, the discovery of chiral phases allowed resolution of enantiomer volatile terpenoids. Separation of enantiomers using these columns is based on three principles (Schurig, 2001; 2002):

(i) Chiral amino acid derivates via hydrogen-bonding
(ii) Chiral metal coordination compounds via complexation
(iii) Cyclodextrin derivatives via inclusion

The columns most commonly used for the separation of enantiomeric terpenes are based on inclusion in cyclodextrin. A wide range of commercial cyclodextrin phases has been developed and used for separation of enantiomeric mono- and sesquiterpenes (Lockwood, 2001). The derivatized cyclodextrins are usually physically incorporated into appropriate polysiloxanes, which are then coated on the column walls or onto an inert support. In certain stationary phases, the polysiloxanes contain a proportion of phenyl groups which renders the mix more thermally stable and thus, can be used at higher temperatures. The level of the cyclodextrin component in the stationary phase varies between 8 and 15%, largely depending on the solubility of the cyclodextrin based material in the particular polysiloxane which can be enhanced by introducing sufficient methylsiloxane into the polymer mix (Beesley and Scott, 1998). Ruiz del Castillo *et al.* (2004) studied the enantiomeric separation capacity of 2 chiral columns in *Mentha*: permethylated β-cyclodextrin (Chirasil-Dex, Chrompack) and 2,3-di-acetoxy-6-*O*-*tert*-butyl dimethylsilyl-γ-cyclodextrin. They found that dimethylsilyl-γ-cyclodextrin yielded the best results, i.e. separation of the main compounds, menthol and its derivates. In a comparative study of chiral phases, Betts (1994) found great increase in relative retention time with a α-cyclodextrin phase than with β- and γ-phases. Retention times of acyclic monoterpenes increased by 60% and that for bicyclic monoterpene increased by 150%. More information on chiral phases is provided by Beesley and Scott (1998).

6. DETECTORS

The detector, most frequently used for terpene analysis, is the flame ionization detector (FID). Its advantages are wide linear dynamic range, very stable response, high sensitivity in the order of 10^{-13} g, speed response and low background noise (high relationship signal/noise). The FID detector is sensitive to the molecular mass and the signal depends on the number of carbon atoms entering by time unit. It is less sensitive to carbonyl groups, amines, alcohols and not sensitive to non-flammable compounds such as H_2O, CO_2, SO_2, NOx. The most important disadvantage of this detector is the sample destruction.

The system most commonly used for the identification and characterization of terpenes is the GC coupled to MS. GC-MS can involve either quadrupole MS or the relatively cheap ion trap MS. This technique is based on the ionization capacity of a molecule (on vapour phase) by electrons bombardment (electronic impact).

The electrons energy must be higher than the ionization potential of the molecule. Energy of ionization generally is 70 ev. This impact yields molecular ions and positive charge fragments, which will then be selected by their mass to charge ratio (m/z) in a quadruple ion trap or a quadruple mass filter. This fragmentation depends on the structure of the molecule and is characteristic for each compound. The identification of the various compounds is achieved by both the retention indices obtained in different polarity columns and the mass spectra. The full scan mode allows selection of peaks in the chromatogram, which are of interest in the mass to charge ratio. The detection limits are in picogram range for the full scan mode. Selection of the scan range will depend on the objective of the study. Successful identification of the compounds depends on the selected range. Most of the studies on essential oils recorded the mass spectra within 40-350 m/z. For monoterpenes, the scan range is within 40-250 m/z while it can be widen for greater weight compounds. Thus, Hamm *et al.* (2003) used a scan range within 29-400 m/z in order to study mono, sesqui- and diterpenes in various olibanum samples. Tolnai *et al.* (2000) used a scan range within 30-350 m/z for studies of organic volatile hydrocarbons. Ojala *et al.* (1999) used a scan range within 50-210 m/z for the study of mono and sesquiterpenes in water samples. Zhao *et al.* (2005) studied essential oils and selected a scan range of 30-500 m/z to identify the compounds up to 30 carbon atoms.

As mentioned in previous sections of this chapter, the GC-MS system is selected by majority of the researchers for terpene analysis. However, its utilization for isomers of very similar structure can generate ambiguous results because the obtained fragmentation patterns and the retention indices can be similar. For this reason, new methods have been developed for terpene study. GC-MS with chemical ionization is one such method. The chemical ionization obtains the ionization and fragmentation from the compounds by the bombardment with ammonium or methane gas. The fragmentation obtained with this method is different from that obtained with electronic ionization. For example, Paolini *et al.* (2005) demonstrated the complementarity of electron impact and chemical ionization in GC-MS for the analysis of a complex mixture, mainly for differentiation of monoterpene esters with similar EI-mass spectra. Furthermore, they conclude that chemical ionization with different gases is an effective method for confirmation of the molecular mass and the determination of acid and alcohol parts of unknown compounds.

GC coupled to infrared spectrometry can be complementary in the use of GC-MS for the analysis of terpenes in complex samples. This technique offers structural information and isomers can be differentiated. This technique produces excellent spectroscopic information and is best suited for non-routine analysis when positive identification of compounds is required. Ferary *et al.* (1996) described a coupled technique of GC-MS with direct deposition to GC-FTIR. A balance of sensitivity level between both techniques was obtained by gas cell system interfaces and the identification of compounds was achieved in the order of picogram. The use of

GC-MS and GC-FTIR enhances reliability in qualitative analysis of terpenes (Cai *et al.*, 2006).

GC-GC is also available for the analysis of terpenes and allows higher peak capacities than conventional GC because each successive small fraction eluting from the conventional size first-dimension column is subjected, in real time, to a second, orthogonal separation, on a relatively short (ca. 0.5 m) second-dimension column with different separation characteristics. In most instances, a non-polar first column is combined with a more polar second column, and the time span of each effluent fraction from the first column that is trapped, refocused and, next, transported to the second-dimension column, is in the order of 3–6 seconds (Korytar *et al.*, 2003).

7. SORBENTS

The choice of a system for collection and analysis of plant volatiles varies with the research objective (Marsili, 2002; Tholl *et al.*, 2006). Methods most commonly used for volatile trapping are:

I. **Solid phase extraction (SPE):** Used to isolate and concentrate selected compounds from liquid, fluid or gas by interaction (sorption) and transfer to solid phase.

II. **Solid phase microextraction (SPME):** An inert fiber coated with a liquid (polymer), a solid (sorbent), or a combination of both. The fiber coating removes the compounds from the sample by absorption (liquid coatings) or adsorption (solid coatings). SPME is one of the methods most commonly used for terpene analysis, allowing collection of sample amounts in parts per billion (ppb). Following equilibration between the fiber and the volatile sample, the fiber is retracted into the needle and can be transferred to the injector block of a GC or a GC–MS for direct thermal desorption. This technique includes speed and sensitivity but has the disadvantage of poor precision for quantitative analysis and long equilibration times for less volatile analytes (Vereen *et al.*, 2000).

After adsorption, compounds are recovered by liquid elution (usually for SPE) or by thermal desorption (for SPME).

Volatile trapping capacity depends on adsorbent material, time of equilibration between the trapping material and the volatile sample and adsorption temperature. Table 4 shows adsorbents and its principal characteristics study of terpenes and volatile organic compounds (VOCs). In the case of SPME, careful selection of polarity and thickness of the fiber coating allow sampling of substances with different polarity and volatility ranging from high-boiling or semivolatiles to VOCs. While thin coatings ensure a fast diffusion and release of semivolatile compounds, thicker coatings may better retain highly volatile compounds until desorption. The amount of compound adsorbed by the SPME fiber depends not

Table 4 Common adsorbents for trapping of plant VOCs (Tholl et al., 2006)

Sample	Identified terpenes	Sorbent	Max. temp. (°C) of sorbent*	Approx. analyte sampling range of sorbent*	Adsorptive features of sorbent*
Volatiles from leaves. Traps with 50 mg of Super Q (Röse et al., 1996)	α-pinene, camphene, β-pinene, myrcene, limonene, β-ocimene, linalool, β-caryophyllene, α-bergamotene, α-humulene, (E)-β-farnesene, α-farnesene, γ-bisabolene, β-bisobolol	Porapak Q Super Q (high purity version)	250	C_5 to C_{12} bp 50-200°C	High affinity for lipophilic to medium polarity organic compounds of intermediate molecular weight; suitable for a wide range of VOCs including oxygenated compounds. Low affinity for polar and/or low molecular weight compounds (H_2O); frequently used in VOC analysis
Artificial air samples. The sample tubes of quartz contained Tenax TA (300 mg/tube) (Hollender et al., 2002)	3-camphene, camphor, Δ-carene, 1,8-cineol, limonene, linalool, α-pinene, β-pinene, α-terpinene, γ-terpinene, fenchyl alcohol	Tenax TA (2,6-diphenylene oxide polymer)	350	C_7 to C_{26} bp 100-400°C	High affinity for lipophilic to medium polarity organic compounds of intermediate molecular weight; not suitable for highly volatile organics. Low affinity for polar and/or low molecular weight compounds (H_2O). Preferred for terpenes; frequently used in VOC analysis
Plant volatiles. Tenax-GC tubes (Padhy and Varshney, 2005)	Isoprenes, pinenes	Tenax GC	350	C_7 to C_{26} bp 100-400°C	ND
Microbial volatile organic compounds of	Aromadendrene, 2-methylenebomane,	Tenax GR (30% graphite carbon	350	C_7 to C_{26} bp 100-400°C	Compared to Tenax TA, it has a higher retention

Application/Reference	Compounds	Adsorbent		bp range	Notes
environmental fungi (Fischer et al., 1999). Artificial air samples. The sample tubes of quartz (Hollender et al., 2002)	2-methyl-2-bornene, 2-methyl-2-bornene isomer, germacrene A, germacrene B, β-elemene, χ-cadinene, camphene, 8-4-carene2, caryophyllene, elemol, χ-curcumcne, bicyclo-elemene, α-farnesene. 3-camphene, camphor, Δ-carene, 1,8-cineol, limonene, linalool, α-pinene, β-pinene, α-terpinene, γ-terpinene, fenchyl alcohol.	and 70% Tenax TA) Tenax GR (300 mg/tube)			volume for most compounds and is twice as dense. Like Tenax TA, it has a low affinity for water or methanol and has an upper temperature limit of 350°C
Volatile organic compounds of surface waters. Tubes with 5,350 mg of Chromosorb-106 (Bianchi and Varney, 1998)	α-pinene, camphene, limonene, naphthalene	Chromosorb 102 Chromosorb 106 (styrene-divinyl-benzene polymer)	250	bp 50-200°C	Wide range of VOCs incl. oxygenated compounds. Used to trap small molecules
Volatile organic hydrocarbons. Multilayer-bed tube (Tolnai et al., 2000).	α-pinene, β-pinene	Carbotrap	>400	C_5 to C_{12}	Wide range of VOCs; ketones, alcohols, aldehydes (bp>75°C), non-polar compounds
Artificial atmosphere SPME (Hollender et al, 2002)	3-camphene, camphor, Δ-carene, 1,8-cineol, limonene, linalool, α-pinene, β-pinene, α-terpinene, γ-terpinene, fenchyl alcohol	Carbosieve SIII or Carboxen 100	>400	bp 60-80°C C_2 to C_5	For small hydrocarbons; not suitable for reactive hydrocarbons (1,3-butadiene, isoprene)

Source	Analytes	Adsorbent			Notes
ND	ND	Activated charcoal	>400	C5 to C16 on CSLA traps	Less efficient than Tenax in trapping aromatic aldehydes; rarely used for thermal desorption
Flowers volatiles SPME (Mac Tavish *et al.*, 2000).	α-pinene, camphene, β-pinene, myrcene, limonene, hotrienol, β-ionone, caryophyllene, humulene, bicyclogermacrene, cis-β-ocimene	PDMS (polydimethylsiloxane)	<300	Volatiles and semivolatiles dependent on thickness of fiber coating MW 60-275	Non-polar absorbent for mostly non-polar volatiles; used as fiber coating for SPME analysis
Resin of trees (olibanum) (Hamm *et al.*, 2003).	α-pinene, β-myrcene, α-phellandrene and limonene α-cubebene, α-copaene, β-elemene, β-caryophyllene, γ-humulene, β-muurolene, β-eudesmene and caryophyllene oxide γ-humulene, β-muurolene, β-eudesmene and caryophyllene oxide	PDMS/DVB (polydimethyl-siloxane/divinyl-benzene)	<300	MW 50-300	Bi-polar adsorbent for polar volatiles used as fiber coating for SPME analysis
		PDMS/Carbo-xen (polydimethyl-siloxane/Carbo-xen)	<300	MW 30-225	Bi-polar adsorbent for trace level volatiles recommended for low molecular weight analytes (MW<90); used as fiber coating for SPME analysis.
		CW/DVB (car-bowax/divi-nilbenzene)	<300	MW 40-275	Polar adsorbent for polar alcohols and polar analytes
		CW	<300	ND	ND
		DVB/CAR/PDMS (divinylbenzene/Carboxen/poly-dimethyl-siloxane)	<300	MW 40-275	Volatiles and semivolatiles, C3-20

ND = No Data

only on the thickness of the fiber coating but also on the distribution constant of the compound, which generally increases with its molecular weight and boiling point (Tholl *et al.*, 2006). On the other hand, the equilibration time is a critical factor. Monoterpenes are highly volatile compounds (limonene: bp_{763} 175.5–176.5°C) and leave the solid matrix very quickly. Thus, to obtain reproducible quantitative results, the fiber and sample should reach the equilibrium. Careful control of sampling parameters (sample volume, temperature, and time) and the use of appropriate standard mixtures for calibration are crucial for quantitative analysis (Tholl *et al.*, 2006).

Traditional fibers, such as polydimethyl-siloxane (PDMS) and polyacrylate (PA), have been used for trapping of terpenes. However, these fibers have poor sensitivity for polar compounds. Mixed coating fibers containing divinylbenzene (DVB), PDMS and carboxen (CAR) or carbowax (CW), increase tramping ability of the fiber due to the synergic effect of adsorption and distribution within the stationary phase, producing higher sensitivity than PDMS and PA fibers (Harmon and Marsili, 2002 cited by Sánchez-Palomo, 2005).

8. EXPERIMENTAL METHODS FOR TERPENE ANALYSIS

Terpenes, mainly mono- and sesquiterpenes, are frequently found among the VOCs emitted by plants to the environment (Weidenhamer *et al.*, 1993; Fuentes *et al.*, 2000). The design of headspace techniques has improved the analysis of these VOCs and has provided a more representative volatile profile of living plants than other methods such as solvent extraction or steam distillation (Tholl *et al.*, 2006). Static headspace analysis is achieved when air surrounding the plant material does not circulate in the chamber. In the case of dynamic headspace analysis, a continuous air stream flows through the sample container (Tholl *et al.*, 2006). In both cases volatiles are trapped on an adsorbent matrix prior to the analysis by GC. Collection of volatile terpenes can be obtained from intact plants or their detached parts. Moreover, detached plant parts can be supplied with water or glucose solution after detachment (ex situ—wet) or not (ex situ—dry). In general, VOCs obtained from intact plant material are different from those obtained from plant cuttings (Jakobsen, 1997). VOCs can also differ depending if either static or dynamic headspace analysis is performed.

9. EXPERIMENTAL PROCEDURES

The experiments described below are for GC and GC-MS analysis of terpene allelochemicals. The experiments are presented in accordance with the components previously mentioned.

9.1 Terpenes

Experiment 1: Essential Oils Extraction by Steam-distillation and GC and GC-MS Analysis

Essential oils are widely distributed in the plant kingdom. Species rich in essential oils include annual, biannual or perennial herbaceous plants, and evergreen or deciduous shrubs and trees. Essential oils can be extracted from plant tissues using the steam distillation method. The high vapor pressure of the oil constituents facilitates their vaporization with water steam at 100°C. Then, oil and water are condensed and the oil layer (upper layer) is easily recovered.

Materials and equipments required

Funnels; Erlenmeyer flasks; Eppendorf tubes; fresh or dry crushed plant material: leaves (young, mature, and senescent), fruits, flowers, stems, roots; clevenger-type apparatus (Fig. 3); rotary evaporator (Fig. 4); GC and GC-MS equipment.

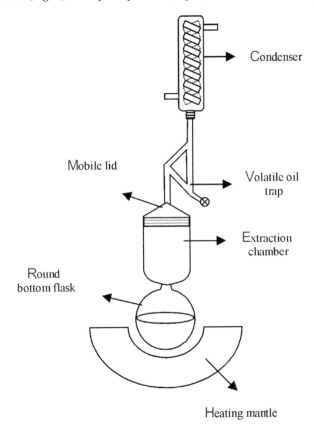

Figure 3 Clevenger-type apparatus.

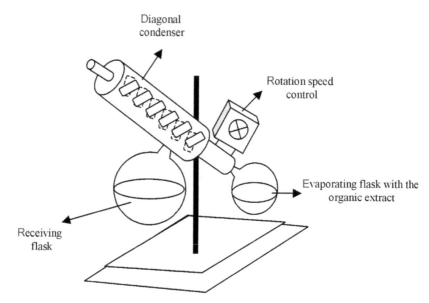

Figure 4 Rotary evaporator showing the evaporating flask maintained under vacuum through a tube connected to a condenser. The evaporating flask rotates while it is heated in the water bath. The solvent vapors leaving the flask are condensed in the diagonal condenser. The condensed vapors drain into the receiving flask where they are collected.

Reagents

Anhydrous sodium sulfate (Na_2SO_4); dichloromethane (Cl_2CH_2); tap water; standard terpenes commercially available.

Procedure

One hundred g of plant material are placed in the chamber of a clevenger-type apparatus (Fig. 3). Tap water is added into a distillation flask (round bottom flask) and the plant material is steam-distilled for 1 hour. Then, the steam-distilled is partitioned with Cl_2CH_2 (200 mL) and dried over anhydrous Na_2SO_4. The Cl_2CH_2 phase is filtered and then evaporated at reduced pressure in a rotary evaporator. Terpenes from the essential oil obtained are identified by GC-MS and quantified by GC.

Calculations

The volume or weight of essential oil obtained is measured and expressed on a dry weight biomass basis.

The terpenes obtained are identified by GC-MS using the retention index and a library (NIST, WILEY, ADAMS) of terpenes and quantified by GC (refer Table 1, row Nos, 1, 2, 8, 16).

Observations

(i) The collected plant material should be distilled as soon as possible to avoid oxidation and evaporation of oil constituents.

(ii) Fresh materials yield greater amounts of essential oil than dry materials.

Precautions

Dichloromethane is harmful if swallowed, inhaled, or absorbed through the skin, and may be carcinogenic. Wear appropriate protective gloves, clothing and goggles (to prevent skin and eye exposure) and respiratory protective mask for solvents.

Experiment 2: Terpenes from Glandular Trichomes

Plant families such as Labiateae, Verbenaceae, Apiaceae and Asteraceae typically contain high concentrations of essential oils in glandular trichomes and other secretory structures. Glandular trichomes on the leaf surface of aromatic plants exude oily drops or terpenoid blends. Phyllospheres often accumulate remarkable amounts of secondary metabolites released by aromatic plants, which in turn can be biotransformed by epiphytic microbes. The objective of this experiment is to analyze and compare the types of terpenes stored in the secretory structures with those lying on the leaf surface. The glandular trichomes are easily separated from the leaves by mechanical abrasion. Then, terpenes stored in the isolated glandular trichomes as well as those on the leaf surface can be extracted with organic solvents according to their solubilities. The organic extracts are analyzed by GC and GC-MS.

Materials and equipments required

Leaves from aromatic plants; Erlenmeyer flasks; filter paper; 250, 105 and 20 μm nylon membranes; glass beads; quartz; rotary evaporator; orbital shaker; light microscopy; GC-MS equipment.

Reagents

Separation solution: 0.4 M mannitol and 50 mM sodium L-ascorbate adjusted to pH 6.8; ethanol (70%); tap water; dichloromethane; gaseous N_2.

Procedure

(a) **Terpenes from the glandular trichomes:** Excised leaves from aromatic plants are rapidly washed, first with 70% ethanol to rinse out exudates from glandular trichomes and dust, and then with running tap water to remove the remaining ethanol. Washed leaves are placed in a 500 mL Erlenmeyer flask, containing 200 mL of the separation solution and 10 g of clean quartz and glass beads. The Erlenmeyer flask is placed on an orbital shaker for 30 minutes at room temperature. Following abrasion, clusters of glandular trichomes are obtained by filtering the contents of the Erlenmeyer flask sequentially through 250, 105 and 20 μm nylon membranes. The isolation of pure glandular trichomes is confirmed by light microscopy. Isolated trichomes are washed several times with the separation solution and transferred immediately to a 50 mL Erlenmeyer flask to which 30 mL dichloromethane is added. The terpenes are extracted by shaking the dichloromethane in an orbital shaker for 10 minutes. The dichloromethane is filtered, carefully concentrated to 0.1 mL and directly analyzed by GC-MS.

(b) **Terpenes from leaf surface:** Leaves of aromatic plants (15 g) are dipped into 100 mL of dichloromethane for 50 seconds to dissolve volatiles on the leaf surface. The leaves are removed and the dichloromethane is filtered through a 0.45 μm Millipore filter and concentrated to a final volume of 1mL using a rotary evaporator. Then, the solvent is evaporated under a N_2 stream until a final volume of 0.1 mL is reached. The samples are analyzed by GC-MS.

Calculations

The terpenes obtained are identified by GC-MS using the retention index and a library of terpenes (NIST, WILEY, ADAMS) (Table 1, row No. 19). The chemical compositions from both the glandular trichomes and from the leaf surface are compared.

Precautions

Dichloromethane is harmful if swallowed, inhaled, or absorbed through the skin, and may be carcinogenic. Wear appropriate protective gloves, clothing and goggles (to prevent skin and eye exposure) and respiratory protective mask for solvents.

9.2 Terpene Release and Accumulation in the Environment

Experiment 3: Terpene Content in Soil Samples

Terpenes released from aromatic plants often accumulate in the soil. They are often extracted from the soil matrix with organic solvents. Then, terpenes are separated by thin layer chromatography (TLC) and subsequently quantified by GC and identified by GC-MS.

Materials and equipments required

Sieve (2 mm mesh); filter paper; volumetric flasks; funnels; spatula; centrifuge tubes; pipettes; Eppendorf tubes; small funnels; Gas Chromatograph FID detector; centrifuge; rotary evaporator; TLC plates; chambers for TLC; iodine vapor chamber/UV lamp; soil samples.

Reagents

Anhydrous sodium sulfate (Na_2SO_4); silica gel 60 G plates (thickness of gap 0.25 mm for analytical TLC and of 1.00 mm for preparative TLC); standard of the allelopathic terpene (commercially available or isolated and purified in laboratory from the plant material); chloroform-methanol azeotrope (87:13); chromatographic grade methanol; benzene.

Procedure

Soil samples are sieved and c.a. 50-80 g of fresh sieved soil is weighed. Then the soil sample is poured into an appropriate volumetric flask (e.g. 250 mL) and extracted with 250 mL of chloroform-methanol (87:13) azeotrope for 24 hours in darkness. The extract is filtered trough filter paper and dried with anhydrous sodium sulfate (Na_2SO_4). The solvent is evaporated at reduced pressure to 1.5 mL. Aliquots of each sample and the standard are spotted at the bottom of an analytical TLC plate. The TLC is developed using benzene as mobile phase. If spots from the extract are detected with the same *Rf* as those of the standard (visualized with the aid of a UV lamp or iodine vapors), repeat TLC using a preparative TLC plate. The band corresponding to the terpene of interest is scrapped off with a spatula into an assay tube. The band is eluted with methanol and filtered through filter paper. The solvent is evaporated in the rotary evaporator until the desired volume is obtained.

Calculations

The extract is analyzed by GC-MS and GC for quantification (Table 1, row No. 15). A standard calibration curve is prepared with the terpene standards for quantification (Fig. 5).

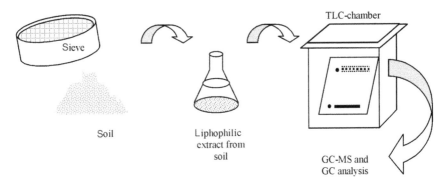

Figure 5 Extraction and isolation of terpenoids from soil samples. Sieved soil is extracted with an organic solvent. Then, the organic extract is spotted in a TLC plate and developed using an appropriate mobile phase. When spots corresponding to the terpene of interest are detected, the samples are processed in preparative TLC. The bands of interest are eluted in an appropriate solvent and after reducing the solvent in a rotary evaporator, the extract is analyzed by GC-MS and GC.

Statistical analysis

If factors such as time (years, months, days or hours) or site (beneath the plant and control sites) are included in the experimental design, ANOVA [Analysis of variance] and the Fisher's LSD [least significance difference] test could be used to detect significant differences among treatments. In case of non-parametric data, use the Kruskal-Wallis test. Presence or absence of the terpene in each sample can be detected from the analytical TLC study. Combine these data for frequency analysis and analyze them by contingency tables and chi-squared test.

Observations

(i) To avoid the spatial heterogeneity in the sampling areas, a common procedure is to mix sub-samples. Sample the first 5 cm of soil [for further details, refer to Chapter 2, Section 1.1].

(ii) More than one extraction of soil samples can be carried out for exhausting the component. Add fresh solvent each time and then mix the extracts.

Precautions

Chloroform-methanol azeotrope is harmful if swallowed, inhaled, or absorbed through the skin. Wear appropriate protective gloves, clothing and goggles (to prevent skin and eye exposure) and respiratory protective mask for solvents.

Experiment 4: Monoterpenes in Soil Solution and Exchangeable Soil Surfaces

Organic compounds incorporated in the soil are subjected to biotic and abiotic factors that define the bioavailability of these substances in the edaphic environment. The amount of bioavailable monoterpenes in soil water can be identified and measured through GC and GC-MS. Thus, solubility of terpenes in the soil aqueous phase and adsorption of terpenes into soil components are the principle of this method.

Materials and equipments required

Centrifuge tubes; funnels; automatic pipettes; filter paper; soil samples; centrifuge apparatus; GC and GC-MS equipment; stove; rotary evaporator.

Reagents

High quality standard monoterpenes commercially available; dichloromethane.

Procedure

Soil samples are sieved (2 mm mesh) and 20 g of each sample are centrifuged at $13,000 \times g$ for 30 minutes. The aqueous soil extract (the supernatant) is twice partitioned with 50 mL of dichloromethane and dried with anhydrous Na_2SO_4. The extract is filtrated and the solvent is evaporated in a rotary evaporator. The dry extract is suspended in 10 µl dichloromethane and analyzed by GC. The centrifuged soil (the pellet) is twice extracted with dichloromethane (200 mL). The extract is filtrated and after drying with anhydrous Na_2SO_4 the solvent is evaporated in a rotary evaporator. Finally the extract is suspended in 10 µl dichloromethane and analyzed by GC (Table 1, row No. 15).

Calculations

(i) Calculate the monoterpene concentration in the aqueous phase.

(ii) Assume that the monoterpene concentration in the pellet water is the same as in the aqueous phase.

(iii) Calculate the monoterpene concentration in the centrifuged soil as the difference between the total content obtained for the centrifuged soil (pellet) and the concentration in the pellet water.

Statistical analysis

Data are analyzed by ANOVA and means are compared using the Fisher's LSD test at $p < 0.05$.

Observations

(i) As the concentration in the aqueous phase needs to be expressed on a soil water content basis, it is necessary to determine the original soil water content. Thus, some samples are heated at 60°C until constant dry weight. On the other hand, because the pellet is wet, the water content of the pellet should also be determined. Thus, some centrifuged soils (pellets) are heated at 60°C until constant dry weight.

(ii) This method can also be applied to determine the behavior of different monoterpene concentrations exogenously added to soil samples collected near the area where the aromatic species grow. For this case, 10 mL of different concentrations of monoterpenes (0, 5, 10, 15, 20 µg/mL) in water are applied to 20 g of each soil sample. Calculate the percent of recovery.

Precautions

(i) Do not pour dichloromethane into plastic centrifuge tubes.

(ii) Dichloromethane is harmful if swallowed or inhaled and may be harmful by skin contact. It is an eye and skin irritant.

(iii) Use safety glasses and gloves; good ventilation; do not inhale vapor; void contact with skin; avoid contact with eyes; wear suitable protective clothing and suitable gloves.

Experiment 5: Mono and Sesquiterpenes from Soil by SPME and GC-MS Analysis

Solid phase microextraction (SPME) allows analysis of VOCs adsorbed to a fiber sorbent. Terpenes, mainly mono and sesquiterpenes, are frequently found among these VOCs because they are compounds with high vapour pressure at ambient temperature. The application of SPME together with GC-MS analysis is a very useful and versatile tool for qualitative composition studies of VOCs from soil samples.

Materials and equipments required

Sieve (2 mm mesh); 50 mL glass vials; PTFE-line septum; aluminium crimp seals with open centre; soil samples; SPME fiber (CAR-PMDS phase); GC-MS equipment.

Reagents

Sodium chloride salt; distilled water; high quality standard terpenes commercially available.

Procedure

Soil samples are sieved (2 mm mesh) and 20 g of sieved soil is placed in a 50 mL vial. 5 g of NaCl salt and an adequate amount of distilled water are added to cover the soil sample. The vials are kept at 60°C for 30 minutes with simultaneous capture of volatiles in the SPME. The adsorbed components are desorbed by introducing the SPME fiber into the GC-MS injection port.

Observations

The terpenes desorbed are identified by GC-MS using the retention index and library of terpenes (Table 1, row Nos. 3, 13, 14, 17, 18).

10. BIOLOGICAL EFFECT OF TERPENES

Experiment 6: Test for Terpene Phytotoxicity

Biological procedures can be utilized directly on the TLC layer in order to detect substances which have physiological activity. Thus, terpenes from EO are separated by TLC and phytotoxic activity can be detected over the chromatographic bands (Inoue *et al.*, 1992).

Materials and equipments required

Commercial radish or lettuce seeds; filter paper; spatula; volumetric flasks; glass tubes; funnels; glass plates (20 cm × 20 cm); TLC chambers; iodine vapor chamber/UV lamp; GC and GC-MS equipment; rotary evaporator; vortex apparatus.

Reagents

Silica gel 60 G plates; agar; essential oil obtained by steam distillation or organic fractions containing terpenes from a litter extract/leachate; azeotrope chloroform-methanol extract from soil; benzene or chloroform (developing solvent); organic solvent (methanol, hexane, dichloromethane) for elution of the bioactive fraction.

Procedure

A TLC of the EO (or organic fraction) is performed on silica gel glass plates with a thickness of 0.25 mm using benzene or chloroform as developing solvent. Then, the TLC plate is covered with agar (0.5% v/v, cooled at 40°C) and the seeds are sown, in a row perpendicular to the bands, over the chromatogram. The plates are incubated for one to three days in a moisture saturated chamber at 24-26°C. The identified bioactive fraction is scrapped off from a matching chromatogram with a spatula into a conveniently sized glass tube. An appropriate solvent is added for the elution of terpenes (MeOH for oxygenated terpenes and n-hexane for hydrocarbons). Then, the fraction is thoroughly mixed in a vortex apparatus and filtered. The solvent volume is reduced in a rotary evaporator and the fraction analyzed by GC-MS (Table 1, row No. 19).

Calculations

Seed germination and growth are examined visually. The *Rf* of the inhibitory band is documented on a chromatogram not treated with agar by detection in short or long wave UV light, or iodine vapors.

Precautions

Silica gel is harmful if inhaled or absorbed through the skin. Wear appropriate protective gloves, clothing and goggles (to prevent skin and eye exposure) and respiratory protective mask.

Experiment 7: Monoterpenes in Vapor Phase

Volatile monoterpenes from aromatic plants inhibit seed germination and seedling growth of other plant species (Panasiuk *et al.*, 1986; Bradow and Connick, 1988; Tarayre *et al*,. 1995; Abrahim *et al.*, 2000). These compounds are thought to be important allelopathic agents in hot, dry climates where they act in the vapor phase. This experiment evaluates in a headspace system the effect of terpenes on seedlings growth. The low boiling point of terpenes allows their volatilization saturating the surrounding atmosphere of the target plants in a closed system. The terpenes in the vapor phase can be recovered by a gastight syringe and then analyzed by GC.

Materials and equipments required

Filter paper; 5 mL glass flask; maize seeds; 3 L glass flask; gastight syringe; GC and GC-MS equipment.

Reagents

High quality standard terpenes commercially available; distilled water.

Procedure

Seeds of maize (test plant) are rolled in the upper 3 cm of 20 cm length paper-towel scrolls. The individual scrolls are moistened with 25 mL of distilled water and placed upright in 3 L flasks. This method allows straight growth of seedling's root and shoot. The scrolled seeds are germinated for 3 days at $27 \pm 1°C$ in darkness. Then, seedlings with uniform root length are harvested and transferred for bioassay.

Bioassays: Thirty maize seedlings, selected as indicated above, are placed in 3 L desiccator flasks on Whatman # 1 filter paper containing 15 mL of distilled water (Fig. 6). In the central area of the 3 L flask a 5 mL glass beaker with the terpene (volatile source) is placed. The quantity of terpene in the glass beaker should be enough to saturate the headspace (2 mL of liquid or 2 g of solid terpene is appropriate to keep the headspace saturated). The flasks are placed in darkness at $27 \pm 1°C$ and at a pressure of 1 Atm. Then, the plants are harvested at 24, 48 and 96 hours and the roots and/or shoot are dissected, and the different experimental parameters are measured.

Volatile compounds analysis: Volatiles from headspace of glass desiccators are trapped by using a 10 mL gastight syringe (Fig. 6) and then analyzed by GC and GC-MS (Table 1, row Nos. 9, 10).

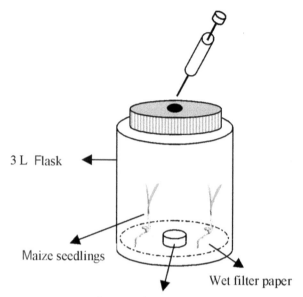

3 L Flask

Maize seedlings

Wet filter paper

Source of terpenes in vapor
phase: 5 ml flask with terpenes in
solid or liquid phase

Figure 6   Static headspace system for collection of an air sample containing exogenous monoterpenes as phytotoxic agents against maize seedlings, in a closed flask. A gastight syringe is inserted through a septum at the flask tape and the air sample is aspirated. Then, the air sample is injected in a GC or a GC-MS.

Calculations
Root and shoot lengths and dry weight are measured. The mean of the 30 seedlings' lengths of each glass flask constitutes one replication.

Statistical analysis
The data are analyzed using ANOVA test and by Duncan's multiple range test.

Observations
(i) Humidity and heat airspace increases with seedlings density and time period of treatment. Hence, the use of 30 seedlings per flask and up to 96 hours of treatment is recommended.
(ii) To study the effect of concentrations lower than saturation, the vapor pressure of terpenes must be considered.

Experiment 8: Biotransformation of Terpenes by Target Seedlings
Principle
Plants are able to detoxify harmful molecules by absorption and conversion into biologically non-active compounds. The incorporated terpenes could be modified either by oxidation, hydroxylation or dealkylation. The objective of this experiment

is to study the biotransformation of terpenes by maize seedlings exposed to the vapor phase of terpenes by GC-MS.

Materials and equipments required

Sterile filter paper; Tween 20; 5 mL glass flask; maize seeds; 3 L glass flask; 200 mL glass vials with a PTFE-line septum; aluminium crimp seals with open centre; SPME fiber; GC and GC-MS equipment; autoclave.

Reagents

Sterile distilled water; 70% ethanol; hypochlorite; high quality standard terpenes commercially available.

Procedure

Seeds are sterilized with 70% ethanol for 10 seconds followed by 1% hypochlorite containing 0.1% Tween 20 for 10 minutes. Then, they are rinsed three times with sterile distilled water. After the seeds are disinfected, they are rolled in the upper 3 cm of 20 cm length autoclaved paper-towel scrolls. The individual scrolls are moistened with 25 mL of sterile distilled water and placed upright in 3 L sterile flasks. The scrolled seeds are germinated for 3 days at 27 ± 1°C in darkness. At the end of this period, seedlings with uniform root lengths are selected and transferred for bioassay.

Bioassay: Proceed as indicated in Experiment 7 but under sterile conditions. Then, seedlings are harvested and placed on 200 mL glass vials. These vials are heated at 30 °C in a water bath for 30 minutes with simultaneous capture of volatiles in a SPME fiber inserted throughout the PTFE line septum. The volatiles are then analyzed by GC and GC-MS (Table 1, row Nos. 5, 14, 17, 18).

Observations

Identification of the headspace volatiles is carried out by using the retention index and library of terpenes (NIST, WILEY, ADAMS).

SUGGESTED READINGS

Abrahim, D., Braguini, W., Kelmer-Bracht, A. and Ishii-iwamoto, E. (2000). Effects of four monoterpenes on germination, primary root growth, and mitochondrial respiration of maize. *Journal of Chemical Ecology* **26**: 611-624.

An, M., Pratley, J. and Haig, T. (2001). Phytotoxicity of Vulpia Residues: IV. Dynamics of allelochemicals during decompostion of Vulpia residues and their corresponding phytotoxicity. *Journal of Chemical Ecology* **27**: 395-409.

Angellini, L., Carpanece, G., Cioni, P., Morelli, I., Macchia, M. and Flamini, G. (2003). Essential oils from Mediterranean Lamiaceae as with germination inhibitors. *Journal of Agricultural and Food Chemistry* **51**: 6158-6164.

Azevedo, N., Campos, I., Ferreira, H., Portes, T., Santos, S., Seraphin, J., Paula, J. and Ferri, P. 2001. Chemical variability in the essential oil of *Hyptis suaveolens*. *Phytochemistry* **57**: 733-736.

Banchio, E., Zygadlo, J. and Valladares, G. (2005). Effects of mechanical wounding on essential oil composition and emission of volatiles from *Minthostachys mollis*. *Journal of Chemical Ecology* **31**: 719-727.

Beesley, T. and Scott, R. (1998). *Chiral Chromatography*. John Wiley and Sons, London.

Betts, T. (1994). Use of esterified and unesterified dipentylated γ-, β- and α-cyclodextrins as gas chromatographic stationary phases to indicate the structure of monoterpenoid constituents of volatile oils. *Journal of Chromatography A* **672**: 254-260.

Bianchi, A. and Varney, M. (1998). Volatile organic compounds in the surface waters of a British estuary. Part 1. Occurrence, distribution and variation. *Water Research* **32**: 352-370.

Bowman, J., Braxton, M., Churchill, M., Hellie, J., Starrett, S., Causby, G., Ellis, D., Ensley, S., Maness, S., Meyer, C., Sellers, J., Hua, Y., Woosley, R. and Butcher, D. (1997). Extraction Method for the Isolation of Terpenes from Plant Tissue and Subsequent Determination by Gas Chromatography. *Microchemical Journal* **56**: 10-18.

Bradow, J. and Connick, Jr. W. (1988). Volatile methyl ketone seed-germination inhibitors from *Amaranthus palmeri* S. Wats. residues. *Journal of Chemical Ecology* **14**: 1617-1631.

Cai, J., Lin, P., Zhu, X. and Su, Q. (2006). Comparative analysis of clary sage (*S. sclarea* L.) oil volatiles by GC-FTIR and GC-MS. *Food Chemistry* **99**: 401-407.

Demyttenaere, J. and De Pooter, H. (1998). Biotransformation of citral and nerol by spores of *Penicillium digitatum*. *Flavour and Fragance Journal* **13**: 173-176.

Dudai, N., Weinberg, Z.G., Larkov, O., Ravid, U., Ashbell, G. and Putievsky, E. (2001). Changes in essential oil during enzyme-assisted ensiling of lemongrass (*Cymbopogon citratus* Stapf.) and lemon eucalyptus (*Eucalyptus citriodora* Hook). *Journal of Agricultural and Food Chemistry* **49**: 2262-2266.

Dudareva, N., Pichersky, E. and Gershenzon, J. (2004). Biochemistry of Plant Volatiles. *Plant Physiology* **135**: 1893-1902.

Dunkel, F. and Sears, L. (1998). Fumigant properties of physical preparations from mountain big sagebrush, *Artemisia tridentata* Nutt. ssp. *vaseyana* (Rydb.) beetle, for stored grain insects. *Journal of Stored Products Research* **34**: 307–321.

Ferary, S., Augera, J. and Touchkb, A. (1996). Trace identification of plant substances by combining gas chromatography-mass spectrometry and direct deposition gas chromatography-Fourier transform infrared spectrometry. *Talanta* **43**: 349-357.

Fischer, G., Schwalbe, R., Möller, M., Ostrowski, R. and Dott, W. (1999). Specific production of microbial volatile organic compounds species- (Mvoc) by airborne fungi from a compost facility. *Chemosphere* **39**: 795-810.

Fisher, N. (1986). The function of mono and sesquiterpenes as plant germination and growth regulators. In: *The Science of Allelopathy* (Eds., A.R. Putnam and C.S. Tang). John Wiley & Sons, New York, USA.

Fisher, N. (1991). Plant terpenoids as allelopathic agents. In: *Ecological Chemistry and Biochemistry of Plant Terpenoids* (Eds., J.B. Harborne and F.A. Tomas-Barberan). Clarendon Press, Oxford, UK.

Fuentes, J., Lerdau, M., Atkinson, R., Baldocchi, D., Bottenheim, J., Ciccioli, P., Lamb, B., Geron, C., Gu, L., Guenther, A., Sharkey, T. and Stockwell, W. (2000). Biogenic hydrocarbons in the atmospheric boundary layer. A review. *Bulletin of the American Meteorological Society* **10**: 1537-1575.

Gil, A., Ghersa, C. and Leicach, S. (2000). Essential oil yield and composition of *Tagetes minuta* accessions from Argentina. *Biochemical Systematics and Ecology* **28**: 261-274.

Graven, E., Webber, L., Benians, G., Venter, M. and Gardner, J. (1991). Effect of soil type and nutrient status on the yield and composition of *Tagetes* oil (*Tagetes minuta* L.). *Journal of Essential Oil Research* **3**: 303-307.

Hamm, S., Lesellie, E., Bleton, J. and Tchapla, A. (2003). Optimization of headspace solid phase microextraction for gas chromatography/mass spectrometry analysis of widely different volatility and polarity terpenoids in olibanum. *Journal of Chromatography A* **1018**: 73-83.

Hollender, J., Sandner, F., Möller, M. and Dott, W. (2002). Sensitive indoor air monitoring of monoterpenes using different adsorbents and thermal desorption gas chromatography with mass-selective detection. *Journal of Chromatography A* **962**: 175-181.

Hudaib, M., Speroni, E., Di Pietra, A. and Cavrini, V. 2002. GC/MS evaluation of thyme (*Thymus vulgaris*) oil composition and variations during the vegetative cycle. *Journal of Pharmaceutical and Biomedical Analysis* **29**: 691-700.

Inderjit and Duke, S. (2003). Ecophysiological aspects of allelopathy. *Planta* **217**: 529-539.

Inderjit and Nilsen, E. (2003). Bioassays and field studies for allelopathy in terrestrial plants: progress and problems. *Critical Reviews in Plant Sciences* **22**: 221-238.

Inoue, M., Nishimura, H., Li, H. and Mizutani, J. (1992). Allelochemicals from *Polygonum sachalinenes* Fr. Schm. (Polygonaceae). *Journal of Chemical Ecology* **18**: 1833-1840.

Jakobsen, H. (1997). The preisolation phase of in situ headspace analysis: methods and perspectives. In: *Plant Volatile Analysis. Methods of Plant Analysis* (Eds., H.F. Linshens and J.F. Jackson). Vol. 19. Springer, Berlin.

Korytar, P., Stee, L., Leonards, P., de Boer, J. and Brinkman, U. (2003). Attempt to unravel the composition of toxaphene by comprehensive two-dimensional gas chromatography with selective detection. *Journal of Chromatography A* **994**: 179-189.

Larsen, T. (1998). Volatile Flavour Production by *Penicillium caseifulvum. International Dairy Journal* **8**: 883-887.

Lee, B., Choi, W., Lee, S. and Park, B. (2001). Fumigant toxicity of essential oils and their constituent compounds towards the rice weevil, *Sitophilus oryzae* (L.). *Crop Protection* **20**: 317-320.

Lockwood, G. (2001). Techniques for gas chromatography of volatile terpenoids from a range of matrices. *Journal of Chromatography A* **936**: 23-31.

Mac Tavish, H., Davies, N. and Menary, R. (2000). Emission of volatiles from brown Boronia flowers: some comparative observations. *Annals of Botany* **86**: 347-354.

Manninen, A., Pasanen, P. and Holopainen, J. (2002). Comparing the VOC emissions between air-dried and heat-treated Scots pine wood. *Atmospheric Environment* **36**: 1763-1768.

Marsili, R. (2002). *Flavor, Fragrance and Odour Analysis.* Marcel Dekker, New York, USA.

Ojala, M., Ketola, R., Mansikka, T., Kotiaho, T. and Kostiainen, R. (1999). Determination of mono- and sesquiterpenes in water samples by membrane inlet mass spectrometry and static headspace gas chromatography. *Talanta* **49**: 179-188.

Orio, O., López, A., Herrero, E., Pérez, C. and Anunziata, O. (1986). *Gas Phase Cromatography.* Edigem S.A. Bs. As. Argentina. (In Spanish).

Padhy, P. and Varshney, C. (2005). Emission of volatile organic compounds (VOC) from tropical plant species in India. *Chemosphere* **59**: 1643-1653.

Panasiuk, O., Bills, D. and Leather, G. (1986). Allelopathic influence of *Sorghum bicolor* on weeds during germination and early development of seedlings. *Journal of Chemical Ecology* **12**: 1533-1543.

Paolini, J., Costa, J. and Bernardini, A. (2005). Analysis of the essential oil from aerial parts of *Eupatorium cannabinum* subsp. *corsicum* (L.) by gas chromatography with electron impact and chemical ionization mass spectrometry. *Journal of Chromatography A* **1076**: 170-178.

Peña-Alvarez, A., Díaz, L., Medina, A., Labastida, C., Capella, S. and Vera, L. (2004). Characterization of three Agave species by gas chromatography and solid-phase microextraction–gas chromatography–mass spectrometry. *Journal of Chromatography A* **1027**: 131–136.

Rohloff, J. and Bones, A. (2005). Volatile profiling of *Arabidopsis thaliana* – Putative olfactory compounds in plant communication. *Phytochemistry* **66**: 1941-1955.

Röse, U., Manukian, A., Heath, R. and Tumlinson, J. (1996). Volatile semiochemicals released from undamaged cotton leaves. *Plant Physiology* **111**: 487-495.

Ruiz del Castillo, M., Blanch, G. and Herraiz, M. (2004). Natural variability of the enantiomeric composition of bioactive chiral terpenes in *Mentha piperita*. *Journal of Chromatography A* **1054**: 87-93.

Sánchez-Palomo, E., Díaz-Maroto, M. and Pérez-Coello, M. (2005). Rapid determination of volatile compounds in grapes by HS-SPME coupled with GC–MS. *Talanta* **66**: 1152-1157.

Sangwan, N., Farooqi, A., Shabih, F. and Sangwan, R. (2001). Regulation of essential oil production in plants. *Plant Growth Regulation* **24**: 3-21.

Schurig, V. (2001). Separation of enantiomers by gas chromatography. *Journal of Chromatography A* **906**: 275-299.

Schurig, V. (2002). Chiral separations using gas chromatography. *Trends in Analytical Chemistry* **21**: 70-71.

Tarayre, M., Thompson, J., Escarré, J. and Linhart, Y. (1995). Intra-specific variation in the inhibitory effects of *Thymus vulgaris* (Labiatae) monoterpenes on seed germination. *Oecologia* **101**: 110-118.

Tholl, D., Boland, W., Hansel, A., Loreto, F., Rose, U. and Schnitzler, J.P. (2006). Practical approaches to plant volatile analysis. *The Plant Journal* **45**: 540-560.

Tolnai, B., Lavay, H., Möller, D., Prümke, H., Becker, H. and Dostler, M. (2000). Combination of canister and solid adsorbent sampling techniques for determination of volatile organic hydrocarbons. *Microchemical Journal* **67**: 163-169.

Vaughn, S. and Spencer, G. (1991). Volatile monoterpenes as potential parent structures for new herbicides. *Weed Science* **41**: 114-119.

Velasco-Negueruela, A., Pérez-Alonso, M., Pérez de Paz, P., Palá-Paúl, J. and Sanz, J. (2003). Analysis by gas chromatography–mass spectrometry of the essential oils from the aerial parts of *Rutheopsis herbanica* (Bolle) Hans. & Kunk., gathered in Fuerteventura (Canary Islands). *Journal of Chromatography A* **984**: 159-162.

Vereen, D., McCall, J. and Butcher, D. (2000). Solid phase microextraction for the determination of volatile organics in the foliage of Fraser fir (*Abies fraseri*). *Microchemical Journal* **65**: 269-276.

Walker, T., Bais, H., Grotewold, E. and Vivanco, J. (2001). Root exudation and rhizosphere biology. *Plant Physiology* **132**: 44-51.

Weidenhamer, J., Macías, F., Fisher, N. and Williamson, B. (1993). Just how insoluble are monoterpenes? *Journal of Chemical Ecology* **19**: 1799-1807.

Zhao, C., Liang, Y., Fang, H. and Li, X. (2005). Temperature-programmed retention indices for gas chromatography–mass spectroscopy analysis of plant essential oils. *Journal of Chromatography A* **1096**: 76-85.

Zunino, M. and Zygadlo, J. (2005). Changes in composition of phospholipid fatty acids and sterols of maize roots in response to monoterpenes. *Journal of Chemical Ecology* **31**: 1269-1283.

Zunino, M., López, M. and Zygadlo, J. (2005). Tagetone induces changes in lipid composition of *Panicum mileaceaum* roots. *Journal of Essential Oil Research* **8**: 239-249.

9

GC and GC-MS for Non-volatile Compounds

Marta S. Maier

1. INTRODUCTION

Gas chromatography (GC) is an analytical technique widely applied to the analysis of mixtures of organic compounds (Rouessac and Rouessac, 2003). The partition of the mixture takes place between a stationary liquid phase and a mobile gas phase. As analysis in the gaseous phase requires volatile derivatives, most of the GC procedures on non-volatile compounds use pre-chromatographic derivatization of the samples (Blau and Halket, 1993). GC is commonly used for qualitative analysis by comparing the retention times of the components of a mixture with those of standards. In those cases where standards are not available, the use of GC coupled to a mass spectrometer (GC-MS) is the method of choice. The mass spectrometer provides the mass spectrum of each of the compounds of the mixture. Interpretation of the spectra and comparison with mass spectra libraries allow identification of each analyte in the mixture.

Fatty acids, sterols and monosaccharides are among natural products usually analyzed by GC and GC-MS. In the next sections the derivatization methods for obtaining volatile derivatives of each of these natural product classes is presented, as well as their characteristic mass fragmentations.

Author's address: Departamento de Química Orgánica, Facultad de Ciencias Exactas y Naturales, Universidad de Buenos Aires, Pabellón 2, Ciudad Universitaria (1428), Ciudad Autónoma de Buenos Aires, Argentina.
E-mail: maier@qo.fcen.uba.ar

2. ANALYSIS OF FATTY ACIDS BY GC AND GC-MS

Fatty acids are structural components of lipids and their compositions and biological functions vary widely in tissues. GC and GC-MS are the most common techniques for routine analysis of fatty acids in lipids. Due to their low volatility and their tendency to tail when they interact with the stationary liquid phase, fatty acids must be derivatized prior to their analysis by GC and GC-MS. As esters are non-polar and more volatile derivatives, esterification is the first choice for derivatization for GC, particularly the conversion into fatty acids methyl esters (FAMEs). Free fatty acids can be obtained by saponification of the lipid (refer Experiment 1 in this chapter) in an alkaline medium, and further conversion into the FAMEs by reaction with diazomethane (refer Experiment 2), alcoholic HCl (refer Experiment 3), boron trifluoride/methanol (refer Experiment 4), or by transesterification in acid (refer Experiment 5) or alkaline media (refer Experiment 6).

Fatty acids can also be converted into their silyl esters for analysis by GC and GC-MS. Treatment of the free fatty acids with N,N-bis(trimethylsilyl) trifluoroacetamide (BSTFA) or bis(trimethylsilyl)-acetamide (BSA) is a quick and easy way of esterifying an acid in high yield. BSTFA is usually preferred for GC and GC-MS because excess reagent and reaction products are very volatile. If an alcohol group is present in the molecule it will be also silylated. It must be taken into account that silylation reagents and their derivatives are sensitive to the hydrolytic effects of moisture (Blau and Halket, 1993). Therefore, silyl derivatives must be analyzed immediately after their synthesis. If more stable derivatives are desired, FAMEs are the best choice.

Fatty acid methyl esters give the molecular ion peak as well as an easily recognizable peak at m/z [M–31] due to the formation of $[R–C=O]^+$. The most characteristic peak in the mass spectrum results from the McLafferty rearrangement (Silberstein and Webster, 1997). Thus, a methyl ester of an aliphatic acid unbranched at the α carbon gives a strong peak at m/z 74, as shown in Fig. 1.

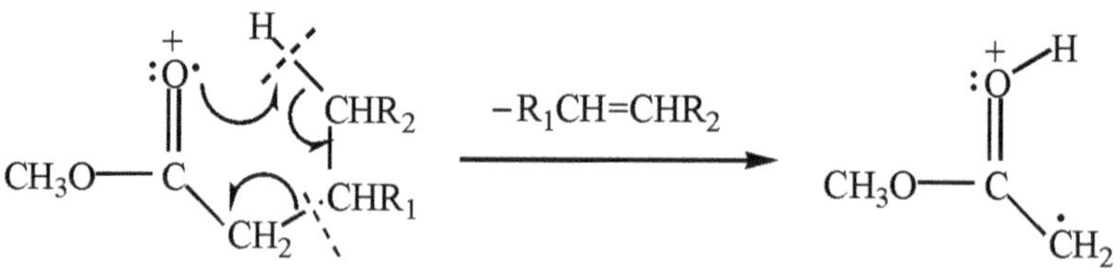

$$m/z = 74$$

Figure 1 Formation of molecular ion from a methyl ester of an aliphatic acid unbranched at α-carbon.

Cleavage at each C–C bond gives an alkyl ion (m/z 29, 43, 57) and an oxygen-containing ion $C_nH_{2n-1}O_2^+$ (m/z 59, 73, 87). Thus, the fragmentation pattern is characterized by clusters of peaks that are 14 mass units apart. In aliphatic FAMEs the largest peak in each cluster represents a C_nH_{2n+1} fragment accompanied by C_nH_{2n} and C_nH_{2n-1} fragments. The most abundant fragments are at C_3 and C_4. In unsaturated FAMEs, the C_nH_{2n-1} and C_nH_{2n} peaks are more intense than the C_nH_{2n+1} peaks.

Experiment 1: Saponification of Acylglycerides

Triacylglycerides, together with minor amounts of mono- and diglycerides, are the most important components of oils and fats. When these are reacted with a base, usually potassium hydroxide in an aqueous/alcoholic medium, a mixture of salts of fatty acids (soap) and glycerol is obtained (Fig. 2). This reaction is called saponification. Acidification and extraction of the mixture of fatty acids in a non-polar solvent allows separation of the fatty acids from the glycerol which remains in the aqueous phase.

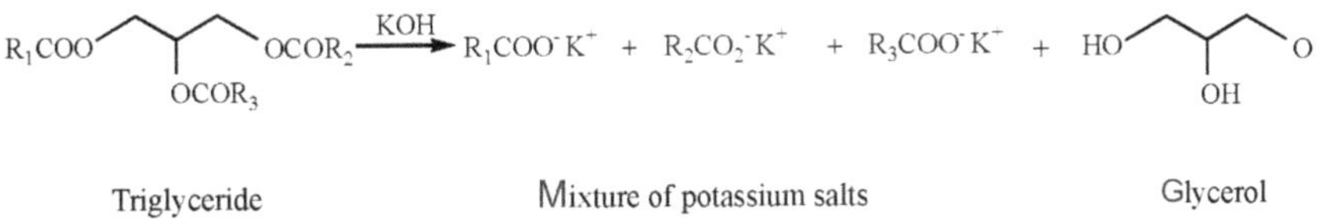

| Triglyceride | Mixture of potassium salts | Glycerol |

Figure 2 Saponification reaction.

Materials and equipments required

Fat, oil or material containing them; 10% KOH in ethanol/H_2O (1:1); 2N HCl; diethyl ether, analytical grade; 2-mL screw top glass tubes and caps.

Procedure

  (i) Weigh 25 mg of the fat or oil in a 2-mL screw top glass tube and add 1 mL of a 10% KOH in EtOH/H_2O (1:1).
 (ii) Heat at 60°C for 2 hours in an oven and cool to room temperature.
(iii) Acidify with 2 N HCl and extract with diethyl ether (3 × 0.75 mL).
(iv) Separate the diethyl ether extract and transfer to a glass tube. Evaporate to dryness under a nitrogen stream.

Experiment 2: Synthesis of Fatty Acids Methyl Esters (FAMEs) by Reaction with Diazomethane

This is one of the quickest esterification reactions to obtain a FAME in high yield, mild conditions and with minimal side reactions. The drawback of the reaction is the preparation of the reagent, its carcinogenity and instability. The high toxicity of diazomethane requires that the sample should be handled in a well-ventilated hood and restricted to microscale. Diazomethane is a yellow gas usually used as an ethereal solution. It is prepared by shaking *N*-nitrosomethylurea at 5°C with aqueous 40% KOH solution and diethyl ether (Vogel, 1963). Separation of the diethyl ether layer and drying with KOH pellets renders the yellow solution ready for use. The esterification reaction is driven by the elimination of gaseous nitrogen as the only by-product (Fig. 3).

$$RCO_2H \quad + \quad CH_2N_2 \quad \longrightarrow \quad RCO_2CH_3 \quad + \quad N_2$$

Figure 3 Esterification reaction with diazomethane.

Materials and equipments required

Mixture of free fatty acids; methanol containing not more than 0.5% water; diethyl ether, analytical grade; diazomethane solution in diethyl ether; 2-mL screw top glass tubes and caps.

Procedure

(i) Weigh 20 mg of the fatty acid mixture in a 2-mL screw top glass tube and add 1 mL of anhydrous diethyl ether containing 10% of methanol.

(ii) Add diazomethane solution in diethyl ether with a Pasteur pipette until the yellow colour just persists. Put the cap on the tube and screw tightly. Leave for 3 hours at room temperature.

(iii) Evaporate the excess of diazomethane and the diethyl ether/MeOH under a nitrogen stream in a hood.

Experiment 3: Synthesis of FAMEs by Reaction with Alcoholic HCl

The esterification of a carboxylic acid with an alcohol requires the use of an acid catalyst. Hydrogen chloride has been the favoured catalyst because of its acid strength and because it is easily removed. Methanolic HCl is readily prepared by dropwise addition of the amount of acetyl chloride required for preparing the desired HCl concentration, to the stirred alcohol, with cooling. The small amount of methyl acetate formed is evaporated at the end of the esterification process along with the excess of methanol.

Materials and equipments required

Mixture of free fatty acids; HCl 3 M in methanol; 2-mL screw top glass tubes and caps.

Procedure

(i) Weigh 20 mg of the fatty acid mixture in a 2-mL screw top glass tube and add 1 mL of HCl 3M in methanol.

(ii) Heat at 60°C in an oven for 2 hours.

(iii) Evaporate the methanol under a nitrogen stream.

Experiment 4: Synthesis of FAMEs by Reaction with Boron Trifluoride

The boron trifluoride (BF_3) catalyst is commercially available as a 14% solution in methanol for preparing methyl esters. The advantage of BF_3-catalyzed esterification is that the reaction is complete after a few minutes on a boiling water bath.

Materials and equipments required

Mixture of free fatty acids; a 14% F_3B-MeOH solution; *n*-hexane, analytical grade; 2-mL screw top glass tubes and caps.

Procedure

(i) Weigh 20 mg of the fatty acid mixture in a 2 mL-screw top glass tube and add 1 mL of BF_3-MeOH.

(ii) Heat on a boiling water bath for two minutes.

(iii) Extract the FAMEs into *n*-hexane (3 × 0.5 mL) and transfer to a glass tube. Evaporate to dryness under a nitrogen stream.

Experiment 5. Synthesis of FAMEs by Transmethylation of Acylglycerides with Methanolic HCl

Transesterification or ester interchange occurs when an ester is heated with an alcohol in the presence of acids or bases (Fig. 4). This technique is very useful to obtain FAMEs by reaction of acylglycerids with excess of methanol. It has the advantage of avoiding losses due to incomplete hydrolysis or side reactions (i.e. polymerization) or the alteration of unsaturated fatty acids, as in the saponification of acylglycerides, and further esterification by one of the methods described above. Methanolic HCl is the preferred acid catalyst although boron trifluoride, H_2SO_4 and perchloric acid have been used.

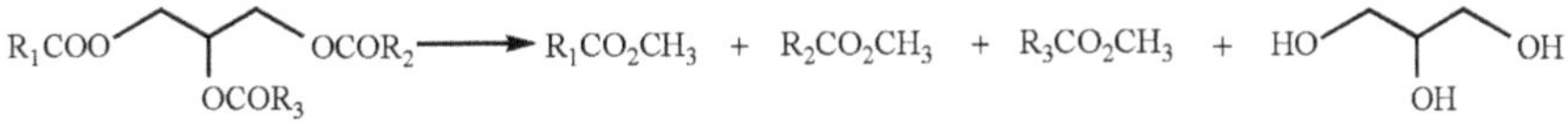

R_1COO ... $OCOR_2$ ⟶ $R_1CO_2CH_3$ + $R_2CO_2CH_3$ + $R_3CO_2CH_3$ + HO ... OH
$OCOR_3$ ... OH

Figure 4 Transesterification reaction.

Materials and equipments required

Fat, oil or material containing them; HCl 2% in methanol; *n*-hexane, analytical grade; 2-mL screw top glass tubes and caps.

Procedure

(i) Weigh 10 mg of the fat or oil in a 2-mL screw top glass tube and add 1 mL of 2% HCl/MeOH.

(ii) Put the cap on the tube, screw tightly and heat at 60°C for 2 hours.

(iii) After cooling, add 0.5 mL of 0.9% NaCl solution and extract with *n*-hexane (2 × 0.75 mL).

(iv) Separate the *n*-hexane layers and transfer to a glass tube. Evaporate to dryness under a nitrogen stream.

Experiment 6: Synthesis of FAMEs by Transmethylation of Acylglycerides with Methanolic KOH

The transmethylation reaction can be achieved in alkaline media at room temperature by a very simple procedure.

Materials and equipments required

Fat, oil or material containing them; a 2 N solution of potassium hydroxide in methanol; *n*-hexane, analytical grade; 2-mL screw top glass tubes and caps.

Procedure

(i) Weigh 20 mg of the fat or oil in a 2-mL screw top glass tube and dissolve in 1 mL of *n*-hexane.

(ii) Add 0.1 mL of a 2 N solution of potassium hydroxide in methanol.

(iii) Put the cap on the tube, screw tightly and shake the tube vigorously for 15 seconds.

(iv) Leave to stratify until the upper layer becomes clear.

(v) Separate the *n*-hexane layer and transfer to a glass tube. Evaporate to dryness under a nitrogen stream.

Experiment 7: Identification and Quantification of FAMEs by Gas-Liquid Chromatography (GC)

Fatty acids are first converted into their volatile FAMEs and these are identified by GC by comparing their retention times with a set of standard esters. Quantification of the components in the mixture can be performed by the method of area normalization or by the use of an internal standard (Rouessac and Rouessac, 2003). The latter is more accurate because it overcomes variable evaporation losses from the syringe needle. The choice of the internal standard is crucial: it must be as similar to the analytes as possible and be resolved in the chromatogram from all the other components of the mixture.

Materials and equipments required

FAMEs prepared as described previously; standards: authentic samples of methyl esters of various fatty acids; chloroform, analytical grade; GC: GC fitted with a capillary column (30 m × 0.25 mm i.d.) packed with crosslinked 5% phenyl methyl silicone. Detector used is flame ionization detector; carrier gas: nitrogen

Procedure

(i) Operate GC as indicated in the operator manual.

(ii) Dissolve the mixture of FAMEs to be analyzed as well as the standards in chloroform.

(iii) Inject an aliquot (1–2 µl) into preconditioned GC in which the temperature of injection port is 250°C and the detector temperature is 290°C. Temperature programme: 1 minute of isothermal at 100°C and then 100 to 290°C at 10°C min^{-1} followed by a 10-minutes hold at 290°C.

(iv) After all the components of the sample have been eluted, inject the standard methyl esters separately and compare their retention times with those of the components of the sample.

Experiment 8: Determination of the Position of Double Bonds in FAMEs by Synthesis of their DMDS (Dimethyl Disulfide) Derivatives and Analysis by GC-MS

Methyl ester derivatives of unsaturated fatty acids have proved to be unsuitable for identification by mass spectrometry, since double bonds migrate when ionized. To overcome this problem, the double bond has to be "fixed" in the structure, generally by preparing derivatives suitable for GC-MS analysis, as for example the dimethyl disulfide (DMDS) derivative shown in Fig. 5 (Vincentini *et al.*, 1987).

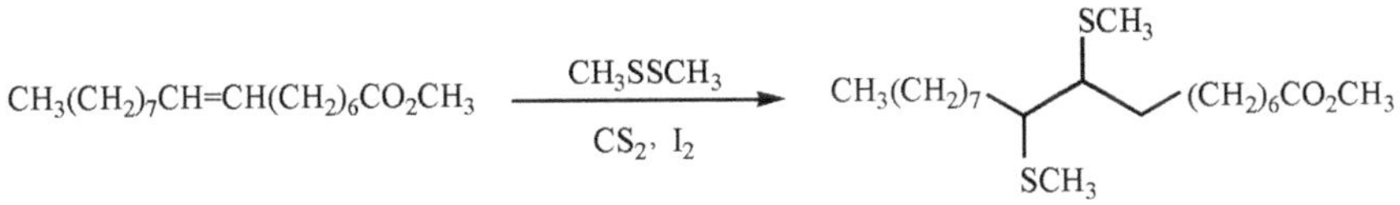

$$CH_3(CH_2)_7CH=CH(CH_2)_6CO_2CH_3 \xrightarrow[\text{CS}_2,\ I_2]{\text{CH}_3\text{SSCH}_3} CH_3(CH_2)_7 \overset{\text{SCH}_3}{\underset{\text{SCH}_3}{\diagup\!\!\diagdown}} (CH_2)_6CO_2CH_3$$

Figure 5 Synthesis of a methyl ester derivative from an unsaturated fatty acid.

Analysis of the fragment ion peaks in the mass spectrum due to cleavage of the simple bond between the carbons attached to the methyl sulfide groups allows determination of the double bond position in the fatty acid chain.

Materials and equipments required
FAMEs; carbon disulfide, analytical grade; dimethyl disulfide, analytical grade; iodine; aqueous 5% $Na_2S_2O_3$ solution; cyclohexane, analytical grade; chloroform, analytical grade; 2-mL screw top glass tubes and caps; GC-MS: mass spectrometer coupled to a GC fitted with a capillary column (25 m × 0.20 mm i.d.) packed with crosslinked 5% diphenyl 95% dimethyl polysiloxane; carrier gas: helium

Procedure
(i) Weigh 5 mg of the FAMEs in a 2-mL screw top glass tube and dissolve in 0.5 mL of carbon disulfide.
(ii) Add 0.5 mL of dimethyl disulfide and 2.5 mg of iodine.
(iii) Put the cap on the tube, screw tightly and heat at 60°C for 40 hours.
(iv) Quench the reaction mixture with aqueous 5% $Na_2S_2O_3$ solution and extract with 1 mL cyclohexane.
(v) Separate the cyclohexane layer and transfer to a glass tube. Evaporate to dryness under reduced pressure.
(vi) Dissolve in chloroform and inject an aliquot (1–2 μl) into preconditioned GC coupled to the mass spectrometer. Temperature programme: 100 to 280°C at 10°C minute^{-1}.

3. ANALYSIS OF STEROLS BY GC AND GC-MS

Naturally occurring sterols usually have 27 to 30 carbon atoms, a hydroxyl group at C-3 and a side chain of at least 7 carbons at C-17 of the perhydro-cyclopentanophenanthrene ring system (Goad and Akihisa, 1997). Sterols can be differentiated from each other by nuclear variations, such as stereochemistry at C-5 (α or β) and the number and location of double bonds as well as C-24- alkyl substituents and number and position of double bonds in the side chain. Usually a double bond at C-5 is present, as in cholesterol and related phytosterols, but saturated sterols such as cholestanol and coprostanol have also been isolated (Fig. 6). Some marine organisms, such as starfish and sea cucumbers, are rich in Δ^7-sterols and most sponges and coelenterates have extremely complex and diverse mixtures of sterols.

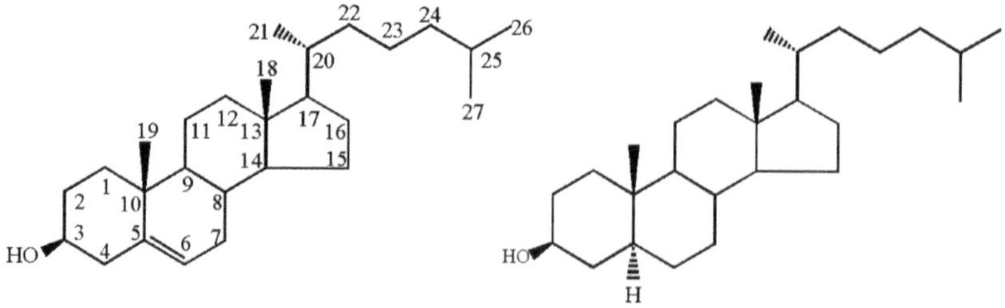

Cholesterol   Cholestanol

Figure 6 Chemical structures of cholesterol and cholestanol.

GC and GC-MS are the analytical techniques most frequently used for analyzing complex sterol mixtures. The various structural features (alkyl group or double bond) in the steroid nucleus and the side chain have individual effects on the retention times of sterols. On the other hand, retention times in a GC column are dependent on the operating conditions and, therefore, they are routinely calculated relative to a reference standard, such as cholesterol, allowing comparison of data tabulated in the same conditions (Lorentz *et al.*, 1989).

The availability of high-temperature stationary phases of different polarities for capillary columns permits the analysis of sterols by GC and GC-MS without previous preparation of derivatives. However, the analysis of derivatives can improve the resolution of sterol mixtures and the stability of some thermally labile compounds such as $\Delta^{5,7}$-sterol dienes. In some cases, the preparation of steryl acetates is part of the purification procedure and it is most practical to analyze them directly by GC without further hydrolysis.

Sterols and their derivatives are usually analyzed by MS with electron impact ionization in the positive mode at an ionizing voltage of 70 eV. The molecular ion is obtained for many free sterols and their derivatives (with the exception of Δ^5- and $\Delta^{5,7}$-steryl acetates) together with characteristic fragmentation ions which are diagnostic for compound identification (Fig. 7). For example, an ion at $[M-18]^+$ is produced for the loss of the 3β-hydroxy group as water in free sterols. This ion is often strong in Δ^5-sterols but it can be quite weak in some sterols with a double bond at a different location in the steroid nucleus.

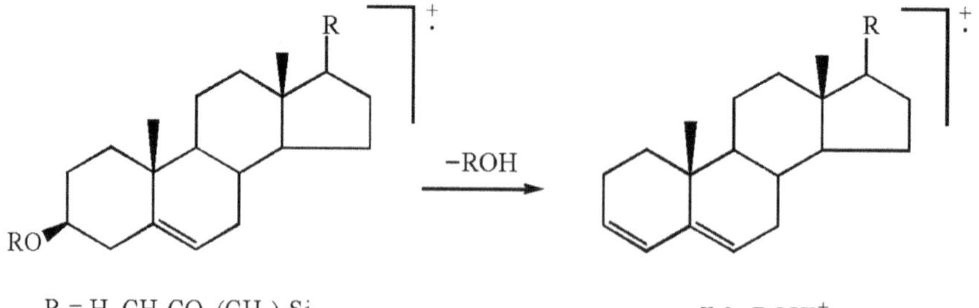

$R = H, CH_3CO, (CH_3)_3Si$ $[M-ROH]^+$

Figure 7 Characteristic fragmentation ions for sterols and their derivatives.

The Δ^5-steryl acetates do not yield a significant molecular ion but instead give a prominent $[M-60]^+$ ion due to the loss of acetic acid. This ion can serve as a diagnostic indicator for the presence of a Δ^5-steryl acetate because most other steryl acetates yield a molecular ion.

Sterol trimethylsilyl ethers show characteristic fragmentations, such as an ion at m/z $[M-90]^+$, for the loss of trimethylsilanol. A common feature in the mass spectra of all sterol derivatives is the presence of an ion or ions for the loss of a methyl group: $[M-15]^+$, $[M-18-15]^+$, $[M-60-15]^+$, $[M-90-15]^+$. Loss of the C-19 methyl group has been shown to be the origin of these ions in some sterols.

An excellent survey of characteristic mass spectral fragmentations of sterol rings with different unsaturation patterns as well as fragmentations of saturated and unsaturated side chains has been referred to by Goad and Akihisa (1997).

Experiment 9: Synthesis and Analysis of Steryl Acetates by GC-MS

Acetylation in sterols involves the introduction of an acetyl group by replacement of a hydrogen atom of a hydroxyl group. Acetylation with acetic anhydride and pyridine is one of the standard acetylation procedures. Pyridine promotes smooth reactions, has great solvent power and acts as an acceptor for the acetic acid formed in the reaction.

Materials and equipments required

Sterol mixture; acetic anhydride, analytical grade; anhydrous pyridine, analytical grade; chloroform, analytical grade; 2-mL screw top glass tubes and caps; GC-MS: mass spectrometer coupled to a GC fitted with a capillary column (30 m × 0.20 mm i.d.) packed with crosslinked 50% phenyl 50% phenylmethyl polysiloxane; carrier gas: helium.

Procedure

(i) Weigh 20 mg of the sterol mixture in a 2-mL screw top glass tube.
(ii) Add 0.5 mL of a (1:1) mixture of acetic anhydride and pyridine.
(iii) Put the cap on the tube, screw tightly and leave at room temperature for 12 hours.
(iv) Evaporate excess acetic anhydride in vacuum.
(v) Dissolve in chloroform and inject an aliquot (1–2 μl) into preconditioned GC coupled to the mass spectrometer. Temperature programme: 80 to 270°C at 12°C min^{-1} and then isothermic at 270°C for 20 minutes.

Experiment 10: Synthesis of Silylated Sterols and Analysis by GC-MS

Silyl derivatives of sterols are formed by the displacement of the active proton in hydroxyl groups, generally by the trimethylsilyl $((CH_3)_3Si)$ group. The trimethylsilylation of commonly occurring sterols containing an unhindered 3β-hydroxyl group is readily achieved by treating the sterol mixture with either N,O-Bis(trimethylsilyl)trifluoroacetamide (BSTFA) containing 1% trimethyl-chlorosilane (TMCS) or equal volumes of hexamethyldisilazane (HMDS) and trimethylchlorosilane (TMCS) (Fig. 8).

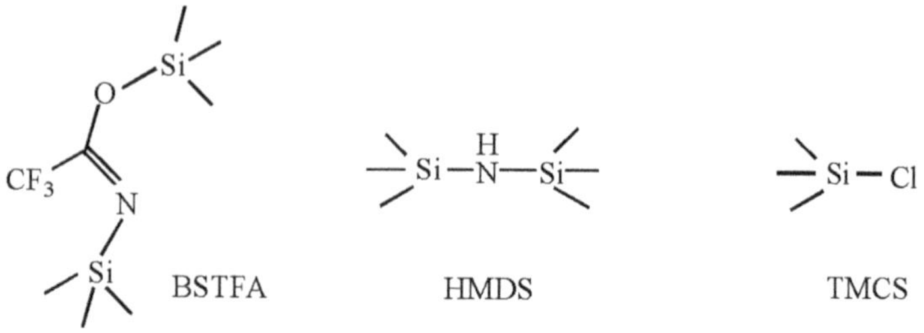

Figure 8   Reagents used for trimethylsilylation of sterols.

Materials and equipments required

Sterol mixture: anhydrous pyridine, analytical grade; hexamethyldisilazane (HMDS); trimethylchlorosilane (TMCS); 2-mL screw top glass tubes and caps; GC-MS: mass spectrometer coupled to a GC fitted with a capillary column (30 m × 0.20 mm i.d.) packed with crosslinked 50% phenyl 50% phenylmethyl polysiloxane; carrier gas: helium.

Procedure

   (i) Weigh 5 mg of the sterol mixture and dissolve in 0.3 mL of pyridine.
   (ii) Add 0.5 mL of HMDS and 0.5 mL of TMCS.
   (iii) Stir vigorously for 5 minutes.
   (iv) Centrifuge and analyze the supernatant by GC-MS under the same conditions as in Experiment 9 (in this chapter).

4. ANALYSIS OF CARBOHYDRATES IN STEROIDAL AND TRITERPENOIDAL GLYCOSIDES (SAPONINS) BY GC AND GC-MS

Steroidal and triterpenoidal saponins are typical metabolites of plant origin but extensive research on echinoderms has demonstrated that these polar compounds are present as major secondary metabolites in starfish and sea cucumbers (Maier and Murray, 2006). Figure 9 shows the structure of Patagonicoside A, the major saponin isolated from the sea cucumber *Psolus patagonicus* as an example of a triterpenoidal tetraglycoside sulfated at C-4 and C-6 of the xylose and glucose units, respectively (Murray *et al.*, 2001).

Due to the complex nature of saponins, structure elucidation requires a combination of spectroscopic techniques (^{1}H- and ^{13}C-NMR and mass spectrometry with soft ionization methods such as FAB, ESI and MALDI) and chemical methods. The structure elucidation of aglycones is most often derived from spectroscopic data since hydrolysis of intact saponins in basic or acid media results in the production of artifacts of the original aglycones due to migration of double bonds, retroaldol and dehydration reactions.

Characterization of the oligosaccharide chain in saponins comprises the identification of each monosaccharide unit and its anomeric configuration, as well as the interglycosidic linkages and their sequence. Identification of each

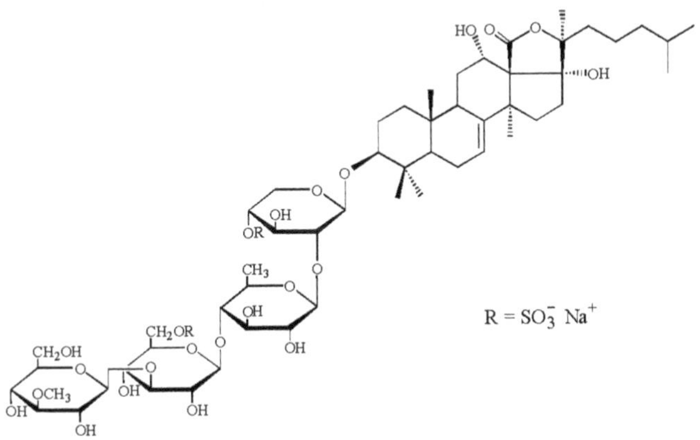

Figure 9 Patagonicoside A, the major saponin isolated from *Psolus patagonicus*.

monosaccharide is easily accomplished by acid hydrolysis of the saponin, derivatization of each monosaccharide and further analysis by GC and comparison with standards. Recent progress in two-dimensional NMR experiments allows the assignment of interglycosidic linkages and monosaccharide sequence in the oligosaccharide chain. The linkage positions can also be obtained by methylation of the saponin followed by acid hydrolysis and GC-MS of the partially methylated alditol acetates derived from the monosaccharides. Determination of the absolute configuration of a monosaccharide can be accomplished by reaction with a quiral amine, reduction to the corresponding 1-amino-1-desoxialditol, acetylation and further analysis by GC and comparison with standards.

Experiment 11: Acid Hydrolysis of Saponins

Saponins are quite fragile molecules. Acid hydrolysis of an intact saponin generally yields an artifact of the aglycone and a mixture of free monosaccharides. The aglycone or its artifact can be removed by extraction with a non-polar solvent. Aqueous solutions of sulfuric acid or trifluoroacetic acid have been used for the hydrolysis. After the reaction is complete, excess of H_2SO_4 is removed by neutralization with $Ba(OH)_2$ and centrifugation of the $BaSO_4$ formed. If trifluoroacetic acid is used, the procedure is simpler because TFA can be removed by evaporation.

Materials and equipments required

Saponin; 2 N trifluoroacetic acid; dichloromethane, analytical grade; 2-mL screw top glass tubes and caps.

Procedure

(i) Weigh 3-10 mg of the saponin in a 2-mL screw top glass tube and add 1 mL of a 2 N TFA solution.

(ii) Put the cap on the tube, screw tightly and heat in an oven at 120°C for 1 hour.

(iii) Cool and evaporate to dryness under vacuum.

(iv) Add 1 mL of a mixture of $H_2O:Cl_2CH_2$ and separate the Cl_2CH_2 layer.

(v) Wash the water layer with Cl_2CH_2 (0.5 mL) and evaporate the aqueous phase to dryness under vacuum.

Experiment 12: Synthesis of Alditol Peracetates of Monosaccharides and Analysis by GC

Sugars are reduced to their alditols and then peracetylated prior to their analysis by GC (Fig. 10). In comparison to other procedures, this methodology is advantageous for aldoses because a single derivative from each sugar is obtained.

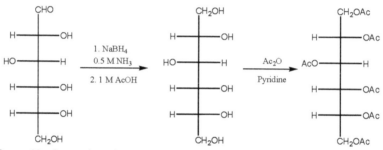

Figure 10 Conversion of a monosaccharide to its peracetylated alditol derivative.

Materials and equipments required

Monosaccharide or monosaccharide mixture; 0.5 M NH_3; $NaBH_4$; 1 M CH_3CO_2H; methanol, analytical grade; acetic anhydride, analytical grade; anhydrous pyridine, analytical grade; $NaHCO_3$ saturated solution; 2-mL screw top glass tubes and caps; chloroform, analytical grade; standards: authentic samples of alditol peracetates of various monosaccharides; GC: GC fitted with a capillary column (30 m × 0.20 mm i.d.) packed with crosslinked 50% phenyl 50% phenylmethyl polysiloxane. Detector used is flame ionization detector; carrier gas: nitrogen.

Procedure

(i) Weigh 5 mg of the monosaccharide or monosaccharide mixture in a 2-mL screw top glass tube.

(ii) Add 0.5 mL of 0.5 M NH_3 and 5 mg of $NaBH_4$.

(iii) Put the cap on the tube, screw tightly and leave at room temperature for 18 hours.

(iv) Acidify with 1 M CH_3CO_2H, add 0.5 mL of MeOH and evaporate under vacuum.

(v) Add 0.5 mL of acetic anhydride and 0.5 mL of pyridine to the residue and heat in an oven at 100°C for 45 minutes.

(vi) Cool and extract with 1 mL of $CHCl_3:H_2O$ (1:1).

(vii) Extract the aqueous phase with $CHCl_3$ and wash the combined chloroform extracts with H_2O (0.5 mL), saturated $NaHCO_3$ solution (0.5 mL) and H_2O (0.5 mL).

(viii) Evaporate to dryness under nitrogen.

(ix) Dissolve in chloroform and inject an aliquot (1-2 µL) into preconditioned GC. Temperature programme: 160 to 210°C at 4°C minute^{-1} and then isothermic at 210°C for 20 minutes.

Experiment 13: Methylation of Saponin, Hydrolysis and Synthesis of Partially Methylated Alditol Peracetates of Monosaccharides

The free hydroxyl groups of the monosaccharide units in the oligosaccharide chain attached to the aglycon can be methylated by Williamson synthesis of ethers. Acid hydrolysis of the methylated saponin following the procedure described in Experiment 11 yields a mixture of monosaccharides methylated at those positions not involved in the glycosidic linkages. These partially methylated monosaccharides are further derivatized as their methylated alditol peracetates and analyzed by GC-MS as described in Experiment 12.

Materials and equipments required

Saponin; anhydrous dimethysulfoxide (DMSO); NaOH; methyl iodide; chloroform, analytical grade; 2-mL screw top glass tubes and caps.

Procedure

(i) Weigh 5 mg of saponin in a 2-mL screw top glass tube and dissolve in 1.3 mL of DMSO.
(ii) Add 63 mg of NaOH and stir at room temperature for 20 minutes.
(iii) Then, add CH_3I and stir for a further 30 minutes.
(iv) Add 4 mL of H_2O and extract with $CHCl_3$ (5 mL).
(v) Separate the chloroform layer, concentrate to 1 mL under a nitrogen stream and analyze by GC-MS.

Experiment 14: Determination of the Absolute Configuration of Monosaccharides by GC

One of the most practical methodologies for determining the absolute configuration of monosaccharides is the GC analysis of 1-[(S)-N-acetyl-(2-hydroxypropylamino)]-1-desoxyalditol acetate derivatives. These are obtained by reaction with a chiral amine to obtain the 1-amino-1-desoxialditol, which is further transformed into the peracetilated alditol (Fig. 11, refer Experiment 12). Cases *et al.* (1995) have found that chiral 1-amino-2-propanol gives the best results to transform a mixture of sugar enantiomers into the corresponding diastereomers. The derivatives are analyzed by GC and comparison with the corresponding standards obtained for the D- and L-sugars.

Materials and equipments required

Monosaccharide or mixture of monosaccharides; (S)-1-amino-2-propanol; glacial CH_3CO_2H; Na[BH$_3$CN]; 3 M aqueous CF_3CO_2H; methanol, analytical grade; 2-mL screw top glass tubes and caps; GC: GC fitted with a capillary column (50 m × 0.20 mm i.d.) packed with crosslinked 5% diphenyl 95% dimethyl polysiloxane. Detector used is flame ionization detector; carrier gas: nitrogen.

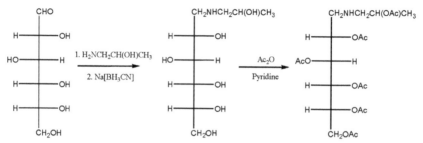

Figure 11 Conversion of a monosaccharide to its 1-[N-acetyl-(2-hydroxypropylamino)]-1-desoxialditol acetate derivative.

Procedure

(i) Weigh 5 mg of monosaccharide in a 2-mL screw top glass tube and add the following solutions: (a) 20 µL of a (1:8) solution of (S)-1-amino-2-propanol in MeOH, (b) 17 µL of a (1:4) solution of glacial acetic acid in MeOH, (c) 13 µL of a 3% solution of Na[BH₃CN] in MeOH.

(ii) Heat at 65°C for 1.5 hours.

(iii) Cool and add dropwise 3 M CF₃CO₂H until the pH drops to 1-2.

(iv) Evaporate the mixture and further co-evaporate with H_2O (2 × 0.5 mL) and MeOH (0.5 mL).

(v) The residue is acetylated as described in Experiment 12.

(vi) Dissolve the derivative in chloroform and inject an aliquot (1-2 µL) into preconditioned GC. Temperature programme: 180 to 220°C at 4°C minute⁻¹ and then 220 to 250°C at 1°C minute⁻¹ and then isothermic at 250°C for 10 minutes.

SUGGESTED READINGS

Blau, K. and Halket, J. (1993). *Handbook of Derivatives for Chromatography*. 2nd edition. John Wiley & Sons, Chichester, England.

Cases, M.R., Cerezo, A.S. and Stortz, C.A. (1995). Separation and quantitation of enantiomeric galactoses and their mono-O-methylethers as their diastereomeric acetylated 1-deoxy-1-(2-hydroxypropylamino) alditols. *Carbohydrate Research* **269**: 333-341.

Goad, L.J. and Akihisa, T. (1997). *Analysis of Sterols*. Chapman and Hall, London.

Lorenz, R.T., Fenner, G., Parks, L.W. and Haeckler, K. (1989). Analysis of Steryl Esters. In: *Analysis of Sterols and Other Biologically Significant Steroids* (Eds., W.D. Nes and E.J. Parish). Academic Press, New York, USA.

Maier, M.S. and Murray, A.P. (2006). Secondary metabolites of biological significance from echinoderms. In: *Biomaterials from Aquatic and Terrestrial Organisms* (Eds., M. Fingerman and R. Nagabhushanam). Science Publishers, New Hampshire, USA.

Murray, A.P., Muniain, C., Seldes, A.M. and Maier, M.S. (2001). Patagonicoside A, a novel antifungal disulfated triterpene glycoside from the sea cucumber *Psolus patagonicus*. *Tetrahedron* **57**: 9563-9568.

Rouessac, F. and Rouessac, A. (2003). *Chemical Analysis. Modern Methods and Instrumental Techniques. Theory and Solved Exercises*. McGraw Hill/Interamericana de España, S.A.U., Madrid. (In Spanish).

Silberstein, R.M. and Webster, F.X. (1997). *Spectrometric Identification of Organic Compounds*. 6th edition. John Wiley & Sons, New York, USA.

Vincentini, M., Guglielminetti, G., Cassani, G. and Tonini, C. (1987). Determination of double bond position in diunsaturated compounds by mass spectrometry of disulphide derivatives. *Analytical Chemistry* **59**: 694-699.

Vogel, A.I. (1963). *A Textbook of Practical Organic Chemistry*. 3rd edition. Longmans, London, England.

Spectrometry: Ultraviolet and Visible Spectra

Cristina B. Colloca and Virginia E. Sosa*

1. INTRODUCTION

Ultraviolet-visible (UV-Vis) spectroscopy is based on the absorption of UV-Vis radiation by a molecule, causing the promotion of some electrons from a basal to an excited energy state. The absorption UV-Vis spectra of plant constituents can be measured in very diluted solutions against a solvent blank using an automatic recording spectrophotometer. Absorbance is measured at wavelengths between 190 and 800 nm. Although UV-Vis spectroscopy provides limited information about a compound, it proves electron distribution in a molecule and is particularly useful when conjugated π electron systems are found in a compound. Qualitative and quantitative methods involving UV-Vis spectroscopy are often used in chemical and clinical laboratories for analysis of aromatic and unsaturated compounds. The following sections describe applications of UV-Vis spectroscopy for analysis of natural products.

2. QUALITATIVE ANALYSIS OF NATURAL PRODUCTS

2.1 Phenols and Phenolic Acids

Free phenols and phenolic acids are usually identified together during the plant analysis (Prynce, 1972). Most phenolic acids in plant tissues are found as lignin

Authors' address: *Departamento de Química Orgánica, Facultad de Ciencias Químicas, Universidad Nacional de Córdoba, Instituto Multidisciplinario de Biología Vegetal - IMBIV (CONICET-UNC) Haya de la Torre y Medina Allende. Edificio de Ciencias II. Ciudad Universitaria, Córdoba, Argentina.
Corresponding author: E-mail: vesosa@fcq.unc.edu.ar

bound constituents or glycosides. Identification of simple phenols and phenolic acids is readily achieved by comparison of their UV spectra in alcoholic and alkaline solutions (Table 1).

Table 1 Spectral properties of simple phenolics

Phenolic	λ_{max}^{EtOH} nm	$\lambda_{max}^{EtOH-NaOH}$ nm
Simple phenols		
Catechol	279	Decomposition
2-Methylresorcinol	275, 280	288
4-Methylresorcinol	282	291
Orcinol	276, 282	294
Phloroglucinol	269, 273	350
Pyrogallol	266	Decomposition
Resorcinol	276, 283	293
Phenolic acids		
Gallic acid	272	Decomposition
Gentisic acid	237, 335	308
p-Hydroxybenzoic acid	265	278
Protocatechuic acid	260, 295	240, 283
Salicylic acid	235, 305	225, 297
Syringic acid	271	298
Vanillic acid	260, 290	285, 297

2.2. Phenylpropanoids

The phenylpropanoids include hydroxycoumarins, phenylpropenes and lignans. Identity of phenylpropanoids can be confirmed by spectral measurements (Table 2). For example, caffeic acid and its derivates have characteristic absorption bands at 243 and 326 nm, with a distinctive shoulder at 300 nm (Fig. 1). Hydroxycoumarins absorb at longer wavelengths than cinnamic acids; aesculetin, the coumarin related to caffeic acid, has absorption bands at 230, 260, 303 and 351 nm. Measurements of the magnitudes of bathochromic shifts (shifts to longer wavelengths) in the presence of alkali are also useful to distinguish the different cinnamic acids and coumarins.

Table 2 Spectral properties of hydroxycinnamic acids

Phenolic	λ_{max}^{EtOH} nm	$\lambda_{max}^{EtOH-NaOH}$ nm
3,4,5-Trimethoxycinnamic acid	232, 303	293
Caffeic acid	243, 326	Decomposition
Ferulic acid	235, 324	344
Isoferulic acid	295, 323	345
o-Coumaric acid	227, 275, 325	390
p-Coumaric acid	227, 310	335
p-Methoxycinnamic acid	274, 310	298
Sinapic acid	239, 325	350

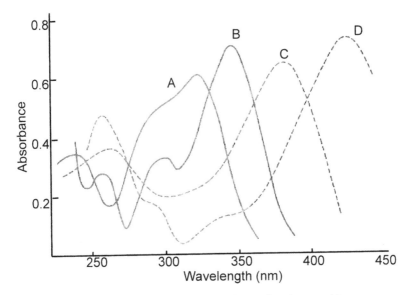

Figure 1 UV-Vis absorption spectra of two phenylpropanoids.
(A) Caffeoylquinic acid (chlorogenic acid) in 95% EtOH.
(B) Aesculetin in 95% EtOH
(C) Caffeoylquinic acid in EtOH-NaOH
(D) Aesculetin in EtOH-NaOH
(Source: Harborne, 1984).

2.3 Coumarins and Furanocoumarins

Coumarin itself shows absorption bands at 274 (log ε 4.03, short band) and 311 nm (log ε 3.72, long band), which have been attributed to the benzene and pyrone rings, respectively.

Furanocoumarins can be further identified by their UV absorption spectra. Unlike hydroxycoumarins, they do not exhibit bathochromic spectral shifts in alkaline solution. Linear furanocoumarins (psoralens) show four absorption bands at 205-225 (log ε 4.00-4.50), 260-290 (4.18-4.80) and 298-320 nm (3.85-4.50) (Nielsen, 1970; Makayama *et al.*, 1971; Tada *et al.*, 2002; Lee *et al*, 2003).

2.3.1 Substituent Effects on Coumarins and Furanocoumarins

Alkyl substitution in coumarins at C-3 leads to a small hypsochromic shift (shift to a lower wavelength) of the long wavelength band, leaving the short wavelength band essentially unchanged. Alkyl substitution at C-5, C-7 or C-8 leads to a bathochromic shift of the short wavelength band but no change in the long wavelength band (Nielsen, 1970). The introduction of a hydroxyl group into the coumarin nucleus causes a bathochromic shift of the main absorption band. The position of the new maximum depends on the ability of the hydroxyl group to conjugate with the chromophoric system (Nielsen, 1970; Murray, 1995).

A useful phenomenon for diagnostic purposes is that acetylated phenolic coumarins show UV spectra similar to those of the corresponding parent hydrocarbons (Stanley and Vannier, 1967; Dreyer, 1970).

Alkaline recorded spectra of 5- and 7-hydroxycoumarins show increased intensities of maxima just to mark bathochromic shift (Govindachari *et al.*, 1967). Bathochromic shifts for 6- and 8-hydroxycoumarins also occur but with a simultaneous fall in the long-wavelength band coefficient absorption (Fig. 2 and Table 3) (Dean, 1963; Sariaslani and Rosazza, 1983).

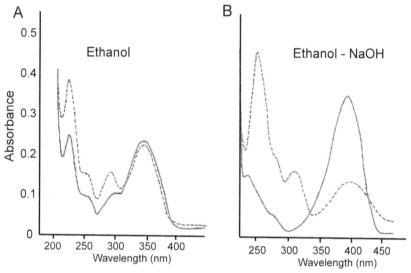

Figure 2 Comparison of the UV absorption spectra of 7-hydroxy-6-hydroxy-7-methoxycoumarin (broken line) in EtOH (A), and ethanolic NaOH (B) (Source: Sariaslani and Rosazza, 1983).

Table 3 Spectral properties of 6- or 7-hydroxycoumarin

Hydroxycoumarin	λ_{max}^{EtOH} nm (log e)	$\lambda_{max}^{EtOH-NaOH}$ nm (log ε)
6-Hydroxy-7-methoxycoumarin	344 (3.60)	400 (3.50)
7-Hydroxycoumarin	325 (4.15)	372 (4.23)
7-Hydroxy-6-methoxycoumarin	344 (4.00)	395 (4.40)

2.3.2 Shift in the Wavelength due to Reagents

Spectral changes induced by the addition of acid and alkali reagents are particularly useful in deducing the orientation of the acyl groups in 4,6,8-trisubstituted-5,7-dihydroxycoumarins (Djerassi *et al.*, 1958 ; Crombie *et al.*, 1966a;

1966b; 1967a; 1967b; Govindachari *et al.*, 1967; Carpenter *et al.*, 1970; 1971). The presence of a free hydroxyl group at C-5 or C-7 is detected by the addition of different bases. The addition of NaAcO produces a marked bathochromic shift in the UV maximum, but the addition of NaMeO produces little or no effect (Horowitz and Gentili, 1960; Livingston *et al.*, 1964). UV spectra of 4-hydroxycoumarin show a hypsochromic shift (343 to 330) by the addition of NaAcO. Spectral shifts produced when other inorganic reagents are added to ethanolic solutions of certain coumarins are also of diagnostic value. Bathochromic shift in the maxima after the addition of H_3BO_3, $AlCl_3$ and NaAcO is shown in Fig. 3 and confirm the presence of *o*-dihydroxy groups (Horowitz and Gentili, 1960).

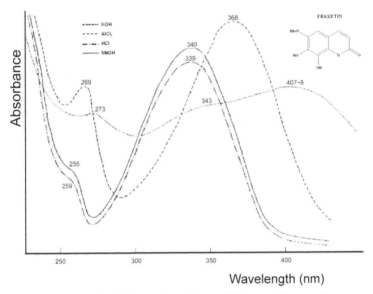

Figure 3 Bathochromic shift produced by addition of inorganic reagents in *o*-dihydroxycoumarin (Source: Murray *et al.*, 1982).

2.4 Lignans

All lignans show a basic UV absorption pattern typical of aromatic compounds with three bands in the region of 210 nm (ε about 5×10^4), 220-240 nm ($\varepsilon = 10$ to 20×10^3) and about 280 nm ($\varepsilon = 2$ to 10×10^3) which correspond to the singlet excited states. In lignans, where there is no conjugation to or between the aryl groups, the molar extinction coefficients corresponding to absorption at the longest wavelength is near to the sum of their molar extinction coefficients (Table 4). Conjugation with a C=C, as in guairetic acid, shifts the absorption maximum towards the visible spectrum. When the system is extended to include the carbonyl group, as in the butyrolactones, a further bathochromic shift occurs (Fig. 4).

Table 4   Ultraviolet absorption of typical lignans

Lignan	λ_{max} (nm) (ε, 10^3)
Austrobailingnan-5	236 (8.20), 288 (7.60)
Dimethoxybenzene	274 (3.929)
Dimethyl lyoniresinol	234 (17.20), 273 (1.98)
Dimethylcycloolivil	230 (17.60), 283 (7.10)
Di-O-methyldihydroxythu-japlicatin methyl ether	240 (18.60), 280 (5.63)
Guaiacol	277 (2.73), 289
Methylenedioxybenzene	282 (3.24)
β-Peltatin	276 (1.77)
Podophyllotoxin	236 (13.60), 292 (4.88)
Sesamin	238 (12.70), 287 (10.10)
Trimethoxybenzene	267 (0.69)
Yangambin	232 (10.80), 270 (1.20)

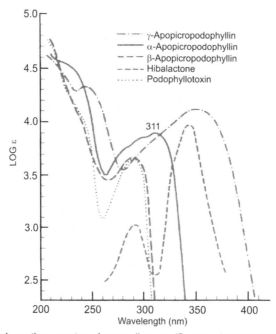

Figure 4   UV absorption spectra of some lignans (Source: Ayres and Loike, 1990).

2.4.1   Shift in the Wavelength due to Reagents

The UV spectra of free phenolic lignans differ slightly from those of fully etherified compounds, but the phenols may be distinguished by a shift of longer wavelength absorption when they are converted in the corresponding anions, as shown in

Table 5 (Rao and Wu, 1978; Chandel and Rastogi, 1980; Le Quesne *et al,.* 1980; Evcim *et al.*, 1986).

Table 5 Ultraviolet absorption of some typical lignans

Lignan	λ_{max} (nm) (ε, 10^3)	$\lambda_{max}^{+ NaOH}$ (nm) (ε, 10^3)
Furan nectandrin A[b]	231 (17.00), 278 (5.60)	294 (18.00)
Haplomyrfolin	230 (15.10), 284 (10.00)	242 (18.20), 290 (11.70)
Lirionol	242 (21.40), 291 (11.00)	262 (12.00), 371 (21.70)

[b] In MeOH

2.5 Quinones

2.5.1 Benzoquinones

The UV-Vis spectra of *p*-benzoquinones (Fig. 5) show an intense absorption band near 240 nm (ε_{max} 26,000), a medium band near 285 nm (ε_{max} 300) and a weaker absorption band in the visible region of the spectrum (Morton, 1965).

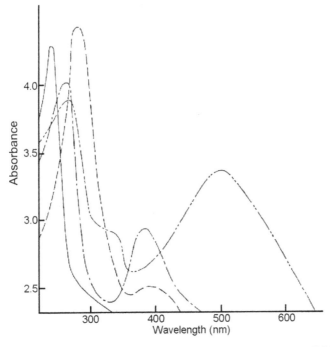

Figure 5 UV-Vis absorption spectra of 1,4-benzoquinone (—) in EtOH; 2,5-dihydroxy-1,4-benzoquinone (— — —) in EtOH; 2-hydroxy-5-methyl-1,4-benzoquinone (• — • — •) in EtOH, and 2-hydroxy-5-methyl-1,4-benzoquinone in ethanolic NaOH (··—··—··). (Source: Thompson, 1971).

2.5.1.1 Substituent Effects on p-benzoquinones

Natural p-benzoquinones can be substituted by alkyl, hydroxyl and alkoxyl groups (Mahmood *et al.*, 2002; Nassar *et al.*, 2002). Exceptionally, aryl group may also be present. Introduction of a substituent into the benzoquinone nucleus produces a small bathochromic shift of the short wavelength band in the spectrum (< 10 nm) but the long wavelength band undergoes a more significant red shift in the order Me- (27 nm), MeO- (69 nm) and HO- (81 nm) in chloroform solution (Table 6).

Table 6 Ultraviolet-visible absorption of 1,4-benzoquinones

Quinone	λ_{max} nm (in CHCl$_3$)		
Parent	246 (4.42)	288 (2.50)	439 (1.35)
	242[a] (4.26)	285[a]sh (2.6)	434[a] (1.26)
			454[a] (1.22)
2,3-Dimethoxy-	254 (4.17)	398 (3.17)	
2,3-Dimethyl-	250 (4.26)	337 (3.05)	425 (1.55)
	257sh (4.20)		
2,5-Dihydroxy-	279 (4.35)	393 (2.43)	
	286sh (4.34)		
2,5-Dimethoxy	278 (4.37)	370 (2.48)	
	284 (4.38)		
2,5-Dimethyl-	254 (4.37)	316 (2.42)	430 (1.43)
	261sh (4.32)		
2-Ethyl-	248 (4.30)	318 (2.95)	437 (1.53)
	256sh (4.29)		
2-Hydroxy-3, 5, 6-trimethyl-	272 (4.29)	409 (2.65)	
	280 (4.30)		
2-Hydroxy-3, 6-dimethyl-	266 (4.24)	402 (2.94)	
2-Hydroxy-3-methyl-	255 (4.17)	396 (3.16)	
2-Hydroxy-6-methyl	268 (4.21)	380 (2.87)	
2-Methoxy-3-methyl-[b]	254 (4.12)	374 (3.13)	
2-Methoxy-6-methyl-[b]	263 (4.23)	355 (2.97)	
2-Methyl-	249 (4.33)	315 (2.80)	436 (1.38)
	255sh (4.27)		
2-Methyl-5, 6-dimethoxy-	264 (4.16)	402 (2.94)	
3,6-Dihydroxy-thymoquinone[a]	293 (4.31)	435 (2.36)	
3-Hydroxythy-moquinone[a]	267 (4.16)	404 (3.01)	
Hydroxy-	256 (4.14)	369 (3.07)	
Methoxy-	254 (4.26)	357 (3.21)	
Tetramethyl-	262 (4.30)	342 (2.34)	430sh
	269 (4.31)		(1.50)

Contd.

Table 6 *Contd.*

Quinone	λ_{max} nm (in $CHCl_3$)		
Trimethoxy-	291 (4.29)	418 (2.65)	
Trimethyl-	258 (4.30)	340 (2.61)	425 (1.53)
	263 (4.27)		430sh

[a] In EtOH; [b] In CCl_4; sh: shoulder.

2.5.2 Shift Reagents Effects on Quinones

In alkaline alcoholic solution, there is a marked shift of the visible absorption. Only instantaneous shifts, reversible on acidification, are important because other changes may take place more slowly, such as notable hydroxylation of the ring or nucleophilic replacement of methoxyl by hydroxyl groups. Addition of sodium borohydride to an ethanolic solution of a benzoquinone affords quinol accompanied by a sharp change to benzenoid absorption with λ_{max} ca. 290 nm and a marked reduction in ε_{max}.

Only two natural *o*-benzoquinones have been found, mansononone and phlebiarubrone, both highly substituted. Relatively few *o*-benzoquinone spectra have been recorded (Table 7). They show triple absorption peaks and are easily distinguished from the *p*-isomers by the relatively low intensity and marked bathochromic displacement. Band 2 is frequently shifted into the visible light range and the absorption may extend as far as 600 nm.

Table 7 Ultraviolet-visible absorption of some benzoquinones

Quinone	λ_{max}^{EtOH} nm (log ε)		
Ardisiaquinone (8)	289 (4.60)	420 (2.83)	
Fumigatin (9)	265 (4.14)	450 (2.96)	
Hydroxyperezone	295 (4.26)	425 (2.41)	
Lagopodin A (1)	257 (4.25)	310 (2.55)	435 (1.55)
Plastoquinone (2)	254 262sh		
Primin (4)	267 (4.33)	365 (2.54)	
Shanorellin (7)	272 (4.05)	406 (2.07)	
Ubiquinone (10)	275	405	
α-Tocopherol-	261 (4.38)		
quinone (3)	269 (4.29)		
o-Benzoquinones			
Parent		375 (3.23)	568 (1.48)
3,5-Dimethyl-[a]	260 (3.19)	410 (3.20)	558 (1.66)
4,5-Dimethoxy-[b]	283 (4.09)	406 (2.82)	504sh (1.60)
5-Ethyl-3-methoxy-[a]		470 (3.19)	570sh (2.60)
Mansonone A (12)	260sh (3.50)	432 (3.20)	
	280sh (3.10)		

Contd.

Table 7 *Contd.*

Quinone	λ_{max}^{EtOH} nm (log ε)		
4-Methyl-[b]	249sh (3.32)	387 (3.23)	544sh (1.48) 570 (1.52)
3-Methoxy-[b]	269sh (2.66)	465 (3.26)	545sh (1.78) 575sh (1.56)
Phlebiarubrone (13)	268 (4.48)	332 (3.64)	465 (3.54)
3,4,5-Trimethyl-[a]	265 (3.33)	425 (3.11)	545 (1.75)

[a] In $CHCl_3$; [b] In CH_2Cl_2; sh: shoulder.

2.5.3 Naphthoquinones

Naphthoquinones possess an additional benzenoid ring in the quinone structure. Both 1,4- and 1,2-naphthoquinones can be found among naturally occurring naphthoquinones.

2.5.3.1 1,4-Naphthoquinones

These spectra are more complex than those of benzoquinones. These compounds show both benzenoid and quinonoid absorptions and one or both rings may be substituted. Alkyl, hydroxyl and alkoxyl groups are frequent in the natural

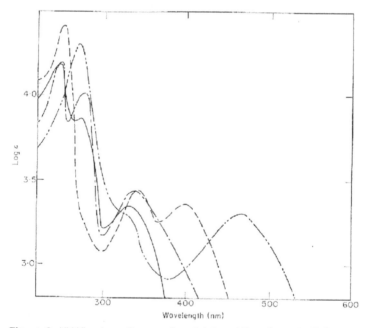

Figure 6 UV-Vis absorption spectra of 1,4-naphthoquinone in EtOH (—);
1,2-naphthoquinone in EtOH (——);
2- hydroxy-1,4-naphthoquinone in EtOH (. — . — .), and
2-hydroxy-1,4-naphthoquinone in ethanolic NaOH (.. — .. — ..).
(Source: Thompson, 1971).

compounds; acyl groups or conjugated double bonds may also be present. The quinonoid ring may be fused to a furan or pyrone ring system.

The spectrum of the parent compound (Fig. 6) shows two intense absorption bands due to the quinone ring in the region 240-290 nm and a medium intensity band due to the benzenoid ring at 335 nm. A broad weak band at 425 nm ($\varepsilon = 32$) is discernible in *iso*-octane solution but it is not detected in more polar solvents.

2.5.3.2 Substituent Effects on 1,4-Naphthoquinones

The +I and +M substituents in the quinone ring produce a bathochromic shift of the band at 257 nm, whereas the benzenoid absorption at 245 and 251 nm is usually scarcely affected. In benzene-substituted 1,4-naphthoquinones the bands associated with benzenoid and quinonoid rings, in the region 240-290 nm, frequently coalesce, and the prominent benzenoid absorption near 335 nm is shifted towards the red (Table 8).

Table 8 Ultraviolet-visible absorption of 1,4-naphthoquinones

1,4-Naphthoquinone	$\lambda_{max}^{CHCl_3}$ nm (log ε)		
Parent	245 (4.34)	257sh (4.12)	
	251 (4.37)		
2,3,5,7-Tetrahydroxy-		320 (3.89)	470 (3.16)
2,3,5,8-Tetrahydroxy-		288 (3.66)	478sh (3.77) 524 (3.64)
		302 (3.66)	511 (3.65)
2,3-Dihydroxy-	262 (4.25)	274sh (4.23)	439 (3.17)
		288sh (4.16)	
2,3-Dimethyl-	243 (4.26)	260 (4.28)	
	249 (4.26)	269 (4.28)	
2,5-Dihydroxy-	240 (4.00)	286 (4.10)	418 (3.56)
			430 (3.56)
2-Hydroxy-	249 (4.20)	275.5 (4.23)	
		282 (4.24)	
2-Hydroxy-3-iso-α-pentenyl		265 (4.41)	419 (3.30)
2-Methoxy-	242 (4.22)	274 (4.21)	
	248 (4.25)	280 (4.21)	
2-Methoxy-3-isopenthyl	252 (4.41)	279 (4.32)	382sh (3.10)
2-Methyl-	245.5 (4.27)	258.5 (4.22)	
	251 (4.30)		
3,5-Dihydroxy-	240 (3.89)	283 (4.15)	419 (3.64)
5,6-Dihydroxy-[a]		263 (4.01)	461 (3.51)
5-Methyl-	252 (4.29)	260sh (4.07)	
Representative naturally occurring 1,4-naphthoquinones			
Alkannin (24)		280 (3.84)	480 (3.74)
			510 (3.78)
			546 (3.60)
Chimaphilin (15)	248 (4.19)	265sh (4.02)	338 (3.19)
	254.5 (4.19)		

Contd.

Table 8 *Contd.*

1,4-Naphthoquinone	$\lambda_{max}^{CHCl_3}$ nm (log ε)			
Dehydro-α-lapachone (17)		267 (4.35) 276 sh (4.17)	333 (3.40)	434 (3.21)
Maturinone (26)	251 (4.39)	266 (4.00) 287sh (3.74)	355 (3.58)	
Menaquinone-1 (14)	245 (4.46) 249 (4.44)	262 (4.34) 273 (4.31)	331 (3.58)	
Solaniol[b](25)			304 (3.97)	472sh (3.84) 500 (3.91) 536sh (3.72)
α-Lapachone (16)	251 (4.45)	282 (4.22)	332 (3.44) 375sh (3.15)	

[a]In EtOH; [b] In dioxan; sh: shoulder

Peri-substitution by a hydroxyl group is exceptional, shifting the benzenoid band almost 100 nm into the visible light range to give a peak at 429 nm (or 422 nm in MeOH), characteristic of simple juglones. The effect is less marked in juglone methyl ether (λ_{max} 396 nm), while in the acetate (-M effect of the ester carbonyl group) the spectrum reverts to that of the parent 1,4-naphthoquinone (Fig. 7).

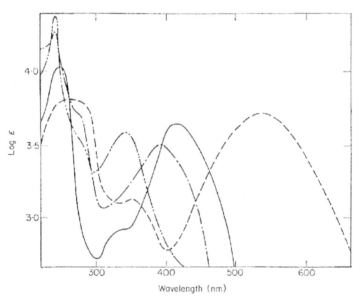

Figure 7 UV-Vis absorption spectra of 5-hydroxy-1,4-naphthoquinone in EtOH (—); 5-hydroxy-1,4-naphthoquinone in ethanolic NaOH (——); 5-methoxy-1,4-naphthoquinone in EtOH (• — • —•), and 5-acetoxy-1,4-naphthoquinone in ethanolic NaOH (•• — •• — ••). (Source: Thompson, 1971).

2.5.3.3 Shifts Reagents Effects on 1,4-Naphthoquinones

The alkali shifts shown by hydroxynaphthoquinones are of diagnostic value. The anion of 2-hydroxy-1,4-naphthoquinone is orange, that of 5-hydroxy-1,4-naphthoquinone is violet, while those of 2,3-, 5,6- and 5,8-dihydroxy-1,4-naphthoquinones are blue. Addition of anhydrous $AlCl_3$ to an ethanolic solution of a naphthazarin gives a spectrum showing characteristic triplet absorption (Fig. 8); this is not shown for juglones.

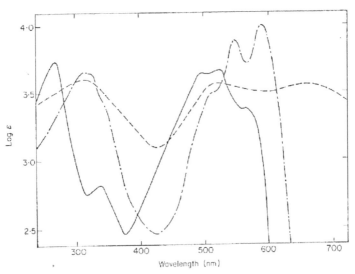

Figure 8 UV-Vis absorption spectra of 5,8-dihydroxy-1,4-naphthoquinone: neutral (—), gaffter addition of OH^- (-----) and after addition of $AlCl_3$ (•—•—•). Compounds were dissolved in EtOH (Source: Thompson, 1971).

2.5.4 Anthraquinones

Anthraquinones show intense benzenoid absorption bands within the range of 240-260 nm and medium absorption at 320-330 nm, and quinonoid band(s) at 260-290 nm. There is a weak quinonoid absorption at 405 nm (Table 9, Fig. 9). The substitution produces little or no effect on the parent absorption in the UV region.

2.5.4.1 Substituent Effects on Anthraquinones

The long wavelength quinonoid absorption band is the most valuable in the structural elucidation of natural anthraquinones. Substitution by hydroxyl or alkoxyl invariably intensifies this band, and in general there is a bathochromic shift. However, there is no shift in the case of 1-hydroxyanthraquinone. Methylation of the 1-hydroxyl group shifts the band at 402 nm to 378 nm due to the elimination of the hydrogen bond contribution (Fig. 9). Exceptionally, 2-hydroxyanthraquinone absorbs at 378 nm and its methyl ether at 363 nm. Absorption above 360 nm is dominated by the number of α-hydroxyl groups, being the influence of much weaker β-hydroxyls (Fig. 10). Absorption at the visible light region is of medium intensity (ε aprox. 10,000) and may obscure the benzenoid band at 320-330 nm. Tri- and tetra-α-hydroxylated derivates frequently display fine structure in the long wavelength band.

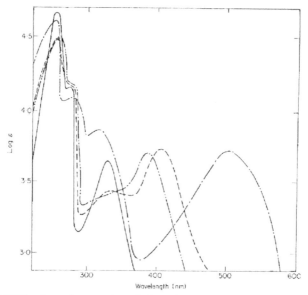

Figure 9   UV-Vis absorption spectra. Anthraquinone (—) and 1-hydroxyanthraquinone (----) were dissolved in EtOH. 1-hydroxyanthraquinone (• — • — •) and 1-methoxyanthraquinone (•• — •• — ••) were dissolved in Ethanolic NaOH (Source: Thompson, 1971).

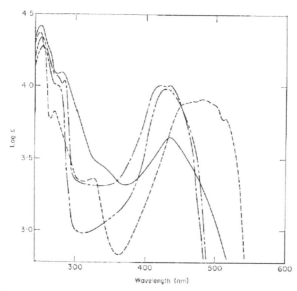

Figure 10   UV-Vis absorption spectra of 1,2-dihydroxyanthraquinone (—), 1,4-dihydroxyanthraquinone (---), 1,5-dihydroxyanthraquinone (• — • — •) and 1-methoxyanthraquinone (•• — •• — ••) dissolved in EtOH (Source: Thompson, 1971).

Table 9 Ultraviolet-visible absorption of anthraquinones

Anthraquinone	$\lambda_{max}^{CHCl_3}$ nm (log ε)			
Parent	243 (4.52) 252 (4.71)	263 (4.31) 272 (4.31)	322 (3.75)	405 (1.95)
1,2-Dihydroxy-	247 (4.45)	278 (4.13)	330 (3.46)	434 (3.70)
1,3,6,8- Tetrahydroxy-	253 (4.17) 262 (4.17)	291 (4.48)	318 (3.96)	372 (3.47) 452 (4.01)
1,3,8-Trihy- droxy-6-methyl-	253 (4.31)	266 (4.29) 289 (4.36) 287 (3.81)		436 (4.14) 487 sh (3.91) 477 sh (3.81)
1,4,5-Trihy- droxy-	250 (4.21)	284 (3.94)		459sh (4.02) 478sh (4.08), 510sh (4.00) 489 (4.12) 523 (3.93)
1,4,5,8-		299sh		509 (4.07) 521 (4.14), 560 (4.19)
Tetrahydroxy-[a]		(3.79)		546.5 (4.14) 601sh (3.25)
1,5-Dihydroxy-	253 (4.33)	275sh (4.05) 284.5 (4.03)		418 (4.00) 432 (4.00)
1,8-Dihydroxy-	251 (4.27)	273sh (4.00) 283 (3.99)		429 (3.98)
1-Hydroxy-	252 (4.46)	277 (4.44) 262 (4.36)	327 (3.52)	402 (3.74)
1-Methoxy-	254 (4.52)	270 (4.18)	328 (3.46)	378 (3.72)
1-Methyl-	252 (4.66)	263 (4.26) 272 (4.16) 265 (4.32)	331 (3.68)	415 (2.18)
2-Hydroxy-	241 (4.31)	271 (4.55) 283 (4.46)	330 (3.55)	378 (3.55)
2-Methoxy-	246 (4.21)	267 (4.51) 280 (4.38)	329 (3.56)	363 (3.60)
2-Methyl-	255 (4.65)	274 (4.24) 266 (4.18)	324 (3.66)	

[a] In EtOH; sh: shoulder

2.6 Flavonoids

The flavonoids are polyphenolic compounds with two benzene rings joined by a linear three carbon chain. Various subgroups of flavonoids are classified according to the substitution patterns of ring C. Both the oxidation state of the heterocyclic ring and the position of ring B are important in the classification. Examples of the seven subgroups are mentioned: flavones, flavonols, dihydroflavones, dihydroflavonols, isoflavones, chalcones and aurones.

2.6.1 Flavones and Flavonols

The UV-Vis spectra of flavones and flavonols generally have two maximum absorption bands. Band I is considered to be associated with the absorption due the B-ring cinnamoyl system and Band II with the absorption involving the A-ring benzoyl system (Fig. 11 and Table 10). The λ_{max} values for Band I and Band II provide information about the type of flavonoids as well as their oxidation patterns (Jurd, 1962).

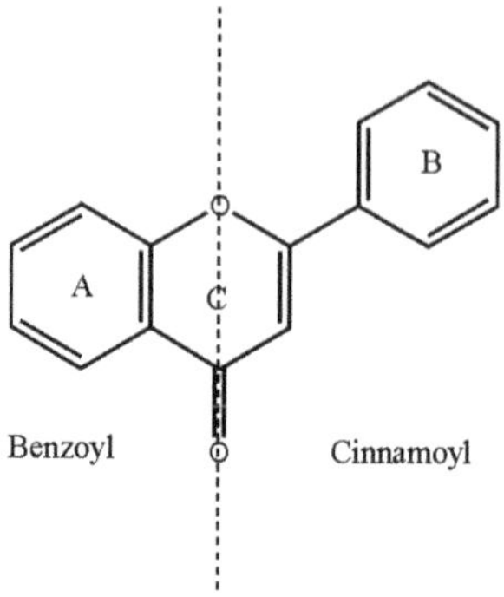

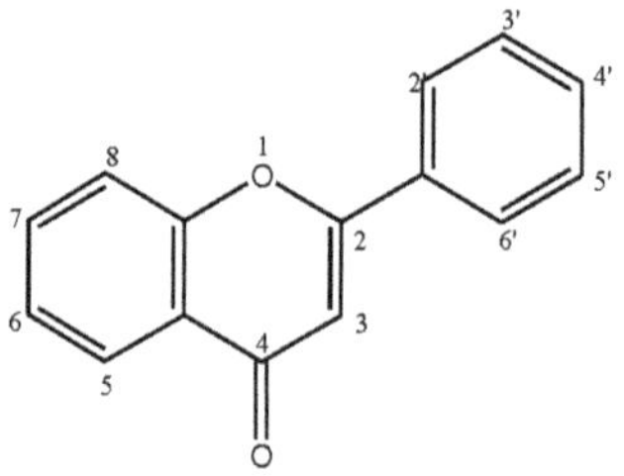

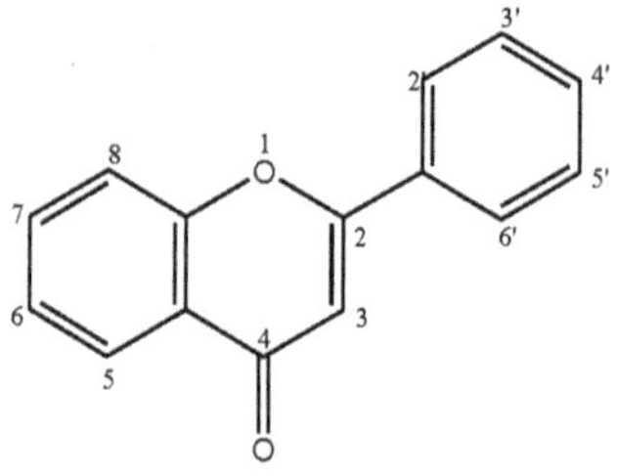

Figure 11 Flavones and flavonols skeletons.

Table 10 Ultraviolet-visible absorption of flavones and flavonols

Flavonoids	EtOH λ_{max} nm (log ε)
Flavones and Flavonols	
3,2'-Dihydroxyflavone	353 (4.21), 303 (3.92), 244 (4.28)
3,4'-Dihydroxyflavone	361 (4.39)
3,5,7,2'-Tetrahydroxyflavone	360 (3.99), 262.5 (4.14)
3,5,7-Trihydroxyflavone (galangin)	360 (4.07), 267.5 (4.23)
3,7,3',4'-Tetrahydroxyflavone (fisetin)	370 (4.43), 315 (4.22), 252.5 (4.33)
3,7,3'-Tri-O-methylquercetin	360 (4.31), 268 (4.24), 257 (4.32)
3',4'-Dihydroxyflavone	342.5 (4.50), 244 (4.46)
3-Methoxyflavone	320, 299 (4.21), 246 (4.25)

Contd.

Table 10 *Contd.*

Flavonoids	EtOH λ_{max} nm (log ε)
4',7-Di-O-ethylvitexin	326 (4.21), 270 (4.26)
5,7,4'-Trimethoxyflavone	325 (4.33), 265 (4.25)
5-Hydroxyflavone	337 (3.88), 272 (4.35)
7-Hydroxyflavone	308 (4.50), 250 (4.33)
Flavone	297 (2.20), 250 (4.06)
Gossypetin	386 (4.15), 341, 278 (4.23), 262 (4.26)
Isoquercitrin	360 (4.32), 258 (4.41)
Luteolin	350 (4.17), 268, 255 (4.13)
Myricetin	378 (4.29), 255 (4.21)
Quercetin 3-L-arabinoside	360 (4.24), 260 (4.32)
Isoflavones	
Isoflavone	307 (3.82)
7,4'-Dihydroxy-5-methoxyisoflavone	256 (4.51)
5,6,7,2'-Tetramethoxyisoflavone	304(4.10), 281 (4.29), 247 (4.57)
5,7,2'-Trimethoxy-8-methylisoflavone	259 (1.48), 249 (1.46)
Flavanones and Dihydroflavonols	
5,7-Dihydroxyflavanone (pinocembrin)	314 (3.78), 288 (4.35)
Dihydrokaempferol	330 (3.75), 292 (4.29), 252 (3.61)
Flavanone	320 (3.37), 250 (3.86)
Hesperetin	330, 289 (4.27)
Naringin	330, 284 (4.28)
Prunin	330, 283 (3.44)
5,7,4'-Trihydroxyflavanone (naringenin)	325, 288 (4.23)
Chalcones and Aurones	
Chalcone	312 (4.35), 230 (3.91)
2'-Hydroxychalcone	366, 316 (4.36), 221 (4.11)
2',4',5',3,4-Pentahydroxychalcone	393 (4.37), 320 (4.03), 268 (4.08)
Aurone	379 (4.06), 316.5 (4.27), 251 (4.10)
3',4',4-Trihydroxyaurone	416 (4.47), 310 (3.92), 276 (4.02), 256 (3.91)
3',4'-Dihydroxyaurone	415.5 (4.43), 330, 277, 259
3'-Hydroxyaurone	381 (4.29), 316 (4.21), 252 (4.03)
4',6,7-Trihydroxyaurone	407 (4.39), 355 (4.22), 241 (4.12)
4'-Hydroxyaurone	405 (4.47), 346 (4.07), 260 (4.32)

2.6.1.1 Substituent Effects on Flavones and Flavonols

Increasing oxidation of the B-ring in flavones and flavonols produce a bathochromic shift in Band I as a consequence of oxygen incorporation. The Band II may appear as either one or two peaks (designated IIa and IIb, with IIa being the peak at longer wavelength), depending on the B-ring oxidation pattern (Table 11).

Table 11 Examples of Band I position according to the B-ring oxidation pattern

Flavonol	Oxidation pattern A- and C-ring	B-ring	Band I
Galangin	3,5,7	–	359
Kaempferol	3,5,7	4'	367
Myricetin	3,5,7	3',4',5'	374
Quercetin	3,5,7	3',4'	370

Increasing hydroxylation of the A-ring in flavones and flavonols causes a bathochromic shift in Band II and a smaller effect on Band I (Table 12).

Table 12 Examples of Band II position according to the A-ring oxidation pattern.

Flavone	A-ring pattern oxidation	Band II nm
Flavone	–	250
Baicalein	5,6,7	274
5,7-Dihydroxyflavone	5,7	268
5-Hydroxyflavone	5	268
7-Hydroxyflavone	7	252
Norwogonin	5,7,8	281

The hydroxylation at C-5 produces a bathochromic shift of both the Band I and Band II for a flavone or a flavonol; a shift of 3-10 nm in Band I and 6-17 nm in Band II (Table 13).

Table 13 The effect of hydroxylation of position 5 in flavones or flavonols

Flavonoids	$\lambda_{max}nm$
3,4'7-Trihydroxyflavone	258, 280sh, 318, 356
3',4',7-Trihydroxyflavone	235, 250sh, 309, 343
3',4',7-Trihydroxyflavone 7-O-rhamnoglucoside	247sh, 255sh, 305, 341
4',7-Dihydroxyflavone	253sh, 312sh, 328
4',7-Dihydroxyflavone 7-O-rhamnoglucoside	255sh, 311sh, 325
5,7-Dihydroxy 3',4'-dimethoxyflavone	240, 248sh, 269, 291sh, 340
5-Deoxyvitexin (Bayin)	255sh, 312sh, 328
7-Hydroxy 3',4'-dimethoxyflavone	239, 262sh, 330
7-Hydroxy 4'-methoxyflavone (Pratol)	253, 314sh, 323
7-Hydroxyflavone	252, 268, 307
Acacetin	269, 303sh, 327
Apigenin	267, 296sh, 336
Apigenin 7-O-neohesperidoside	268, 333
Chrysin	247sh, 268, 313
Fisetin	248, 262sh, 307sh, 319, 362

Contd.

Table 13 *Contd.*

Flavonoids	λ_{max} nm
Kaempferol	253sh, 266, 294sh, 322sh, 367
Luteolin	242sh, 253, 267, 291sh, 349
Luteolin 7-O-rutinoside	255, 265sh, 349
Quercetin	255, 269sh, 301sh, 370
Vitexin	270, 302sh, 336

sh: shoulder

The methylation or glycosylation of 3, 5 or 4'-hydroxyl groups on both the flavone or flavonol nucleus produces a hypsochromic shift, especially in Band I. The substitution at other positions has little or no effect on the UV spectrum (Table 14).

Table 14 The effect of the methylation or glycosylation.

Flavonoids	λ_{max} nm (in MeOH)
3',4'-Dihydroxyflavone	242, 308sh, 340
3',4'-Dimethoxyflavone	242, 314sh, 333
4',7-Dihydroxyflavone	253sh, 312sh, 328
5,7-Dihydroxy 3',4'-dimethoxyflavone	240, 248sh, 269, 291sh, 340
7-Hydroxy 4'-methoxyflavone (Pratol)	253, 314sh, 323
Acacetin	269, 3003sh, 327
Apigenin	267, 296sh, 336
Chrysoeriol	241, 249sh, 269, 347
Diosmetin	240sh, 252, 267, 291sh, 344
Galangin	267, 305sh, 359
Galangin 3-methyl ether	266, 312sh, 340sh
Isorhamnetin	253, 267sh, 306sh, 326sh, 370
Isorhamnetin 3-O-galactoside	255, 268sh, 303sh, 357
Kaempferol	253sh, 266, 294sh, 322sh, 367
Kaempferol 3-O-robinoside 7-O-rhamnoside (Robinin)	244sh, 265, 315sh, 350
Luteolin	242sh, 253, 267, 291sh, 349
Quercetin	255, 269sh, 301sh, 370
Quercetin 3',4',5,7-tetramethyl ether	252, 270sh, 304sh, 362
Quercetin 3-methyl ether	257, 269sh, 294sh, 358
Quercetin 3-methyl ether 4'-O-glucoside 7-O-diglucoside	254, 269, 349
Quercetin 3-O-glucoside 7-O-rhamnoside	257, 269sh, 358
Rutin	259, 266sh, 299sh, 359
Tamarixetin 7-O-neohesperidoside	255, 269sh, 369
Tamarixetin 7-O-rutinoside	255, 271sh, 291sh, 367

sh: shoulder

2.6.1.2 Shifts Reagents Effects on Flavones and Flavonols

The addition of NaOMe to flavones or flavonols, in MeOH solution, is diagnostic for detection of 4'-hydroxyl group. The presence of this group produces a bathochromic shift of Band I of about 40-65 nm, without a decrease in intensity. A similar effect is observed in flavonols with a free 3-hydroxyl group (Table 15).

Table 15 Bathochromic shift by addition of NaOMe in flavones and flavonols (in MeOH)

Flavonoid	Band I bathochromic shift (nm)
Flavones	
2″-O-xylosylvitexin	60
3',4',7-Trihydroxyflavone	52
3',4',7-Trihydroxyflavone 7-O-rhamnoglucoside	64
3',4'-Dihydroxyflavone	64
4',7-Dihydroxyflavone	58
4',7-Dihydroxyflavone	60
7-O-rhamnoglucoside 5-Deoxyvitexin (Bayin)	62
Amentoflavone	47
Apigenin	56
Apigenin 7-O-glucoside	53
Apigenin 7-O-neohesperidoside	53
Chrysoeriol	58
Fisetin 3-O-glucoside	68
Flavonols	
Galangin	53
Isoorientin	57
Isorhamnetin 3-O-rutinoside	58
Isorhamnetin 3-O-galactoside	58
Isovitexin	62
Jaceidin	61
Jacein	49
Kaempferol	49
Kaempferol 3-O-robinoside 7-O-rhamnoside (Robinin)	39
Kaempferol 7-O-neohesperidoside	61
Lucenin-1	59
Luteolin	52
Luteolin 7-O-glucoside	46
Luteolin 7-O-rutinoside	45
Morin	47
Orientin	59
Patuletin 3-O-rutinoside	55
Patuletin 3–O-glucoside	55
Penduletin	48
Quercetin	43
Quercetin 3,7-O-diglucoside	41

Contd.

Table 15 *Contd.*

Flavonoid	Band I bathochromic shift (nm)
Quercetin 3-galactoside	47
Quercetin 3-methylether	49
Quercetin 3-O-glucoside 7-O-rhamnoside	38
Quercetin 3-O-glucoside 7-O-rutinoside	38
Rhamnosylvitexin	60
Rutin	51
Saponarin	53
Scoparin	61
Tricen	66
Violanthin	63
Vitexin	59
Xanthomicrol	59

The addition of NaOAc (not fused) to flavones or flavonols, which have a 4'-hydroxyl group and no free 3- or 7-hydroxyl groups, show a pronounced shoulder on the long wavelength side of Band I. A similar effect is observed when 7-hydroxyl group is free whether or not a 4'-hydroxyl group is present in the flavonoid. Band I appears as a peak similar to that observed with NaOMe when fused (HOAc-free) NaOAc is used with flavones and flavonols containing a 4'-hydroxyl group.

Some oxygenation patterns are alkali-sensitive. The presence of oxygenation patterns in flavones or flavonols produces decomposition by addition of NaOAc after several minutes (Jurd and Horowitz, 1957). The most common alkali-sensitive oxygenation patterns in this type of flavonoids are those containing 5, 6, 7; 5, 7, 8 or 3, 3', 4' trihydroxyl groups. Flavonols with 3',4'-dihydroxyl system can suffer decomposition in a few minutes by addition of NaOMe or NaOAc, as shown in Table 16 (Jurd and Horowitz, 1957).

Table 16 Effect of NaOMe and NaOAc in flavonols containing a 3',4'-dihydroxyl system (in MeOH)

Flavonols	Alkali[a]	
	NaOMe	NaOAc
Fisetin	Dec	Dec
Gossypetin	Dec	Dec
Gossypin	Dec	Dec
Gossypitrin	Dec	Dec
Herbacetin 8-methyl ether	Dec	Dec
Isorhamnetin	Dec	Dec
Kaempferol	Slow dec	No dec
Kaempferol 7-O-neohesperidoside	Slow dec	No dec

Contd.

Table 16   *Contd.*

Flavonols	Alkali[a]	
	NaOMe	NaOAc
Morin	Slow dec	No dec
Myricetin	Dec	Dec
Patuletin	Dec	Dec
Patulitrin	Dec	Dec
Quercetin	Dec	Dec
Quercetin 7-O-rhamnoside	Dec	Dec
Rhamnetin	Dec	Dec

[a]Dec = Spectrum decomposed determined by a comparison of the spectrum in alkali measured immediately, with that measured 5-10 min later.

For 7-hydroxyl derivatives of flavones and flavonols, the addition of NaOAc produces a bathochromic shift of Band II (5-20 nm). In 7-hydroxyl, 6,8-dioxygen derivatives of flavones, this bathochromic shift is often small or imperceptible (Table 17).

Table 17   Bathochromic shift of Band II in 7-hydroxyflavones and 7-hydroxyflavonols with added NaOAc

Flavonoids	Bathochromic shift (nm)[a]
Flavones	
2″-O-Xylosylvitexin	10
4′,7-Dihydroxyflavone	8
5,6,7-Trihydroxyflavone (Baicalein)	Dec
5,7,8-Trihydroxyflavone (Norwogonin)	Dec
5,7-Dihydroxy-2′-methoxyflavone	5
5-Deoxyvitexin (Bayin)	13
7-Hydroxy-4′-methoxyflavone (Pratol)	17
7-Hydroxyflavone	14
Acacetin	7
Apigenin	7
Chrysin	7
Chrysoeriol	30
Isoorientin	21
Isovitexin (Saponaretin)	8
Lucenin-1	25
Luteolin	16
Orientin	23
Rhamnosylvitexin	11
Violanthin	7
Vitexin	10

[a]Dec = Spectrum decomposed determined by a comparison of the spectrum in NaOAc measured after 2-5 min with that measured 5-10 min later.

The presence of *ortho*-dihydroxyl group in flavones and flavonols can be determined by the addition of H_3BO_3 in the presence of NaOAc. The chelate formed between H_3BO_3 and *ortho*-dihydroxyl groups in the B-ring produces a bathochromic shift (12-30 nm) of Band I (Table 20).

$AlCl_3$ forms a complex with *ortho*-dihydroxyl groups or with 3- or 5-hydroxyl groups. The formation of these complexes produces a bathochromic shift in relation to the MeOH spectrum. The complexes formed with the former type of hydroxyl groups are acid labile. The UV spectra of the complexes formed with *ortho*-dihydroxyl groups of B-ring suffer a hypsochromic shift (30-40 nm) in Band I by addition of HCl (Table 18). The presence of three adjacent hydroxyl groups in B-ring show only a 20 nm hypsochromic shift. For *ortho*-dihydroxyl groups of A-ring the same effect is observed, except for 5-hydroxyl group (Markham and Mabry, 1968).

Table 18 Effect of $NaOAc/H_3BO_3$ and $AlCl_3$ on Band I of the UV spectrum of 3',4'-dihydroxyflavones and 3',4'-dihydroxyflavonols

Flavonoids	Bathochromic shift with $NaOAc/H_3BO_3$ (nm) relative to MeOH spectrum	Bathochromic shift with $AlCl_3$ (nm) relative to $AlCl_3$/HCl spectrum
Flavones		
3',4',7-Trihydroxyflavone 7-O-rhamnoglucoside	24	39
3',4',7-Trihydroxyflavone	17	31a
3',4'-Dihydroxyflavone	25	36
Isoorientin	28	45
Lucenin-1	33	46
Luteolin	21	41
Luteolin 3-O-glucoside	24	45
Luteolin 3-O-rutinoside	21	43
Orientin	29	45
Flavonols		
3,3',4'-Trihydroxyflavone	22	39
Fisetin	19	35
Fisetin 3-O-glucoside	25	b
Gossypetin	21	45
Gossypin	20	11
Gossypitrin	14	21
Myricetin	18	22
Patuletin	22	32
Patuletin 3-O-glucoside	27	31
Patuletin 3-O-rutinoside	25	31
Patulitrin	21	31
Quercetin	18	30

Contd.

Table 18 *Contd.*

Flavonoids	Bathochromic shift with NaOAc/H₃BO₃ (nm) relative to MeOH spectrum	Bathochromic shift with AlCl₃ (nm) relative to AlCl₃/HCl spectrum
Quercetin 3,7-O-diglucoside	25	38
Quercetin 3-methyl ether	20	41
Quercetin 3-O-galactoside	15	33
Quercetin 3-O-glucoside 7-O-rhamnoside	22	37
Quercetin 3-O-glucoside 7-O-rutinoside	22	37
Quercetin 7-O-rhamnoside	14	32
Quercitrin	17	29
Rhamnetin	18	28
Robinetin	18	21
Rutin	28	31

a = Based on major absorption bands at 340 and 371 nm.
b = In the presence of AlCl₃, fisetin 3-O-glucoside rapidly hydrolyzes to fisetin.

The complexes formed with 3- or 5-hydroxyl groups are stable (Jurd and Geissman, 1956), hence the effect indicated above after addition of HCl is not observed (Table 19).

Table 19 Effect of AlCl₃ on Band I of the UV spectrum of 5-hydroxyflavones and 3-substituted flavonols

Flavonoids	Bathochromic shift (nm) of Band I (in MeOH) to Band Ia (in the presence of AlCl₃/HCl)
2″-O-Xylosylvitexin	47
5- Hydroxyflavones	
5,7-Dihydroxy-2′-methoxyflavone	53
5,7-Dihydroxy-3′,4′,5′-trimethoxyflavone	51
5,7-Dihydroxy-3′,4′-dimethoxyflavone	41
5-Hydroxyflavone	60
Amentoflavone	50
Apigenin	45
Apigenin 7-O-glucoside	49
Apigenin 7-O-neohesperidoside	47
Chrysoeriol	39
Diosmetin	39
Flavonols with 3-hydroxyl substituted	
Galangin 3-methyl ether	51

Contd.

Table 19 *Contd.*

Flavonoids	Bathochromic shift (nm) of Band I (in MeOH) to Band Ia (in the presence of AlCl₃/HCl)
Isoorientin	35
Isorhamnetin 3-O-galactoside	46
Isorhamnetin 3-O-rutinoside	43
Isovitexin	44
Kaempferol 3-O-robinoside 7-O-rhamnoside (Robinin)	48
Luteolin	36
Luteolin 7-O-glucoside	39
Luteolin 7-O-rutinoside	40
Orientin	38
Quercetin 3-methyl ether	44
Quercetin 3-methyl ether 4'-O-glucoside 7-O-diglucoside	50
Quercetin 3-O-galactoside	43
Quercetin 3-O-glucoside 7-O-rhamnoside	46
Quercetin 3-O-glucoside 7-O-rutinoside	46
Quercetin 3,7-O-diglucoside	47
Quercitrin	51
Rhamnosylvexin	47
Rutin	43
Saponarin	42
Scoparin	37
Tricin	36
Violanthin	48
Vitexin	47

2.6.2 *Isoflavones, Dihydroflavones and Dihydroflavonols*

Isoflavones, flavanones and dihydroflavonols have similar UV spectra since there is little or no conjugation between the A- and B-rings (Fig. 12).

Isoflavone skeleton   Flavanone skeleton   Dihydroflavonol skeleton

Figure 12 Isoflavones, dihydroflavones and dihydroflavonols skeletons.

2.6.2.1 Substituent Effects on Isoflavones

The effect of oxidation and substitution patterns on isoflavone nucleus can be summarized as follows:

- Absorption Band II of isoflavones usually occurs in the region of 245-270 nm. This is relatively unaffected when the hydroxylation of the B-ring increases, but increasing oxygenation in the A-ring produces a bathochromic shift (Table 20).
- The UV spectra of 6,7-dioxygenated isoflavones, show an abnormal intense Band I and the spectra are similar to those observed for flavones such as texasin (325 nm) and afrormosin (320 nm).
- The substitution of the 5-hydroxyl group (methylation or glycosidation) causes a 5-10 nm hypsochromic shift and the loss of this group causes a 7-17 nm hypsochromic shift of Band II (Table 21).
- The substitution of 7- or 4'-hydroxyl groups has little or no effect on the UV spectrum (Table 22).

Table 20   Effect produced by increasing oxygenation in the B-ring of isoflavones.

Isoflavones	λ_{max} nm
6-Hydroxygenistein	254sh, 270, 350sh
Iridin	268, 331sh
Irigenin	268, 336sh
Irisolidone	265, 335sh
Tectoridin	266,331
Tectorigenin	267, 330sh

sh: shoulder

Table 21   Effect produced by substitution of 5-hydroxyl group

Isoflavones	λ_{max} nm
5,7-Dihydroxyisoflavone	259, 303sh, 315sh
5,7-Dimethoxyisoflavone	251, 308sh
Daidzein	238sh, 249, 259sh, 303sh
Genistein	261, 328sh
Genistein 5-methyl ether	256, 283sh, 317sh

sh: shoulder

Table 22   Effect produced by substitution of 7- or 4'-hydroxyl groups.

Isoflavones	λ_{max} nm
Biochanin A	261, 330sh
Genistein	261, 328sh
Lanceolarin	262, 325sh
Prunetin	262, 327sh
Sphaerobioside	262, 327sh
Sophoricoside	261, 324sh

sh: shoulder

2.6.2.2 Substituent Effects on Dihydroflavones and Dihydroflavonols

Both flavanones and dihydroflavonols have their major absorption peak (Band II) in the range of 270-295 nm and are, therefore, easily distinguished from the isoflavones spectra (245-270 nm). The effect of the oxidation and substitution pattern for flavanones and dihydroflavonols can be summarized as follows:

- Loss of the 5-hydroxyl group causes a 10-15 nm hypsochromic shift of Band II (Table 23).
- Increasing oxygenation in the B-ring has no noticeable effect on the UV spectra (Table 24).

Table 23   Effect produced by loss of 5-hydroxyl group in MeOH

Flavonoids	λ_{max} nm
Dihydrokaempferol	291, 329sh
Garbanzol	276, 311
Liquiritigenin	276, 312
Naringenin	289, 326sh

sh: shoulder

Table 24   Effect produced by increasing oxygenation in the B-ring in MeOH

Flavonoids	λ_{max} nm
Dihydrofisetin	277, 310
Dihydrorobinetin	275, 308
Eriodictyol	289, 324sh
Garbanzol	276, 311
Naringenin	289, 326sh

2.6.2.3 Shifts Reagents Effects on Isoflavones, Dihydroflavones and Dihydroflavonols

The addition of NaOMe or NaOAc to isoflavones, dihydroflavones and dihydroflavonols produces a bathochromic shift of Band I, Band II or both Bands. The isoflavones carrying hydroxyl groups in A-ring show a bathochromic shift for both Band I and Band II. The bathochromic shift observed for 3',4'-dihydroxyisoflavones spectra shows a reduction of intensity in a few minutes. This effect indicates sample decomposition. The decomposition rate helps in distinguishing between 4'-monohydroxyisoflavones (slow decomposition) and 3',4'-dihydroxyisoflavones (fast decomposition).

Dihydroflavones and dihydroflavonols with A-ring hydroxylation show a bathochromic shift in Band II (Table 26). However, under alkaline conditions, some flavanones (particularly those lacking a free 5-hydroxyl group) will isomerize to chalcones, which have entirely different UV spectra.

The addition of NaOAc to 7-hydroxyisoflavones produces a bathochromic shift of Band II (6-20 nm) (Table 25). Bathochromic shifts of 34-37 nm for 5,7-dihydroxyflavanones and 5,7-dihydroxydihydroflavonols are observed, while for their 5-deoxy-equivalents, the observed shifts are about 51-58 nm (Table 26). The UV spectra of 5,6,7-trihydroxyflavanone show degeneration few minutes after the addition of NaOAc. The presence of *ortho*-dihydroxyl groups in the A-ring can be detected by a 10 nm bathochromic shift of Band I with the addition of NaOAc/H_3BO_3.

Table 25 Band II shifts in the UV spectra of 7-hydroxyisoflavones with NaOAc

Flavonoids	Bathochromic shift (nm)
2-Carboxy-5,7-dihydroxyisoflavone	14
2-Carboxy-6,7-dihydroxy-4′-methoxyisoflavone	13
3′,4′,7-Trihydroxyisoflavone	8
5,7-Dihydroxyisoflavone	14
6-Hydroxygenistein	Dec
7-Hydroxyisoflavone	21
Baptigenin	8
Biochanin A	11
Daidzein	4
Formononetin	6
Genistein	10
Genistein 5-methyl ether	8
Irigenin	5
Irisolidone	8
Orobol	8
Pratensein	9
Pseudobaptigenin	8
Sophoricoside	11
Tectorigenin	6

Table 26 The shift of Band II in the UV spectra of 7-hydroxyflavanones and 7-hydroxydihydroflavonols with NaOAc

Flavonoids	Band II [a] Bathochromic shift (nm)
Flavanones	
(+)-Fustin 3-O-glucoside	58
5,6,7-Trihydroxyflavanone	Dec
Astilbin	37
Dihydrofisetin	57
Dihydroflavonols	
Dihydrokaempferol	36
Dihydrorobinetin	58
Engeletin	36
Eriodictyol	36
Garbanzol	58
Liquiritigenin	51
Naringenin	34
Pinocebrin	34
Taxifolin	37

[a] Major absorption band; Dec = decompose

The presence of 3′,4′-dihydroxyl groups in isoflavones, flavanones and dihydroflavonols is not detectable by means of the addition of AlCl₃ because the B-ring has little or no conjugation with the major chromophore.

The compounds containing *ortho*-dihydroxyl groups at positions 6, 7 or 7, 8, which do not involve the C-5 hydroxyl group, exhibit bathochromic shift of both Bands I and II. 5-hydroxyisoflavones and 5-hydroxydihydroflavonols suffer a bathochromic shift of 20-26 nm (Table 27).

Table 27 The shift of Band II in the UV spectra of 5-hydroxyisoflavones, 5-hydroxyflavanones and 5-hydroxydihydroflavonols in the presence of $AlCl_3/HCl$

Flavonoids	Band II Bathochromic shift (nm)
Isoflavones	
2-Carboxy-5,7-dihydroxyisoflavone	10
5,7-Dihydroxyisoflavone	14
6-Hydroxygenistein	11
Biochanin A	12
Genistein	21
Genistin	12
Lanceolarin	12
Prunetin	12
Sphaerobioside	11
Flavanones	
5,6,7-Trihydroxyflavanone	22
5,6,7-Trihydroxyflavanone 7-O-glucuronide	26
Eriodictyol	20
Hesperidin	23
Naringenin	22
Pinocebrin	20
Dihydroflavonols	
Dihydrokaempferol	21
Engeletin	21
Taxifolin	22

2.6.3 Chalcones and Aurones

The UV spectra of both chalcones an aurones (Fig. 13) are characterized by an intense absorption in Band I and a diminished absorption in Band II.

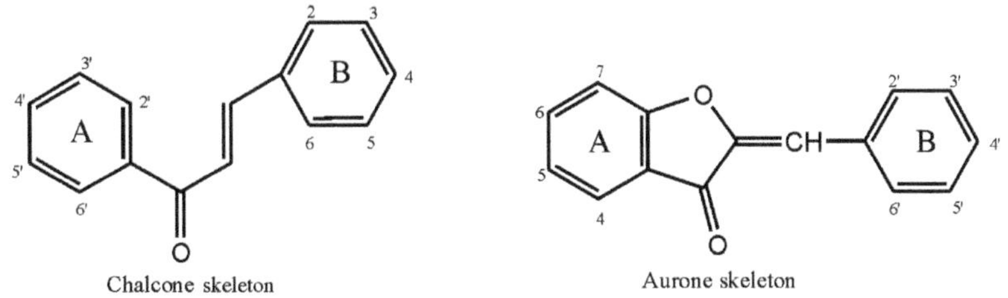

Chalcone skeleton   Aurone skeleton

Figure 13 Chalcones and aurones skeletons.

2.6.3.1 Substituent Effects on Chalcones and Aurones

As with flavones and flavonols, increased oxygenation of either the A- or B-ring usually results in bathochromic shifts of Band I. The aurones exhibit a long wavelength absorption band between 370-430 nm; although some of the simpler aurones such as 6-hydroxyaurone and 5,7-dihydroxyaurone absorb at shorter wavelengths. Methylation or glycosylation of a hydroxyl group on the aurone nucleus has little effect on the UV spectrum, with the exception of the 18 nm hypsochromic shift observed when the 7-hydroxyl group in a 6,7-dihydroxyaurone is methylated.

2.6.3.2 Shifts Reagents Effects on Chalcones and Aurones

The UV-Vis spectra of chalcones show a bathochromic shift of Band I after addition of NaOMe. The chalcones that contain a free 4-hydroxyl group show a bathochromic shift of 60-100 nm with an increase in the intensity. Chalcones with either a free 2- or 4'-hydroxyl group also suffer a bathochromic shift of 60-100 nm, but without an increase in the intensity.

Aurones with a free 4'-hydroxyl group show a bathochromic shift of Band I of 80-95 nm. For those with a free 6-hydroxyl group the effect is of *ca.* 70 nm. Aurones containing 6- and 4'-hydroxyl groups show a reduced bathochromic shift. Chalcones containing free 4' and/or 4-hydroxyl group and aurones with 4'- and/or 6-hydroxyl group show a bathochromic shift in Band I. Chalcones with three adjacent hydroxyl groups may decompose in NaOAc.

The complexes formed between H_3BO_3/NaOAc and chalcones or aurones carrying *ortho*-dihydroxyl groups in B-ring show a bathochromic shift of 28-36 nm. For A-ring *ortho*-dihydroxyl groups a small bathochromic shift is observed (Table 28).

Chalcones or aurones with *ortho*-dihydroxyl groups in B-ring suffer a bathochromic shift (40-70 nm) in Band I by the addition of $AlCl_3$. Meanwhile,

Table 28 Band I shifts in the UV spectra of *ortho*-dihydroxychalcones and *ortho*-dihydroxyaurones in the presence of NaOAc/H_3BO_3 and $AlCl_3$

Flavonoids	Bathochromic shift with NaOAc/H_3BO_3 (nm) relative to MeOH spectrum	Bathochromic shift with $AlCl_3$ (nm) relative to $AlCl_3$/HCl spectrum
Chalcones		
2',3,4,4'-Tetrahydroxychalcone	36	63
2',3,4-Trihydroxychalcone	30	67
2',3',4'-Trihydroxychalcone	10	22
3,4-Dihydroxychalcone	36	48
Aurones		
3',4',6,7-Tetrahydroxyaurone	33	48
3',4'-Dihydroxyaurone	32	50
6,7-Dihydroxyaurone	22	39
Leptosidin	28	44

those with *ortho*-dihydroxyl groups in A-ring show a smaller bathochromic shift (Table 29). 2-Hydroxychalcones show a large bathochromic shift (48-64 nm) for Band I in presence of $AlCl_3/HCl$ and 2',3',4'-trihydroxychalcones and its derivatives show a bathochromic shift of about 40 nm (Table 29).

Table 29 Band I shifts in the UV spectra of 2'-hydroxychalcones in the presence of $AlCl_3/HCl$

Flavonoids	Band I Bathochromic shift (nm)
2,2',4-Trihydroxychalcone	62
2,2'-Dihydroxychalcone	64
2',3,4,4'-Tetrahydroxychalcone	48
2',3,4-Trihydroxychalcone	63
2',3',4'-Trihydroxychalcone	39
2',4,4'-Trihydroxychalcone	54
2',4-Dihydroxychalcone	58
2'-Hydroxy 4'-methoxychalcone	64

3. EXPERIMENTAL PROCEDURES

3.1 Structural Analysis by UV-Vis Spectroscopy (Mabry *et al.*, 1970)

Experiment 1: Use of Shifts Reagents

Principle

Substances with chromophore groups can be identified according to their UV-Vis spectra. This can be greatly enhanced by repeating the measurements of compounds dissolved in solution either at different pH values or in the presence of particular reagents. In these cases, shifts to longer wavelengths (bathochromic shifts) or to shorter wavelengths (hypsochromic shifts) may occur. The following procedure is applied to flavonoid analysis. Nevertheless, shift reagent effects described above are for the analysis of coumarins, furanocoumarins and naphthoquinones, but can also be applied using standards of these compounds.

Materials and equipments required

Analytical balance; glass stoppered bottle; volumetric flask; pipettes; UV-Vis light (200 to 700 nm) spectrophotometer; 1-cm path length cuvette.

Reagents

Sodium methoxide (NaOMe): Freshly cut metallic sodium (2.5 g) is added cautiously in small portions to dry spectroscopic methanol (100 mL); the solution is stored in a glass container with a tightly fitting plastic stopper.

Aluminum chloride ($AlCl_3$): 5 g of fresh anhydrous reagent grade $AlCl_3$ (which appears yellow-green and reacts violently when mixed with water) is added cautiously to 100 mL of spectroscopic methanol.

Hydrochloric acid (HCl): 50 mL of concentrated HCl is mixed with 100 mL of distilled water; the solution is stored in a glass stoppered bottle.

Powdered anhydrous sodium acetate (NaOAc).

Boric acid (H_3BO_3): For procedure 1, anhydrous H_3BO_3 is used; for procedure 2, spectroscopic methanol (100 mL) is saturated with anhydrous H_3BO_3.

Stock solutions described above can be used upto six months after preparation. To avoid excessive exposure of stock solutions to the atmosphere, four 30 dropping bottles, each containing about 15 mL of one of the stock solutions, should be kept near the spectrophotometer. The solutions in the dropping bottles are used for the spectral analyses and are always replaced monthly.

Procedure

   (i) A stock solution of the flavonoid is prepared dissolving 0.1 mg compound in 10 mL of spectroscopic methanol. The concentration is then adjusted so that absorbance of the major absorption peak, comprised between 250 and 400 nm region, is 0.6 to 0.8. When the flavonoid (or any compound) has been purified by paper chromatography, the zone of the chromatogram where the compound was detected, is cut into small pieces which are then shaken for 10 minutes or less with 50 to 100 mL of the highest quality spectral grade methanol in a 250 mL Erlenmeyer flask (reagent grades of methanol contain traces of non-volatile substances which absorb in the 280-220 nm range). The solution is filtered and is then evaporated to dryness on a rotary evaporator; the obtained residue is dissolved in 10 mL spectral grade methanol and directly used (or further diluted if needed) for spectral analysis.

  (ii) The methanol spectrum is measured at normal scan speed using 2-3 mL of the sock solution of the flavonoid.

 (iii) The methanol spectrum is rerun at slow scan speed in the regions of the peak maxima to determine the wavelength (λ) of each maximum more accurately.

 (iv) The NaOMe spectrum is measured immediately after the addition of three drops of the NaOMe stock solution to the solution used for steps 2 and 3. After 5 minutes the spectrum is rerun to check for flavonoid decomposition. The solution is then discarded.

  (v) The $AlCl_3$ spectrum is measured immediately after the addition of six drops of the $AlCl_3$ stock solution to 2-3 mL of fresh stock solution of the flavonoid. For few isoflavones and dihydroflavonols, $AlCl_3$ requires about a minute to produce its maximum effect on the UV spectrum.

 (vi) The $AlCl_3$/HCl spectrum is recorded immediately after the addition of three drops of the stock HCl solution to the cuvette containing the $AlCl_3$ (from step 5). The solution is then discarded.

(vii) To obtain NaOAc spectrum, coarsely powdered anhydrous NaOAc is added, by shaking the cuvette containing 2-3 mL of fresh stock solution of the flavonoid. About a 2 mm layer of NaOAc remains on the cuvette. The NaOAc spectra should be recorded within 2 minutes after the addition of NaOAc to the solution (with decomposing compounds, the time factor is critical). A second spectrum is run after 5-10 minutes to check for decomposition.

(viii) For obtaining $NaOAc/H_3BO_3$ spectrum, two methods can be used depending on whether or not decomposition of the compound is observed during the recording of the NaOMe spectrum. If no decomposition is observed when the NaOMe spectrum is rerun after 5 minutes, proceed as indicate below in **procedure 1.** When decomposition of the flavonoid occurs after addition of NaOMe, **procedure 2** is applied.

Procedure 1: Powdered anhydrous H_3BO_3 is added, shaking the cuvette, to the solution containing NaOAc (from step 7). The solution is discarded after spectrum recording.

Procedure 2: Five drops of the H_3BO_3 stock solution are added to 2-3 mL of fresh stock solution of the flavonoid. The solution is then quickly saturated with coarsely powdered NaOAc and the $NaOAc/H_3BO_3$ spectrum is recorded.

3.2 Quantitative Analysis

3.2.1 Derivative Spectrophotometry

Experiment 2: Total Flavonoids Quantification by Derivative Spectrophotometry

Principle

Derivative spectrophotometry (Rolim *et al.*, 2006) consists of differentiating the normal spectrum by mathematical transformation of spectral curve into a derivative (first- or higher derivatives). This technique usually improves the resolution bands, eliminates the influence of background or matrix and provides fingerprints more defined than traditional or direct absorbance spectra, since it enhances the detectability of minor spectral features (Kazemipour *et al.*, 2002; Ojeda and Rojas, 2004). Derivative transformation allows the discrimination against broad band interferents, arising from turbidity or non-specific matrix absorption and it tends to emphasizes subtle spectra features, allowing the enhancement of the sensitivity and specificity in mixture analysis (Karpinska, 2004).

Materials and equipments required

Analytical balance, beakers, 25 and 50 mL volumetric flasks, pipettes, Beckman DU-640 UV-Vis spectrophotometer capable of taking first to fourth-derivative spectra by an internal data processing system, 1-cm quartz cuvette.

Reagents

Standard stock solution and serial dilutions: 25.0 mg of rutin are dissolved in 50 mL of 99.5% ethanol in a volumetric flask. Serial dilutions are prepared to concentrations of 10.0, 20.0, 30.0, 40.0, 50.0 and 60.0 $\mu g \ mL^{-1}$, transferring appropriate amounts of the stock solution to 25 mL volumetric flasks.

Sample solutions: An oil/water (O/W) emulsion is developed with a self-emulsifying agent that allows emulsification at room temperature. The emulsifying agent is made with the following components of pharmaceutical grade: polyacrylamide, C13-14 isoparaffin, laureth-7 (Seppic, France), octyldodecanol (Cognis, Germany), isopropyl palmitate (Croda, UK), phenoxyethanol, methylparaben, ethylparaben, propylparaben, butylparaben (Croda, UK), cyclomethicone (Dow Corning, United States), glycerine (LabSynth, Brazil), disodium edentate (LabSynth, Brazil), *Trichilia castigua* Adr. Juss (Meliaceae), and *Ptychopetalum olacoides* Bentham (Olacaceae) extract (Chemyunion, Brazil), and distilled water. Qualitative and quantitative composition (% w/w) of O/W emulsion is presented in Table 30.

Table 30 Qualitative and quantitative (% w/w) composition of O/W emulsion

Chemical name (% w/w)	Composition
Cyclomethicone	1
Disodium edentate	0.1
Distilled water	83.4
Glycerine	3
Isopropyl palmitate	1
Octyldodecanol	3
Phenoxyethanol, methylparaben, ethylparaben, propylparaben and butylparaben	0.5
Polyacrylamide, C13-14 isoparaffin and laureth-7	3
Trichilia castigua Adr. Juss (Meliaceae) and *Ptychopetalum olacoides* Bentham (Olacaceae) extract	5

O/W emulsion is weighed (1.0, 1.5 and 2.0 g) and transferred to 25 mL volumetric flasks. Volumes are completed with ethanol 99.5%. Solutions are centrifuged at $863 \times g$ for 5 minutes at room temperature and supernatants are discarded.

Procedure

(i) The O/W emulsion analyte-free samples are spiked with a known appropriate amount of standard rutin (0.05% w/w).

(ii) Spiked samples are diluted to concentrations of 20.0, 30.0 and 40.0 µg mL^{-1} of rutin in 99.5% ethanol.

(iii) Spectra are recorded in a wavelength range of 300 to 450 nm, in nine replications (i.e. Fig. 14).

(iv) The first and second derivate of normal spectrum is obtained for each replicate as sigmoid functions (i.e. Fig. 15).

(v) Amplitude of the sigmoid function is plotted against concentration of rutin (i.e. Fig. 16)

(vi) The total flavonoids concentration is determined by extrapolation from the calibration curve.

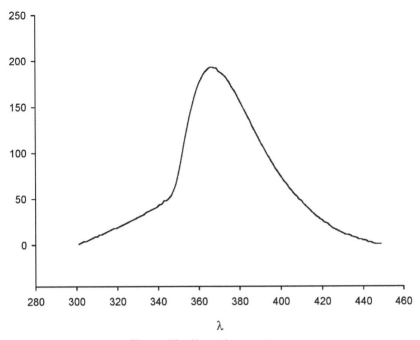

Figure 14 Absorption spectrum.

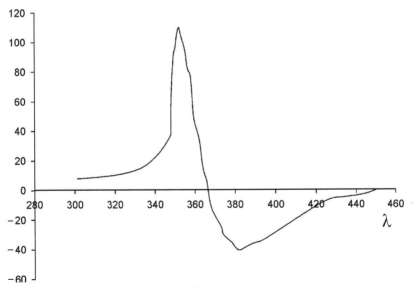

Figure 15 Sigmoid function.

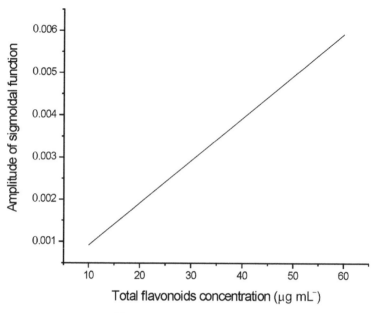

Figure 16 Calibration curve.

3.2.2 Spectrophotometric Methods Based on the Borntrager's Reaction

Experiment 3: Anthraquinones Quantification by Spectrophotometry

Materials and equipments required

Analytical balance, reflux equipment, rotary evaporator, water bath, beakers, pipettes, UV-Vis (200 to 700 nm) spectrophotometer, 1-cm path length cuvette.

Borntrager's reaction is commonly used to identify the anthraquinone derivatives (Barrese *et al.*, 2005). A known amount of the sample material is boiled in dilute, aqueous potassium hydroxide for a few minutes. This hydrolyzes glycosides and also oxidizes anthrones or anthranols to anthraquinones. The alkaline solution is cooled, acidified and extracted with ethyl ether. When the ethyl ether phase is separated and shaken with dilute alkali, the alkaline phase becomes red, indicating presence of quinones. If partially reduced anthraquinones are present, the original solution does not turn red immediately in alkaline medium but turns yellow with green fluorescence and then gradually becomes red as oxidation occurs. Oxidation may be hastened by adding a small amount of 3% hydrogen peroxide. The Borntrager's reaction can be used for quantitative colorimetric determinations.

Procedure

(i) 5 g of sample is weighed. 100 mL of water is added. The mixture is refluxed for 15 minutes.

(ii) Two aliquots of 20 mL are taken. 20 mL of 0.1 M HCl and 1.2 g of $FeCl_3$ are added to one aliquot (solution I). 20 mL of 0.1 M HCl are added to the other (solution II).

(iii) The solutions (solution I and solution II) are refluxed for 1 hour.

(iv) The solutions are extracted with ethyl ether.

(v) The organic phase is washed twice with 10 mL of water and is then evaporated.

(vi) The residues are dissolved in 100 mL of 0.1 M KOH (solution I' and solution II').

(vii) Two aliquots of 20 mL are taken from each solution.

(viii) 5 drops of $KMnO_4$ are added to aliquots warming up in a water bath. It is continued until the oxidizing agent become colourless.

(ix) These are extracted with ethyl ether. Organic phases are washed twice with water (10 mL) and evaporated.

(x) The residues are dissolved with 20 mL of KOH 0.1 M (solution I'' and solution II'').

(xi) The quantitative determination can be made by:
    (A) Direct spectrophotometry ($COCl_2$ as reference compound)
    (B) Derivative spectrophotometry

(A) Direct Spectrophotometry ($COCl_2$ as Reference Compound)

Experiment 4: Anthraquinones Quantification by Direct Spectrophotometry

Reagents
Standard stock solution and serial dilutions: The stock solution is prepared by dissolving 200 mg of $COCl_2$ with 50 mL of methanol:water (65:35) in a volumetric flask. Serial dilutions are prepared to concentrations of 150, 300, 400, 500, 600 and 800 µg mL^{-1} transferring appropriate amounts of the stock solution to 25 mL volumetric flasks.

Sample solutions

Procedure
Refer to Section 3.2.2 (in this chapter)

(i) Spectra are recorded in wavelengths between 420 and 500 nm, in nine replicates.

(ii) The first and second derivates of normal spectrum are obtained for each replicate as sigmoid functions.

(iii) Amplitude of the sigmoid function is plotted against concentration of anthraquinone.

(iv) The anthraquinone concentration is determined by extrapolation from the calibration curve.

(B) Derivative Spectrophotometry

Experiment 5: Anthraquinones Quantification by Derivative Spectrophotometry

Reagents

Standard stock solution and serial dilutions: Stock solution is prepared dissolving 150 mg of kermesic acid with 50 mL of methanol:water (65:35) in a volumetric flask. Serial dilutions are prepared to concentrations of 100, 250, 400, 500, 600 and 800 µg mL^{-1} transferring appropriate amounts of the stock solution to 25 mL volumetric flasks.

Sample solutions

Procedure

Refer to Section 3.2.2 (in this chapter)

(i) Spectra are recorded in wavelengths between 420 and 500 nm, in nine replicates.
(ii) The first and second derivate of normal spectrum is obtained for each replicate in the form of sigmoid functions.
(iii) Amplitude of the sigmoid function is plotted against concentration of kermesic acid.
(iv) The anthraquinone concentration is determined by extrapolation from the calibration curve.

SUGGESTED READINGS

Ayres, D. and Loike, J. (1990). *Lignans: Chemical, Biological and Clinical Properties.* Cambridge University Press, Cambridge, United Kingdom.

Barrese, Y., Hernández, P. and Garcia, O. (2005). Development of an analytical technique to quantify anthraquinones present in *Senna alata* (L.). *Roxb. Revista Cubana de Plantas Medicinales* **10**: 3-4. (In Spanish).

Carpenter, I., McGarry, E. and Scheinmann, F. (1970). The neoflavonoids and 4-alkylcoumarins from *Mammea africana* G. Don. *Tetrahedron Letters* **11**: 3983-3986.

Carpenter, I., McGarry, E. and Scheinmann, F. (1971). Extractives from Guttiferae. Part XXI. The isolation and structure of nine coumarins from the bark of *Mammea africana* G. Don. *Journal of the Chemical Society* **3C**: 3783-3790.

Chandel, R. and Rastogi, R. (1980). Pygeoside a new lignan xyloside from *Pygeum acuminatum*. *Indian Journal of Chemistry* **19B**: 279-282.

Crombie, L., Games, D. and McCormick, A. (1966a). Isolation and structure of Mammea A/BA, A/AB and A/BB: A group of 4-aryl-coumarin extractives of *Mammea americana* L. *Tetrahedron Letters* **7**: 145-149.

Crombie, L., Games, D. and McCormick, A. (1966b). Isolation and structure of Mammea B/BA, B/BB, B/BC and C/BB: A group of 4-n-propyl- and 4-n-amyl-coumarin extractives of *Mammea americana* L. *Tetrahedron Letters* **7**: 151-155.

Crombie, L., Games, D. and McCormick, A. (1967a). Extractives of *Mammea americana* L. Part I. The 4-n-Alkylcoumarins. Isolation and structure of Mammea B/BA, B/BB, B/BC and C/BB. *Journal of the Chemical Society* **2C**: 2545-2552.

Crombie, L., Games, D. and McCormick, A. (1967b). Extractives of *Mammea americana* L. Part II. The 4-Phenylcoumarins. Isolation and structure of Mammea A/AA, A/A, cycloD, A/BA, A/AB and A/BB. *Journal of the Chemical Society* **5C**: 2553-2559.

Dean, F. (1963). *Naturally Occurring Oxygen Ring Compounds*. Butterworth, London.

Djerassi, C., Eisenbraun, E., Gilbert, B., Lemin, A., Marfey, S. and Morris, M. (1958). Naturally occurring oxygen heterocyclics II. Characterization of an insecticidal principle from *Mammea americana* L. *Journal of the American Chemical Society* **80**: 3686-3691.

Dreyer, D.L. (1970). Extractives of *Angelica genuflexa* Nutt. *Journal of Organic Chemistry* **35**: 2294-2297.

Evcim, U., Gozler, B., Freyer, A. and Shamma, M. (1986). Haplomyrtin and (–)-haplomyrfolin: two lignans from *Haplophyllum myrtifolium*. *Phytochemistry* **25**: 1949-1951.

Govindachari, T., Pai, B., Subramaniam, P., Rao, U. and Muthumaraswamy, N. (1967). Constituents of *Messua ferrea* L.-H. Ferruol A, a new 4-alkylcoumarin. *Tetrahedron* **23**: 4161-4165.

Harborne, J. (1984). *Phytochemical Methods. A Guide to Modern Techniques of Plant Analysis*. 2nd edition. Chapman and Hall, London.

Horowitz, R. and Gentili, B. (1960). Flavonoids of Citrus IV. Isolation of soe aglycones from Lemon (*Citrus limon*). *Journal of Organic Chemistry* **25**: 2183-2187.

Jurd, L. and Geissman, T. (1956). Absorption spectra of metal complexes of flavonoid compounds. *Journal of Organic Chemistry* **21**: 1395-1401.

Jurd, L. and Horowitz, R. (1957). Spectral studies on flavonols – the structure of Azalein. *Journal of Organic Chemistry* **22**: 1618-1622.

Jurd, L. (1962). Spectral properties of flavonoid compounds. In: *The Chemistry of Flavonoid Compounds* (Ed., T.A. Geissman). Pergamon Press, Oxford, United Kingdom

Karpinska, J. (2004). Derivative spectrophotometry - recent applications and directions of developments. *Talanta* **64**: 801-822.

Kazemipour, M., Noroozian, E., Tehrani, M. and Mahmoudian, M. (2002). A new second-derivate spectrophotometric method for determination of permethrin in shampoo. *Journal of Pharmaceutical and Biomedical Analysis* **30**: 1379-1384.

Le Quesne, P., Larrahondo, J. and Raffauf, R. (1980). Anti-tumour plants, part X : Constituents of *Nectandra rigida*. *Journal of Natural Products* **43**: 353-359.

Lee, S., Li, G., Kim, H., Kim, J., Chang, H., Jahng, Y., Woo, M., Song, D. and Son, J. (2003). Two new furanocoumarins from the roots of *Angelica dahurica*. *Bulletin of Korean Chemical Society* **24**: 1699-1701.

Livingston, A.L., Bickoff, E., Lundin, R. and Jurd, L. (1964). Trifoliol, a new coumestan from ladino clover. *Tetrahedron* **20**: 1963-1970.

Mabry, T., Markham, K. and Thomas, M. (1970). *The Systematic Identification of Flavonoids*. Springer-Verlag, Berlin.

Mahmood U., Kaul, V. and Jirovetz, L. (2002). Alkylated benzoquinones from *Iris kumaonensis*. Phytochemistry **61**: 923-926.

Makayama, M., Fujimoto, K. and MacLeod, J. (1971). The isolation, structure, and synthesis of Halkendin. *Australian Journal of Chemistry* **24**: 209-211.

Markham, K. and Mabry, T. (1968). A procedure for the UV spectral detection of ortho-dihydroxyl groups in flavonoids. *Phytochemistry* **7**: 1197-1200.

Morton, R. (Ed). (1965). *Biochemistry of Quinones*. Academic Press, New York.

Murray, R. (1995). Coumarins. *Natural Product Reports* **12**: 477-505.

Murray, R., Mendéz, J. and Brown, S. (1982). *The Natural Coumarins. Occurrence, Chemistry and Biochemistry*. John Willey & Sons, Chichester, New York.

Nassar, M., Abdel-Razik, A., El-Khrisy, Eel-D, Dawidar, A., Bystrom, A. and Mabry, T. (2002). A benzoquinone and flavonoids from *Cyperus alopecuroides*. *Phytochemistry* **60**: 385-387.

Nielsen, B. (1970). Coumarins of umbelliferous plants. *Dansk Tidsskrift for Farmaci* **44**: 111-286.

Ojeda, C. and Rojas, E. (2004). Recent developments in derivative ultraviolet/visible spectrophotometry. *Analytica Chimica Acta* **518**: 1-24.

Prynce, R. (1972). Gallic acid as a natural inhibitor of flowering in *Kalanchoe blossfeldiana*. *Phytochemistry* **11**: 1911-1918.

Rao, K. and Wu, W. (1978). Glycoside of Magnolia, Part III. Structural elucidation of magnolenin C. *Lloydia* **41**: 56-62.

Rolim, A., Oishi, T., Maciel C., Zague, V., Pinto, C., Kaneko, T., Consiglieri, V. and Velasco, M. (2006). Total flavonoids quantification from O/W emulsion with extract of Brazilian plants. *International Journal of Pharmaceutics* **308**: 107-114.

Sariaslani, F. and Rosazza, J. (1983). Novel Biotransformations of 7-Ethoxycoumarin by *Streptomyces griseus*. *Applied and Environmental Microbiology* **46**: 468-474.

Stanley, W. and Vannier, S. (1967). Psoralens and Substituted Coumarins from Expressed Oil of Lime. *Phytochemistry* **6**: 585-596.

Tada, Y., Shikisshima, Y., Takaishi, Y., Shibata, H., Higuti, T., Honda, G., Ito, M., Takeda, Y., Kodzhimatov, O., Ashurmetov, O. and Ohmoto, Y. (2002). Coumarins and γ-pyrone derivatives from *Prangos pabularia*: antibacterial activity and inhibition of cytokine release. *Phytochemistry* **59**: 649-654.

Thompson, R. (1971). *Naturally Occurring Quinones*. 2nd edition. Academic Press, London.

Spectrometry: Infrared Spectra

Adriana del V. Pacciaroni and Virginia E. Sosa*

1. INTRODUCTION

Electromagnetic radiation (EM) is an invaluable tool for structural elucidation of organic compounds. The EM radiation with frequencies between 4,000 and 400 cm^{-1} (wave numbers), is termed infrared (IR) radiation and its application to organic chemistry is known as IR spectroscopy. The IR radiation allows determination of organic structures because it is absorbed by the covalent bonds of organic compounds. Since different chemical bonds absorb at different frequencies with different intensities, the IR spectrum of an organic compound is constituted by a collection of absorption intensities at different frequencies. The frequencies at which IR radiation is absorbed ("peaks" or "signals") can be correlated directly to bonds found within the compound under analysis. Although the IR spectrum is characteristic of the entire molecule, certain groups of atoms give rise to absorptions at or near the same frequency regardless of the structure of the rest of the molecule. The persistence of these characteristic bands provides useful structural information.

IR radiation is the part of the electromagnetic spectrum comprised between the visible and microware radiations. Band positions in the IR spectra are presented as wave numbers (v) expressed in reciprocal centimeters (cm^{-1}). This unit is proportional to the energy of vibration. Wavelength (λ) was used in older literature and expressed in micrometer units (μm = 10^{-6} m). Wave numbers are reciprocally related to wavelength.

$$cm^{-1} = 10^4/\mu m$$

Authors' address: *Departamento de Química Orgánica, Facultad de Ciencias Químicas, Universidad Nacional de Córdoba, Instituto Multidisciplinario de Biología Vegetal IMBIV (CONICET-UNC). Haya de la Torre y Medina Allende. Edificio de Ciencias II. Ciudad Universitaria. Córdoba, Argentina.
Corresponding author: E-mail: vesosa@fcq.unc.edu.ar

Organic chemists usually report the intensity of the signal in quantitative terms as follows: (*vs*) very strong, (*s*) strong, (*m*) medium, (*w*) weak, and (*vw*) very weak (Pasto *et al.*, 1992).

2. CHARACTERISTIC ABSORPTION BANDS

The wave number for a stretching vibration is related to both the constant force between the atoms (K) and the mass of the two atoms (m_1 and m_2) according with Hooke's Law equation:

$$\bar{v} = \frac{1}{2\pi c}\left[K\left(\frac{m_1 + m_2}{m_1 - m_2}\right)\right]^{\frac{1}{2}}$$

This relationship allows deduction of two important trends in wave number for stretching vibrations:

A. As the bond strength increases, the wave number increases. For example:

Bond stretch	C–C	C=C	C≡C	C–H	=C–H	≡C–H
Wave number of absorption (Cm^{-1})	1,200	1,650	2,150	2,900	3,100	3,300

CC stretch   CH stretch

B. As the mass of one of the two atoms in the bond increases, the wave number decreases (assuming relatively small changes in bond strength). For example:

Bond stretch	C–H	C–D	C–C	C–O	C–CI	C–I
Wave number of absorption (cm^{-1})	3,000	2,100	1,200	1,100	800	500

These trends in absorption can be summarized into the following categories:

Wave number regions (cm^{-1})	Bond stretch
3,600 – 2,700 cm^{-1}	X–H
2,700 – 1,900 cm^{-1}	X≡Y
1,900 – 1,500 cm^{-1}	X=Y
1,500 – 500 cm^{-1}	X–Y

In an IR spectrum the intensities of the bands are different and this characteristic, together with the frequency of absorption, is useful in identifying the functional groups present in a molecule.

The absorption intensity registered in the IR spectrum is related to the change in the dipole moment that occurs during vibration. Consequently, vibrations that produce a large change in the dipole (e.g. C=O stretch) give a more intense absorption than those that result in a relatively small change in the dipole (e.g. C=C). Vibrations that do not produce change in the dipole moment (e.g. a symmetrical alkyne C=C stretch) will show little or no absorption.

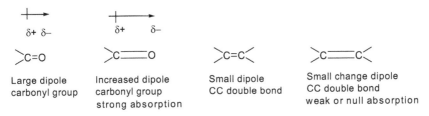

| Large dipole carbonyl group | Increased dipole carbonyl group strong absorption | Small dipole CC double bond | Small change dipole CC double bond weak or null absorption |

Interpretation of peaks in the fingerprint region (1,300-600 cm^{-1}) is complicated by the large number of different vibrations that occur there. These include single bond stretching and a wide variety of bending vibrations. This region gets its name since nearly all molecules (even very similar ones) have a unique pattern of absorption in this region (Silverstein *et al.*, 1981).

Organic chemists have recorded and catalogued the types and locations of IR absorptions produced by a wide variety of chemical bonds in various chemical environments. These data are quickly referenced through tables of IR absorption ranges and compared to the spectrum under consideration. As a general rule, the most important factors determining the location where a chemical bond will absorb are the bond order in the molecule and the types of atoms joined by the bond. Conjugation and nearby atoms shift the frequency to a lesser degree. Therefore, the same or similar functional groups in different molecules will typically absorb within the same, specific frequency ranges. Consequently, tables of IR absorptions are arranged by functional groups, in some versions these may be further subdivided to give more precise information.

Upon first inspection, a typical IR spectrum is visually divided into two regions (Table 1). The left half, above 1,600 cm^{-1}, usually contains relatively few peaks and provides extremely valuable diagnostic information. First, alkane C–H stretching absorptions just below 3,000 cm^{-1} demonstrate the presence of saturated carbons, and signals just above 3,000 cm^{-1} demonstrate unsaturation (olefinic C-H bond). A very broad peak in the region comprised between 3,100 and 3,600 cm^{-1} indicates the presence of exchangeable protons, typically from alcohol, amine, amide or carboxylic acid groups. Alkyne or nitrile groups can be easily detected in the frequencies from 2,300 to 2,100 cm^{-1} since this region is normally void of other absorptions. The carbonyl group typically gives rise to a very strong peak around 1,700 cm^{-1}. Due to its intensity and also because it occurs in a region of the spectrum usually void of other signals, a strong absorption between 1,800-1,650 cm^{-1} is perhaps the most reliable diagnosis for the presence of a C=O group.

In contrast, the right half of the spectrum, below 1,600 cm^{-1}, normally contains many peaks of different intensities, many of which are not readily identifiable. Only the C–O bond that displays one or two strong peaks between 1,250 and 1,000 cm^{-1} is considered here. Since this latter signal(s) appears in the usually very complex region known as the "fingerprint region" (1,300-600 cm^{-1}), owing to which almost every organic compound produces a unique pattern in this area, it is convenient to support the assignment by comparison of this region to a known spectrum.

Although Table 1 and similar IR absorption tables provide a good starting point for IR spectra analysis, additional details of IR properties are often needed. The following topics cover the most important IR principles.

Table 1   Important regions of the IR spectrum (Rubio, 1974; Chapman and Hall, 1975)

X-H stretch region: 3,600-2,700 cm^{-1}		
3,600-3,300 cm^{-1}	Alcohol O–H Amine or Amide N–H Alkyne C–H	Alcohol OH stretch: usually a broad and strong absorption near 3,400. NH stretch: typically not as broad or strong as the OH. NH$_2$ it may appear as two peaks. Terminal alkyne: C–H may be confirmed by a weak CC triple bond stretch near 2,150 cm^{-1}.
3,300-2,500 cm^{-1}	Acid O–H	Normally a very broad signal centered near 3,000 cm^{-1}.
2,850-2,750 cm^{-1}	Aldehyde C–H	Two medium intensity peaks on the right hand shoulder of the alkyl C–H's.
3,200-3,000 cm^{-1}	Aromatic (sp^2) =C–H Alkene (sp^2) =C–H	Aromatic CH's: usually appear as a number of weak absorptions. Alkene C–H is one or a couple stronger absorptions.
3,000-2,800 cm^{-1}	Alkyl (sp^3) C–H	Most of organic compounds have alkyl CH's: not usually too informative. The intensity of these peaks relative to other peaks gives a hint as to the size of the alkyl group.
C≡X stretch region: 2,700-1,900 cm^{-1}		
2,260-2,210 cm^{-1}	Nitrile C≡N	A sharp, medium intensity peak. Carbon dioxide in the atmosphere may also result in an absorption in this area if not subtracted out.
2,260-2,100 cm^{-1}	Alkyne C≡C	Peak's intensity varies from medium to nothing. Symmetrical alkynes: little or no absorption.
C=X stretch region: 1,900-1,500 cm^{-1}		
1,850-1,750 cm^{-1}	Anhydride C=O	Anhydrides have two absorptions: 1,830-1,800 and 1,775-1,740.

Contd.

	3-6 membered ring C=O	The absorption frequency increases as the ring size decreases: cyclohexanone = 1,715, cyclopentanone = 1,745, cyclobutanone = 1,780, cyclopropanone = 1,850.
1,750-1,700 cm^{-1}	Aldehyde C=O Ketone C=O Ester C=O Acid C=O	Usually the most intense absorption in the spectrum. The carbonyl stretching absorption is one of the strongest IR absorptions, and is very useful in structure determination as one can determine not only the number of carbonyl groups (assuming peaks do not overlap) but also estimate which types are present.
1,700-1,640 cm^{-1}	Amide C=O Conjugated C=O	Due to resonance, amides and conjugated carbonyls come slightly lower than "normal" C=O. Conjugation lowers the absorption by 20-50 cm^{-1}.
1,680-1,620 cm^{-1}	Alkene C=C	Absorption not as intense as that for C=O. Variable and may be fairly small in symmetrical, or nearly symmetrical cases.
1,600-1,400 cm^{-1}	Aromatic C=C	Multiple sharp, medium peaks. The pattern of peaks varies depending upon the substitution pattern. Usually one peak around 1,600 and several others at lower wave numbers.

Fingerprint region: 1,500-500 cm^{-1}

1,300-1,000 cm^{-1}	C–O	A strong absorption.
1,500-500 cm^{-1}	Various	Interpretation of peaks in the fingerprint region is complicated by the large number of different vibrations that occur here. These include single bond stretches and a wide variety of bending vibrations. Nearly all molecules have a unique pattern of absorptions in this region.

2.1 Effects of Mass on Frequency

As mentioned previously, one of the major factors influencing the IR absorption frequency of a bond is the identity of the two atoms involved. To be more precise, as the masses of the two atoms increase, the vibration frequency decreases.

2.2 Free vs Hydrogen-bonded Hydroxyl Groups

One of the most distinct and easily recognizable peaks in an IR spectrum is the broad O–H absorption of alcohols and phenols. However, it is important to understand why this broadening takes place and to consider the situations in which the peak may not have this characteristic shape. First, note that any significant quantity of a compound will contain a very large number of individual molecules, and each molecule may be hydrogen bonded to a slightly different extent. Thus, IR absorptions will occur at varying frequencies for each of these bonds when IR spectrum is acquired. The final result is that the IR peak appears broadened as a consequence of these slightly different absorptions.

IR spectra of hydroxyl-containing compounds can be acquired without seeing this broad signal. In very dilute solution of the sample or in the gas phase, hydrogen bonding is prevented through lack of molecular contact. Even in concentrated solution, larger compounds may sterically hinder hydrogen bonding, preventing exchange. In these situations the broad O–H peak is replaced by a sharp signal around 3,600 cm^{-1}.

In IR absorption tables, signal intensities (height) are usually denoted by the following abbreviations: w = weak, m = medium, s = strong, v = variable. A broad signal shape is sometimes indicated by br. Occasionally, absorption frequency is given as a single approximation denoted with a '~' rather than a range.

2.3 Near IR Region

Near IR spectroscopy is based on molecular overtones and combination vibrations. Such transitions are forbidden by the selection rules of quantum mechanics. As a result, the molar absorptivity of the signals appearing in the near IR region is typically quite small. One advantage is that near IR typically penetrates into a sample further than mid infrared radiation. Near IR spectroscopy is, therefore, not a particularly sensitive technique, but it can be very useful in analysis of bulk material with little or no sample preparation.

The near IR spectrometer provides a resolution of 5 A° allowing determining positions of bands at ± 0.001 µ; in contrast, resolution of IR instruments is in the order of ± 0.02 µ and ± 0.01 µ for the best instruments.

2.4 Far IR Region

The far IR region contains absorption bands coming from deformations of tension and flexion of the heavy atom connections, deformations of skeleton, of functional group, stretching deformations of the ring and ways of vibration of the network.

3. INTERPRETATION OF IR SPECTRA

Interpretation of IR spectra involves the correlation of absorption bands in the spectrum of an unknown compound with the known absorption frequencies for each type of bond. Table 2 will help users to become more familiar with the process. **Intensity** (weak, medium or strong), **shape** (broad or sharp), and **position** (cm⁻¹) are significant parameters for identification of an absorption band in the spectrum. Examples of intensity and shape of absorption bands are shown in Table 2.

Table 2 Characteristic infrared absorption frequencies

Bond	Compound type	Frequency range (cm⁻¹)
C–H	Alkanes	2,960-2,850 (s) stretch
		1,470-1,350 (v) scissoring and bending
C–H	CH₃ Umbrella deformation	1,380 (m–w) – Doublet – isopropyl, t-butyl
C–H	Alkenes	3,080-3,020 (m) stretch
		1,000-675 (s) bend
C–H	Aromatic Rings	3,100-3,000 (m) stretch
	Phenyl Ring Substitution Bands	870-675 (s) bend
	Phenyl Ring Substitution Overtones	2,000-1,600 (w) – fingerprint region
C–H	Alkynes	3,333-3,267 (s) stretch
		700-621 (b) bend
C=C	Alkenes	1,680-1,640 (m, w) stretch
C≡C	Alkynes	2,260-2,100 (w, sh) stretch
C=C	Aromatic Rings	1,600-1,500 (w) stretch
C–O	Alcohols, Ethers, Carboxylic Acids, Esters	1,260-100 (s) stretch
C–F	Alkyl Fluorides	1,000-1,400 (s)
C–Cl	Alkyl Chlorides	600-800 (s)
C–Br	Alkyl Bromides	500-600 (s)
C–I	Alkyl Iodides	500 (s)
C=O	Aldehydes, Ketones, Carboxylic Acids, Esters	1,760-1,670 (s) stretch
O–H	Monomeric: Alcohols, Phenols	3,640-3,160 (s, br) stretch
	Hydrogen Bonded: Alcohols, Phenols	3,600-3,200 (b) stretch
	Carboxylic Acids	3,000-2,500 (b) stretch
N–H	Amines	3,500-3,300 (m) stretch
		1,650-1,580 (m) bend
C–N	Amines	1,340-1,020 (m) stretch
C≡N	Nitriles	2,260-2,220 (v) stretch
NO₂	Nitro Compounds	1,660-1,500 (s) asymmetrical stretch
		1,390-1,260(s) symmetrical stretch

References: v = variable; m = medium; s = strong; b = broad; w = weak; sh = sharp

4. QUICK PROCEDURE FOR IR ANALYSIS

Absorption bands can be looked for as follows:

I. **C-H absorption(s) between 3,100 and 2,850 cm^{-1}:** An absorption above 3,000 cm^{-1} indicates C=C–H, either vinyl or aromatic. The presence of the aromatic ring should be confirmed by the presence of peaks at 1,600 and 1,500 cm^{-1} and the strong C–H out-of-plane deformations below 900 cm^{-1} which indicate the substitution pattern on the aromatic ring. Alkenes should be confirmed by absorption at 1,640-1,680 cm^{-1}. C–H absorptions between 3,000 and 2,850 cm^{-1} are due to aliphatic hydrogens.

II. **O–H or N–H absorption appears between 3,200 and 3,600 cm^{-1}:** This indicates either an alcohol, N–H containing amine or amide, or carboxylic acid. For –NH$_2$ a doublet will be observed.

III. **Carbonyl (C=O) absorption between 1,690 and 1,760 cm^{-1}:** This group gives rise to a strong absorption that indicates either an aldehyde, ketone, carboxylic acid, ester, amide, anhydride or acyl halide. Aldehydes also show a distinctive absorption of moderate intensity at *ca.* 2,720 cm^{-1} due to the aldehyde C–H which is usually accompanied by another band at 2,820 cm^{-1}.

IV. **C–O absorption between 1,050 and 1,300 cm^{-1}:** These peaks are normally rounded like the O–H and N–H signals and are prominent. Carboxylic acids, esters, ethers, alcohols and anhydrides contain this peak.

V. The structure of aromatic compounds may also be confirmed from the pattern of the weak overtone and combination tone bands found from 2,000 to 1,600 cm^{-1}.

VI. **C=C and C=N absorptions at 2,100-2,260 cm^{-1}:** These bands are small but exposed.

VII. A methyl group may be identified by a signal at 1,380 cm^{-1} owing to the umbrella deformation: this band is split into a doublet for isopropyl and *gem*-dimethyl groups.

It should be noted that absence of typical absorption bands in the diagnosis region is also useful to rule out the presence of a functional group.

4.1 Sample Preparation

Infrared spectra may be obtained from gases, liquids, or solids. The spectra of gases or low-boiling liquids may be obtained by expansion of the sample into an evacuated cell. Gas cells are available in different lengths, from few cm to 40 m.

Liquids may be examined pure or in solution. Pure liquids are examined between salt plates, usually without a spacer. Samples between flat plates produce a liquid film of approximately 0.02 mm thickness. The minimum weight of sample required is approximately 1.0 mg. Silver chloride plates should be used for samples that dissolve sodium chloride. Solids are usually examined as a mull, as a pressed disk, or deposited as a glassy film.

Mulls are prepared by thoroughly grinding 2-5 mg of a solid in a smooth agate mortar. Grinding is continued after the addition of 1 or 2 drops of the mulling oil.

The pellet (pressed-disk) technique depends on the fact that dry, powdered potassium bromide (or other alkali metal halide) can be compacted under pressure in vacuo to transparent disks. The sample (0.5-1.0 mg) is intimately mixed with approximately 100 mg of dry, powdered KBr and the mixture thus obtained is used to prepare the pellet.

4.2 Materials for Sample Preparation

Typical IR materials for various applications are shown in Table 3. These IR materials are used as IR windows or diluents. Note that hygroscopic materials are suitable for organic samples, while non-hygroscopic materials are used for water-containing samples.

4.3 Sample Preparation for Transmission Analysis

Infrared spectrum is typically acquired with the sample either mounted on or suspended in KBr or NaCl. This is because neither of these compounds has an IR-active stretch in the region typically observed for organic and several inorganic molecules (Williams and Fleming, 1995).

4.3.1 Solid Samples

Thin plate samples: You can use a standard magnetic plate sample holder as shown in Fig. 1.

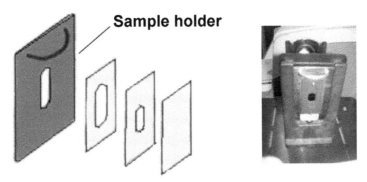

Sample holder

Figure 1 Magnetic sample holder.

Powder samples: Powder samples are pressed into a pellet. By this method, different diluents (matrixes) for various applications can be selected. For mid IR frequency range, KBr, KCl or diamond dust can be used. For far IR testing, high-density polyethylene (HDPE) or diamond dust is suitable. For near IR analysis,

Table 3 Typical materials for IR experiments

Materials	NaCl	KBr	CsI	AgCl	ZnSe	Diamond
Transmission range (cm^{-1})	40,000 -625	40,000-400	40,000-200	40,000-360	40,000-454	40,000-2,500
% trans. (window thickness)	91.5 (4 mm)	90.5 (4 mm)	92.0 (2 mm)	84.0 (3 mm)	65.0 (1 mm)	70.0 (1 mm)
Water sol (g/100 g, 25°C)	35.7	53.5	44.4	Insol.	Insol.	Insol.
Cleaning solvents	Anhydrous solvents	Anhydrous solvents	Anhydrous solvents	Acetone CH_2Cl_2	Acetone alcohol H_2O	Acetone alcohol
Attacking materials	Wet solvents	Wet solvents	Wet solvents	Ammonium salts	Acids, strong alkalies	$K_2Cr_2O_2$ conc. H_2SO_4

CsI or KBr can be selected. Pellets of KBr are prepared as follows:

(i) The powder sample and KBr must be ground to reduce the particle size to less than 5 mm in diameter. Otherwise, large particles scatter the IR beam and cause a slope baseline of spectrum.

(ii) Add a spatula full of KBr into an agate mortar and grind it to a fine powder until no crystals are seen and sticks to the mortar.

(iii) Take a small amount of powder sample (about 0.1-2% of the KBr amount, or just enough to cover the tip of spatula) and mix with the KBr powder. Subsequently grind the mixture for 3-5 minutes.

(iv) Assemble the die-set as shown in Fig. 2. When assembling the die, add the powder to the 7 mm collar. Put the die together with the powder into the press. Press the powder for 2 minutes to form a pellet. A good KBr pellet is thin and transparent. Opaque pellets give poor spectra, because little infrared beam passes through them. White spots in a pellet indicate that the powder is not ground well enough, or is not properly dispersed in the pellets.

(v) Disassemble the die set and take out the 7 mm collar. Put the collar together with the pellet onto the sample holder.

(vi) Place the sample holder in the IR chamber and scan.

(vii) Clean the dye after the experiment is complete. Note that Br⁻ from the KBr can often replace ligands in the compound whose spectrum is desired.

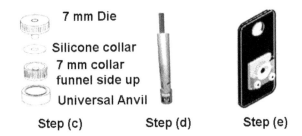

7 mm Die

Silicone collar

7 mm collar funnel side up

Universal Anvil

Step (c)   **Step (d)**   **Step (e)**

Figure 2   Steps to prepare a KBr disk: (c) die-set parts, (d) spatula, (e) sample holder.

4.3.2   Liquid Samples

Making a sandwich: To prepare a liquid sample for IR analysis, place a drop of the liquid on the face of a highly polished salt plate (such as NaCl, AgCl or KBr), then place a second plate on top of the first plate to spread the liquid in a thin layer between the plates and clamps the plates together (Fig. 3). Finally, wipe off

Figure 3   Salt plates assembly for liquid samples.

the liquid out of the edge of the plate. You can mount the sandwich plate onto the sample holder. After finishing the experiment, clean the plates with isopropanol or other appropriate solvent (not water) and return them to a desiccator.

Caution: Volatile liquids cannot be determined using this method, because they will evaporate during spectrum acquisition. Do not use this method when liquid samples are toxic. In addition, NaCl and KBr are soluble in water, and thus they should not be used for aqueous samples.

Using a liquid cell: A demountable path length cell is designed for liquid sample. Assemble the cell as shown in Fig. 4. Then, use a syringe to fill the liquid into the cell. Finally seal the cell.

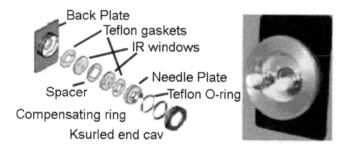

Figure 4 Thermo Nicolet demountable path length cell.

4.4 Suitable Solvents for IR Analysis

The solvent selected must be transparent in the region of interest. When the entire spectrum is of interest, several solvents must be used. Some of the useful solvents are carbon tetrachloride (CCl_4) and carbon disulfide (CS_2). CCl_4 is relatively free of absorption at frequencies above 1,333 cm^{-1} but CS_2 shows little absorption below 1,333 cm^{-1}. CS_2 cannot be used as a solvent for primary or secondary amines. Cl_3CH is also a suitable solvent for IR spectra in liquid solution.

5. RAMAN SPECTROSCOPY

Raman spectroscopy is used in condensed matter physics and chemistry to study vibrational, rotational, and other low-frequency modes in a system. It relies on inelastic scattering, or Raman scattering of monochromatic light, usually from a laser in the visible, near IR or near ultraviolet range. Phonons or other excitations in the system are absorbed or emitted by the laser light, resulting in the energy of the laser photons being shifted up or down. The shift in energy gives information about the phonon modes in the system. Infrared spectroscopy yields similar, but complementary information.

Typically, a sample is illuminated with a laser beam. Light from the illuminated spot is collected with a lens and sent through a monochromator. Wavelengths close to the laser line (due to elastic Rayleigh scattering) are filtered out and those in a certain spectral window away from the laser line are dispersed onto a detector.

Spontaneous Raman scattering is typically very weak. As a result, the main difficulty of Raman spectroscopy is separating the weak inelastically scattered light from the intense Rayleigh scattered laser light. Raman spectrometers typically use holographic diffraction gratings and multiple dispersion stages to achieve a high degree of laser rejection. A photon-counting photomultiplier tube (PMT) or, more commonly, a ccd camera is used to detect the Raman scattered light. Raman spectroscopy has a stimulated version, analogous to stimulated emission, called stimulated Raman scattering.

The Raman effect occurs when light impinges upon a molecule and interacts with the electron cloud of the bonds of that molecule (Figure 5). The amount of deformation of the electron cloud is the polarizability of the molecule. The amount of the polarizability of the bond will determine the intensity and frequency of the Raman shift. The molecule must be symmetric to observe Raman shift. The photon (light quantum), excites one of the electrons into a virtual state. When the photon is released, the molecule relaxes back into vibrational energy state. The molecule will typically relax into the first vibration energy state, and this generates Stokes Raman scattering. If the molecule was already in an elevated vibrational energy state, the Raman scattering is then called Anti-Stokes Raman scattering.

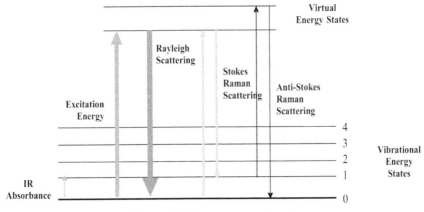

Figure 5 Raman energy levels.

6. APPLICATIONS OF IR SPECTROSCOPY

Compounds from numerous structural classes have been implicated in allelopathic interactions. The major focus is on compounds isolated in the last 20 years (Vyvyan, 2002). Their structures have been established, mainly by spectroscopic methods, including IR spectroscopy. The IR spectroscopic data reported for selected classes of allelochemicals are summarized in Table 4. They may be classified as benzoquinones, coumarins, flavonoids and terpenoids.

Table 4 IR diagnostic bands for selected allelochemicals

Compound	Diagnostic IR bands (cm⁻¹)	References
1	3,380 (OH) other signals not reported	Chang *et al.*, 1986
2	3,345 (OH), 1,664 C=O, 1,642 (C=C), 1,601, 1,463 (C=C aromatic)	Rimando *et al.*, 1998
3	1,644 br (C=O and C=C), 1,644, 1,601 (C=C aromatic)	
4	3,430 (OH), 1,645 (C=O)	Macías *et al.*, 1997
5	3,420 (OH), 1,650 (C=O)	
6	3,420 (OH), 1,652 (C=O)	
7	1,714 (C=O, cyclohexanone), 1,698 (C=O), 1,271 (C–O-C)	Macías *et al.*, 1998a
8	1,715 (C=O, cyclohexanone), 1,672 (C=O), 1,630 (C=C), 1,257 (C–O–C)	
9	3,417 (OH), 1,697 (C=O), 1,649 (C=C), 1,130 (C-O-C)	
10	3,406 (OH), 1,680 (α,β-unsaturated C=O), 1,649 (C=C)	
11	3,488 (OH), 1,759 (α,β-unsaturated-γ-lactone), 1,715 (α,β-unsaturated ester)	Macías *et al.*, 1999
12	3,418 (OH), 1,767 (α,β-unsaturated-γ-lactone), 1,717 (α,β-unsaturated ester)	
13	3,312 (OH), 1,693 (C=O, carboxylic acid), 1,693 (α,β-unsaturated aldehyde)	Macías *et al.*, 1998b
14	3,401 (OH), 1,710 (C=O), 1,699 (C=O, carboxylic acid)	
15	3,450 (OH), 1,712 (C=O), 1,699 (C=O, carboxylic acid)	
16	3,340 (OH), 1,714 (C=O), 1,693 (C=O, carboxylic acid)	
17	3,431 (OH), 1,708 (C=O), 1,657 (C=C)	
18	3,427 (OH), 1,705 (C=O)	

For coumarins, diagnostic bands are those caused by the γ-pyrone ring, giving rise to characteristic signals at 1,715-1,745 and 1,130-1,160 cm⁻¹, which should be analyzed with those produced by the aromatic ring at 1,625-1,640 cm⁻¹.

In the IR spectra of flavonoids, a broad band at 3,300-3,400 cm⁻¹, indicates phenolic groups and C=O signals show up with medium intensity at 1,660-1,650 cm⁻¹, while at 1606 and 1,450 cm⁻¹, the aromatic C–C stretching are found.

For terpenoids usually there are no diagnostic bands since they belong to several structural types and contain a variety of functional groups. Table 4 summarizes several structural classes of allelopathic natural products whose IR data have been reported.

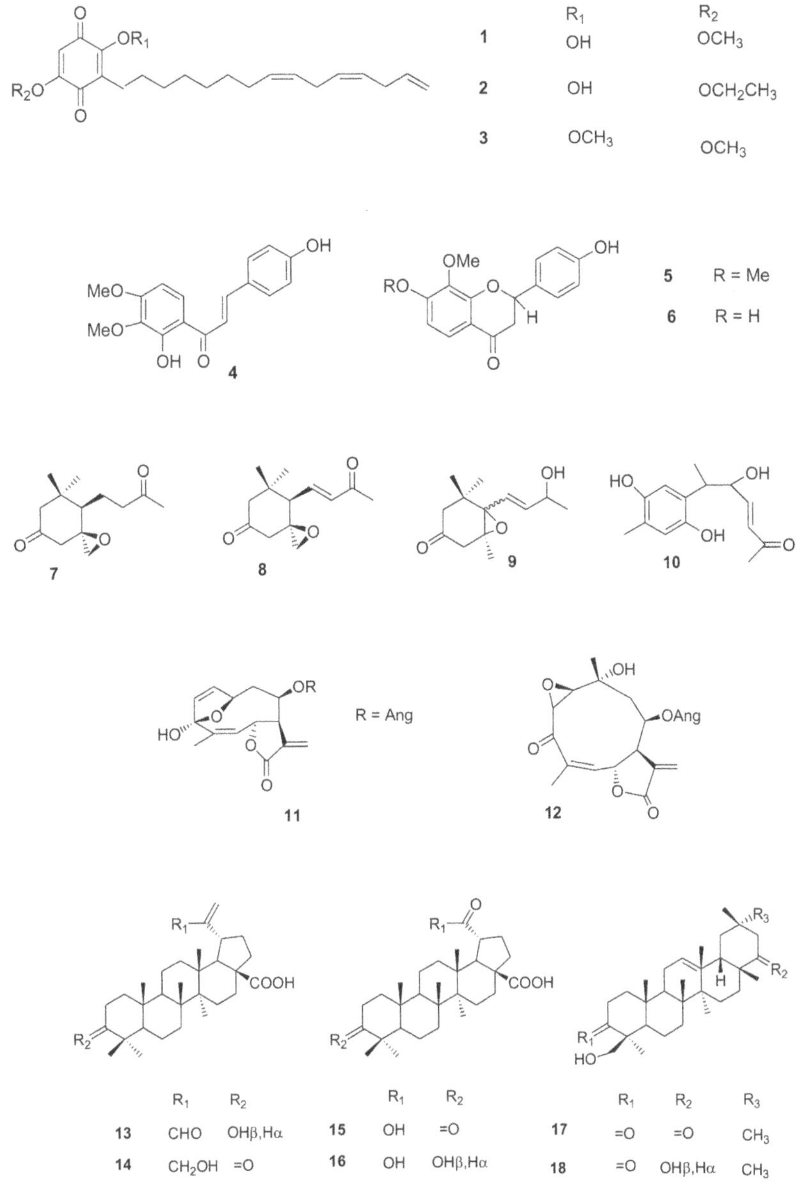

Figure 6 Structures of selected allelochemicals.

Among the benzoquinones, special attention is paid to sorgoleone-358 (**1**), 5-ethoxy-sorgoleone (**2**) (Chang *et al.*, 1986), and 2,5-dimethoxy-sorgoleone (**3**) (Fig. 6) isolated from *Sorghum tricolor* (Rimando *et al.*, 1998). The bioactive flavonoids heliannone A (**4**), heliannone B (**5**), and heliannone C (**6**) were isolated from *Helianthus annuus* cultivars (Macías *et al.*, 1997) and the two more characteristic IR absorptions were reported (Table 4).

Among the terpenoids, numerous structural types have been obtained, for example norsesquiterpenes annuionones A-C (**7-9**), and the norbisabolene helinorbisabone (**10**) (Macías *et al.*, 1998a). Also, some sesquiterpene lactones are potential allelochemicals. These latter compounds are easily identified from the IR spectra due to the characteristic α,β-unsaturated-γ-lactone signal at 1,750-1,780 cm^{-1}. Typical examples are those of helivypolide D (**11**), and helivipolide E (**12**) (Macías *et al.*, 1999) showing signals at 1,759 and 1,767 cm^{-1}.

A series of polar bioactive triterpenoids (**13-18**) have been isolated from *Melilotus messanensis* and the IR data have been reported (Macías *et al.*, 1998b). These are good examples for the different carbonyl groups in the molecules.

6. INFRARED SPECTRA OF SELECTED COMPOUNDS WITH ASSIGNMENT OF THE DIAGNOSIS ABSORPTIONS

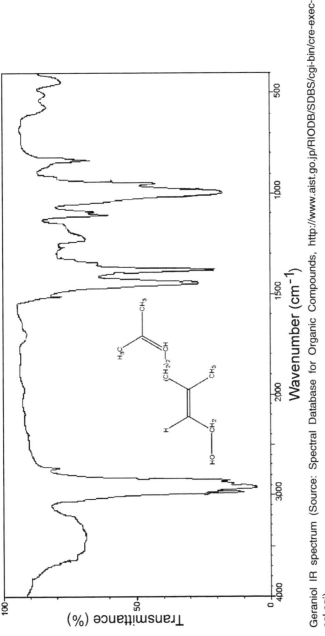

Figure 7 Geraniol IR spectrum (Source: Spectral Database for Organic Compounds, http://www.aist.go.jp/RIODB/SDBS/cgi-bin/cre-exec-sql.cgi)

$3,620\text{-}3,200$ cm^{-1}—broad, hydrogen bonded OH
$2,970\text{-}2,850$ cm^{-1}—CH aliphatic asymmetric and symmetric vibrations
$1,450$ cm^{-1}—$-CH_3$ group asymmetric bending vibration and CH_2 scissoring vibration
$1,382$ cm^{-1}—$-CH_3$ group symmetric bending vibration

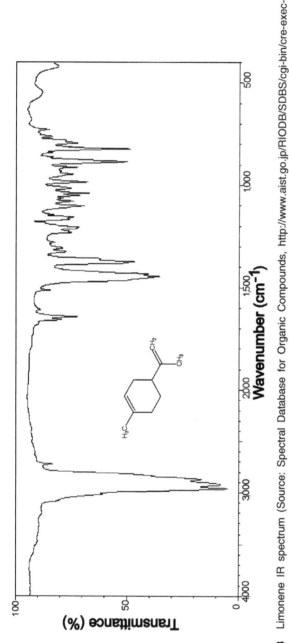

Figure 8 Limonene IR spectrum (Source: Spectral Database for Organic Compounds, http://www.aist.go.jp/RIODB/SDBS/cgi-bin/cre-exec-sql.cgi)

3,040 cm^{-1}—vinyl hydrogen stretching absorption

2,960-2,850 cm^{-1}—CH aliphatic asymmetric and symmetric vibrations

1,453 cm^{-1}—aliphatic CH$_2$ scissoring deformation

1,376 cm^{-1}— –CH$_3$ group symmetric bending vibration (umbrella deformation)

895 cm^{-1}—C–H bending vibration of gem-disubstituted alkene (R$_1$R$_2$C=CH$_2$)

798 cm^{-1}—C–H bending vibration of trisubstituted alkene (R$_1$R$_2$C=CHR$_3$)

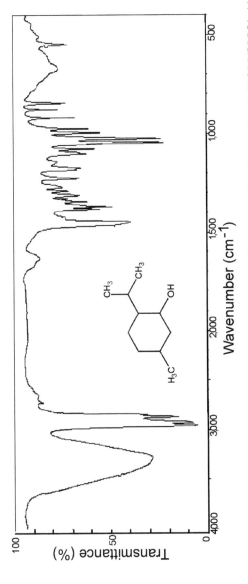

Figure 9 Menthol IR spectrum (Source: Spectral Database for Organic Compounds, http://www.aist.go.jp/RIODB/SDBS/cgi-bin/cre-exec-sql.cgi).

3,294 cm⁻¹—hydrogen bonded OH

2,960-2,850 cm⁻¹—asymmetric and symmetric absorptions of aliphatic CH

1,466 cm⁻¹—aliphatic CH₂ scissoring deformation

1,057 cm⁻¹—C–O stretching absorption

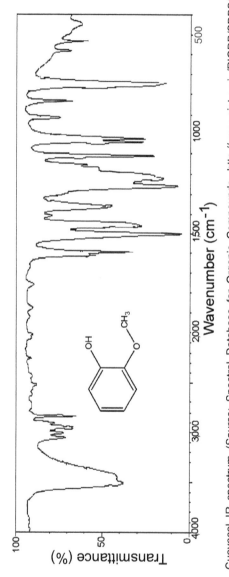

Figure 10 Guayacol IR spectrum (Source: Spectral Database for Organic Compounds, http://www.aist.go.jp/RIODB/SDBS/cgi-bin/cre-exec-sql.cgi).

3,511 cm^{-1}—OH stretching vibration
3,010-3,100 cm^{-1}—aromatic CH absorptions
2,964 cm^{-1}—CH$_3$ asymmetric stretching vibration
2,870 cm^{-1}—CH$_3$ symmetric stretching vibration
1,615-1,500 cm^{-1}—aromatic nucleus
1,250 cm^{-1}—phenol C–O stretching
725 cm^{-1}—four adjacent hydrogens on aromatic ring (*ortho* substitution)

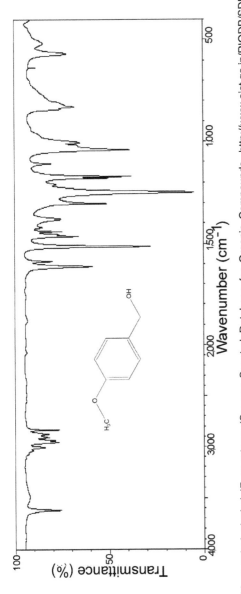

Figure 11 p-anisyl alcohol IR spectrum (Source: Spectral Database for Organic Compounds, http://www.aist.go.jp/RIODB/SDBS/cgi-bin/cre-exec-sql.cgi).

$3,010$ cm^{-1}—aromatic CH absorption

$2,960$-$2,850$ cm^{-1}—CH aliphatic asymmetric and symmetric stretchings

$1,500$ cm^{-1} and $1,600$ cm^{-1}—aromatic nucleus

$1,250$ cm^{-1}—Asymmetric C—O—C stretch

$1,000$ cm^{-1}—Symmetric C—O—C stetch

800 cm^{-1}—two adjacent hydrogens on aromatic ring (*para* substitution)

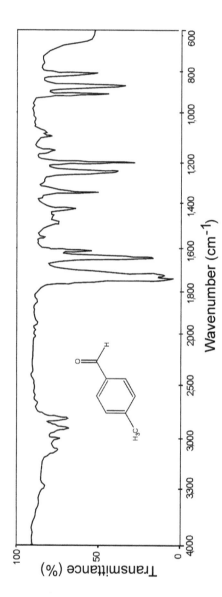

Figure 12 P-tolualdehyde, 4-methylbenzaldehyde IR spectrum (Source: Spectral Database for Organic Compounds, http://www.aist.go.jp/ RIODB/SDBS/cgi-bin/cre-exec-sql.cgi).

3,080 cm⁻¹—aromatic CH absorption
2,970-2,850 cm⁻¹—asymmetric and symmetric absorptions of aliphatic CH
2,760 cm⁻¹—aldehyde CH
1,735 cm⁻¹—C=O absorption
1,615 cm⁻¹—carbonyl conjugated with aromatic ring. The 1,600 and 1,500 bands for the aromatic nucleus are variable
1,380 cm⁻¹— –CH₃ group symmetric bending vibration
820 cm⁻¹—two adjacent hydrogens on aromatic ring (*para* substitution)

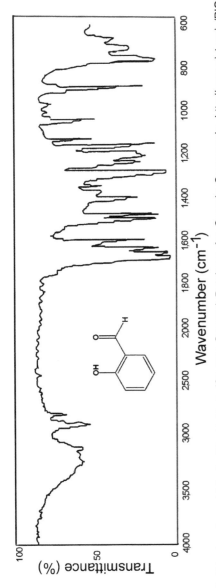

Figure 13 2-hydroxybenzaldehyde IR spectrum (Source: Spectral Database for Organic Compounds, http://www.aist.go.jp/RIODB/SDBS/cgi-bin/cre-exec-sql.cgi).

3,250 cm^{-1}—hydrogen bonded OH.
3,120 cm^{-1}—aromatic CH.
2,820 cm^{-1}—aldehyde CH
1,690 cm^{-1}—C=O
1,600 and 1,500 cm^{-1}—aromatic nucleus
1,200 cm^{-1}—phenol C–O absorption. The C–O absorption for alcohols is: primary, near 1,050 cm^{-1}; secondary, near 1,100 cm^{-1}; and tertiary near 1150 cm^{-1}
760 and 720 cm^{-1}—four adjacent hydrogens on an aromatic ring (*ortho* substitution)

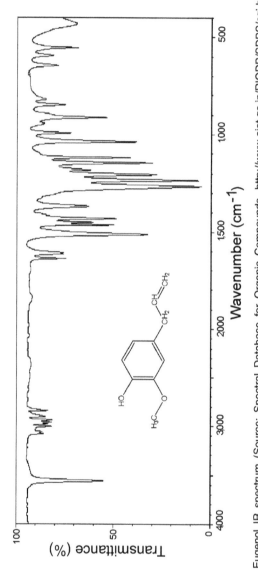

Figure 14 Eugenol IR spectrum (Source: Spectral Database for Organic Compounds, http://www.aist.go.jp/RIODB/SDBS/cgi-bin/cre-exec-sql.cgi)

3,669 cm^{-1}—OH, intramolecular hydrogen bond

2,970-2,850 cm^{-1}—asymmetric and symmetric absorptions of aliphatic CH

1,230 cm^{-1}—asymmetric C–O–C stretch.

1,004 cm^{-1}—symmetric C–O–C stetch

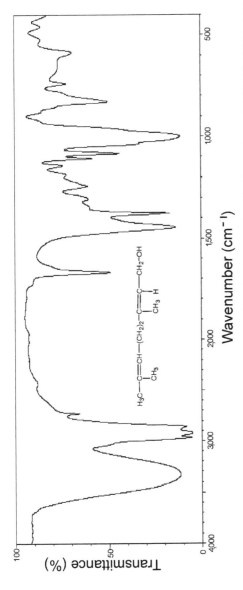

Figure 15 Nerol IR spectrum (Source: Spectral Database for Organic Compounds, http://www.aist.go.jp/RIODB/SDBS/cgi-bin/cre-exec-sql.cgi).

3,327 cm^{-1}—hydrogen bonded OH

3,030 cm^{-1}—shoulder; vinyl hydrogen stretching absorption

2,970-2,850 cm^{-1}—asymmetric and symmetric absorptions of aliphatic CH

1,380 cm^{-1}— –CH$_3$ group asymmetric bending vibration

1,000 cm^{-1}—C–O stretching absorption (primary alcohol)

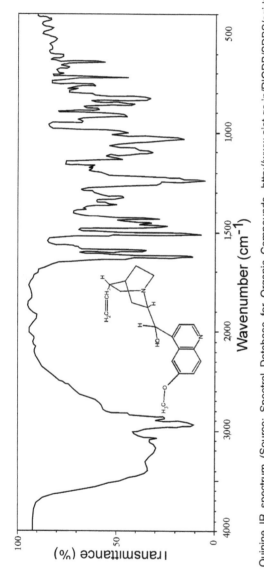

Figure 16 Quinine IR spectrum (Source: Spectral Database for Organic Compounds, http://www.aist.go.jp/RIODB/SDBS/cgi-bin/cre-exec-sql.cgi)

3,400 cm^{-1} and 3,200 cm^{-1}—OH stretching
3,090 cm^{-1}—vinyl C–H stretching (=CH$_2$)
3,015 cm^{-1}—aromatic C–H
2,960-2,850 cm^{-1}—aliphatic CH symmetric and asymmetric stretching vibrations
1,509 cm^{-1} and 1,622 cm^{-1}—C=C, C=N ring stretching

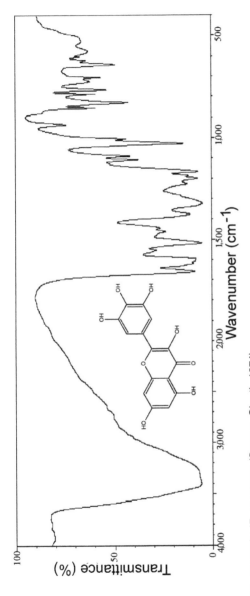

Figure 17 Myricetin IR spectrum (Source: Oberti, 1974).

3,000-3,600 cm⁻¹—very broad, hydrogen bonded OH

1,670 cm⁻¹—C=O stretch

1,020 cm⁻¹ and 1,600 cm⁻¹—aromatic ring

710-900 cm⁻¹—absorptions associated with aromatic CH out-of-plane vibrations

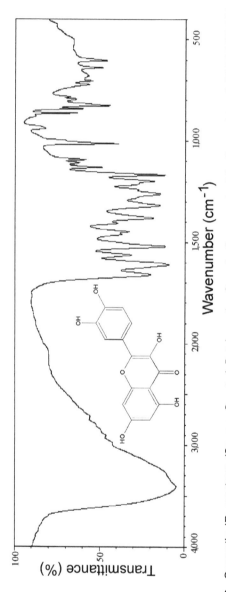

Figure 18 Quercetin IR spectrum (Source: Spectral Database for Organic Compounds, http://www.aist.go.jp/RIODB/SDBS/cgi-bin/cre-exec-sql.cgi).

3,400 cm^{-1}—very broad, hydrogen bonded OH

1,680 cm^{-1}—C=O stretch.

1,180 cm^{-1} and 1,600 cm^{-1}—aromatic ring

700-900 cm^{-1}—absorptions associated with aromatic CH out-of-plane vibrations

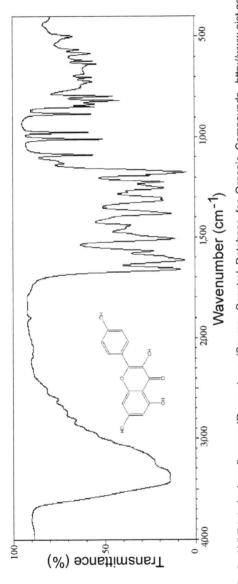

Figure 19 3, 4′,5,7-tetrahydroxyflavone IR spectrum (Source: Spectral Database for Organic Compounds, http://www.aist.go.jp/RIODB/ SDBS/cgi-bin/cre-exec-sql.cgi).

$3{,}389 \text{ cm}^{-1}$—broad, hydrogen bonded OH

$1{,}660 \text{ cm}^{-1}$—C=O stretch

$1{,}150 \text{ cm}^{-1}$ and $1{,}600 \text{ cm}^{-1}$—aromatic ring

$760\text{-}900 \text{ cm}^{-1}$—absorptions associated with aromatic CH out-of-plane vibrations

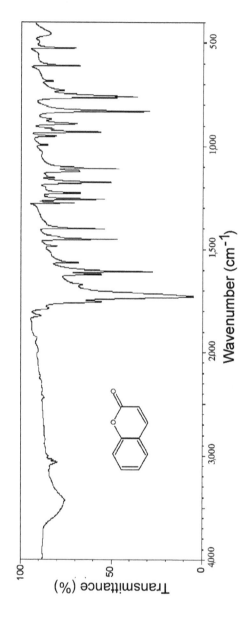

Figure 20 Coumarin IR spectrum (Source: Spectral Database for Organic Compounds, http://www.aist.go.jp/RIODB/SDBS/cgi-bin/cre-exec-sql.cgi).

3,090-3,010 cm^{-1}—aromatic CH absorption
1,729 cm^{-1}—C=O absorption.
1,115 and 1,610 cm^{-1}—aromatic ring
760 cm^{-1}—aromatic CH out-of-plane deformation, four adjacent-free hydrogen atoms

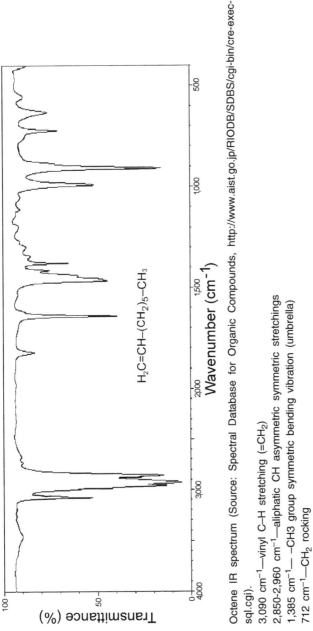

Figure 21 Octene IR spectrum (Source: Spectral Database for Organic Compounds, http://www.aist.go.jp/RIODB/SDBS/cgi-bin/cre-exec-sql.cgi).

3,090 cm^{-1}—vinyl C–H stretching (=CH$_2$)
2,850-2,960 cm^{-1}—aliphatic CH asymmetric symmetric stretchings
1,385 cm^{-1}—CH3 group symmetric bending vibration (umbrella)
712 cm^{-1}—CH$_2$ rocking

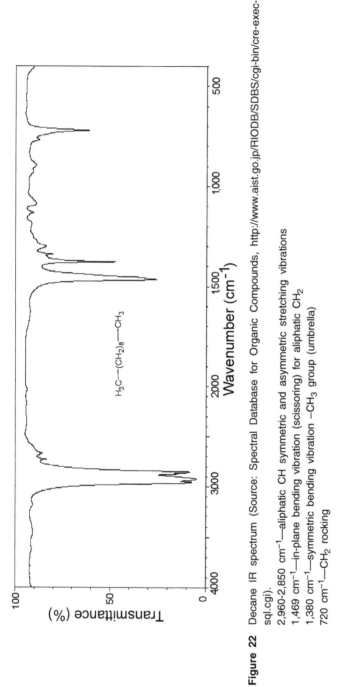

Figure 22 Decane IR spectrum (Source: Spectral Database for Organic Compounds, http://www.aist.go.jp/RIODB/SDBS/cgi-bin/cre-exec-sql.cgi).

2,960-2,850 cm^{-1}—aliphatic CH symmetric and asymmetric stretching vibrations

1,469 cm^{-1}—in-plane bending vibration (scissoring) for aliphatic CH$_2$

1,380 cm^{-1}—symmetric bending vibration –CH$_3$ group (umbrella)

720 cm^{-1}—CH$_2$ rocking

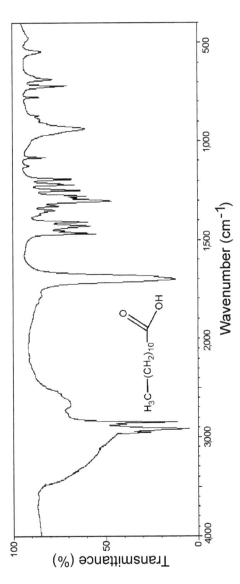

Figure 23 Lauric acid IR spectrum (Source: Spectral Database for Organic Compounds, http://www.aist.go.jp/RIODB/SDBS/cgi-bin/cre-exec-sql.cgi).

$2,800-3,400$ cm^{-1}—typical broad absorption of carboxylic acid due to OH stretching (bonded)

$2,500-2,700$ cm^{-1}—weak, two or more bands due to OH stretching (bonded) of carboxylic acid

$2,964-2,850$ cm^{-1}—aliphatic C–H symmetric and asymmetric stretching (mounted on the carboxylic acid absorption)

$1,710$ cm^{-1}—C=O stretch.

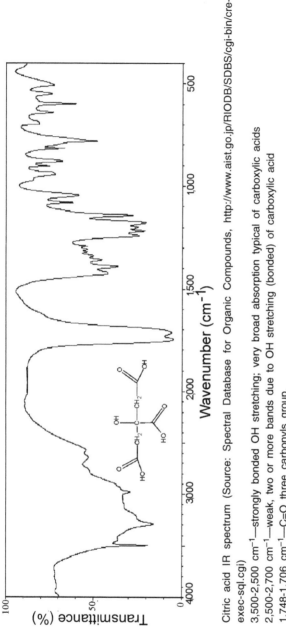

Figure 24 Citric acid IR spectrum (Source: Spectral Database for Organic Compounds, http://www.aist.go.jp/RIODB/SDBS/cgi-bin/cre-exec-sql.cgi)

$3,500-2,500$ cm^{-1}—strongly bonded OH stretching; very broad absorption typical of carboxylic acids $2,500-2,700$ cm^{-1}—weak, two or more bands due to OH stretching (bonded) of carboxylic acid $1,748-1,706$ cm^{-1}—C=O three carbonyls group.

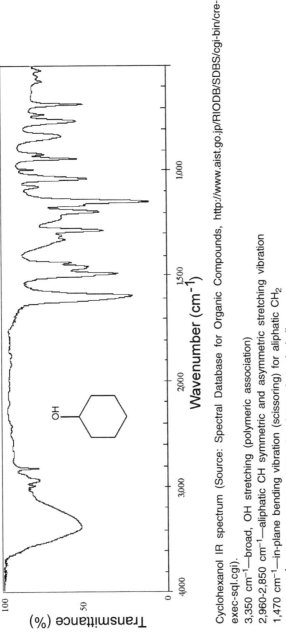

Figure 25 Cyclohexanol IR spectrum (Source: Spectral Database for Organic Compounds, http://www.aist.go.jp/RIODB/SDBS/cgi-bin/cre-exec-sql.cgi).

$3,350~cm^{-1}$—broad, OH stretching (polymeric association)

$2,960$-$2,850~cm^{-1}$—aliphatic CH symmetric and asymmetric stretching vibration

$1,470~cm^{-1}$—in-plane bending vibration (scissoring) for aliphatic CH_2

$1,065~cm^{-1}$—C—O stretching absorption (secondary alcohol)

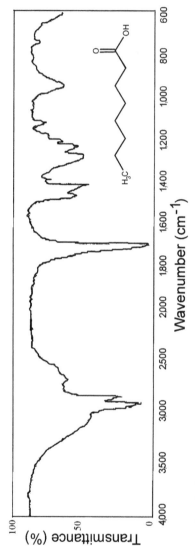

Figure 26   Octanoic acid IR spectrum (Source: Spectral Database for Organic Compounds, http://www.aist.go.jp/RIODB/SDBS/cgi-bin/cre-exec-sql.cgi)

3,400-2,500 cm^{-1}—typical broad absorption of carboxylic acids (dimmer)

2,500-2,700 cm^{-1}—weak, two or more bands due to OH stretching (bonded) of carboxylic acid 2,850-2,960 cm^{-1}—aliphatic C–H symmetric and asymmetric stretchings (mounted on the carboxylic acid absorption)

1,700 cm^{-1}—C=O group

950 cm^{-1}—OH out-of-plane bending of the dimmer

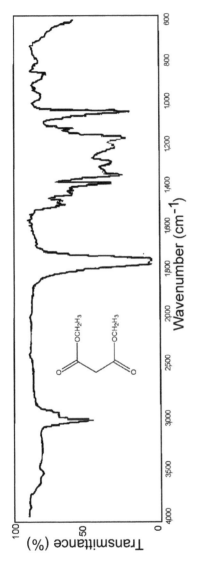

Figure 27 Diethyl malonate IR spectrum (Source: Spectral Database for Organic Compounds, http://www.aist.go.jp/RIODB/SDBS/cgi-bin/cre-exec-sql.cgi)

2,960-2,850 cm^{-1}—aliphatic CH symmetric and asymmetric stretching vibrations

1,765 and 1,740 cm^{-1}—C=O; note the two minima of this band (two carbonyls)

1,380 cm^{-1}—symmetric bending vibration –CH$_3$ group (umbrella)

1,280 cm^{-1}—asymmetric stretch for C–O–C of ester

1,045 cm^{-1}—symmetric stretch for C–O–C of ester

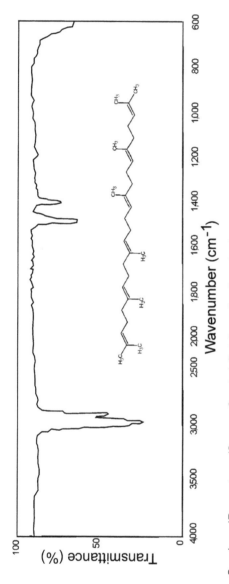

Figure 28 Squalane IR spectrum (Source: Spectral Database for Organic Compounds, http://www.aist.go.jp/RIODB/SDBS/cgi-bin/cre-exec-sql.cgi)

3,020 cm⁻¹—vinyl C–H stretching
2,960-2,850 cm⁻¹—aliphatic CH asymmetric and symmetric stretching
1,480 cm⁻¹—in-plane bending vibration (scissoring) of aliphatic CH₂
1,395 cm⁻¹—symmetrical bending vibration –CH₃ group (umbrella)

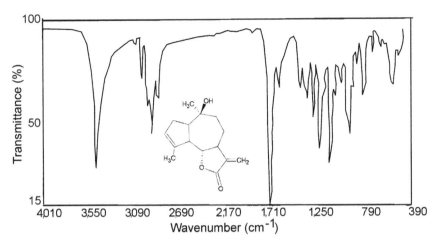

Figure 29 10-Epi-8-desoxycumambrin B IR spectrum (Source: Gil, 1989).

3,487 cm^{-1}—O–H stretching

3,023 cm^{-1}—vinyl C–H stretching

2,960-2,850 cm^{-1}—aliphatic C–H symmetric and asymmetric stretching

1,741 cm^{-1}—C=O stretch γ-lactone

1,660 cm^{-1}—C=C stretching

1,154 cm^{-1}—C–O bending of tertiary alcohol (near 1100 cm^{-1} for secondary alcohols and near 1050 cm^{-1} for primary alcohols)

SUGGESTED READINGS

Chapman and Hall. (1975). *The Infrared Spectra of Complex Molecules*. 3rd edition, London, England.

Chang, M., Netzly, D.H., Butler, L.G. and Lynn, D.G. (1986). Chemical regulation of distance: characterization of the first natural host germination stimulant for *Striga asiatica*. *Journal of the American Chemical Society* **108**: 7858-7860.

Fayer, M.D. Ed. (2001). *Ultrafast Infrared and Raman Spectroscopy*, Marcel Dekker, Inc. New York, Basel.

Gil, R.R. (1989). *Secondary Metabolites from Species of Stevia genus, Tribu Eupatorieae (Compositae)*. Ph.D. Dissertation, National University of Cordoba, Argentina. (In Spanish).

Macías, F.A., Molinillo, J.M.G., Torres, A., Varela, R.M. and Castellano, D. (1997). Bioactive flavonoids from *Helianthus annuus* cultivars. *Phytochemistry* **45**: 683-687.

Macías, F.A., Oliva, M.R., Varela, R.M., Torres, A. and Molinillo, J.M.G. (1999). Allelochemicals from sunflower leaves cv. Peredovick. *Phytochemistry* **52**: 613-621.

Macías, F.A., Simonet, A.M., Galindo, J.C.G., Pacheco, P.C. and Sánchez, J.A. (1998b). Bioactive polar triterpenoids from *Melilotus messanensis*. *Phytochemistry* **49**: 709-717.

Macías, F.A., Varela, R.M., Torres, A., Oliva, M.R. and Molinillo, J.M.G. (1998a). Bioactive norsesquiterpenes from *Helianthus annuus* with potential allelopathic activity. *Phytochemistry* **48**: 631-636.

Oberti, J.C. (1974). *Main Flavonoids and Alkaloids from Bark of* Prosopis uscifolia Griseb. Ph.D. dissertation. National University of Cordoba, Argentina (In Spanish).

Pasto, D., Johnson, C. and Miller, M. (1992). *Experiments and Techniques in Organic Chemistry.* Prentice-Hall, Englewood-Cliffs, New Jersey, USA.

Rimando, A.M., Dayan, F.E., Czarnota, M.A., Weston, L.A. and Duke, S.O. (1998). A new photosystem II electron transfer inhibitor from *Sorghum bicolor. Journal of Natural Products* **61**: 927-930.

Rubio, J.M.. (1974). *Infrared Spectroscopy.* Chemical Serie. AEO. Regional Program of Scientific and Technological Development. (In Spanish).

Silverstein, R.M., Bassler, G.C. and Morrill, T.C. (1981). *Spectrometric Identification of Organic Compounds.* 4th edition. John Wiley and Sons, New York, USA.

Spectral Database for Organic Compounds. Available from: http://www.aist.go.jp/RIODB/ SDBS/cgi-bin/cre_exec_sql.cgi.; *20 October 2006.*

Vyvyan, J.R. (2002). Allelochemicals as leads for new herbicides and agrochemicals. *Tetrahedron* **58**: 1631-1646.

Williams, D.H. and Fleming, I. (1995). *Spectroscopic Methods in Organic Chemistry.* 5th edition. McGraw-Hill, London.

12

NMR

Pedro C. Rossomando

1. INTRODUCTION

There has been an explosive growth in nuclear magnetic resonance (NMR) spectroscopy in recent years. The advent of 1D and 2D techniques such as DEPT, COSY, NOESY, ROESY, 2D *J*-resolved, one-bond hetero-correlations such as HETCOR and long range correlations COLOC and more, have provided the organic chemist powerful new tools for structural elucidation of complex natural products. Inverse measurement techniques, such as HMBC and HMQC or HSQC, have allowed substantial enhancements of sensitivity in the heteronuclear shift correlation experiments. These methods have been employed in structure elucidation of new natural products from terrestrial and marine life forms.

2. INTRODUCING HIGH-RESOLUTION NMR

A nucleus placed in a static magnetic field of a flux density B_0 may suffer NMR, if it has an angular moment p known as nuclear spin (Derome, 1987; Claridge, 1999). The p component in direction of B_0 (Fig. 1), p_0, only takes multiple values of m according to:

$$p_0 = m(h/2\pi); \quad m = \pm n(1/2); \quad n = 0,1,2,...$$

Values of m are limited by the spin quantum number I:

$$m = I, I - 1, ..., -I$$

I is a constant characteristic of many nuclei in the fundamental state and its

Author's address: Química Orgánica, Facultad de Química, Bioquímica y Farmacia, Universidad Nacional de San Luis, Chacabuco 917, D5700BWS, San Luis, Argentina. E-mail: prosso85@gmail.com

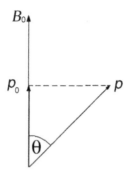

Figure 1 Component p_0 of the vector p in the direction of the B_0 field.

magnitude depends on the atomic number z and the mass number a of an atom (Table 1).

Table 1 Atomic number z, mass number a and spin quantum number I of selected nuclei.

z	a	I	$^a X$
Even	Even	0	^{12}C, ^{16}O
Odd	Odd	½, 3/2, 5/2,...	^{1}H, ^{15}N, ^{19}F, ^{31}P
Even	Odd	½, 3/2, 5/2,...	^{13}C, ^{17}O
Odd	Even	1, 2, 3,...	^{2}H, ^{14}N

Nuclei with $I \neq 0$ interact with the magnetic field due to their magnetic moments μ_0. μ relates with p through the constant of proportionality γ:

$$\mu = \gamma p$$

Constant γ, called magnetogyric ratio, is a characteristic of each nucleus. The component of μ in direction of B_0, μ_0:

$$\mu \cos \theta = \mu_0 = \gamma I \, (h/2\pi)$$

When it exposes a spinning nucleus to B_0, it behaves as a gyroscope in a gravitational field (Fig. 2). Vector μ precesses around B_0, with a precession frequency v_0 known as Larmor's frequency of the mentioned nucleus. v_0 is proportional to B_0.

$$\boxed{v_0 = \gamma B_0 / 2\pi}$$

The energy of the magnetic moment μ in the magnetic field B_0 is the product of B_0 and v_0. Thus

$$E = -\mu_0 B_0 = -\gamma \, (h/2\pi) \, I \, B_0$$

Vector μ cannot process aleatory, but it can to certain angles with regard to B_0. Quantum number I determines the number of possible orientations.

$$2I + 1 = \text{orientations}$$

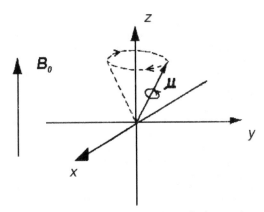

Figure 2 Larmor precession of a spinning nucleus.

For nuclei with spin 1/2 like ^1H, ^{13}C, ^{15}N, ^{19}F, ^{31}P two possible alignments arise with regard to B_0 (Fig. 3).

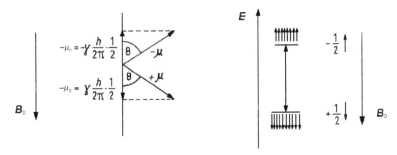

Figure 3 Energy levels and spin populations of nuclear spins with $I = 1/2$ in B_0.

As was seen, the absorption frequency in an NMR experiment is given by Larmor's equation. The value of v_0 falls inside the radio frequency [r.f.] range:

B_0	v_0 (^1H)	v_0 (^{13}C)
9.4 T	400.13 MHz	100.6 MHz
11.75 T	500.13MHz	125.8 MHz
14.1 T	600.13 MHz	150.9 MHz
18.8 T	800.13 MHz	201.2 MHz

The NMR experiment affects protons which have $+1/2$ spin (parallel to B_0) and they absorb energy when the inversion to the $-1/2$ spin is produced (anti-parallel to B_0). The difference of $+1/2$ and $-1/2$ spins populations is small. For a $B_0 = 4.23$ T (180 MHz)

$$\Delta E = (h_{\gamma H}/2\pi)\, B_0 = 9.5 \cdot 10^{-14} \cdot 26753 \cdot 42300/(2 \cdot 3.14)\ [=]\ \text{Kcal/mol}$$
$$\Delta E = 17 \cdot 10^{-7}\ \text{Kcal/mol} = 0.017\ \text{cal/mol}$$

In the equilibrium, the difference of populations is given by Boltzmann's distribution as:

$$N_\alpha/N_\beta = e^{0.017/RT} = 1.00003$$

For every 100,000 β spins there are only 100,003 α spins for that field. The obtained value shows: (a) NMR is not a trace spectroscopy, and (b) the importance of having magnets of high field.

So far the precessing of an isolated nucleus has been reviewed. A group of nuclei show, as was seen, a small excess of nuclei in the lowest energy level (Fig. 4).

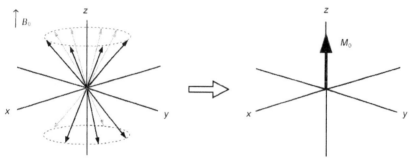

Figure 4 Collection of spins. The tiny population excess parallel to the z-axis gives rise to M_0.

The objective of NMR is to measure the precessional movement of the nuclei being studied, which is possible by disturbing the system in equilibrium with r.f. pulses, applied perpendicularly to B_0. The r.f. pulse behaves like a time dependent magnetic field and it is denoted as B_1. The analysis of this disturbance, in the static system of coordinates (laboratory frame) is complicated. The problem can be resolved by observing the experiment from a perspective in which the precessional movement vanishes. Take for example that the system of coordinates rotates around axis z, with a similar frequency to v_0 (rotating frame) (Fig. 5).

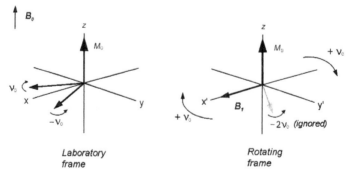

Laboratory
frame

Rotating
frame

Figure 5 Representations of the laboratory and rotating frame. In the rotating frame it rotates at a rate equal to the applied r.f. frequency, v_0.

In this way, the influence of B_0 (precession) disappears, and an important field B_1 is created (which is small in comparison with B_0). In Fig. 6, the effect of an r.f. pulse (B_1) on the magnetization in the equilibrium M_0 is observed. The action of the r.f. pulse over the sample disturbs the nuclear spins thermal equilibrium in the fundamental state.

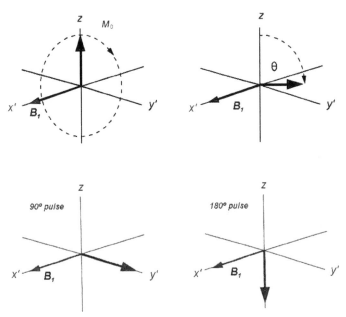

Figure 6 A radio frequency pulse applies a torque to M_0 magnetization and drives towards the x'-y' plane.

After the disturbance is completed, the system relaxes recovering the original equilibrium.

It is considered that this relaxation occurs mainly through two mechanisms, which give origin to two different relaxation times, the longitudinal relaxation time T_1, and the transversal relaxation time T_2. T_1 is the time during which recovering of magnetization along axis z (Fig. 7) is produced, this is the reestablishment of the equilibrium of populations associated to the loss of the energy of the spins. T_2 is the time during which the individual spins lose their phase coherence. This produces a dispersion of the individual magnetization vectors because the experienced field for each of them is lightly different. This leads to the absence of pure magnetization in the transverse plane (Fig. 8). The lifetimes of the excited nuclear spins are extremely long compared to the excited electronic states of the optic spectroscopy. This is a consequence, as was seen, of the small transition energies, associated to nuclear resonance. The advantages of having extensive lifetimes are that resonance signals are obtained rather narrower than those found in rotational, vibrational or optical transitions. Besides, there is the possibility of having time to manipulate the spins system after having been initially excited, which permits to perform a wide variety of alterations to the mentioned system, which later remain evidenced in the final spectrum.

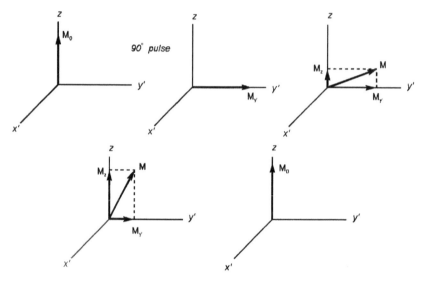

Figure 7 Longitudinal relaxation. The recovery of *M* re-establishes the longitudinal (M_z) component.

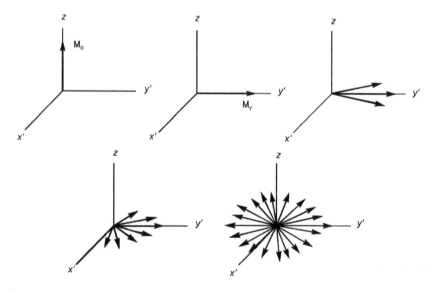

Figure 8 Transverse relaxation. Field inhomogeneity associated with spin-spin dipolar interactions may dephase the individual spins without exchanging energy with the surrounding.

It is important to highlight that even in the simplest case, such as the experiment of one pulse (Fig. 9), the relaxation speeds influence the resolution as in the sensibility of the spectrum (Braun *et al.*, 1998). The effect of an r.f. pulse applied to the sample can now be visualized. The "pulse" consists of lighting a radiation of r.f. of a determined largeness during a time t_p, being extinguished later. The effect on the magnetization M_0 is that of applying a torque in a perpendicular direction to B_1, which makes the vector rotate from axis z to the plane x-y. The angle made by the vector will depend on the amplitude and duration of the pulse. If the angle crossed by the vector is of 90°, then it falls on the y axis representing a 90° pulse. In the case of the example where B_1 was applied over the axis x, the pulse is of $90°_x$. The response to the disturbance generated by the pulse is detected as a signal that exponentially decays with time, a Free Induction Decay or FID (Fig. 10) which shows how the excited spins system relaxes towards thermal equilibrium. This type of spectrum in *time domain* can be converted into a spectrum in the *frequency domain*, through a mathematical operation known as Fourier Transformation. The FID can be mathematically manipulated to increase sensibility (e.g. NMR routine ^{13}C) at the expense of the resolution, or reverse, to increase resolution (often important in ^1H NMR) at the expense of sensibility. Besides, it is possible to develop pulses sequences, which, after an adequate mathematical manipulation, result in great value data (as 2D NMR). The vast utility of NMR spectroscopy is based on two secondary phenomena based on electronic density that surrounds the nucleus, the *chemical shift* and the *spin-spin coupling*. As in the case of chemical shift, the spin-spin coupling reflects the chemical environment in which the nuclear spin is involved.

Under the influence of the applied magnetic field, electrons that surround the magnetic nucleus move generating a magnetic field that opposes the applied magnetic field, in the region of the nucleus. Thus, the applied magnetic field necessary to produce the resonance of the mentioned nucleus, must be greater than the existing one in absence of electrons. It could be said that electrons shield the nucleus.

This shielding is proportional to B_0. Any effect that alters density or spatial distribution of the electrons around the nucleus will alter the grade of shielding

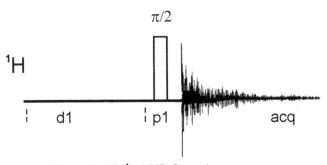

Figure 9 1D ^1H NMR One-pulse sequence.

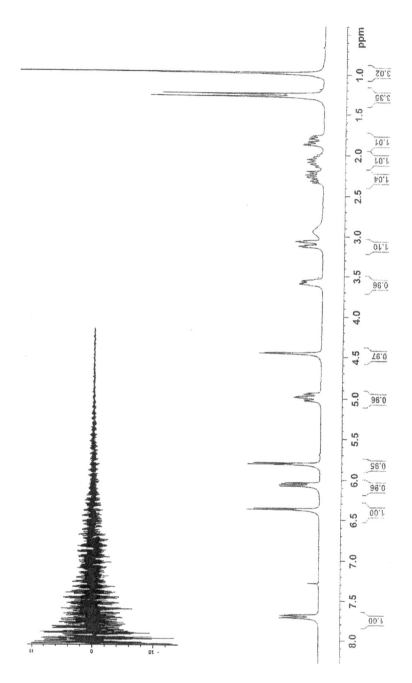

Figure 10 FID signal and ^{1}H spectrum of helenalin.

and this will be seen reflected in the nucleus resonance frequency in the constant applied magnetic field.

The local field B_1 that the nucleus "feels" is:

$$B_i = B_0 (1 - \sigma_i) \qquad \sigma_i = \text{shielding constant}$$

Therefore, its frequency v_i:

$$v_i = (\gamma/2\pi) B_0 (1 - \sigma_i)$$

Nuclei located in different chemical (electronic) environments experience different protection phenomena giving place to different Larmor's frequencies. This is known as *chemical shift*.

The positions of the different signals are usually referred to the resonance of a standard substance, tetramethylsilane (TMS), used for measurement of ^1H and ^{13}C NMR chemical shifts. Hence, the chemical shift (δ_i) is an adimensional parameter which does not depend on the frequency of the spectrometer v_0 and is expressed in units of 10^{-6} (ppm):

$$\delta_i = (v_i - v_{\text{ref}})/v_0$$

Any structural change taking place will be reflected in the chemical shift: if the environment of two nuclei (for example, two hydrogen or two carbons) is not identical, their chemical environments will be in the nature of any of the two nuclei, unless that *accidental equivalence* occurs. When the nuclei have identical chemical environments and, therefore, the same chemical shift, it is said they are *chemically equivalent*. As hybridization changes in the electronic density due to the different substituents, greater effects are produced in the carbon nucleus than in the hydrogen nucleus bonded to it. In fact, the range of chemical shifts of ^{13}C in typical organic compounds comprises 200 ppm approximately, while the corresponding range in ^1H shifts is only of approximately 12 ppm, as can be seen in Table 2.

Table 2 ^1H and ^{13}C chemical shifts (source: Pretsch *et al.*, 1998)

Compounds	Functional groups	^1H (ppm)	^{13}C (ppm)
Alkanes	CH_3	0.6–1.2	15–30
	CH_2	1.2–1.5	22–45
	CH	1.4–1.8	30–58
Cycloalkanes	3-ring CH_2	−0.2–0.2	−2.9
	4-ring CH_2	1.95	22.3
	5-ring CH_2	1.50	26.5
	6-ring CH_2	1.44	27.3
Various CH_3	CH_3-C-C-**G**		
	(**G** = X, OH, OR, N ..)	0.8–1.4	27–29
	CH_3-C-**G**		
	(**G** = C=C, Ar)	1.05–1.20	15–30

Contd.

Table 2   *Contd.*

Compounds	Functional groups	^1H *(ppm)*	^{13}C *(ppm)*
	CH$_3$-C–**G**		
	(**G** = X, OH, OR, C=O)	1.0–2.0	25–30
	CH$_3$ –C=C	1.5–2.0	12–25
	CH$_3$–COR, CH$_3$-Ar	2.1–2.4	20–30
	CH$_3$–C? C	1.7	5–30
	CH$_3$–**G**		
	(**G** = N, X)	2.2–3.5	25–35
	CH$_3$–**G**		
	(**G** = OR, OAr)	3.2–3.8	56–60
Various CH$_2$	R-CH$_2$–**G**, **G** = C=O	2.3–2.6	32–45
	G = C=C	1.9–2.3	32–35
	G = Ar	2.4–2.7	38–40
	G = F	4.3	88
	G = Cl	3.4	51
	G = Br	3.3	40
	G = I	3.1	13
	G = OH, OR	3.5	67–69
	G = NH$_2$	2.5	47–49
	G = NR$_2$	2.5	60–62
	G = CO$_2$H	2.4	39–41
	G = CN	2.5	25–27
Various CH	R$_2$CH-**G**, **G** = C=O	2.5	40
	G = C=C	2.2	Variable
	G = Ar	2.8	32
	G = F	4.6	83
	G = Cl	4.0	52
	G = Br	4.1	45
	G = I	4.2	20
	G = OH, OR	3.9	57–58
	G = NH$_2$	2.8	43
	G = NR$_2$	2.8	56
	G = CO$_2$H	2.6	Variable
	G = CN	2.7	23
Alkenes	=CH$_2$	4.5–5.0	115
	=CH$_2$ (conjugated)	5.3–5.8	117
	=CHR	5.1–5.8	120–140
	=CHR (conjugated)	5.8–6.6	130–140
	C=C=CH$_2$	4.4	75–90
	C=C=C	–	210–220
Alkynes	R**C**≡**C**H	2.4 – 2.7	65–70
	R**C**≡CH	–	85–90

Contd.

Table 2 *Contd.*

Compounds	Functional groups	^1H (ppm)	^{13}C (ppm)
Benzenes	General Ranges	6.5 – 8.5	115–160
Specific examples:	$PhNO_2$, ipso-	–	148.5
	ortho-	8.2	123.5
	meta-	7.4	129.4
	para-	7.6	134.3
	$PhOCH_3$, ipso-	–	159.9
	ortho-	6.8	114.1
	meta-	7.2	129.5
	para-	6.7	120.8
	PhBr, ipso-	–	123.0
	ortho-	7.5	131.9
	meta-	7.1	130.2
	para-	6.7	126.9
	$PhCH_3$, ipso-	–	137.8
	ortho-	7.4	129.3
	meta-	7.2	128.5
	para-	7.1	125.6
Carbonyl Groups (aldehydes)	RCHO	9.4–9.7	200
	ArCHO	9.7–10.0	190
(Ketones)	R_2CO	–	205–215
	5-ring C=O	–	214
	6-ring C=O	–	209
	ArCOR	–	190–200
(Carboxyls)	RCO_2H, $ArCO_2H$	–	170–180
(Esters)	RCO_2R, $ArCO_2R$	–	165–172
(Acid chlorides)	RCOCl, ArCOCl	–	168–170
(Amides)	$RCONH_2$, $ArCONH_2$	–	170
Nitriles	RC≡N	–	115-125
Exchangeable (Acidic) Hydrogens	ROH (free)	0.5–1.0	–
	ROH (H-bonded)	4.0–6.0	–
	ArOH (free)	4.5	–
	ArOH (H-bonded)	9.0–12.0	–
	CO_2H (H-bonded)	9.6–13.3	–
	NH, NH_2 (free)	0.5–1.5	–
	ArNHR, $ArNH_2$ (free)	2.5–4.0	–
	R_3NH^+, $R_2NH_2^+$, RNH_3^+ (in CF_3CO_2H)	7.0–8.0	–
	Ar_3NH^+, etc. (in CF_3CO_2H)	8.5–9.5	–
	RSH	1.0–1.6	–
	ArSH	3.0–4.0	–

A nucleus chemical shift can also be affected by the presence of *magnetically anisotropic* groups in its neighborhood. These groups generate, under the influence of the applied magnetic field, a shielding/deshielding phenomenon, which has different effects on the neighboring nuclei, *through the medium of space*. For instance, aromatic rings shield nuclei placed above or beneath the plane of the ring, and deshield nuclei placed in their plane. This last effect is the main cause of the deshielding observed for benzene protons.

The value of chemical shift of any magnetic nucleus in a given molecule is unique, and it depends on all the other groups present in it.

An organic molecule generally contains more than one magnetic nucleus (i.e. ^1H). The energy associated to the spin flip of the observed nucleus can be affected by the different quantum energetic states of the other nuclei. This gives origin to a fine additional structure (multiplicity), which provides an enormous amount of structural information when it is correctly interpreted.

A nucleus "feels" different orientations ($+1/2$ o $-1/2$) of another nucleus through the electrons of the chemical bonds that separate them. This is called spin-spin coupling.

The magnitude of the spin-spin coupling falls rapidly as the number of bonds that separate the interacting nuclei increases. This interaction is usually negligible when the nuclei that interact are more than three bonds separate. The spin-spin interaction magnitude is the difference of the nuclear transition energy caused by the coupling with another spin. It is expressed by the coupling constant J. Since J depends only on the number, type and spatial arrangement of the bonds that separate both nuclei, it does not depend on the applied magnetic field. Thus, J values provide valuable structural information.

3. ONE-DIMENSIONAL TECHNIQUES

3.1 One-Pulse Techniques

In Fig. 10 the spectrum of helenalin, a sesquiterpene resulting from some *Helenium* genus species (Herz *et al.*, 1962) is observed. It was taken using the sequence of pulses shown in Fig. 9.

3.1.1 ^{13}C NMR

^{13}C is a rare isotope (Breitmaier and Voelter, 1987). The most obvious consequence of its low natural abundance (1.06%) is the extremely low chance of observing a coupling of a ^{13}C atom with another ^{13}C atom. Complications resulting from spin-spin coupling in spectra of ^{13}C emerge only from their coupling with ^1H. Direct carbon-proton coupling constants (through a bond) have values between 125 and 250 Hz, while the coupling constants CH through two or more bonds are much smaller (generally under 10 Hz). The simplest ^{13}C spectrum emerges from applying the pulses sequence from Fig. 11, where ^{13}C is observed while all protons decouple using *broadband* or *noise decoupling*. In the ^{13}C spectra with broadband decoupling (Fig. 12), individual resonances appear like singlets, making these spectra very easy to interpret.

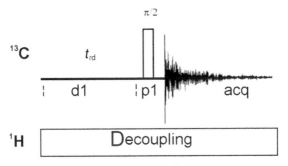

Figure 11 ^{13}C One-pulse sequence with ^{1}H decoupling.

3.1.2 DEPT

DEPT (Distortionless Enhancement by Polarization Transfer), is a technique that involves polarization transfer, used to observe nuclei with small magnetogyric constant, that are scalar coupled (J-coupled) to ^{1}H, a nucleus with big magnetogyric constant (Bendall *et al.*, 1981; Dodrell *et al.*, 1982). DEPT is the most widely used experiment in ^{13}C spectroscopy. DEPT makes use of the generation and manipulation of quantum multiple coherences to differentiate different ^{13}C signals. Thus, DEPT experiments allow differentiation of methyl (–CH$_3$), methylene (–CH$_2$) and methine (–CH) signals. Quaternary carbons disappear from DEPT spectra because the one-bond heteronuclear coupling constant (J_{XH}) is used to generate polarization transfer. DEPT may be run with or without ^{1}H decoupling, and it is relatively insensitive to a precise correspondence between delays and coupling constants. Thus, DEPT is easier to use than INEPT [Insensitive Nucleus Enhanced by Polarization Transfer], a pulse sequence closely related to DEPT (Morris and Freeman, 1979). However, DEPT is more sensitive to imperfections in pulses than INEPT.

To follow the events that occur during the pulse sequence (Fig. 13), take into consideration the pair ^{1}H-^{13}C and notice that the action of the two pulses of 180° is to refocus chemical shifts, where necessary. The sequence starts with a pulse of 90° (H) after which magnetization evolves under the carbon-proton coupling influence such as, after a time period of 1/2J, the two satellite vectors of the proton are in anti phase. Application at this point of a 90° (C) pulse produces a new state of matters in which a transverse magnetization proton, as in the case of carbon, evolves coherently. This new state of matters is known as heteronuclear multiple-quantum coherence (HMQC), which can not be visualized with the vector model. However, HMQC can be imagined as a container where proton and carbon magnetizations live, influencing each other, waiting to be separated at some later point. HMQC evolves not only during a determined period under the influence of the proton chemical shift, but also, simultaneously, under the influence of the carbon chemical shift up to the application of the θ pulse. In order to remove the effects of the second proton chemical shifts, a second time period of 1/2J evolution is adjusted and a 180° (H) pulse is applied between the

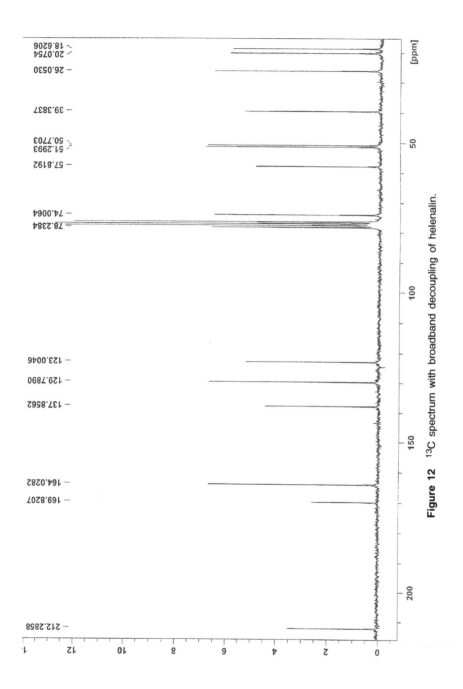

Figure 12 ^{13}C spectrum with broadband decoupling of helenalin.

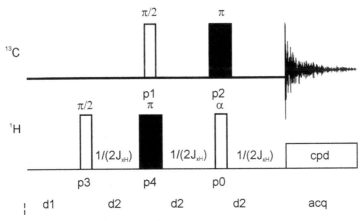

Figure 13 DEPT pulse sequence.

two delays, coincident with the 90° (C) pulse. The action of the θ pulse is to transfer HMQC in the C transverse magnetization, regenerating observable magnetization. The result of this transference process depends, above all, on the multiplicity of carbon resonance. In other words, methyl, methylene and methine groups respond in a different manner when facing θ pulse, and this provides the basis of the edition with DEPT. Simultaneous application of the 180° (C) pulse with the θ pulse (H) refocuses carbon shifts during the 1/2J final delay. The final result is the polarization transfer from H to C, combined with the potential of the spectra edition.

In DEPT experiment, edition is a product of the variable angle of the θ pulse which does not depend on J. As a result, having found an adequate range of J values, the experiment is carried out easily, even with small errors in the adjustments of these J values. The determination in the resonance multiplicities with DEPT starts with the collection of three spectra with θ = 45, 90 and 135°, taking into account that 90° DEPT requires at least double the number of scans to make the same signal to noise relation than the others. The signs of the signals are as shown in Table 3.

Table 3 Signs of the DEPT signals

	DEPT 45	DEPT 90	DEPT 135
CH	+	+	+
CH_2	+	NS	−
CH_3	+	NS	+

This pattern is trivial to determine multiplicity of each of the signals through direct comparison (Fig. 14). Signals of quaternary C are distinguished because they appear only in the normal ^{13}C spectrum.

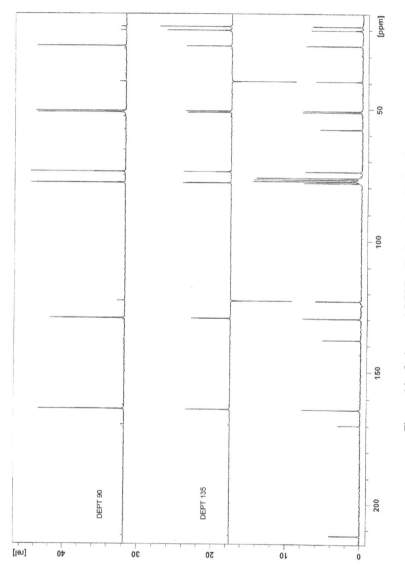

Figure 14 Carbon and DEPT edited spectra of helenalin.

In practice, it is not necessary to acquire DEPT 45 spectrum because the information it offers is summarized in DEPT 135, saving machine time. The incorrect adjustment of the proton pulses, especially θ pulse, leads to the appearance of small peaks in DEPT 90. Due to its low intensity, these spurious signals are easily recognized.

If the pulses are correctly measured, errors may emerge when delay values are very different from those of $^1J_{CH}$. A typical value for $^1J_{CH}$ is 145 Hz, for which $1/2\,J$ is 3.45 ms. With this $1/2J$ value some carbons such as of an alkyne group can present unusually weak signals and sometimes they can be considered as quaternary carbon signals of an alkyne group due to the exceptionally big value of $^2J_{CH}$.

4. CORRELATIONS VIA INDIRECT (SCALAR) COUPLINGS

4.1 Two-dimensional Methods

The pulse sequence of a 2D NMR experiment can be divided into the following two blocks:

(1) Preparation – Evolution (t_1), (2) Mixing – Detection (t_2)

The detection period corresponds exactly to the one seen for 1D NMR spectroscopy. Time t_2 provides, after Fourier transformation, the f_2 frequencies axis in the 2D NMR spectrum. What distinguishes 2D NMR spectroscopy is that a second variable time, t_1 evolution time, has been introduced. Evolution time is increased step by step. For each value of t_1 a different FID is detected in t_2. Thus, the signal obtained is a function of two variable times t_1 and t_2: S (t_1, t_2). It is obtained that way after Fourier transformation of each of the FID, i.e. a series of f_2 spectra. These spectra differ between each other in the intensities and/or the phases of the individual signals, according to the different increases in t_1. A second Fourier transformation on t_1, "perpendicular" to the f_2 dimension results in a spectrum in function of two frequencies f_1 and f_2 (i.e. Fig. 15). Basically two spin-spin interactions can be used in the transference of coherence (see DEPT). An interaction is the scalar spin-spin coupling through bond (J coupling). This interaction is effective, as has been seen, between spins separated by a few bonds. It is very useful in the elucidation of connectivity (bonds), which helps in the investigation of the constitution of organic molecules, and also in spectral assignment. The magnitude of the scalar coupling is sensitive to the torsion angle around simple and double bonds, giving important information about the spatial structure (conformation and configuration) (Karplus, 1959; 1963; Bystrov *et al.*, 1978).

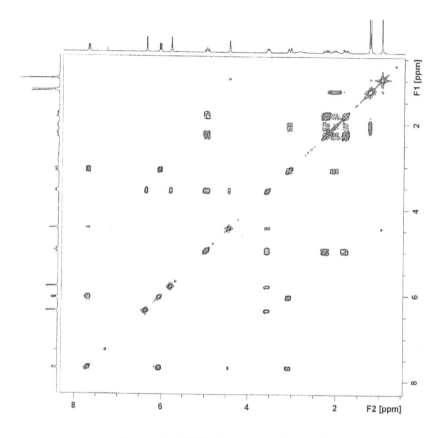

Figure 15 COSY-45 spectrum of helenalin.

The other interaction is dipolar coupling of two spins, which occurs through the space. This coupling is responsible of the dominant relaxation mechanism in solution, although it only manifests in 1D NMR spectrum as line broadening. Dipole coupling also generates mutual relaxation between two close nuclei in space, known as cross relaxation.

Intermolecular cross relaxation initiates a variation in the intensity of a signal as a result of the alteration in the population of another spatially close nuclei. This phenomenon is known as Nuclear Overhauser Enhancement (NOE) (Neuhaus and Williamson, 1989).

4.2 Homonuclear Shift Correlation

4.2.1 ¹H-¹H COSY

Every discussion about structural elucidation of natural products using NMR spectroscopy includes the treatment of bidimensional COSY (Correlation

SpectroscopY) experiments, to establish the relationship between scalar coupled protons. Homonuclear correlation through bidimensional NMR methods is responsible for the development of the present day massive field of 2D NMR spectroscopy. COSY is a 2D homonuclear technique which is used to correlate ^{1}H nuclei chemical shifts that are J-coupled ones to others (Kessler *et al.*, 1988; Martin and Zektzer, 1988). Two types of COSY sequences will be discussed, magnitude COSY and COSY with double quantum filtered, with and without field pulsed gradients.

COSY pulse sequence (Fig. 16) consists of the application of two pulses ^{1}H of 90°, separated between each other by an evolution period systematically increased (t_1), followed by an acquisition time t_2. The bases that operate in the COSY are considerably complex and have been the subject of various treatments with density matrix, as well as with formalisms operating product (Sorensen *et al.*, 1983; Ernst *et al.*, 1987).

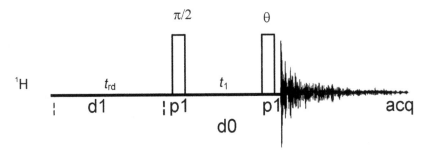

Figure 16 COSY pulse sequence.

In the COSY basic sequence, the first pulse creates transverse magnetization, whose components evolve during the evolution time t_1 according to their chemical shifts and homonuclear J couplings. The second pulse mixes the magnetization components among all transitions that belong to the same coupled spins system. The final distribution of the magnetization components is detected measuring its precession frequencies during the detection period t_2. The COSY spectrum is processed by a 2D Fourier transformation with regard to t_1 and t_2, and cross peaks (peaks outside the diagonal) indicate which ^{1}H are J coupled.

Various simple programmes of two pulses can be used to obtain a COSY spectrum in magnitude mode, i.e. COSY 90 and COSY 45. These are differentiated based on the angle of the final pulse. However, a pulse angle of 45° is recommended because the greatest signal-noise relation is obtained, together with a simpler cross peak structure in the final spectrum. In some cases it is possible to differentiate geminal from vicinal couplings.

The selection of parameters to acquire a COSY is relatively simple. It is only needed to establish two parameters, assuming that the 90° ^{1}H pulse has been determined. These are the dwell time (DW) and the interpulse delay (t_{rd}). It should also be decided about digitalization of the second domain of the frequencies

(f_1). The experiment COSY being autocorrelated, a square data matrix with equivalent frequency ranges is desirable. It may also be necessary that the data matrix is symmetrized prior to the plotting.

In a very simple way, it can be remembered that DW is reciprocal to the spectral width (SW) and that DW multiplied by the number of acquired points determines the time of acquisition t_2. Then, once the SW is determined, the DW can be automatically calculated. In practice, a standard ^{1}H spectrum is acquired and processed.

The SW and the carrier frequency f_1 are optimized in such a way that ^{1}H signals cover all the SW. The interpulse delay in the experiment allows the system to go back to the magnetic equilibrium. As a starting point it can be added to t_2, 1 sec. (t_{rd}).

Digitalization of the second domain of the f_1 frequencies is the third consideration in the selection of parameters. A spectral resolution of 3 Hz/pt is enough to solve big scalar couplings, in a well-resolved spectrum. Fine digital resolution, necessary to solve small scalar couplings, notably increases experimental time. A good starting point is to choose 512 data points in f_2, and 128 data points in f_1.

Once the COSY spectrum is acquired, zero filling is applied in f_1 up to 256 real data points, in order to obtain a symmetric matrix of 256×256 real points.

Acquired signals in one of these experiments have absorptive and dispersive contributions, as in f_1 and f_2. This means that it is impossible to place the spectrum on phase, with all the peaks purely absorptive. As a consequence, the spectrum must be presented in magnitude mode (Fig. 15). Finally, data can be symmetrized.

Symmetry (Baumann *et al.*, 1981) of the matrix improves the presentation of the data, eliminating through an algorithm those signals outside the diagonal which do not appear symmetric on both sides of the mentioned diagonal.

The DQF [Double Quantum Filter]–COSY pulses sequence (Fig. 17) consists of three pulses, where the third pulse transforms part of the multiple quantum coherence in simple quantum coherence, which is observable and it is detected during the period of acquisition t_2 (Derome *et al.*, 1990; Rance *et al.*, 1984).

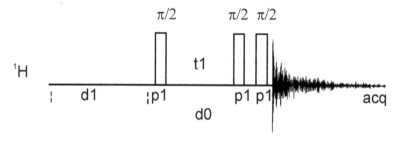

Figure 17 DQF-COSY pulse sequence.

A DQF–COSY advantage is its sensibility to the phase, i.e. the cross peaks can be seen as pure absorption signals as in f_1 and f_2. In general, the spectrum sensitive to the phase has a greater resolution than the equivalent magnitude spectrum, because in the magnitude spectrum, the signal is wider than the pure absorption signal. Another advantage of the DQF–COSY experiment is the partial cancellation of the diagonal peaks. Thus, the diagonal crest is much less pronounced in a DQF–COSY than in a normal COSY. A third advantage of the DQF is the elimination of big signals, as those of the solvent, which do not experience scalar coupling at all. The DQF–COSY sequence is sensitive to high repetition speeds, therefore, it is important to choose a long recycling time t_{rd}, with the purpose of avoiding multiple–quantum devices in the spectrum.

4.3 Heteronuclear Shift Correlation

4.3.1 HMQC

The Heteronuclear Multiple-Quantum Correlation (HMQC) spectroscopy is an inverse chemical shift correlation experiment which provides the same information as XHCORR [heteronuclear (X, H) shift correlation spectroscopy] (Bax and Morris, 1981; Bax and Subramanian, 1986). The advantage of HMQC is the detection of the nucleus of the greater γ (^{1}H) and, in that way, obtaining the highest sensitivity. The challenge of an inverse experiment of chemical shift correlation is to suppress through a difference experiment, the intense signals that emerge from the ^{1}H ones that are not directly coupled to ^{13}C. The interest signal is that of ^{1}H directly coupled to ^{13}C; however, the detected signal is ruled by the contribution of ^{1}H directly joined to ^{12}C. HMQC minimizes this problem optimizing the sensitivity of the experiment.

The resonance of the spins of low γ can be detected with good sensibility for the creation of multiple quantum coherence. HMQC experiment provides a map of connectivity in which a cross peak correlates two bonded nuclei, as can be seen in the HMQC spectrum of helenalin (Fig. 18). In such a spectrum the main characteristics of the experiment are seen, for instance, known assignments of protons can be transferred to the carbon spectrum, extending the molecule characterization. Besides, the resonance of protons scatters according to the chemical shift of ^{13}C, which helps the interpretation of the ^{1}H spectrum. The final feature is the ability to identify diasterotopic geminal pairs (cross peaks *a* and *b* in the spectrum), not always identified in COSY spectra. The pulses sequence (Fig. 19) is rather simple. The first pulse ^{1}H creates transverse magnetization, part of which evolves in an anti-phase magnetization at the end of the first $1/(2J_{XH})$ delay. This anti-phase magnetization is transformed into multiple-quantum coherence by the pulse **p3** and evolves according to the chemical displacement during $t1$. A $1/(2J_{XH})$ delay is inserted between the last pulse of 90° and the beginning of the acquisition. Without this delay the ^{1}H magnetization components that could be anti-phase would be cancelled with

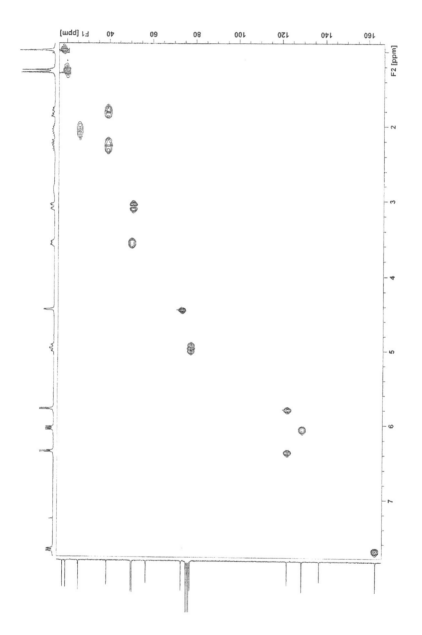

Figure 18 HMQC (with BIRD) spectrum of helenalin.

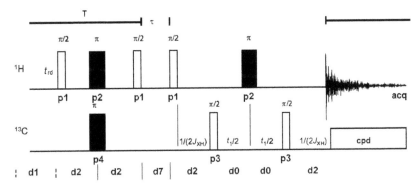

Figure 19 HMQC with BIRD pulse sequence.

the decoupling of ^{13}C. For small molecules, it is useful to use a BIRD [Bilinear Rotation pulse] preparation period in conjunction with the HMQC experiment (Fig. 19). The basic idea of this preparation period is to saturate every ^{1}H not directly bonded to ^{13}C.

4.3.2 HSQC

The Heteronuclear Single-Quantum Correlation (HSQC) experiment differs from HMQC, i.e. only transverse magnetization (single-quantum) of the heteronuclear spin evolves during the period t_1 (Fig. 20). The transverse heteronuclear magnetization is generated by polarization transfer from protons directly bonded via INEPT sequence (Morris and Freeman, 1979; Bodenhausen and Ruben, 1980). While correlations similar to HMQC experiment are produced in the last instance, cross peaks do not contain ^{1}H-^{1}H couplings along the f_1. This is due to the fact that only heteronuclear magnetization evolves during t_1. This results in a better resolution in that dimension, becoming the main advantage of HSQC over HMQC in small molecules.

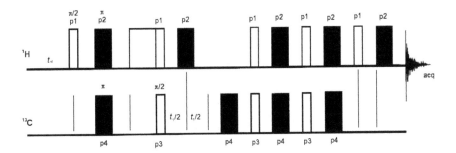

Figure 20 Basic HSQC pulse sequence.

4.3.3 HMBC

Heteronuclear Multiple Bond Correlation (HMBC) spectroscopy is a modified version of the HMQC adequate to determine long-range connectivities $^1H/^{13}C$ (Bax and Summers, 1986; Summers *et al.*, 1986). It allows establishing correlations between carbons and protons placed to more than one bond; hence the name of multiple bond. In most of the cases, this involves proton-carbon correlations through couplings to two or three bonds ($^nJ_{CH}$, n = 2,3), since the interaction tends to disappear at greater distances. It provides the same information as COLOC [correlation spectroscopy for long-range couplings] but its sensitivity is much greater, since it is an inverse experiment, ^1H detection. It allows recovering quaternary carbon resonances not detected in the correlations experiments to a bond like HMQC and HSQC. The HMBC spectrum of helenalin is shown in Fig. 21.

5. CORRELATIONS VIA DIRECT (DIPOLAR) COUPLINGS

5.1 Correlations Through Space: The Nuclear Overhauser Effect

When a resonance in a NMR spectrum is disturbed by saturation or inversion, the pure intensities of other resonances may vary. The change in the signal intensities is because of the spins that are *near in space* to those affected by the disturbance. This phenomenon is known as Nuclear Overhauser Effect (NOE).

In a NOE difference experiment, a ^1H resonance is selectively pre-irradiated until saturation is obtained. During the pre-irradiation period, the NOE is built in another ^1H nucleus near in space. Then, through a $\pi/2$ pulse, an observable magnetization is created, which is detected during the acquisition period. The experiment is repeated using different pre-irradiation frequencies that include one that is off-resonance. This last one is used to get a reference or control spectrum.

Final spectra are shown as the difference between a spectrum obtained with on-resonance pre-irradiation and the reference spectrum. The devices resulting from temperature changes or drift on the magnetic field can be minimized acquiring pre-irradiated and reference data in an alternated way. The NOE's difference pulse sequence (Fig. 22), starts with the d1 recycle delay time, followed by a period of cw irradiation with a duration of L4*d20, where d20 is the irradiation time for a particular frequency, while L4 is a loop counter that determines irradiation total time.

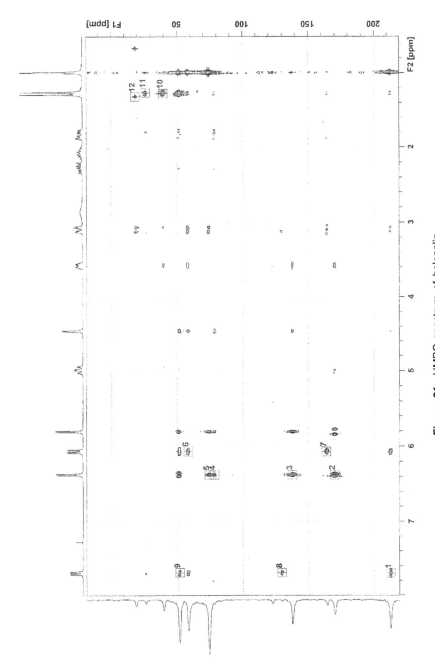

Figure 21 HMBC spectrum of helenalin.

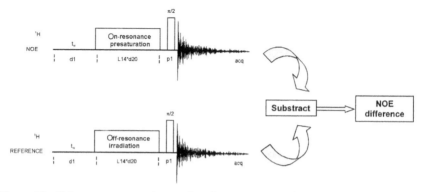

Figure 22 Pulse sequence and procedure for generating NOE difference experiment.

5.2 NOESY

NOESY (Nuclear Overhauser Effect SpectroscopY) is a 2D spectroscopy method, valuable in the spins identification that suffer cross-relaxation (Jeener *et al.*, 1979; Wagner and Wutrich, 1982). In its commonest form, NOESY is used as a H homonuclear technique. In the NOESY experiment, dipolar direct couplings indicate basic cross relaxation, and the spins that suffer cross relaxation are those near in space. Thus, cross peaks in a NOESY spectrum indicate that protons are near one another in space. The basic NOESY sequence (Fig. 23) has three $\pi/2$ pulses. The first pulse creates transverse magnetization. This one precesses during evolution time t_1, which is increased during the course of the 2D experiment. Then, after initial excitement and evolution time t_1, magnetization vector exists in the transverse plane. The second pulse of 90° (Fig. 24) places the component of the orthogonal magnetization vector to the pulse direction, over z-axis, generating the inversion in the populations, allowing the evolution of NOE during the subsequent mixing time (τ_m). Transverse magnetization component that remains in the transverse plane is detected in the COSY experiment. The third pulse $\pi/2$ again creates a transverse magnetization, which is acquired immediately after the pulse.

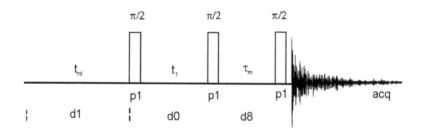

Figure 23 NOESY pulses sequence.

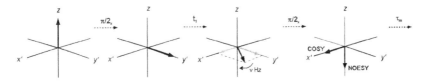

Figure 24 Evolution of magnetization vectors during NOESY.

5.3 ROESY

ROESY (Rotating frame Overhauser Effect SpectroscopY) is an experiment in which homonuclear NOE is measured under spin-lock conditions (Bax and Davis, 1985).

ROESY (Fig. 25) is especially useful for medium size molecules with a mass of around 1,000-2,000 daltons. In these cases, the observed NOE is near zero, but the rotating frame NOE (ROE) is always positive.

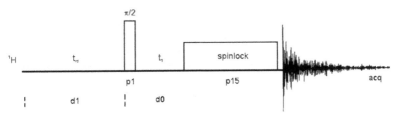

Figure 25 ROESY pulse sequence.

6. IDENTIFICATION OF NATURAL PRODUCTS USING NMR: SOME EXAMPLES

The following paragraphs provide examples of identification of natural products using NMR. Compounds analyzed are helenalin (**1**), 12-*epi*-teupireinin (**2**), 8-hydroxysalviarin (**3**) and 7, 8-didehydrorhyacophiline (**4**) (Fig. 26).

6.1 Structural Determination of Helenalin (1)

Several spectra of compound **1** were previously provided. The assignment of the ^{1}H NMR and ^{13}C NMR signals is discussed here. The ^{13}C NMR spectrum (Figs. 12 and 14) shows signals for 15 carbon atoms, of which 7 are CH groups, 2 are CH$_3$ groups, 2 are CH$_2$ groups, and 4 are as quaternary carbons. In the ^{1}H NMR spectrum (200.13 MHz) (Fig. 10), the signals at δ 7.7 (*dd*, *J* = 6.0, 1.75 Hz) and 6.07 (*dd*, *J* = 6.0, 2.8 Hz) correspond to β and α hydrogens respectively of a α, β-unsaturated carbonyl system, and can be assigned to H-2 and H-3, respectively. The carbonyl group is located at C-4. The H-2 y H-3 signals show

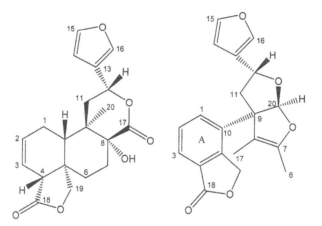

Compound 1. Helenalin

Compound 2. 12-*epi*-teupireinin

Compound 3. 8-hydroxysalviarin   Compound 4. 7,8-didehydroryacophiline

Figure 26 Compounds selected as examples for NMR analysis.

a clear coupling between them in the COSY spectrum (Fig. 15). Furthermore, both signals are scalar connected to a signal at δ 3.1 (*ddd*, *J* = 11.5, 2.8, 1.75 Hz), which could be assigned to H-1. The combination of 1D/2D techniques such as DEPT (Fig. 14) and HMQC (Fig. 18) allows to assign the signals at δ 169.8 (C-2), 129.8 (C-3) and 212.3 ppm (C-4). H-2 shows, among other HMBC correlations (Fig. 21), a clear coupling to C-3 and C-4. H-3 shows correlations with C-2 and C-4 confirming this partial structure. Signals appearing as doublets at δ 6.38 and 5.80 ppm with *J* = 3.1 Hz are consistent with methylene protons of an exocyclic double bond, and can be assigned to H-13 and H-13'. In the COSY spectrum these signals are shown to be coupled to H-7 at 3.58 (*m*). H-7 also shows a coupling to a hydrogen at δ 4.98 (*dt*, *J* = 8.5, 2.5 Hz) which can be assigned to H-8, and with the signal at δ 4.46 (*d*, *J* = 1.75 Hz) which should be assigned to H-6. H-6 also shows a long-range coupling (four bonds) with the hydrogens of C-15 at δ 0.99 (*br s*). Furthermore, H-6 shows HMBC correlations with C-5 (57.8 ppm, *s*), C-7 (50.8, *d*), C-8 (78.2 ppm, *d*) and C-11 (137.8 ppm, *s*). HMBC correlations H-8/C-12 (169.8 ppm, *s*), H-7/C-9 (39.4 ppm, *t*), H-7/C-5 and H-7/C-11 are also observed. The doublet at δ 1.27 (*J* = 6.5 Hz) integrating for three hydrogens showing a coupling with the multiplet centered at 2.07 ppm (H-10) should be assigned to the H-14 methyl group. H-14 shows HMBC correlations with C-10 (26.0 ppm, *d*), C-9, C-1 and C-5.

The NOESY spectrum (Fig. 27) shows a dipolar coupling (through space) between H-1–H-7 and between H-10–H-15, which suggest a relative stereochemistry of **1**.

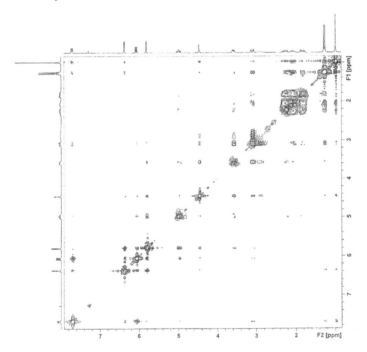

Figure 27 Phase sensitive NOESY spectrum of helenalin.

6.2   Structural Determination of 12-*epi*-teupireinin (2)

Compound **2** was isolated from *Teucrium nudicaule* (Gallardo *et al.*, 1996). The ^{13}C NMR spectrum (Fig. 28) shows signals for 25 carbon atoms, of which 12 are CH/CH$_3$ groups (two CH$_3$ corresponding to acetate have identical chemical shift), 5 are CH$_2$ groups and 8 are quaternary carbons. In the ^{1}H NMR spectrum

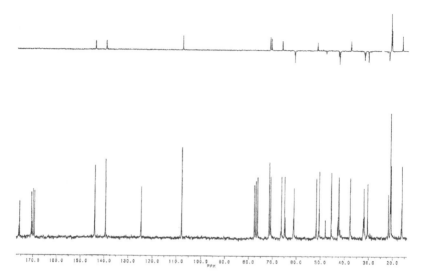

Figure 28 ^{13}C NMR and DEPT spectra of 12-*epi*-teupireinin.

(200.13 MHz, CDCl$_3$) (Fig. 29) signals are observed as multiplets at δ 7.42 and 6.38 ppm which integrate for 2 and 1 hydrogen respectively, and correspond to 2 H-α (H-15 and H-16) and 1 H-β (H-14) of a β-substituted furan ring. A group of signals integrating for 2 hydrogens is found at 5.40 ppm. One of these signals

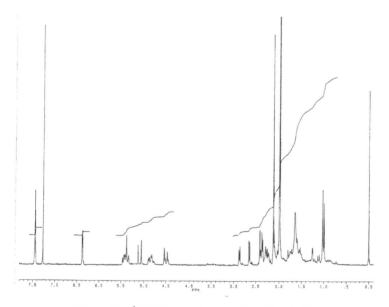

Figure 29 ^{1}H NMR spectrum of 12-*epi*-teupireinin.

corresponds to a triplet which was assigned to the hydrogen base of lactone H-12. In the COSY spectrum (Fig. 30) this hydrogen shows a clear coupling with the signal at 2.38 ppm corresponding to the methylene hydrogens H-11A and H-11B which have an identical chemical shift. The direct heteronuclear correlation experiment, HETCOR [heteronuclear correlation spectroscopy] (Fig. 31), shows connectivity among the aromatic carbons C-14 (δ_C 107.8), C-15 (144.1 ppm), C-16 (139.5 ppm) and the corresponding signals of H-14, H-15 and H-16, which are characteristics of a furan ring. The correlation H-12/C-12 (δ 71.42) is also observed. According to these data, the typical fragment of diterpenic furans isolated from *Teucrium*, as shown in Fig. 32, can be deduced.

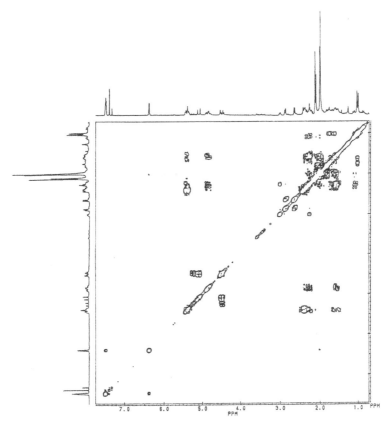

Figure 30   COSY 90 spectrum of 12-*epi*-teupireinin.

The other signal is overlapped to the triplet at 5.38 ppm (H-3, *dd*) geminal to an acetoxyl group.

Two doublets at 2.62 and 2.85 ppm constitute an AB system (*J* = 4.9 Hz) and can be assigned to two geminal protons of an oxirane moiety (H-18A and H-18B). A doublet centered at δ 5.10 (*J* = 13.0 Hz) and a broadened double doublet

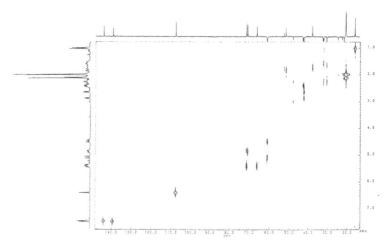

Figure 31 HETCORR (CH one-bond correlation) spectrum of 12-*epi*-teupireinin.

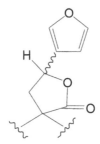

Figure 32 Fragment of diterpenic furans isolated from *Teucrium.*

at 4.50 ppm (J = 13.0, 1.8 Hz) form other AB system corresponding to two methylene hydrogens belonging to an acetoxymethyl group bonded to a quaternary carbon (H-19B and H-19A). The COSY spectrum clearly shows couplings corresponding to both AB systems and, in the case of the second system, the long range W-type coupling with the signal at δ 4.86 (H-6β), which has together with H-19A, a planar conformation allowing the W coupling through four simple bonds. A signal at δ 1.02 (H-17, 3H, d, J = 7 Hz) corresponds to a methyl group axially oriented on C-8. Signals appearing as singlet at 1.98 and 2.00 ppm which integrate for six and three hydrogens respectively, are consistent with acetoxyl groups; one of them belongs to the acetoxymethyl group bonded to C-5 and the other to the α-acetoxyl group on C-6. The third acetoxyl group found in the *trans*-decalin system is geminal to an axial H-3. The NOE difference experiment (Fig. 33) allows establishing the C-12 configuration. This experiment allows assigning the stereochemistry of *neo*-clerodanes 20,12-olide derivatives when only one of the epimers is provided. Irradiation of H-17 protons produced a 1% NOE enhancement of H-14, while no NOE enhance is observed at H-12,

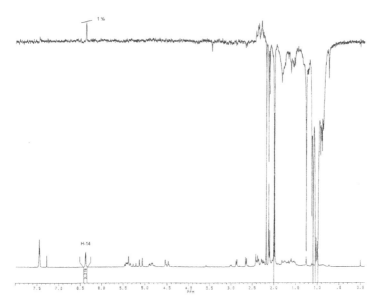

Figure 33 NOE difference experiment of 12-*epi*-teupeirinin. The signal at δ 1.02 (H-17) was irradiated.

indicating that H-12 and H-17 are on opposite sides of the plane defined by the lactone ring C-20, C-12, establishing the configuration of C-12 as *S*.

6.3 Structural Determination of de 8-hydroxysalviarin (3)

Compound **3** is a neoclerodane diterpene isolated from leaves of *Salvia reflexa* (Nieto *et al.*, 1996). The ^{1}H NMR spectrum (500.13 MHz) (Fig. 34; see expanded spectrum in Fig. 36) showed signals at δ 7.42 (*dd*, *J* = 2.0, 1.0 Hz), 7.39 (*d*, 1.0 Hz) and 6.40 (*dd*, 2.0, 1.0 Hz) ascribable to a β-substituted furan ring. Three one-proton signals clearly coupled in the COSY spectrum (Fig. 35) at δ 5.95 (*ddd*, *J* = 10.0, 4.0, 3.0 Hz), 5.60 (*ddd*, *J* = 10.0, 8.0, 3.0 Hz) and 2.80 ppm (*ddd*, *J* = 8.0, 1.5, 1.0 Hz) were assigned, respectively, to H-2, H-3 and H-4.

The one-proton signal at δ 5.25 (*dd*, *J* = 13.0, 4.0 Hz) together with signals as double doublets at 2.40 (*J* = 15.0, 4.0 Hz) and 1.80 (*J* = 15.0, 13.0 Hz) ppm, were consistent with an ABX system formed by H-12 and the C-11 methylene group of **3**. The angular methyl group appeared as a singlet at δ 0.98 and the C-19 protons appeared at δ 4.19 (*d*, *J* = 8.8 Hz) and 4.15 (*dd*, *J* = 8.8, 1.5 Hz). The latter signal was assigned to the H-19 *pro-S* diasterotopic proton, which showed long-range W coupling (^4J = 1.5 Hz) with H-6β. These observations were indicative of an axial orientation for C-19 and of the absence of a β-substituent at C-6. The

C-1 allylic protons appeared as a complex multiplet centered at 2.10 ppm coupled through five bonds with H-4 (Fig. 35).

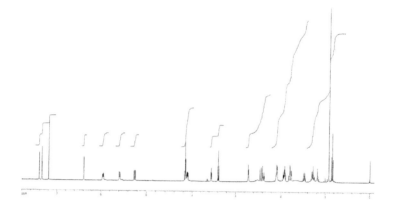

Figure 34 ¹H NMR (500.13 MHz) spectrum of 8-hydroxysalviarin.

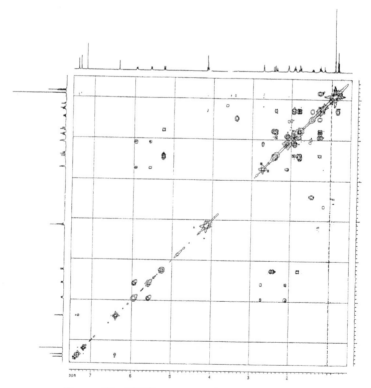

Figure 35 COSY spectrum of 8-hydroxysalviarin.

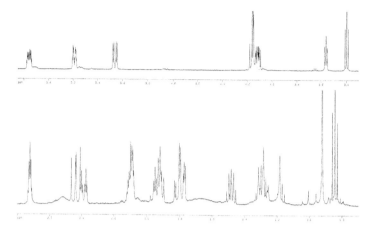

Figure 36 ¹H NMR spectrum (expanded) of 8-hydroxysalviarin.

The ¹³C NMR spectrum (125.7 MHz) (Fig. 37) of **3** showed signals for 20 carbon atoms, nine of which are CH/CH_3, five are CH_2 and six are quaternary carbons.

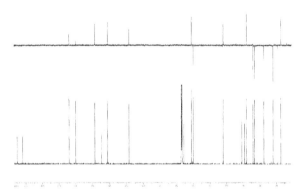

Figure 37 ¹³C NMR and DEPT spectra of 8-hydroxysalviarin.

All unambiguous ¹³C NMR assignments of **3** were resolved by a combination of 1D- and 2D-NMR techniques comprising DEPT (Fig. 37), HMBC and HMQC. The HMQC spectrum (Fig. 38) provided ¹³C assignments of C-2 (128.6 ppm), C-3 (120.9 ppm), C-4 (51.9 ppm), C-12 (71.1 ppm), C-14 (108.5 ppm), C-15 (130.9 ppm), C-16 (143.7 ppm), C-20 (17.2 ppm).

The C-8 signal at δ 75.3 showed HMBC correlations (Fig. 39) with signals for H-7α, H-7β and H-20. In addition, the resonance at δ 22.0 (C-1) showed a correlation with the signal at δ 1.95 as multiplet assigned to H-10. These observations confirmed the attachment of the hydroxyl group at C-8.

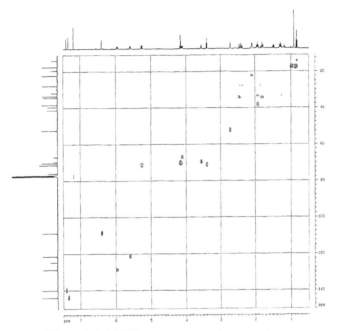

Figure 38   HMQC spectrum of 8-hydroxysalviarin.

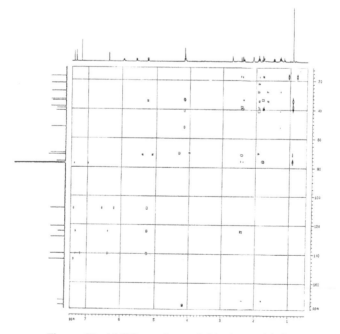

Figure 39   HMBC spectrum of 8-hydroxysalviarin.

The ^{13}C NMR chemical shift of the C-20 methyl group (δ 17.2) was in agreement with an A/B trans ring junction, with C-20 axially oriented. The relative stereochemistry of **3** was established from its NOESY spectrum (Fig. 40). Dipolar interactions among H-20, H-19 (two protons), H-1α and H-11α indicated that these are all on the same face of the molecule. Similar interactions among H-10, H-4, H-6β, H-11β and H-12 supported their close proximity. The NOE correlation between H-10 and H-12 in the NOESY spectrum, together with the coupling constants of the latter proton (J = 13.0, 4.0 Hz) suggested that the C-12 stereocenter has an R configuration.

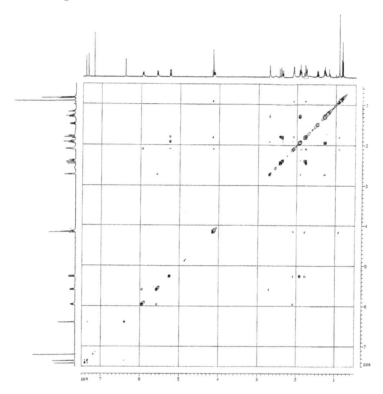

Figure 40 NOESY spectrum of 8-hydroxysalviarin.

6.4 Structural Determination Of 7,8-Didehydrorhyacophiline (4)

Compound **4** isolated from aerial parts of *S. reflexa* is a diterpene with a secoclerodane skeleton (Nieto *et al.*, 1996). The ^{13}C NMR data (DEPT) (Fig. 41) showed that the molecule contained 13 unsaturated carbons; two methyl groups attached to sp^2 carbons, two methylene groups, one of which bearing an

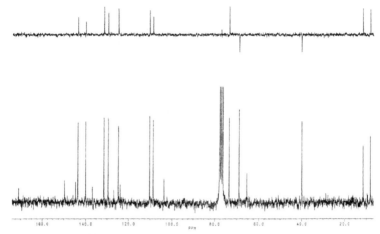

Figure 41 ^{13}C NMR and DEPT spectra of 7,8-dehydroryacophiline.

oxygen atom, and two oxygenated methine groups, with one at δ 73.5 similar to C-12 in compound **3**, and the other at δ 110.2, suggesting an acetal methine carbon (C-20).

The ^{1}H NMR spectrum (200.13 MHz) (Fig. 42) of **4** analyzed with the aid of 2D COSY (Fig. 43), showed resonances for aromatic protons at δ 7.88 (1H, *dd*, *J* = 9.0, 3.0 Hz) assigned to H-3, while H-1 and H-2 appeared as an unresolved complex multiplet centered at 7.58 ppm near to the signals of the furan protons H-16 (δ 7.51, broad singlet) and H-15 (δ 7.45, *dd*, *J* = 2.0, 1.0). The absence of the C-20 methyl signal together with the one-proton singlet at δ 5.90 (H-20), as well as the observed downfield shift of the C-19 methylene protons as compared to **3**, were indicative of a ryacophane skeleton for **4** with an aromatic A ring. The C-19 methylene protons resonated as doublets (*J* = 13.5 Hz) at 5.35 and 5.15 ppm, superimposed with the H-12 resonance (δ 5.12, *dd*, *J* = 10.0, 5.7 Hz), which was clearly coupled with H-11α at δ 2.45 (*dd*, *J* = 12.0, 10.0 Hz) and H-11β at δ 2.60 (*dd*, *J* = 12.0, 5.7 Hz). Two three-proton broad singlets at 1.95 and 1.35 ppm were ascribed to H-6 and H-17 respectively, which showed homoallylic coupling through de C-7, C-8 double bond. The HETCOR and COLOC experiments were in agreement with structure **4** (Table 4).

Table 4 COLOC correlations on compound **4**

Carbon[a]	Correlated protons	Carbon[a]	Correlated protons
4	H-1, H-19	9	H-17, H-20, H-1
5	H-2, H-3, H-19	10	H-2, H-19, H-20
7	H-6, H-17, H-20	13	H-16
8	H-6, H-17	18	H-19

[a]Quaternary carbons only

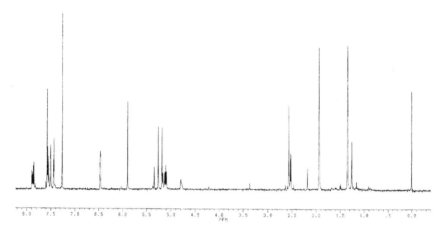

Figure 42 ^{1}H NMR spectrum of 7,8-dehydroryacophiline.

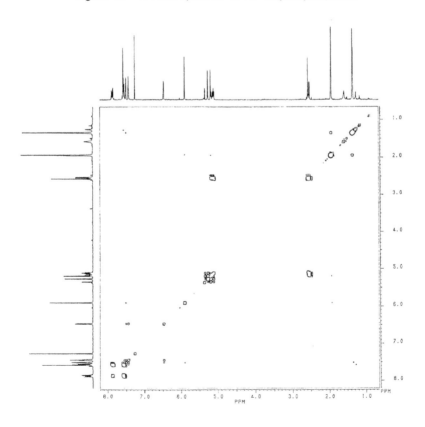

Figure 43 COSY spectrum of 7,8-dehydroryacophiline.

NOESY correlations (Fig. 44) made it possible to determine the relative stereochemistry of **4**. The observed NOE between H-19 and H-20 is only possible if C-20 has an *S* configuration, while interactions between H-14 and H-19 were in agreement with an *R* configuration at C-12. All these NOESY spectral data revealed that the ring junction in the furofuran moiety should be *cis*.

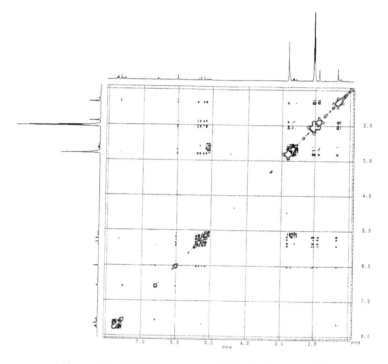

Figure 44   NOESY spectrum of 7,8-dehydroryacophiline.

SUGGESTED READINGS

Baumann, R., Wider, G., Ernst, R.R. and Wüthrich, K. (1981). Improvement of 2D NOE and 2D correlated spectra by symmetrization. *Journal of Magnetic Resonance* **44:** 76-83.

Bax, A. and Davis, D.G. (1985). Practical aspects of two-dimensional transverse NOE spectroscopy. *Journal of Magnetic Resonance* **63:** 207-213.

Bax, A. and Morris, G.A. (1981). An improved method for heteronuclear chemical-shift correlation by two-dimensional NMR. *Journal of Magnetic Resonance* **42:** 501-505.

Bax, A. and Subramanian, S. (1986). Sensitivity-enhanced two-dimensional heteronuclear shift correlation NMR-spectroscopy. *Journal of Magnetic Resonance* **67:** 565-569.

Bax, A. and Summers, M.F. (1986). H-1 and C-13 assignments from sensitivity-enhanced detection of heteronuclear multiple-bond connectivity by 2-D multiple quantum NMR. *Journal of the American Chemical Society* **108:** 2093-2094.

Bendall, M.R., Dodrell, D.M. and Pegg, D.T. (1981). Editing of carbon-13 NMR spectra. 1. A pulse sequence for the generation of subspectra. *Journal of the American Chemical Society* **103**: 4603-4605.

Bodenhausen, G. and Ruben, D.J. (1980). Natural abundance nitrogen-15 NMR by enhanced heteronuclear spectroscopy. *Chemical Physical Letters* **69**: 185-187.

Braun, S., Kalinowski, H.O. and Berger, S. (1998). *150 and More Basic NMR Experiments: A practical course.* 2nd edition. Wiley-VCH, Weinheim, Germany.

Breitmaier, E. and Voelter, W. (1987). *Carbon-13 NMR Spectroscopy. High-Resolution Methods and Applications in Organic Chemistry and Biochemistry.* 3rd edition. Wiley-VCH, Weinheim, Germany.

Bystrov, V.F., Arseniev, A.S. and Gavrilov, Y. (1978). NMR Spectroscopy of Large Peptides and Small Proteins. *Journal of Magnetic Resonance* **30**: 151-154.

Claridge, T.D.W. (1999). *High-Resolution NMR Techniques in Organic Chemistry.* Pergamon, Oxford, U.K.

Derome, A.E. (1987). *Modern NMR Techniques for Chemistry Research.* Pergamon, Oxford, U.K.

Derome, A. and Williamson, M.P. (1990). Rapid-pulsing artifacts in double-quantum-filtered COSY. *Journal of Magnetic Resonance* **88**: 177-179.

Dodrell, D.M., Pegg, D.T. and Bendall, M.R. (1982). Distortionless enhancement of NMR signals by polarization transfer. *Journal of Magnetic Resonance* **48**: 323-327.

Ernst, R.R., Bodenhausen, G. and Wokaun, A. (1987). *Principles of NMR in One and Two Dimensions.* Clarendon Press, Oxford, U.K.

Gallardo, V.O., Tonn, C.E., Nieto, M., Morales, G., Giordarw, O.S. (1996). Bioactive neo-clerodane diterpenoids towards *Tenebrio molitor* larvae from *Teucrium nodicaule* H. and Baocharis spicata (Lam.) *Beill. Nat. Prod. Lett.* **8**: 189-197.

Herz, W., Romo de Vivar, A., Romo, J. and Viswanathan, N. (1963). Constituents of *Helenium* species, XV. The structure of mexicanin. *Journal of the American Chemical Society* **85**: 19-25.

Jeener, J., Meier, B., Bachmann, P. and Ernst, R.R. (1979). Investigation of exchange processes by two-dimensional NMR spectroscopy. *Journal of Chemical Physics* **69**: 4546-4553.

Karplus, M. (1959). Contact Electron-Spin Coupling of Nuclear Magnetic Moments. *Journal of Chemical Physics* **30**: 11-15.

Karplus, M. (1963). Vicinal Proton Coupling in Nuclear Magnetic Resonance. *Journal of the American Chemical Society* **85**: 2870-2873.

Kessler, H., Gehrke, M. and Griesinger, C. (1988). Two-dimensional NMR-Spectroscopy — Background and Overview of the Experiments. *Angewandte Chemie International Edition in English* **27**: 490-536.

Martin, G.E. and Zektzer, A.S. (1988). *Two-dimensional NMR Methods for Establishing Molecular Connectivity.* Wiley-VCH, Weinheim, Germany.

Morris, G.A. and Freeman, R. (1979). Enhancement of Nuclear Magnetic-Resonance Signals by Polarization Transfer. *Journal of the American Chemical Society* **101**: 760-764.

Neuhaus, D. and Williamson, M. (1989). *The Nuclear Overhauser Effect in Structural and Conformational Analysis.*Wiley-VCH, Weinheim, Germany.

Nieto, M., Gallardo, O.V., Rossomando, P.C. and Tonn, C.R. (1996). 8-Hydroxysalviarin and 7,8-Didehydrorhyacophiline, Two New Diterpenes from *Salvia reflexa. Journal of Natural Products* **59**: 880-882.

Pretsch, E., Seibl, J., Simon, W. and Clerc, T. (1998). *Tables for Structural Determination by Spectroscopic Methods*. Springer-Verlag, Ibérica Barcelona, Spain. (In Spanish).

Rance, M., Sorensen, O.W., Bodenhausen, G., Wagner, G., Ernst, R.R. and Wüthrich, K. (1984). Improved spectral resolution in COSY proton NMR spectra of proteins via double quantum filtering. *Biochemical and Biophysical Research Communications* **117**: 479-482.

Sorensen, O.W., Eich, G., Levitt, M.H., Bodenhausen, G. and Ernst, R.R. (1983). Progress in NMR spectroscopy. *Progress in Nuclear Magnetic Resonance Spectroscopy* **16**: 163-166.

Summers, M.F., Marzilli, L.G. and Bax, A. (1986). Complete H-1 and C-13 assignments of co-enzyme B-12 through the use of new two-dimensional NMR experiments. *Journal of the American Chemical Society* **108**: 4285-4294.

Wagner, G. and Wütrich, K. (1982). Amide proton exchange and surface conformation of the basic pancreatic trypsin inhibitor in solution: Studies with two-dimensional nuclear magnetic resonance. *Journal of Molecular Biology* **155**: 347-349.

Biological Activity of Natural Products

Bioassays with Whole Plants and Plant Organs

Diego A. Sampietro*, Emma N. Quiroga, Jose R. Soberón, Melina A. Sgariglia and Marta A. Vattuone

1. INTRODUCTION

Assays of whole plant or plant organs consist in the exposure of plant materials to allelochemicals under controlled conditions. They are done for better understanding of the plant-plant interactions, identification of new growth regulators or characterization of new structure-activity relationships. Some of these assays are inexpensive, less time consuming and provide valuable qualitative and quantitative data. Control of assay conditions allows to separate the allelochemical effects from natural environmental interference and/or to evaluate the influence of specific environmental conditions on allelochemical activity. This chapter describes assays to test the biological effects of allelochemicals on whole plant and plant organs. The following points must be considered before performing one or more of the suggested assays:

I. **Sensitivity and selectivity:** The isolation and purification processes often yield small amounts of an allelochemical. Availability and cost of several commercial allelochemical standards is often restricted. High sensitive assays aid to solve these concerns (Hoagland and William, 2004). Sensitivity and selectivity depend on the assayed plant species or organ and on the biological potency of the allelochemical under study. Algal species assayed

Authors' address: *Instituto de Estudios Vegetales "Dr. A. R. Sampietro"*, Facultad de Bioquímica, Química y Farmacia, Universidad Nacional de Tucumán, España 2903, 4000, San Miguel de Tucumán, Tucumán, Argentina.
Corresponding author: E-mail: dasampietro2006@yahoo.com.ar

in 96 well plates allow to maximize sensitivity (Duke *et al.*, 2000), but results with algae do not always correlate well with growth responses of higher plants. Small aquatic plants, such as those of the *Lemna* genus, have been used in assays with small amounts of allelochemicals (Tanaka *et al.*, 1993). However, these assays often generate an undesirable proportion of false positives. Assays in 24 well plates using small seeded-plants from dicotyledons (i. e. *Lactuca sativa* L.) and monocotyledons (i.e. *Agrostis stolonifera* L.) as model species allow maximum sensitivity and selectivity (Rimando *et al.*, 2001; Sampietro *et al.*, 2006).

II. **Reproducibility and duration:** The assay should be easily reproduced and results should be obtained after short times of exposure to an allelochemical. Long duration assays often suffer from the interference of strange factors that affect assay reproducibility (Hoagland and Williams, 2004). The strange factors are: environmental conditions, allelochemical decomposition and microbial contamination in assay, these lead to irreproducible assays.

III. **Selection of a biological response:** Seed germination is most common assay used in phytotoxic and allelopathic studies (Hoagland and Brandsaeter, 1996). The main limitation of germination assays is that several species are insensitive during the germination to several allelochemicals, but are more sensitive during seedling growth and further growth stages (Hoagland and Williams, 2004). Interpretation of allelochemical effects on plant germination has also limitations, became some allelochemicals delays germination instead of inhibiting it. Therefore, assay to measure seed germination together with radicle and shoot length may lead to misinterpretation of post-germination growth responses. To prevent these concerns, pre-germinated seeds should be used for radicle and shoot elongation assays (Hoagland and Brandsaeter, 1996). Studies in plant physiology often involve molecular/biochemical assays oriented to specific molecular target sites and are often used in bioassay guided isolation of allelochemicals. However, these assays ignore alternative molecular target sites reducing the effectiveness to detect the biological activity of an allelochemical (Duke *et al.*, 2000). Hence, assays on whole plant and plant organs should be first conducted because they provide the global plant growth response to an allelochemical that is the result of several interacting physiological processes (González *et al.*, 1995). After this first approach, molecular/biochemical assays are needed to identify the primary action mechanism of a compound.

IV. **Reduced cost and easy to use:** An efficient assay must need a minimum space, time and equipment (Hoagland and Williams, 2004).

Experiment 1: Seed Germination Assay

Seed germination is the resumption of growth by the embryo within the seed leading to the development of a new plant (Bewley, 1997). Germination comprises those events that begin with water uptake of the quiescent dry seed (imbibition) and finish with the elongation of the embryonic axis. Protrusion of the radicle through the testa allows visual detection of seed germination. In germination assays, seeds of a plant species are placed in contact with different concentrations of an allelochemical, including a proper control without it. After a specific time interval under controlled conditions, the number of germinated seeds are counted and germination is expressed in one or more germination indices (Chiapusio *et al.*, 1997). Conditions needed for the germination assays (i. e. dormancy break, temperature, photoperiod, volume of solution per Petri dish and pH) vary with the plant species and must be well identified (Leather and Einhellig, 1988; Macías *et al.*, 2000). The most common germination assay is performed in polystyrene or glass Petri dishes containing a proper support such as Whatman # 1 filter paper, sand, agar, soil, vermiculite or cellulose sponge.

Materials and equipments required

9 cm Petri dishes, pipettes, 500 mL beakers, 500 mL Erlenmeyer flasks, assay tubes; Whatman # 1 paper discs (9 cm in diameter); filter cartridges with 0.2 μm pore size membranes; tongs; stainless steel strainer (8 cm in diameter or less); autoclave; laminar flow hood; lettuce (*Lactuca sativa* L.) seeds; balance with a sensitivity of 0.01 mg; growth chamber; Parafilm.

Reagents

1% Sodium hypochlorite (add 2 mL of sodium hypochlorite to 198 mL of distilled water); 1 N NaOH (dissolve 20 g of NaOH in 500 mL of distilled water); 5 mM 2-[N-morpholino] ethanesulfonic acid (MES) buffer (dissolve 488 mg of MES in 500 mL of distilled water; adjust to pH 5.7 with 1 N NaOH); solutions of 2-benzoxazolinone (BOA) at pH 5.7 (weigh 25.9 mg of BOA; dissolve in 5 mL of buffer MES; then, dilute with 5 mM MES to prepare 100 mL of 10^{-3}, 10^{-4}, 10^{-6}, 10^{-8} M BOA solutions; in a laminar flow hood, filter solutions with a sterile 0.22 μm-pore filter membrane; receive each sterile solution in a sterile 500 mL Erlenmeyer flask).

Procedure

(i) Put the Whatman # 1 paper discs in a Petri dish. Wrap glass material, tongs and the strainer with paper sheets or aluminium foil. Sterilize all materials in an autoclave at 120°C for 15 minutes before use.

(ii) Weigh 100 seeds of lettuce. Calculate the seed amount to be used as follows:

$$\frac{25 \times DW_{100}\,(g) \times NR \times NT \times 1.2}{100}$$

where, DW_{100} = dry weight of 100 lettuce seeds, NR = number of replications per treatment; NT = number of treatments. Weigh the calculated seed amount.

(iii) Working in the laminar flow hood, place seeds in a 500 mL beaker containing 200 mL of 1% sodium hypochlorite. After 15 minutes, drain seeds in the strainer. Dry seeds using a sterile paper-disc towel.

(iv) Transfer a filter paper disc with a tong to each Petri dish. Then, add 4 mL of a BOA solution or 5 mM buffer MES (control). Place 25 sterile seeds equidistant in each Petri dish. Seal Petri dishes with a plastic bag or Parafilm.

(v) Place Petri dishes in a growth chamber (25°C, 16-h photoperiod at 400 μmol m^{-2} s^{-1} photosynthetically active radiation). Count the number of germinated seeds every 6 hours for two days, until no further seeds are germinated. A seed is considered germinated after rupture of seed coats and radicle emergence.

Calculations

Simultaneous calculation of several germination indices has been proposed to provide a better interpretation of allelochemical activity on seed germination (Chiapusio *et al.*, 1997):

Total germination (GT)

$$G_T = \frac{N_T \times 100}{N}$$

where, N_T = number of germinated seeds for each treatment at the end of the assay and N = total number of seeds used in the assay. This index is the most commonly applied. However, it doesn't detect delays in seed germination.

Speed of germination (S)

$$S = (N_1 \times 1) + (N_2 - N_1) \times \frac{1}{2} + (N_3 - N_2) \times \frac{1}{3} + \cdots + (N_n - N_{n-1}) \times \frac{1}{n}$$

where, N_1, N_2, N_3...N_n = number of seeds germinated at 6 (1), 12 (2), 18 (3)... and (n) hours after the assay beginning.

Speed of accumulated germination (AS)

$$AS = \frac{N_1}{n} + \frac{N_2}{2} + \frac{N_3}{3} + \cdots + \frac{N_n}{n}$$

where, N_1, N_2, N_3..., N_n = cumulative number of seeds which germinated on time 1, 2, 3, ...or n, following set up of the experiment.

Coefficient of the rate of germination (CRG)

$$CRG = \frac{(N_1 + N_2 + N_3 + \cdots + N_n)}{(N_1 \times T_1) + (N_2 \times T_2) + (N_3 \times T_3) + \cdots + (N_n \times T_n)} \times 100$$

where, N_1 = number of germinated seeds on time T_1, N_2 = number of germinated seeds on time T_2 and N_n = number of germinated seeds on time T_n.

Observations

(i) Make at least three independent experiments (N = 3). Each treatment should have at least three replications.

(ii) Osmotic potential of fractions obtained during isolation and separation of natural products sometimes inhibit *per se* germination and post-germination growth (Sampietro *et al.*, 2006). In these cases, osmotic potential should be measured with an osmometer and mannitol solutions with the same osmotic potentials as those determined in the assayed fractions should also be assayed on seed germination of the test plants.

(iii) Ionizable substances, such as BOA (2-Benzoxazolinone) and several phenolic compounds, can drastically change the pH (Dayan *et al.*, 2000). At the same time, the biological activity of several substances is pH dependent (Blum *et al.*, 1984). Hence, solutions should always be buffered. MES [2-(N-Morpholino)ethanesulfonic acid] buffer allow to overcome these concerns.

(iv) Variations in germination response among plant species is often related with seed size. At the same concentration and with the same number of seeds per dish, the effect of an allelochemical may be less pronounced in large-seeded species than in small-seeded ones because higher tissue volume may dilute the effect of a bioactive compound in a seed with large size (Dayan *et al.*, 2000).

Statistical analysis

Analyze data using one way Analysis of the Variance and Mann-Whitney U test at the significant level of 0.05.

Experiment 2: Postgermination Assays

Most general postgermination assay consists in the exposure of pregerminated seeds with uniform radicle elongation to different concentrations of an allelochemical, including a control without it. Shoot and/or root elongation are measured after a specific time interval under controlled conditions, being possible to build a dose response-curve (Fig. 1). Critical parameters such as the inhibition threshold (lowest concentration required to initiate inhibition), the IC_{50}

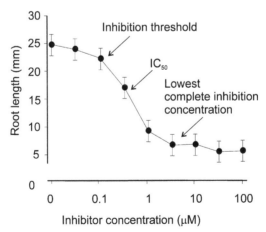

Figure 1 A typical dose–response curve illustrating various important parameters that can be estimated from the curve: inhibition threshold is the lowest concentration required to have an observable inhibition, IC_{50} is the concentration required for 50% inhibition and LCIC is the lowest complete-inhibition concentration (source: Dayan *et al.*, 2000).

(concentration required to obtain 50% inhibition) and the LCIC (lowest complete inhibition concentration) are estimated from these curves, establishing the range of concentrations where an allelochemical exerts its biological activity (Dayan *et al.*, 2000). Allelochemical concentrations for further more expensive and/or time consuming assays can be selected according to these parameters.

Materials and equipments required

Use the same materials and reagents described for Experiment 1.

Procedure

(i) Wrap glass material, tongs and the strainer with paper sheets or aluminium foil. Sterilize all materials in an autoclave at 120°C for 15 minutes before use.

(ii) Calculate the seed amount needed as follows:

$$\text{Seed amount (g)} = \frac{100 \times DW_{100}\,(g) \times NR \times NT \times 1.2}{100}$$

where, DW_{100} = dry weight of 100 lettuce seeds, NR = number of replications per treatment; NT = number of treatments. Weigh the calculated seed amount. Prepare NR × NT Petri dishes. Work in a laminar flow hood.

(iii) Place seeds in a 500 mL beaker containing 200 mL of 1% sodium hypochlorite. After 15 min, drain seeds in the strainer. Dry seeds using a sterile paper-disc towel. Work in a laminar flow hood.

(iv) Place 100 seeds in each Petri dish on discs of Whatman # 1 filter paper. Place the Petri dishes in a growth chamber (25°C, 16-h photoperiod at 400 μmol m^{-2} s^{-1} photosynthetically active radiation) for 24 h.

(v) In the laminar flow hood, transfering a sterile filter paper disc to each one with a tong. Then, add 4 mL of BOA solutions or 5 mM buffer MES (control). Place 25 pre-germinated seeds with uniform radicle length (up to 1 mm) in each Petri dish. Seal Petri dishes with Parafilm.

(vi) Place Petri dishes in a growth chamber (25°C, 16-h photoperiod at 400 μmol m^{-2} s^{-1} photosynthetically active radiation). After 4 days, measure root and shoot elongation with a ruler.

Calculations

Step 1: Calculate root/shoot elogation in mm for each treatment as mean ± standard deviation.

Step 2: Plot root/shoot elongation vs BOA concentration. Determine inhibition threshold, IC_{50} and LCIC from the graphic as indicated in Fig. 1.

Observations

Observations are the same indicated for Experiment 1.

Statistics

Analyze variance of data using general lineal model (GLM) procedure and compare means using Dunnet T3 test at a significant level of 0.05. Sometimes dose–response curves of root/shoot elongation do not fit with linear regression functions, being needed other functions for statistical analysis (Inderjit *et al.*, 2002; Sampietro *et al.*, 2006; Finney, 1979).

Experiment 3: Assay of Volatile Compounds

Plant material is exposed to air containing different concentrations of volatile allelochemicals in a closed chamber (Fig. 2). Humidity is regulated by a cotton layer impregnated with distilled water placed at the bottom of the chamber. A little piece of filter paper soaked in a solution containing the volatile allelochemical is placed in a small container localized in the center of the Chamber (Yun *et al.*, 1993; Connick *et al.*, 1987).

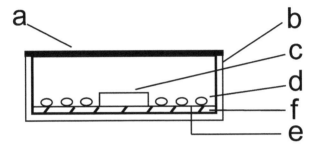

Figure 2   Diagrammatic section through test chamber: a, vinyl wrap; b, Petri dish; c, glass vessel containing a disk impregnated with a solution of the volatile compound; d, seeds; e, filter paper; f, absorbent cotton moistened with water.

Materials and equipments required

Glass material: 2500-mL containers or jars, 5-cm Petri dishes, pipettes; tomato (*Lycopersicum esculentum* Mill.) seeds; sheets and discs of Whatman # 1 filter paper; cotton; vinyl wraps.

Reagents

Heptanone standard.

Procedure

  (i) Prepare four 2500 mL containers per treatment. Each container should be prepared as under: Add a known weight of cotton at the bottom of the container dripped with a known volume of distilled water. Then, place 50 tomato seeds on a layer of filter paper onto the cotton layer. Before seeding, place a Petri dish containing a filter paper disc impregnated with 5 mL of distilled water, in the center of the container.

 (ii) Prepare five treatments. Add 0, 1.2, 2.4, 6.0, or 12.3 μl of heptanone in the Petri dishes corresponding to 0, 0.5, 1 1.0, 2.4 and 4.9 ppm treatments, respectively.

(iii) Close the containers with vinyl wraps and place them in a growth chamber (25°C, 16-h photoperiod at 400 μmol m^{-2} s^{-1} photosynthetically active radiation) for 4 days.

Calculations
Calculate total germination (see *Experiment 1*)

Observations
1. Make at least three independent experiments (N = 3).
2. The assay described above can be transformed in a postgermination assay, using uniform pregerminated tomato seeds and measuring root and shoot elongation after 4 days.

Statistical analysis
See *Statistical analysis* of Experiment 1.

Experiment 4: Allelochemicals on Growth Rates of Duckweed

The duckweeds (family Lemnaceae) are aquatic macrophytes found in slow moving fresh water around the world (Hillman, 1961). These small floating plants consist of a single leaf-like frond and a single root (Fig. 3). Daughter fronds are propagated vegetatively and plants generally appear as rosettes of three or four fronds (Cross, 2002). Researchers have used these plants for basic research in plant development, biochemistry and photosynthesis and in ecotoxicological studies because their unexpensive maintenance, small size, fast grow and high sensitivity (Einhellig *et al.*, 1985). These characteristics allow the use of lemnaceae species to evaluate the biological activity of limited quantities of potential active natural products.

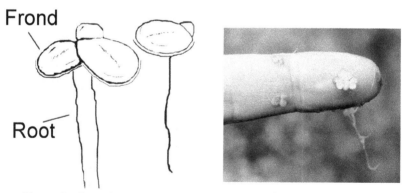

Figure 3 General morphology of *Lemna* species (Source: Cross, 2002).

Materials and equipments required
2 L Erlenmeyer flasks, beakers, pipettes; 24 well tissue culture plate (8.9 × 13.3 cm); growth chamber with a light source; laminar flow hood; fronds of *Lemna minor* L. (or other species from *Lemna* genus); autoclave; stainless steel loop; oven.

Reagents
Medium for *Lemna* growth (KNO_3: 202 mg L^{-1}; KH_2PO_4: 50.3 mg L^{-1}; K_2HPO_4: 27.8 mg L^{-1}; K_2SO_4: 17.4 mg L^{-1}; $MgSO_4 \cdot 7 H_2O$: 49.6 mg L^{-1}; $CaCl_2$: 11.1

mg L^{-1}; Na$_2$-EDTA: 10 mg L^{-1}; FeSO$_4$ · 7 H$_2$O: 6 mg L^{-1}; H$_3$BO$_3$: 5.72 mg L^{-1}; MnCl$_2$ · 4 H$_2$O: 2.82 mg L^{-1}; ZnSO$_4$: 0.6 mg L^{-1}; (NH$_4$)Mo$_7$O$_{24}$ · 4 H$_2$O: 43 µg L^{-1}; CuCl$_2$ 2 H$_2$O: 8 µg L^{-1}; CoCl$_2$ 6 H$_2$O: 54 µg L^{-1}; adjust the pH of the medium to 6.5); solutions of ferulic acid (weigh 1.5 mg of ferulic acid; add 5 mL of methanol and vortex; dilute with methanol to prepare 50 mL of solutions with concentrations of 50, 20 and 10 µg/mL); 10% sodium hypochlorite solution (dilute 10 mL of sodium hypochlorite in 90 mL of distilled water).

Procedure

Stock culture of *Lemna minor*

(i) Prepare 500 mL of inorganic medium.
(ii) Transfer the medium to a 2 L Erlenmeyer and close with a cotton cap. Autoclave the Erlenmeyer flask at 120°C for 1 hour.
(iii) Surface sterilize fronds with 10% sodium hypochlorite solution.
(iv) Open the Erlenmeyer, place 1-3 fronds using a sterile loop and close. Work under aseptic conditions in a laminar flow hood.
(v) Maintain the stock Erlenmeyers in a growth chamber at 28°C under constant light. After one week, a culture flask should contain a sufficient number of plants for a bioassay.

Assay

(i) Prepare 2 L Erlenmeyers with culture medium and autoclave 1 hour at 120°C.
(ii) After cooling, pipette 1.5 mL of medium into wells of a 24 well plate.
(iii) Add 5 µl of each concentration of ferulic acid to different wells. Add 5 µl of methanol to wells that will be used as control. Replicate each treatment six times.
(iv) Working in a laminar flow hood, transfer 3 fronds from the stock culture to each well using a sterile loop. Close the 24 well plates and place in the growth chamber.
(v) Maintain the 24 well plates in the growth chamber at 28°C under constant light for 7 days.
(vi) Count the number of fronds in each well and transfer it to tared containers. Ovendry the fronds for 24 hours at 90°C and weigh them.

Calculations

Calculate the growth rate in each treatment as under:

$$\text{Growth rate} = \frac{\text{Log}_{10}[\text{final frond number}] - \text{Log}_{10}[\text{initial frond number}]}{\text{Number of days}}$$

Observations

Incorporation of sucrose and other organic substances can accelerate *Lemna* growth in the stock culture. These organic additives, however, favor the growth of bacterial contaminants. Medium of stock *Lemna* culture can be changed every seven days to minimize algae contamination.

Statistical analysis

Compare data of each treatment with control using Student's t test at the significant level of 0.05.

General plant biometric parameters

Biometric parameters such as plant height, root length, number of fresh and dry leaves, leaf and root biomass, fresh and dry weight and leaf area are common indices of growth and development in plant physiology (Sánchez-Moreira and Reigosa, 2005). As previously indicated, information provided by these parameters help to choose which physiological or biochemical studies at a molecular level should be further conducted to understand primary allelochemical effect. After 15-day assay, to obtain lettuce seedlings with and without exposition to ferulic acid (Caspersen *et al.*, 1999). In subsequent experiments, biometric parameters on lettuce seedlings are measured.

Experiment 5: Lettuce Hydroponics Assay

Materials and equipments required

500 mL Erlenmeyer flasks, beakers, Petri dishes, pipettes; Cellulose acetate filters for ultracleaning filtration (0.2 μm pore) of up to 100 mL of liquid; paper or aluminium sheets; autoclave; balance with a sensitivity of 0.01 mg; oven; laminar flow hood; growth chamber; sterile glass material (wrap hermetically 9 cm Petri dishes and 1, 5 and 10 mL pipettes with paper or aluminium sheets; autoclave the glass material at 120°C for 20 min); preparation of Petri dishes (prepare tryptic soy agar, TSA, dissolving 50 g of agar in 1 L of tryptic soy broth; boil while stirring until dissolved; transfer to 500 mL Erlenmeyer flasks and close with cotton plugs; autoclave at 120°C for 20 min; before cooling, pipette 20 mL of the 5% TSA into each sterile 9 cm Petri dish; work in a laminar flow hood).

Reagents

1 N HCl (place 80.4 mL of HCl concentrated and make up carefully to 1 L with distilled water; 10% H_2O_2); nutrient solution (prepare the stock solutions as shown in Table 1; take volumes indicated for each stock solution, mix in 1 L beaker and make up to 200 mL; adjust the pH to 6.0 with 1 N HCl and make up to 1 L). Open a 500 mL Erlenmeyer flask (add 200 mL of nutrient solution after passing through a 0.2 μm cellulose acetate filter and closed with a sterile cotton plug). Nutrient solution with ferulic acid (prepare nutrient solution as indicated above. Add a proper weight of ferulic acid to each volume of nutrient solution according to Table 2).

Procedure

(i) Immerse seeds of lettuce (*Lactuca sativa* L. cv. Grand Rapids) in 10% H_2O_2 for 30 min followed by rinsing in sterile distilled water.

Table 1 Stock solutions and volumes of each one needed to prepare 1 L of nutrient solution

	Chemical	Weight	Distilled water (mL)	Volumes for nutrient solution
	KH_2PO_4	2.7 g	20 mL	1 mL
	Na_2HPO_4	2.8 g	20 mL	1 mL
	NH_4Cl	1.3 g	20 mL	1 mL
	KNO_3	10.1 g	100 mL	10 mL
	$Ca(NO_3)_2$	11.5 g	70 mL	4 mL
	$MgSO_4$	4.8 g	40 mL	1 mL
	H_3BO_3	37.1 mg		
Micronutrients	$MnSO_4$	15.1 mg	20 mL	
	$ZnSO_4$	12.9 mg	all	1 mL
	$CuCl_2$	2.0 mg	together	
	FeNa-EDTA	293.6 mg		
	$Na_2MoO_4 \cdot H_2O$	2 mg		

Table 2 Amounts (mg) needed to prepare different concentrations of ferulic acid

Concentration of ferulic acid ($\mu mol\ L^{-1}$)	Amount of ferulic acid (mg)
100	19.4
500	97.1
1000	194.2

(ii) Working in a laminar flow hood, place 10 sterile lettuce seeds in each Petri dish with 5% TSA. Place in a growth chamber (25°C with a 16-h photoperiod at 400 $\mu mol\ m^{-2}\ s^{-1}$ photosynthetically active radiation).

(iii) After six days, transfer each sterile seedling to a sterile 500 mL Erlenmeyer flask containing 200 mL of the nutrient solution (control) or nutrient solution with a concentration of ferulic acid.

(iv) Distribute the culture Erlenmeyer flasks at random on trolleys in a grow chamber.

(v) The atmosphere in each flask needs aireation. An alternative is to use an air filter system, with 0.22 μm pore filters, working at a flow rate of 170–180 mL min^{-1} (see Chapter 14, Section 1).

(vi) After 2 weeks, seedlings are ready to be used for biometric measurements.

(vii) Each treatment should be replicated at least four times and the whole assay should be twice repeated.

Experiment 6: Plant Height and Root Length

Plant height is the length measured from the stem base to the plant shoot tip. Length of primary root is the distance measured between stem base and the primary root tip. These biometric parameters are often measured with a ruler or a tape measure. Several natural products, i.e. ferulic acid, changes the root phytohormone balance expressed as an increase in secondary root number (Sampietro *et al.*, 2006). Hence, you may also count the number of secondary roots.

Procedure

   (i) Take lettuce seedlings from the hydroponics lettuce assay.
  (ii) Measure with the ruler plant height and primary root length of each lettuce seedling.
 (iii) Count number of secondary roots borne on the primary root.

Calculations

Calculate means of plant height and length of primary root ± standard deviation for each treatment. Repeat the experiment twice.

Experiment 7: Leaf Area

Leaf area is often determined using electronic area meters or photoelectric planimeters with a high degree of accuracy (Jonckheere *et al.*, 2004). This kind of determinations, however, are expensive and time consuming. A common approach is to use linear dimensional or dry weight measurements to compute mathematical models for accurace and fast estimation of leaf areas. The use of leaf dry weight involves destructive processing and is rather laborious and time consuming. Models based on measurement of leaf length and width with a ruler are non-destructive and provide a reliable and inexpensive alternative to leaf

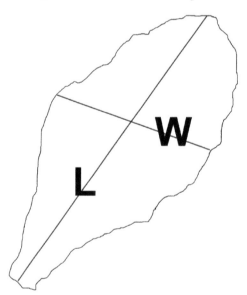

Figure 4   Lettuce leaf showing maximum length (L) and width (W).

area meters (Gameily *et al.*, 1991; Blanco and Folegatti, 2003; Blum and Dalton, 1985). An equation proposed to measure lettuce leaf area is:

$$\text{Leaf area} = 0.7 \times L \times W - 2.4$$

where leaf area is in cm^2 a and length (L) and width (W) values are in cm (Yoshida *et al.*, 1997).

Procedure

Measure maximum length and width from each leaf of the lettuce seedlings obtained in the hydroponics assay. Select full expanded leaves. Then, apply the above equation. Calculate mean leaf area ± standard deviation for each treatment. Repeat the experiment twice.

Experiment 8: Dry and Fresh Weight

Plant exposed to natural products can suffer modifications in pattern of biomass allocations between organs, rates of plant growth, water content and relative water content. These parameters are gravimetrically estimated through the dry and fresh weight of plant organs.

Materials and equipments required

Balance with a sensitivity of 0.01 mg, oven; 0.2 and 1.5 mL Eppendorff tubes; punch of 5 mm diameter; tongs; glass pipettes.

Procedure

A. Plant Water Content

(i) Remove lettuce seedlings from the hydroponics assay. Detach roots from lettuce seedlings of a treatment's replication, weigh immediately 1 g in a tared 1.5 mL Eppendorff tube. Register fresh weight substracting the weight of the Eppendorff tube. Leave roots to dry in the Eppendorff tube at 80°C for 48 h. Repeat this operation with each treatment's replication of the hydroponics lettuce assay. If possible, prepare two Eppendorff tubes from each replication.

(ii) Leave to cool the dry material and weigh again. Register dry weight substracting the weight of the Eppendorff tube.

Calculations

Calculate water content for each replication as follows (Khan *et al.*, 2000):

$$\text{Water content (g)} = \text{fresh root weight (g)} - \text{dry root weight (g)}$$

Calculate water content mean ± standard deviation for each treatment. Repeat the experiment twice.

B. Relative Growth Rate

(i) Remove lettuce seedlings at the beginning of the lettuce hydroponics assay (seedlings transfered from TSA, see steps 1 and 2 of Experiment 5). Detach roots from lettuce seedlings, weigh immediately 1 g in a tared 1.5 mL Eppendorff tube. Register fresh weight subtracting the weight of the

Eppendorff tube. Leave roots to dry at 80°C for 48 hours in an oven. Repeat this operation with each treatment's replication of Experiment 5. If possible, prepare two Eppendorff tubes from each replication.
(ii) Remove lettuce seedlings at the end of Experiment 5. Detach roots from lettuce seedlings, weigh immediately 1 g in a tared Eppendorff tube. Register fresh weight substracting the weight of the Eppendorff tube. Leave roots to dry at 80°C for 48 hours in an oven. Repeat this operation with each treatment's replication of Experiment 5. If possible, prepare two Eppendorff tubes from each replication.

Calculations

Calculate relative growth rate (from time 7 to 22 days) for each replication as under (Barkosky and Einhellig, 2003):

$$\text{Relative growth rate} = \frac{(\ln DW_2 - \ln DW_1)}{(T_2 - T_1)}$$

where, DW_1 and DW_2 are dry weights at the beginning and at the end of the lettuce bioassay, respectively; T_1 and T_2 are initial and harvest days of the lettuce hidroponics assay, respectively. Calculate means of relative growth rate ± standard deviation for each treatment. Repeat the experiment twice.

C. Relative Water Content

The relative water content (RWC) is an indicator of the plant water balance (González and González-Vilar, 2001). RWC is expressed as percent of water in a given time respect to the amount needed to saturate the plant tissues:

$$RWC = \frac{FW - DW}{SW - DW} \times 100$$

where, FW is fresh weight of a plant material at a given time, DW is the dry weight of the same plant material and SW is the weight of the same material after artificial saturation with water (full turgor).

RWC indicates dehydration degree of plant tissues and can be used to measure the induction of water stress by an allelochemical. González and González-Vilar (2001) tabulated the probable effects on plant tissues of increasing dehydration as follows:

RWC 100-90%: Stomatal closure in leaves and reduction in cellular expansion and growth.

RWC 90-80%: Correlates with changes in the composition of plant tissues and some alterations in the relative rates of photosynthesis and respiration.

RWC < 80%: Imply usually water potential of the order of –1.5 MPa or less. This water potentials would be companied with ceasing photosyntesis and increase in respiration, proline content and abscisic acid levels.

Procedure

(i) 2 mL Eppendorff tubes and number them. Label with the same numbers 1.5 mL Eppendorff tubes.
(ii) Fill with cold de-ionised water the 1.5 mL tubes.

(iii) Remove the youngest developed leaf from lettuce seedlings. Obtain four discs from lettuce leaves using the corck borer (avoid borders, and apical and basal parts of the leaf). Put four discs in each 0.2 mL tube. Seal the tubes and place them in ice to prevent growth and evaporation. Weigh the tubes with leaf discs inside to give a value for tissue fresh weight (FW).

(iv) Once every sample has been weighed, transfer the leaf pieces from each 0.2 mL Eppendorff tube to the 1.5 mL tube with the same number, containing the cold de-ionized water. Tubes should be placed in ice, and left for 4 hours in a refrigerator (5°C) to allow the tissue taking up water until saturation.

(v) Take away the leaf discs from the de-ionised water. Remove carefully the excess of water on the leaf surface with filter paper.

(vi) Transfer every sample to their original tubes (0.2 mL) and reweigh. This gives a measure of fully turgid fresh weight (TFW). Open the tubes, and place them in a stove to 80°C during 48 hours.

(vii) After drying reweigh the tissue to obtain a value for dry weight (DW).

Calculations

Calculate RWC (see above) for each treatment ± standard deviation. Repeat the experiments twice. At least you should replicate each treatment three times.

Statistical analysis

Analyze variance of data using one way ANOVA and compare means using Dunnet T3 test at the significant level of 0.05.

Experiment 9: Coleoptile Elongation

Auxins are a group of phytohormones that regulate several growth responses, including coleoptile elongation (Salisbury and Ross, 1992). Coleoptile is a protective sheath enclosing the shoot tip and embryonic leaves of grasses. At low concentrations auxins stimulate elongation of coleoptile cells. In nature, auxins coming from the shoot tip produce coleoptile cell elongation. When natural auxin-producing zones are lacking, i.e. in coleoptile segments, coleoptile elongation can only occur after exogenous hormone supplementation (Park *et al.*, 2001). The ratio of final length/original length for a grass coleoptile is measured after coleoptile segments are floated into incubation mixtures containing different concentrations of an auxin such as 1-naphthyl acetic acid (NAA) (Hoshi-Sakoda *et al.*, 1994). A modification of this assay allows to study the effect of allelochemicals on auxin-induced growth responses. The allelochemical is added to the incubation mixtures at different concentrations, providing a constant concentration of an auxin. Length of coleoptiles subjected to different concentrations of the allelochemical and a control (without allelochemical) are compared, and coleoptile elongation vs allelochemical concentration is plotted.

Reagents

1 N NaOH (dissolve 20 g of NaOH in 250 mL of distilled water and make up carefully to 500 mL with distilled water); 1.9% 1-naphthyl acetic acid, NAA (dissolve

95 mg of NAA in 5 mL of distilled water); incubation mixture (dissolve 1.1 g of 2-(N-morpholino) ethanesulfonic acid (MES) in 800 mL of distilled water; add 20 g of sucrose and 0.1 mL of 0.19% NAA); control of incubation mixture (adjust the pH of the incubation mixture, previously described, to pH 5.7 with 1 N NaOH; make up to 1 L); 16% 6-Methoxy-2-benzoxazolinone, MBOA (dissolve 0.80 g in 5 mL of distilled water); incubation mixtures with MBOA (prepare solution as indicated for incubation mixture; before pH adjustment, add a volume of 16% MBOA according with the desired final concentration (Table 3); then, make up to 1 L).

Table 3 Volumes needed to prepare 1 L of MBOA 1, 0.1 and 0.01 mM

Volume of 16% MBOA	Concentration
1 mL	10^{-3} M
0.1 mL	10^{-4} M
0.01 mL	10^{-6} M

Procedure

(i) Soak maize caryopses and plant them in wet vermiculite. Let grow 4 days at 25°C in the darkness.

(ii) Remove seedling leaves. Then, cut coleoptile segments of 1 cm long at 7 mm from the apical section.

(iii) Transfer 10 coleoptile segments to 5 mL of an incubation mixture (control) or a incubation mixture with MBOA (6-Methohybenzoxazolinone). Incubate at 25°C for 18 h. After incubation, measure the length of each coleoptile segment.

Calculations

For each treatment, calculate mean of coleoptile elongation (mm) ± standard deviation. Graph coleoptile elongation vs MBOA concentration.

Observations

Adventitious roots initiated in the hypocotyl cuttings of mung bean (*Vigna radiata* L.) is also an auxin mediated growth response (Chou and Lin, 1976). In this case, anti-auxin activity is expressed by less number of adventitious roots arised in the base of hypocotyl segments respect to those initiated in a proper control. This assay can be considered complementary to the proposed coleoptile assay.

Statistical analysis

Analyze variance of data using one way ANOVA and compare means using Dunnet T3 test at the significant level of 0.05.

Experiment 10: Cell Metabolic Activity

Principle

Tetrazolium salts, such as 2,3,5-triphenyl tetrazolium chloride (TTC), have been successfully used for measuring dehydrogenase activity in plant tissues.

Dehydrogenase enzymes couple the oxidation of cell reduced substrates or enzyme co-factors [NADH (Nicotinamide adenine dinucleotide), NADPH (Nicotinamide adenine dinucleotide phosphate) or $FADH_2$ (Flavin adenine dinucleotide)] with the reduction of hydrogen acceptors (Baud and Graham, 2006). Dehydrogenases are able to reduce water soluble TTC (a non-biological hydrogen acceptor) yielding intensely colored water-insoluble formazan crystals. Formazan can be extracted from root tissues with organic solvents and its absorbance is read at 485 nm (Sampietro *et al.*, 2006). Formazan production is a measure of cellular energetic metabolism. As mitochondrial dehydrogenases are the main component in cellular dehydrogenase activity, formazan production in root tissues is also an indirect measure of root respiration (Stowe *et al.*, 1995).

Materials and equipments required
Those indicated for Experiment 1; glass funnel; Whatman # 1 filter paper; spectrophotometer or filter photometer capable of measurement at 485 nm, equipped with a 1 cm path length microcuvette; pestle and morter; water bath.

Reagents
1 M sulfuric acid (carefully add 28 mL of concentrated sulfuric acid and make up to 500 mL); ethyl acetate; 50 mM sodium phosphate buffer, pH 7 (dissolve 6.9 g of $NaH_2PO_4 \cdot H_2O$ in 1 L of distilled water; then, dissolve 7.1 g of Na_2HPO_4 in 1 L of distilled water; mix 39 mL of $NaH_2PO_4 \cdot H_2O$ solution with 61 mL of Na_2HPO_4 solution and make up to 200 mL with distilled water); 0.2% 2,3,5-triphenyl tetrazolium chloride, TTC (dissolve 0.4 g of TTC in 200 mL of 50 mM sodium phosphate buffer, pH 7).

Procedure
(i)   Sterilize materials (refer *Procedure*, Experiment 1).
(ii) Germinate lettuce seeds (refer *Procedure*, Experiment 2).
(iii) Obtain lettuce seedlings (refer *Procedure*, Experiment 2).
(iv) Wash lettuce roots (100 mg fresh weight) of seedlings grown with and without exposure to ferulic acid. Then, blot dry, weigh and soak the root sample in 5 mL of 0.2% TTC solution (pH 7) at 37°C for 4 hours in the dark.
(v)   Add 0.5 mL of 1 M sulfuric acid to stop the reaction.
(vi) Wash lettuce roots with distilled water and blot dry with a towel of filter paper. Add roots to a morter. Then, add 3.5 mL of ethyl acetate and ground them with a pestle. Filter the obtained extract through filter paper.
(vii) Make up to 7 mL with ethyl acetate.
(viii) Read formazan content at 485 nm.

Calculations
Calculate formazan production as follows:

$$\text{Formazan content (\%)} = \frac{A_{485} \text{ nm in a treatment}}{A_{485} \text{ nm in the control}}$$

Perform each treatment in triplicate and repeat experiment twice.

Observations

Lettuce excised roots must be weighed as fast as possible to avoid tissue dehydration and loss of metabolic activity.

Statistical analysis

Analyze variance of data using one way ANOVA and compare means using Dunnet T3 test at the significant level of 0.05.

Experiment 11: Chlorophyll Content

Principle

Chlorophylls are pigments that give green colour to plants. They capture sun's light energy as the first step in its transformation into chemical energy (ATP) and reducing power (NADPH) utilized in the reduction of carbon dioxide to carbohydrates (Salisbury and Ross, 1992). The chlorophylls, chlorophyll *a* and *b*, are virtually essential in this process. Several reasons justify measuring leaf Chlorophyll content:

1. Chlorophyll concentration in plant leaves can directly limit photosynthetic potential and hence the production of primary plant metabolites (Richardson *et al.*, 2002).
2. Chlorophyll content gives an indirect measure of nutrient status because much of leaf nitrogen is incorporated in chlorophyll (Moran *et al.*, 2000).
3. Pigmentation can be directly related to stress physiology, as carotenoids concentration increase and chlorophylls generally decrease under stress and during senescence (Richardson *et al.*, 2002).
4. The relative concentrations of pigments are known to change with abiotic factors such as light (i. e. sun leaves have higher chlorophyll *a*: chlorophyll *b* ratio than those expose to darkness). Quantifying these proportions can provide important information about relationships between plants and their environment.

   Chlorophyll content is often determined through spectrophotometry, fluorometry and HPLC. Spectrophotometric methods are most common because they allow fast and inexpensive measurements. These methods need previous chlorophyll extraction with organic solvents. Aqueous acetone (90%), DMSO (dimethyl sulfoxide) and cold methanol are the most common solvents used for chlorophyll extraction from higher plants and algae (Barnes *et al.*, 1992; Hiscox and Israelstam, 1979).

Materials and equipments required

9 cm Petri dishes, pipettes, Pasteur pipettes, Erlenmeyers; growth chamber; tongs; lettuce seeds (*Lactuca sativa* L.); Whatman # 1 filter paper discs (9 cm in diameter); spectrophotometer or filter photometer capable of measure between 600 and 700 nm, equipped with a 1 cm path length microcuvette; vortex; autoclave; filter cartridges with 0.2 µm pore size membranes; laminar flow hood; lettuce (*Lactuca sativa* L.) seeds; balance with a sensitivity of 0.01 mg; growth chamber; Parafilm; vortex.

Reagents

1 N NaOH (dissolve 20 g of NaOH in 500 mL of distilled water); 5 mM 2-[N-morpholino] ethanesulfonic acid (MES) buffer (dissolve 488 mg of MES in 500 mL adjusted to pH 5.7 with 1 N NaOH); solutions of ferulic acid at pH 5.7 (weigh 69.9 mg of ferulic acid; dissolve in 5 mL of buffer MES. Then, dilute with buffer MES to prepare 100 mL of 388, 194, 97 and 19 µg ferulic acid/mL); DMSO.

Procedure

(i) Sterilize materials (refer *Procedure*, Experiment 1).

(ii) Germinate lettuce seeds (refer *Procedure*, Experiment 2).

(iii) Design the assay with four treatments and four replications per treatment. Label Petri dishes in groups of four indicating the control and ferulic acid solutions with concentrations of 388, 194, 97 and 19 µg of ferulic acid mL^{-1}. Working in a laminar flow hood, place a sterile filter paper disc in each Petri dish. Add 3 mL of the appropiate solution to each Petri dish according with its label. Use buffer MES for control dishes. Place 25 pregerminated seeds in each Petri dish.

(iv) Close Petri dishes and seal them with a plastic bag or Parafilm.

(v) Place in a growth chamber (25°C with a 16-h photoperiod at 400 µmol $m^{-2} s^{-1}$ photosynthetically active radiation) for 7 days.

(vi) Remove cotyledons from each Petri dish with a tong and weigh them.

(vii) Place cotyledons from each dish in an assay tube with 3.5 mL of DMSO. Cover the tubes with glass rubbers and incubate at 60°C for 1 h. After cooling, vortex the tubes and stand for 5 min.

(viii) Read absorbance at 645 nm and 663 nm.

Calculations

Calculate the amount of chlorophyll (µg mL^{-1}) by the Nerst equation (Arnon, 1949):

$$Total\ chlorophyll\ (µg\ mL^{-1}) = 20.2\ A_{645} + 8.02\ A_{663}$$
$$Chlorophyll\ a\ (µg\ mL^{-1}) = 12.7\ A_{663} - 2.69\ A_{645}$$
$$Chlorophyll\ b\ (µg\ mL^{-1}) = 22.9\ A_{645} - 4.68\ A_{663}$$

where, A_{645} = absorbance at 645 nm and A_{663} = absorbance at 663 nm. Multiply values obtained from each equation by total DMSO extraction volume (3.5 mL). Then, express amount of chlorophyll per unit of cotyledon fresh weight as follows:

$$Total\ chlorophyll\ (µg\ g^{-1})$$

$$= \frac{chlorophyll\ extracted\ from\ cotyledons\ at\ a\ given\ treatment\ (µg)}{fresh\ weight\ from\ cotyledons\ at\ a\ given\ treatment\ (g)}$$

Also express the amount of chlorophyls *a* and *b* per unit of cotyledon fresh weight. Calculate means ± standard deviation.

Statistical analysis

Analyze variance of data using one way ANOVA and compare means using Dunnet T3 test at the significant level of 0.05.

SUGGESTED READINGS

Arnon, D.I. (1949). Copper enzymes in isolated chloroplasts. Polyphenoloxidase in *Beta vulgaris*. *Plant Physiology* **24**: 1-15.

Barkosky, R.R. and Einhellig, F.A. (2003). Allelopathic interference of plant-water relationships by para-hydroxibenzoic acid. *Botanical Bulletin of Academia Sinica* **44**: 53-58.

Barnes, J.D., L. Balaguer, E., Manrique, Elvira, S. and Davison, A.W. (1992). A reappraisal of the use of DMSO for the extraction and determination of chlorophylls *a* and *b* in lichens and higher plants. *Environmental and Experimental Botany* **32**: 85-100.

Baud, S. and Graham, I.A. (2006). A spatiotemporal analysis of enzymatic activities associated with carbon metabolism in wild-type and mutant embryos of *Arabidopsis* using in situ histochemistry. *The Plant Journal* **46**: 155-169.

Bewley, J.D. (1997). Seed germination and dormancy. *The Plant Cell* **9**: 1055-1066.

Blanco, F.F. and Folegatti, M.V. (2003). A new method for estimating the leaf area index of cucumber and tomato plants. *Horticultura Brasileira* **21**: 666-669.

Blum, U. and Dalton, B.R. (1985). Effects of ferulic acid, an allelopathic compound, on leaf expansion of cucumber seedlings grown in nutrient cucumber. *Journal of Chemical Ecology* **11**: 279-301.

Blum, U., Dalton, B.R. and Rawlings, J.O. (1984). Effects of ferulic acid and some of its microbial metabolic products on radicle growth of cucumber. *Journal of Chemical Ecology* **10**: 1169-1191.

Chou, C.H. and Lin, H.J. (1976). Autointoxication mechanism of *Oryza sativa* I. Phytotoxic effects of decomposing rice residues in soil. *Journal of Chemical Ecology* **2**: 353-367.

Chiapusio, G., Sánchez, A.M., Reigosa, M.J., González, L. and Pellissier, F. (1997). Do germination indices adequately reflect allelochemical effects on the germination process? *Journal of Chemical Ecology* **23**: 2445-2453.

Connick, W.J., Bradow, J.M., Legendre, M.G., Vail, S.L. and Menges, R.M. (1987). Identification of volatile allelochemicals from *Amaranthus palmeri* S. Wats. *Journal of Chemical Ecology* **13**: 463-472.

Cross, J.W. (2002). *The Charms of Duckweed.* http://www.mobot.org/jwcross/duckweed.htm

Dayan, F.E., Hernández, A. and Allen, S.N. (1999). Comparative phytotoxicity of artemisinin and several sesquiterpene analogues. *Phytochemistry* **50**: 607-614.

Dayan, F.E., Romagni, J.G. and Duke, S.O. (2000). Investigating the mode of action of natural phytotoxins. *Journal of Chemical Ecology* **26**: 2079-2094.

Duke, S.O., Dayan, F.E., Romagni, J.G. and Rimando, A. M. (2000). Natural products as sources of herbicides: current status and future trends. *Weed Research* **40**: 99-111.

Einhellig, F.A., Leather, G.R. and Hobbs, L.L. (1985). Use of *Lemna minor* L. as a bioassay in allelopathy. *Journal of Chemical Ecology* **11**: 65-72.

Finney, D.J. (1979). Bioassay and the practice of statistical interference. *International Statistical Review* **47**: 1-12.

Gameily, S.W., Ranole, W.M., Hills, A.A. and Smittle, D.A. (1991). A rapid and non-destructive method for estimating leaf area in onion. *Horticulture Science* **26**: 206-208.

González, L. and González-Vilar, M. (2001). Determination of relative water content. In: *Handbook of Plant Ecophysiology Techniques* (Ed., M.J. Reigosa). Kluwer Academic Publishers, Netherland.

González, L., Souto, X.C., Sánchez, A.M. and Reigosa, M.J. (1995). Response of *Rumex crispus, Plantago lanceolata, Chenopodium album, Amaranthus retroflexus, Cirsium* sp. y *Solanum nigrum* to the action of allelopathic compounds. Congreso 1995 de la Sociedad Española de Matoherbología. (In Spanish).

Hillman, W.S. (1961). Experimental control of flowering in *Lemna*. III. A relationship between medium composition and the opposite photoperiodic response of *L. perpusilla* 6746 and *L. gibba* G3. *American Journal of Botany* **48**: 413-419.

Hiscox, J.D. and Israelstam, G.F. (1979). A method for the extraction of chlorophyll from leaf tissue without maceration. *Canadian Journal of Botany* **57**: 1332–1334.

Hoagland, D.R. and Arnon, D.I. (1950). *The Water-Culture Method for Growing Plants Without Soil*, Univ. Calif. Agric. Exp. Stn., Berkeley, CA, Circular No. **347**: 1-39.

Hoagland, R.E. and Brandsaeter, L.O. (1996). Experiments on bioassay sensitivity in the study of allelopathy. *Journal of Chemical Ecology* **22**: 1845-1859.

Hoagland, R.E. and Williams, R.D. (2004). Bioassays — Useful tools for the study of allelopathy. In: *Allelopathy: Chemistry and Mode of Action of Allelochemicals*. (Eds., F.A. Macías, J.C.G. Galindo, J.M.G. Molinillo and H.G. Cutler), CRC Press, Boca Raton, FL, USA.

Hoshi-Sakoda, M., Usui, K., Ishizuka, K., Kosemura, S., Yamamura, S. and Hasegawa, K. (1994). Structure–activity relationships of benzoxazolinones with respect to auxin-induced growth and auxin-binding protein. *Phytochemistry* **37**: 297-300.

Inderjit, Streibig, J.C. and Olofsdotter, M. (2002). Joint action of phenolic acid mixtures and its significance in allelopathy research. *Physiologia Plantarum* **114**: 422-428.

Jonckheere, I., Fleck, S., Nackaerts, K., Muys, B., Coppin, P., Weiss, M. and Baret, F. (2004). Review of methods for in situ leaf area index determination Part I. Theories, sensors and hemispherical photography. *Agricultural and Forest Meteorology* **121**: 19-35.

Khan, M.A., Ungar, I.A. and Showalter, A.M. (2000). Effects of salinity on growth, water relations and ion accumulation of the subtropical perennial halophyte, *Atriplex griffithii* var. *stocksii*. *Annals of Botany* **85**: 225-232.

Leather, G.R. and Einhellig, F.A. (1988). Bioassay of naturally occurring allelochemicals for phytotoxicity. *Journal of Chemical Ecology* **14**: 1821-1828.

Macías, F.A., Castellano, D. and Molinillo, J.M.G. (2000). Search for a Standard Phytotoxic Bioassay for Allelochemicals. Selection of Standard Target Species. *Journal of Agricultural and Food Chemistry* **48**: 2512-2521

Moran, J.A., Mitchell, A.K., Goodmanson, G. and Stockburger, K.A. (2000). Differentiation among effects of nitrogen fertilization treatments on conifer seedlings by foliar reflectance: a comparison of methods. *Tree Physiology* **20**: 1113–1120.

Park, W.J., Schafer, A., Van Onckelen, H., Kang, B.G. and Hertel, R. (2001). Auxin-induced elongation of short maize coleoptile segments is supported by 2,4-dihydroxy-7-methoxy-1,4-benzoxazin-3-one. *Planta* **213**: 92-100.

Richardson, A.D., Duigan, S.P. and Berlyn, G.P. (2002). An evaluation of noninvasive methods to estimate foliar chlorophyll content. *New Phytologist* **153**: 185-194.

Rimando, A.M., Olofsdotter, M., Dayan, F.E. and Duke, S.O. (2001). Searching for rice allelochemicals: An example of bioassay-guided isolation. *Agronomy Journal* **93**: 16-20.

Salisbury, F.B. and Ross, C.W. (1992). *Plant physiology.* Wadsworth Publishing Company, Belmont.

Sampietro, D.A., Vattuone, M.A. and Isla, M.I. (2006). Plant growth inhibitors isolated from sugarcane (*Saccharum officinarum*) straw. *Journal of Plant Physiology* **163**: 837-846.

Sampietro, D.A., Sgariglia, M.A., Soberón, J.R. and Vattuone, M.A. (2006). Effects of allelochemicals from sugarcane straw on growth and plant physiology of weeds. *Allelopathy Journal* **19**: 351-360.

Sánchez-Moreiras, A.M. and Reigosa, M.J. (2005). Whole plant response of lettuce after root exposure to BOA (2-3H-Benzoxazolinone). *Journal of Chemical Ecology* **31**: 2689-2703.

Stowe, R.P., Koenig, D.W., Mishra, S.K. and Pierson, D.L. (1995) Nondestructive and continuous spectrophotometric measurement of cell respiration using a tetrazolium-formazan microemulsion. *Journal of Microbiological Methods* **22**: 283-292.

Tanaka, T., Abbas, H.K. and Duke, S.O. (1993). Structure-dependent phytotoxicity of fumonisins and related compounds in a duckweed bioassay. *Phytochemistry* **33**: 779-785.

Teseire, H. and Guy, V. (2000). Copper-induced changes in antioxidant enzymes activities in fronds of duckweed (*Lemna minor*). *Plant Science* **153**: 65-72.

Yoshida, S., Kitano, M. and Eguchi, H. (1997). Growth of lettuce plants (*Lactuca sativa* L.) under control of dissolved O_2 concentration in hydroponics. *Biotronics* **26**: 39-45.

Yun, K.W., Kil, B.S. and Han, D.M. (1993). Phytotoxic and antimicrobial activity of volatile constituents of *Artemisia princeps* var. *orientalis*. *Journal of Chemical Ecology* **19**: 2757-2766.

Bioassays on Plants: Plant Cells and Organelles

Osvaldo Ferrarese-Filho*, Maria de Lourdes Lucio Ferrarese and Wanderley Dantas dos Santos

1. INTRODUCTION

Plants release organic compounds into the environment from their aerial or sub-aerial parts as rainwater leachates, exudates, volatiles, and/or decomposition residues. These compounds may accumulate in the soil environment and affect the growth and development of neighboring plants, an interaction called allelopathy (Einhellig, 1995). Allelopathic compounds are released by various mechanisms, which include release in soil by rainwater, root exudation and natural decomposition of plant residues lying above or mixed in soil. Numerous compounds have been referred as allelochemicals [benzoic acid derivatives (e.g., p-hydroxybenzoic, vanillic and salicylic acids), cinnamic acid derivatives (e.g., p-coumaric and ferulic acids), non-protein amino acids (e.g., L-3,4-dyhydroxyphenylalanine) and flavonoids (e.g., naringenin)] etc. In general, allelochemicals belong to many chemical classes with an even larger array of structural complexities (Fig. 1). Membrane perturbations are often suggested the primary site of action of many allelochemicals that trigger further modifications in physiological processes of plant cell (Einhellig and Barkosky, 2003). However, a clear insight into the primary allelochemical action on plant physiology has not been obtained. Several modes of action for allelochemicals are involved in the inhibition and modification of plant growth and development (Fig. 2).

Authors' address: *Department of Biochemistry, University of Maringá, Av. Colombo, 5790, 87020-900, Maringá, PR, Brazil.
Corresponding author: E-mail: oferrarese@uem.br

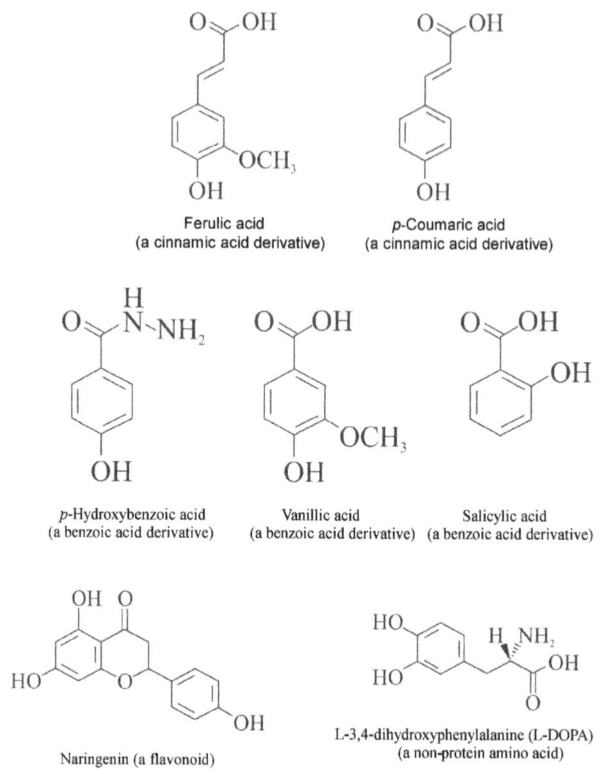

Ferulic acid
(a cinnamic acid derivative)

p-Coumaric acid
(a cinnamic acid derivative)

p-Hydroxybenzoic acid
(a benzoic acid derivative)

Vanillic acid
(a benzoic acid derivative)

Salicylic acid
(a benzoic acid derivative)

Naringenin (a flavonoid)

L-3,4-dihydroxyphenylalanine (L-DOPA)
(a non-protein amino acid)

Figure 1   Structural chemistry of allelochemicals used in this chapter.

Allelochemicals are also active against higher plants. Tobacco, sunflower, sorghum, cucumber, wheat and soybean crops typically suppress seed germination, cause injury to root growth and inhibit seedling growth (Einhellig, 1995). This chapter describes the experiments to evaluate the allelochemical interferences on plant growth. Physiological modifications studied begin with a disruption of normal root membrane function which leads to changes in the specific enzymes of phenylpropanoid pathway and on root lignification. Furthermore, the effects of allelochemicals on oxygen (consumption or production) in organelles, such as mitochondria and chloroplasts are evaluated. The experiments are conducted in a hydroponic system (Fig. 3) containing: 1) a glass container (16 cm of height and 10 cm of diameter); 2) an adjustable acrylic plate with 25 holes (5 mm each) where the seedlings are inserted. The plate is suspended by an adjustable small rod screwed to the support plate in the opening of the container; 3) a short hose; 4) a membrane filter (0.45 µm) coupled to hose and 5) a small aquarium pump for aeration of nutrient solution (200 mL of half-strength Hoagland's solution, pH 6.0), as described in Table 1. The low part of system (until the adjustable acrylic plate) may be wrapped with an

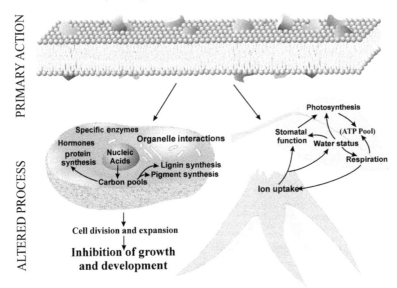

Figure 2 Action model for allelochemicals in plants (Source: Adapted from Einhellig, 1995).

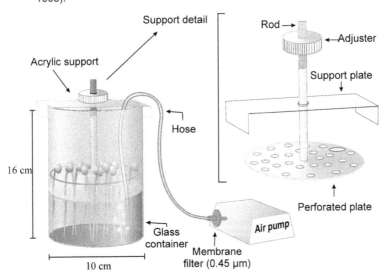

Figure 3 Hydroponic system. Flasks based on hydroponic culture are incubated in a chamber at controlled temperature, humidity and light (Source: Ferrarese *et al.*, 2000).

Table 1 Composition of the nutrient solution (Source: Hoagland and Arnon, 1950)

	Stock solution (g liter^{-1})	Nutrient solution (mL of stock solution per liter)
Macronutrient		
KH_2PO_4	136.0	0.5
KNO_3	101.0	2.5
$Ca(NO_3)_2 \cdot 4H_2O$	236.0	2.5
$MgSO_4 \cdot 7H_2O$	246.5	1.0
Micronutrient	*	0.5
Fe-EDTA	**	0.5

*Micronutrient solution: 2.86 g H_3BO_3; 1.81 g $MnCl_2 \cdot 4H_2O$; 0.22 g $ZnSO_4 \cdot 7H_2O$; 0.08 g $CuSO_4 \cdot 5H_2O$ and 0.02 g $H_2SMoO_4 \cdot H_2O$ are dissolved in one liter of deionized water.
**Fe-EDTA solution: 60.25 g $FeCl_3 \cdot 6H_2O$ and 65.10 g EDTA are separately dissolved in deionized water at 50°C. After, the solutions are mixed and the volume completed to one liter of deionized water. To store in a dark glass flask. The solution should be maintained in refrigerator.

aluminum foil, to avoid light in the roots. The experimental system containing 25 seedlings is maintained in a growth chamber (25°C, 12 hours light/12 hours dark cycle, irradiance of 280 μmol m^{-2} s^{-1}) or in an adequate room. Soybean (or another plant species) seeds, surface-sterilized with 2% sodium hypochlorite for 5 minutes and rinsed extensively with deionized water, are dark-germinated for 72 hours (at 25°C) on two sheets of moistened filter paper. Twenty-five equal-sized germinated seeds are transferred into the glass container filled with nutrient solution containing the allelochemical or without it (control). After 24 hours (or another experimental condition when indicated), the effects of allelochemical on cellular membrane (Experiments 1 to 3), enzymes of phenylpropanoid pathway (Experiments 4 and 5), lignification of cell wall (Experiment 6) and organelles (Experiments 7 and 8) are evaluated.

Experiment 1: Plasma Membrane Integrity

One of the early symptoms of allelochemicals toxicity is the root growth inhibition, which may be accompanied by cells death as a consequence of the loss of plasma membrane integrity. Evans blue, a non-permeating pigment of low phytotoxicity, is a dependable stain for determination of cell death. This selective staining of dead cells depends upon exclusion of this pigment from living cells by the intact plasmalemma, whereas it passes through the damaged plasmalemma of dead cells and accumulates as a blue protoplasmatic stain (Baker and Mock, 1994). The technique allows rapid and reproducible quantification of the stain retained by dead cells. The flavonoid naringenin, an intermediate of the phenylpropanoid metabolism in plants has been reported to suppress the plant growth (Deng *et al.*, 2004). Experiments are conducted to investigate if this allelochemical affects the viability of soybean meristematic cells.

Materials and equipments required

Seeds, fresh plant roots, analytical balance, beakers, dark bottle, germination paper, hydroponic system, pipettes, polypropylene flasks, visible light (400 to 700 nm), spectrophotometer, 1-cm cuvette, pH meter (or pH indicator sticks).

Reagents

0.25% Evans blue solution: dissolve 0.25 g of dye in 100 mL of distilled water and store in a dark bottle; *N,N*-dimethylformamide; nutrient solution (see Table 1); 2% sodium hypochlorite: dilute 16.7 mL of commercial sodium hypochlorite solution (12%) in 83.3 mL of distilled water; KOH: dissolve 28 g of KOH in 100 mL of distilled water and store in a polypropylene flasks; 5 M HCl: dilute 41.5 mL of concentrated HCl (12.06 M) in 58.5 mL of distilled water and store in a dark bottle; 0.4 mM naringenin solution (see *Observation* 2).

Procedure

(i) Sterilize soybean seeds with 2% sodium hypochlorite for 5 minutes, and rinse extensively with deionized water.

(ii) Spread the seeds uniformly on germination paper. Roll up the paper and transfer to a beaker (or another container) with a small amount of distilled water in the bottom. Seeds must be germinated in the dark at 25°C for 72 h.

(iii) Transfer twenty five equal-sized seedlings into the hydroponic system as described in the Introduction.

(iv) After 24 hours of incubation with 0.4 mM naringenin or without it (control), remove the seedlings to determine the loss of plasma membrane integrity using the Evans blue staining spectrophotometric assay (Baker and Mock, 1994) with modifications. Incubate the freshly harvested roots treated or untreated with the allelochemical for 15 minutes with 30 mL of 0.25% Evans blue solution. Wash the roots with distilled water for 30 minutes to remove excess and unbound dye. Soak the excised root tips (2 cm) in 3 mL of *N,N*-dimethylformamide for 50 minutes at room temperature to solubilize the dye bound to dead cells. Measure the absorbance of released Evans blue at 600 nm, using distilled water (or *N,N*-dimethylformamide) as blank.

Observations

(i) Make at least three independent experiments (N = 3).

(ii) Naringenin should be prepared immediately before use. Each hydroponic system contains 200 mL of nutrient solution (see Introduction). Since N = 3, prepare 600 mL of this solution. Prepare 0.4 mM naringenin as follows: a) dissolve 65.3 mg of naringenin in 25 mL of distilled water; b) add drops of 5 M KOH until complete naringenin dissolution (up to pH 10); c) add drops of 5 M HCl until pH 7.5 to avoid naringenin precipitation; d) complete with enough nutrient solution for 600 mL. Finally, adjust to pH 6.0.

Calculations

The loss of plasma membrane integrity may be expressed as Absorbance at 600 nm of treated roots (sample) in relation to untreated roots (control).

Generate a bar graphic by plotting absorbance at 600 nm versus root treatment (Fig. 4). Alternatively, the relative Evans blue uptake may be expressed as % of control (considered as 100%) by the equation:

Relative Evans blue uptake (%) = (A_{600} of naringenin treatment/A_{600} of control) × 100

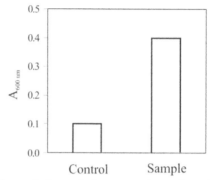

Figure 4 Loss of plasma membrane integrity.

Statistical analysis

Difference between data of treated and untreated (control) roots may be evaluated by Student *t*-test.

Precautions

HCl solution is highly corrosive. *N,N*-dimethylformamide is toxic by inhalation, contact or swallowing. Evans blue is toxic and may cause cancer. Avoid exposure. Obtain special instruction before use.

Experiment 2: Cell Membrane Permeability

Damage to cellular membrane is a virtually ubiquitous response to a wide variety of stresses. The permeability of the membrane is altered leading to leakage of electrolytes from the cells of stressed tissues, which may result in cell death. Electrolyte leakage from cells may be determined by conductivity. This measure is a sensitive and reliable indicator of the severity of the stress in plant cell. Exogenously applied cinnamic and benzoic acids modifies the permeability of cellular membrane in response to stress on plant tissues (Baziramakenga *et al.*, 1995). Ferulic acid, a cinnamic acid derivative, is a plant secondary metabolite that exhibits strong allelochemical properties. This allelochemical rapidly depolarizes the cucumber root cell membrane by causing a generalized increase in membrane permeability (Politycka, 1996). The following experiments are conducted to evaluate the effects of ferulic acid on soybean cell permeability, measured by electrical conductivity.

Materials and equipments required

Seeds, fresh plant roots, analytical balance, beakers, conductivity meter, dark bottle, germination paper, hydroponic system, pH meter (or pH indicator sticks), pipettes, polypropylene flasks, water bath.

Reagents

Nutrient solution (see Table 1); 2% sodium hypochlorite, 5 M KOH and 5 M HCl solutions (see Experiment 1, *Reagents*); ferulic acid (see *Observation* 2).

Procedure

(i) Sterilize soybean seeds with 2% sodium hypochlorite for 5 minutes, and rinse extensively with deionized water.

(ii) Spread the seeds uniformly on germination paper. Roll up the paper and transfer to a beaker (or another container) with a small amount of distilled water in the bottom. Seeds must be germinated in the dark at 25°C for 72 h.

(iii) Rinse the seedlings four times with distilled water to eliminate the electrolytes at the surface.

(iv) Cut twenty-five roots (about 1.0 g) from the seedlings. Immerse in a beaker containing 25 mL of aqueous solution (pH 5.3) of 0.5 mM ferulic acid or without it (control) for 24 hours at room temperature.

(v) Measure the electrical conductivity of the bathing solution by using a conductimeter.

Calculations

Electrical conductivity (EC) may be expressed as % of control (considered as 100%) by the equation:

$$EC\ (\%) = (EC\ of\ ferulic\ acid\ treatment/EC\ of\ control) \times 100.$$

Observations

1. Make at least three independent experiments ($N = 3$).

2. Prepare 0.5 mM ferulic acid immediately before use: a) dissolve 9.7 mg of ferulic acid in 25 mL of distilled water; b) add drops of 5 M KOH until complete dissolution of the compound (up to pH 12); c) add drops of 5 M HCl until pH 6.0 approximately (to avoid precipitation of the compound); d). add 70 mL of water and e) adjust to pH 5.3. Finally, complete with enough water for 100 mL of solution.

Experiment 3: Lipid Peroxidation

Lipid peroxidation may be defined as the oxidative deterioration of lipids containing any number of carbon-carbon double bonds, such as polyunsaturated fatty acids of cell membrane. In plants, lipid peroxidation has been referred as a mechanism of cellular injury. This mechanism involves a process whereby unsaturated lipids are oxidized to form additional radical species as well as toxic by-products that may be harmful to the plant. Polyunsaturated lipids are especially susceptible to this type of damage because in an oxidizing environment they react to form lipid peroxides and, further, malondialdehyde (MDA). This end-product may be found in different biological samples as a result of peroxidation. Determination of MDA using 2-thiobarbituric acid (TBA) is the most common method for estimating the oxidative stress effects on lipids, indicating the extension of the peroxidative injury. The assay is based on the reaction of MDA with TBA forming the MDA-TBA complex, which is measured spectrophotometrically at 532 nm. Phenolic allelochemicals, such as benzoic and cinnamic acids and their derivatives rapidly induce lipid peroxidation, causing a generalized cellular disruption that ultimately leads to cell death (Baziramakenga *et al.*, 1995; Politycka, 1996). The scope of experiment is to investigate the lipid peroxidation in soybean root tips treated with two phenolic allelochemicals, *p*-coumaric and *p*-hydroxybenzoic acids.

Materials and equipments required

Seeds, fresh plant roots, analytical balance, beakers, centrifuge tubes, dark bottle, germination paper, hydroponic system, ice-bath, mortar and pestle, pH meter (or pH indicator sticks), pipettes, polypropylene flasks, refrigerated preparative centrifuge capable of speed of 2,000 × g, visible light (400 to 700 nm) spectrophotometer, 1-cm cuvette, water bath.

Reagents

Nutrient solution (see Table 1); 67 mM phosphate buffer, pH 7.0 (see Table 2); polyvinylpolypyrrolidone (PVP); 0.5% 2-thiobarbituric acid (TBA) in 20% trichloroacetic acid (TCA): dissolve 20 g of TCA in 100 mL of distilled water

Table 2 $Na_2HPO_4 - KH_2PO_4$ buffer solution* (Source: McKenzie and Dawson, 1969)

Na_2HPO_4 solution (mL)	KH_2PO_4 solution (mL)	pH
0.25	9.75	5.288
0.5	9.5	5.589
1.0	9.0	5.906
2.0	8.0	6.906
3.0	7.0	6.239
4.0	6.0	6.468
5.0	5.0	6.643
6.0	4.0	6.813
7.0	3.0	7.168
8.0	2.0	7.381
9.0	1.0	7.731
9.5	0.5	8.043

*These buffer solutions are generally useful, since the range of the mixtures is from pH 5 to 8 at 25ºC.

For preparation of 67 mM phosphate buffer (pH 7.0), proceed as follows: a) Prepare 67 mM Na_2HPO_4 solution: dissolve 9.511 g of Na_2HPO_4 in distilled water and dilute to exactly one liter. Prepare 67 mM KH_2PO_4 solution: dissolve 9.118 g of KH_2PO_4 in distilled water and dilute to exactly one liter. Both solutions should be absolutely clear. Store in dark glass flasks. The solutions should be maintained in refrigerator. b) Prepare one liter of buffer by mixing about 650 mL of 67 mM Na_2HPO_4 solution plus 350 mL of 67 mM KH_2PO_4 solution (choose the adequate proportion in the table). Adjust to pH 7.0.

Note: Similar procedure should be applied for other phosphate buffers described in this chapter.

before addition of 0.5 g of TBA; 2% sodium hypochlorite, 5 M KOH and 5 M HCl solutions (see Experiment 1, *Reagents*); 1.0 mM *p*-coumaric acid and *p*-hydroxybenzoic acid solutions (see *Observation* 5).

Procedure

(i) Sterilize soybean seeds with 2% sodium hypochlorite for 5 minutes, and rinse extensively with deionized water.

(ii) Spread the seeds uniformly on germination paper. Roll up the paper and transfer to a beaker (or another container) with a small amount of distilled water in the bottom. Seeds must be germinated in the dark at 25°C for 72 h.

(iii) Transfer twenty-five equal-sized seedlings into the hydroponic system as described in the Introduction.

(iv) Incubate the seedlings with 1.0 mM phenolic allelochemical (*p*-coumaric or *p*-hydroxybenzoic acids) or without it (control) for 48 h.

(v) Cut the roots for determination of lipid peroxidation (Doblinski *et al.*, 2003). Transfer fresh roots (0.25 g) to a mortar and thoroughly mixed with 2.5 mL of 67 mM phosphate buffer (pH 7.0) and 0.05 g PVP, which adsorbs polyphenols. Centrifuge the extract at 2,000 × g for 15 minutes at 4°C. Use the supernatant to determine the lipid peroxidation, as follows. Add an aliquot (0.75 mL) of supernatant to 3 mL of 0.5% TBA (prepared in 20% TCA). Blank contains 0.75 mL of supernatant and 3 mL of 20% TCA. Place the sample and the blank in a water bath at 90°C (10 minutes). Cool quickly in an ice-bath for 15 minutes. Centrifuge the sample and the blank at 2,000 × g for 5 minutes. Measure the absorbance of supernatant at 532 nm and 600 nm, against blank. After subtracting the non-specific absorbance (600 nm), determine the MDA concentration by using the molar extinction coefficient ($\varepsilon = 155$ mM^{-1} cm^{-1}). Express the results as nmol MDA g^{-1} fresh root weight.

Observations

(i) Roots are exposed to allelochemicals for 48 hours with the nutrient solution completely renewed after the first 24 h.

(ii) After preparation, 20% TCA should be stored at 2–8°C. Since TCA is hygroscopic, the mixture TBA-TCA must be prepared just before use.

(iii) Make at least three independent experiments (N = 3).

(iv) To avoid inconsistent data, the spectrophotometric measures must be done immediately.

(v) Allelochemical solution must be prepared before use in the experiments. Each hydroponic system contains 200 mL of nutrient solution (see Introduction). Since N = 3, prepare 600 mL of this solution. For 1.0 mM *p*-coumaric acid, proceed as follows:

(a) Calculate the mass (m) of *p*-coumaric acid from Molarity (M).

$$M = m/(\text{Molecular Weight} \times \text{Volume})$$

$$\therefore \qquad m = M \times \text{Molecular Weight} \times \text{Volume}$$

If Molarity is 0.001 mol L^{-1}, Molecular Weight is 164.2 g mol^{-1} and volume is 0.6 l, the mass (m) required should be:

$$m = 0.001 \text{ mol } L^{-1} \times 164.2 \text{ g } mol^{-1} \times 0.6 \text{ L}$$
$$m = 0.0985 \text{ g (or 98.5 mg)}$$

(b) dissolve 98.5 mg of *p*-coumaric acid in 25 mL of distilled water; c) add drops of 5 M KOH until complete dissolution of the compound (up to pH 12); c) add drops of 5 M HCl until pH 6.0, approximately (to avoid precipitation of the compound); d) complete with enough nutrient solution for 600 mL. Finally, adjust to pH 6.0.

Note: Similar procedure should be done for any other allelochemical.

Calculations

Step 1. Calculate the sample absorbance (A).

$$A = A_{532nm} - A_{600nm} \quad (1)$$

For example, if $A_{532nm} = 0.200$ and $A_{600nm} = 0.020$, absorbance should be:

$$A = 0.200-0.020 \qquad \therefore A = 0.180$$

Step 2. Calculate the sample concentration (C).

$$C = A/\varepsilon \ (2), \text{ where } \varepsilon = 155 \text{ mM}^{-1} \text{ cm}^{-1}$$

Following example in step 1, concentration should be:

$$C = 0.180/155 \text{ mM}^{-1} \text{ cm}^{-1}$$
$$C = 1160 \text{ nmol } L^{-1}$$

Volume used in the spectrophotometric measurement is 3.75 mL. Then, the concentration in assay is:

$$C = 1160 \text{ nmol} \times 3.75 \text{ mL}/1000 \text{ mL}$$
$$C = 4.35 \text{ nmol}$$

Step 3. The weight of a soybean root sample is 0.25 g. Volumes of the original extract and the enzyme extract are 2.5 mL and 0.75 mL, respectively. Then, calculate root quantity in the assay:

$$\text{Root quantity} = 0.25 \text{ g} \times 0.75 \text{ mL}/2.5 \text{ mL}$$
$$\text{Root quantity} = 0.075 \text{ g}$$

Step 4. Calculate the lipid peroxidation.

$$\text{Lipid peroxidation} = 4.35 \text{ nmol} \times 1 \text{ g}/0.075 \text{ g}$$
$$\text{Lipid peroxidation} = 58 \text{ nmol MDA g}^{-1} \text{ fresh root weight}$$

Note: Similar procedure should be applied to calculate MDA (Malondialdehyde) content in the control (without allelochemical treatment)

Statistical analysis

Difference between data of treated and untreated (control) roots may be evaluated by Student *t*-test.

Precautions

TBA has a strong mercaptan odor. Do not breathe dust. Avoid contact with skin and eyes. TCA is caustic acid. Handle with care.

Experiment 4: Phenylalanine Ammonia-lyase Activity

Principle

The phenylpropanoid pathway is one of the most important plant metabolic pathways, where phenolic compounds and a wide range of secondary products are synthesized. Phenylalanine ammonia-lyase (PAL, EC 4.3.1.5) is regarded as the primary enzyme of the phenylpropanoid biosynthetic pathway. It catalyzes the non-oxidative deamination of L-phenylalanine to *trans*-cinnamic acid which is the precursor of many phenylpropanoids, such as lignins, flavonoids and coumarins (Dixon *et al.*, 2002). Changes in PAL activity have been found as a response of plants against various biotic and abiotic stresses. Increase of PAL activity has been related to root growth inhibition in maize (Devi and Prasad, 1996), cucumber (Politycka, 1999) and soybean (Herrig *et al.*, 2002), when these plant species are exposed to cinnamic and benzoic acids derivatives. These growth reductions are associated with the cell wall stiffening originated in the cross-linking among cell wall polymers and lignin synthesis. In this experiment, high performance liquid chromatography (HPLC) is applied to measure the enzyme activity through the quantification of *trans*-cinnamate, the product of PAL reaction.

Materials and equipments required

Seeds, fresh plant roots, analytical balance, beakers, centrifuge tubes, dark bottle, disposable syringe filter, germination paper, hydroponic system, high performance liquid chromatography (HPLC), mortar and pestle, pipettes, pH meter (or pH indicator sticks), polypropylene flasks, refrigerated preparative centrifuge capable of speed of 2,000 × g, water bath.

Reagents

Nutrient solution (see Table 1); methanol; 0.1 M sodium borate buffer (pH 8.8): dissolve 3.8 g of $Na_2B_4O_7$ in 100 mL of distilled water and adjust to pH 8.8; 10 µM *trans*-cinnamic acid: a) prepare a stock solution of 1.0 mM (dissolve 14.8 mg of *trans*-cinnamic acid in 25 mL of distilled water; add drops of 5 M KOH until complete dissolution of the compound; add drops of 5 M HCl until pH 6.0 to avoid precipitation of the compound, and complete with enough water for 100 mL of solution. b) dilute 0.1 mL of stock solution in 9.9 mL of chromatographic mobile phase (methanol:water, 70:30 v/v); 50 mM L-phenylalanine: dissolve 0.8 g of L-phenylalanine in 100 mL of distilled water; L-3,4-dihydroxyphenylalanine (see *Observation* 2); 1.0 mM ferulic acid and vanillic acid solutions (see Experiment 3,

Observation 5); 2% sodium hypochlorite, 5 M KOH and 5 M HCl solutions (see Experiment 1, *Reagents*).

Procedure

(i) Sterilize soybean seeds with 2% sodium hypochlorite for 5 minutes, and rinse extensively with deionized water.

(ii) Spread the seeds uniformly on germination paper. Roll up the paper and transfer to a beaker (or another container) with a small amount of distilled water in the bottom. Seeds must be germinated in the dark at 25°C for 72 h.

(iii) Transfer twenty-five equal-sized seedlings into the hydroponic system as described in the Introduction.

(iv) Incubate the seedlings with 1.0 mM allelochemical (ferulic acid, vanillic acid or L-3,4-dihydroxyphenylalanine) or without it (control) for 24 h.

(v) Cut the roots for determination of PAL activity as described by Ferrarese *et al.* (2000). Macerate the tissues (2 g) in a mortar (at 4°C) with 5 mL of 0.1 M sodium borate buffer (pH 8.8). Centrifuge the homogenates (2,200 × g, 15 minutes). Use the supernatant as the enzyme preparation. Incubate the reaction mixture (1 mL of sodium borate buffer, pH 8.7, and 0.25 mL of enzyme extract) at 40°C, for 5 minutes, for PAL activity assay. Start the reaction by adding 0.3 mL of 50 mM L-phenylalanine. Stop the reaction after one hour of incubation by adding 50 µL of 5 M HCl. Filter the samples through a 0.45 µm disposable syringe filter. Inject 20 µL in a Liquid Chromatograph equipped with a pump, an injector and an UV detector. A reversed-phase ODS column (150 × 4.6 mm, 5 µm) is used at room temperature. The mobile phase is methanol: water (70:30 v/v) with a flow rate of 0.5 mL min^{-1}. Measure the absorption at 275 nm. Identify *trans*-cinnamate, the product of PAL, by comparing its retention time with that of standard's (at 10 µM). Made parallel controls without L-phenylalanine or with *trans*-cinnamate (added as internal standard in the reaction mixture). Express the PAL activity as nmol *trans*-cinnamate hour^{-1} g^{-1} fresh root weight.

Observations

(i) Make at least three independent experiments (N = 3).

(ii) Since N = 3, prepare 1.0 mM L-3,4-dihydroxyphenylalanine dissolving 118.3 mg in 600 mL of nutrient solution, and adjusting to pH 6.0.

(iii) All material (pestle and mortar, centrifuge tube) must be refrigerated before enzyme extraction.

(iv) Using the isocratic HPLC method, identification of *trans*-cinnamate is possible by comparison of its peak retention time (Rt) in the sample with that of a *trans*-cinnamate standard (Fig. 5). The compound is quantified according with its peak area. For quantification, the external standard method is proposed. Figure 6 shows a curve obtained relating peak area with *trans*-cinnamate concentration. It confirms linear relation between peak areas and *trans*-cinnamate concentrations comprised in the range 5-50 µM.

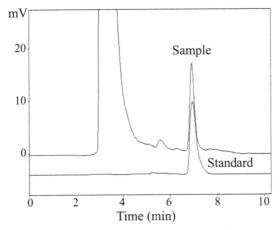

Figure 5 HPLC profiles of a sample and a *trans*-cinnamate standard.

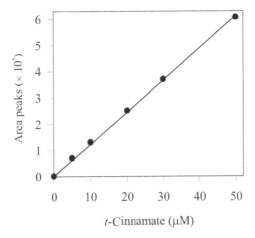

Figure 6 Curve of *trans*-cinnamate.

Moreover, the figure demonstrates that this methodology is highly sensitive to very low PAL activities (Ferrarese *et al.*, 2000). Based on peak areas, PAL activity is determined as described in the *Calculations*.

Calculations
Step 1. Calculate the sample concentration (C).

For example, if the peak area of standard (10 µM *trans*-cinnamate) = 270000 and the peak area of sample = 123456, concentration should be:

$$C = 10 \text{ µM} \times 123456/270000$$
$$C = 4.57 \text{ µM (or 1160 µmol L}^{-1})$$

Volume used in the HPLC measurement is 1.55 mL. Then, the concentration in assay is:

$$C = 4.57 \ \mu mol \times 1.55 \ mL/1000 \ mL$$
$$C = 0.0071 \ \mu mol \ (or \ 7.1 \ nmol)$$

Note: Since reaction time is 1 hour then concentration = 7.1 nmol $hour^{-1}$

Step 2. The weight of a soybean root sample is 2 g. Volumes of the original extract and the enzyme extract are 5 mL and 0.25 mL, respectively. Then, calculate root quantity in the assay:

$$Root \ quantity = 2 \ g \times 0.25 \ mL/5 \ mL$$
$$Root \ quantity = 0.1 \ g$$

Step 3. Calculate the PAL activity.

$$PAL \ activity = 7.1 \ nmol \ hour^{-1} \times 1 \ g/0.1 \ g$$
$$PAL \ activity = 71 \ nmol \ \textit{trans}\text{-cinnamate} \ hour^{-1} \ g^{-1} \ fresh \ root \ weight$$

Note: Similar procedure should be applied to calculate PAL activity in the control (without allelochemical treatment).

Statistical analysis

Difference between data of treated and untreated (control) roots may be evaluated by Student *t*-test.

Precautions

Methanol is flammable, keep away sources of ignition. It is toxic by inhalation, contact or swallowing.

Experiment 5: Soluble and Cell Wall-bound Peroxidases

Plants respond to a variety of environmental stresses through induction of antioxidant enzymes that provide protection against further damage. Peroxidase (POD, EC 1.11.1.7) has been implicated in a broad range of physiological and developmental processes such as suberin and lignin formation, cross-linking of cell wall components, auxin metabolism and response against pathogens and biotic/abiotic stresses. Due to this role, its activity may be detected in plants from germination to senescence. Cross-linking of phenolic monomers in the formation of suberin and the oxidative coupling of lignin subunits, as part of lignin biosynthesis, are related to secondary cell wall formation and have been associated with reduction of extensibility and growth. Reduction of seedling root elongation has been attributed to the enhancement of soluble and cell wall-bound POD activities, which lead to premature lignification of the cell walls (Passardi *et al.*, 2005). Phenolic allelochemicals affect the soluble and cell wall-bound POD in different plant species (Devi and Prasad, 1996; Politycka *et al.*, 2004; Böhm *et al.*, 2006). In this experiment, spectrophotometry is used to measure the soluble and cell wall-bound POD activities in soybean root tips treated with different phenolic allelochemicals.

Materials required

Seeds, fresh plant roots, analytical balance, beakers, centrifuge tubes, dark bottle, germination paper, hydroponic system, pestle and mortar, pipettes, pH meter (or pH indicator sticks), polypropylene flasks, refrigerated preparative centrifuge capable of speed of 2,000 × g, visible light (400 to 700 nm) spectrophotometer, 1-cm cuvette.

Reagents

Nutrient solution (see Table 1); polyvinylpolypyrrolidone (PVP); 67 mM phosphate buffer, pH 7.0, and 25 mM phosphate buffer, pH 6.8 (see Table 2); 1 M NaCl: dissolve 5.85 g of NaCl in 100 mL of distilled water; reaction mixture: 10 µL guaiacol, 35 µL H_2O_2 and 30 mL of 25 mM phosphate buffer (pH 6.8). This preparation is sufficient for ten samples; 1.0 mM vanillic, *p*-coumaric, *p*-hydroxybenzoic and ferulic acids solutions (see Experiment 3, *Observation* 5); 2% sodium hypochlorite, 5 M KOH and 5 M HCl solutions (see Experiment 1, *Reagents*).

Procedure

(i) Sterilize soybean seeds with 2% sodium hypochlorite for 5 min, and rinse extensively with deionized water.

(ii) Spread the seeds uniformly on germination paper. Roll up the paper and transfer to a beaker (or another container) with a small amount of distilled water in the bottom. Seeds must be germinated in the dark at 25°C for 72 h.

(iii) Transfer twenty-five equal-sized seedlings into the hydroponic system as described in the Introduction.

(iv) Incubate the seedlings with 1.0 mM allelochemical (ferulic acid, vanillic acid, *p*-coumaric or *p*-hydroxybenzoic acids) or without it (control) for 24 h.

(v) Cut the roots for determination of POD activity. Transfer the roots (0.5 g fresh weight) to a mortar and mix thoroughly with 5 mL of 67 mM phosphate buffer (pH 7.0) and 0.1 g PVP. Centrifuge the extract (3,600 × g, 5 min, 4°C). Use the supernatant to determine the activity of soluble POD. For cell wall-bound POD isolation, wash the pellet with deionized water until no soluble POD activity is detected in the supernatant. Incubate the pellet in 1 M NaCl (2 mL, 4°C, 60 min). Centrifuge the homogenate (3,600 × g, 5 min). The supernatant contains the cell wall-(ionically)-bound POD. Guaiacol-dependent activities of the soluble and cell wall-bound POD are determined according to Cakmak and Horst (1991) with slight modifications. The reaction mixture (3 mL) contains 25 mM sodium phosphate buffer (pH 6.8), 2.6 mM guaiacol and 10 mM H_2O_2. Start the reaction starts by adding the diluted enzyme extract (0.4 mL) in extraction medium (see *Observation* 4). Guaiacol oxidation must be followed for 5 min, at 470 nm. Enzyme activity must be calculated from the molar extinction coefficient ($\varepsilon = 25.5$ mM^{-1} cm^{-1}) for tetraguaiacol. Blank consists of a reaction mixture without enzyme extract whose absorbance is subtracted from the mixture with enzyme extract. Express the POD activity as µmol tetraguaiacol min^{-1} g^{-1} fresh root weight.

Observations

(i) All material (pestle and mortar, centrifuge tube) must be refrigerated before enzyme extraction.

(ii) Make at least three independent experiments (N = 3).

(iii) Guaiacol is not easy miscible, and the mixture reaction must be vigorously mixed.

(iv) Before spectrophotometric assays, soluble POD extract should be diluted (about 20 times) in 67 mM phosphate buffer (pH 7.0), to yield an appropriate absorbance at 470 nm (0.3 to 0.5). Cell wall-bound POD extract should be diluted (about 10 times) in 1 M NaCl.

Calculations

1. Determination of soluble POD activity

Step 1. The weight of a soybean root sample is 0.5 g. Volumes of the original extract and the diluted enzyme extract are 5 mL and 0.02 mL, respectively. Note that 0.4 mL of diluted enzyme extract (dilution = 20 for soluble POD) corresponds to 0.02 mL (0.4 mL/20 = 0.02 mL). Then, calculate root quantity in the assay:

$$\text{Root quantity} = 0.5 \text{ g} \times 0.02 \text{ mL}/5 \text{ mL}$$
$$\text{Root quantity} = 0.002 \text{ g}$$

Step 2. Calculate the sample concentration (C).

$$C = A_{470}/\varepsilon, \quad \text{where } \varepsilon = 25.5 \text{ mM}^{-1} \text{ cm}^{-1}$$

For example, if A_{470} = 0.400, concentration should be:

$$C = 0.400 / 25.5 \text{ mM}^{-1} \text{ cm}^{-1}$$
$$C = 0.01569 \text{ mM}$$

Step 3. Calculate the soluble POD activity.

$$\text{POD activity} = 0.01569 \text{ mM} \times 1 \text{ g}/0.002 \text{ g}$$
$$\text{POD activity} = 7.84 \text{ mM g}^{-1} \text{ (or 7.84 mmol L}^{-1} \text{ g}^{-1})$$

Volume used in the spectrophotometric measurement is 3 mL. Then:

$$\text{POD activity} = 7.84 \text{ mmol g}^{-1} \times 3 \text{ mL}/1000 \text{ mL}$$
$$\text{POD activity} = 0.023 \text{ mmol g}^{-1} \text{ (or 23 µmol g}^{-1})$$

Reaction time in the spectrophotometric measurement is 5 min. Then, the enzyme activity in the assay is:

$$\text{POD activity} = 23 \text{ µmol g}^{-1}/5 \text{ min}$$
$$\text{POD activity} = 4.6 \text{ µmol min}^{-1} \text{ g}^{-1} \text{ fresh root weight}$$

2. Determination of cell wall-bound POD

The weight of a soybean root sample is 0.5 g. Volumes of the original extract and the diluted enzyme extract are 5 mL and 0.04 mL, respectively. Note that 0.4 mL of diluted enzyme extract (dilution = 10 for cell wall-bound POD) corresponds to 0.04 mL (0.4 mL/10 = 0.04 mL). Then, calculate root quantity in the assay:

Root quantity = 0.5 g × 0.04 mL/5 mL
Root quantity = 0.004 g

Note: Proceed as indicated in step 2 for soluble POD.

Statistical analysis

Difference between data of treated and untreated (control) roots may be evaluated by Student *t*-test.

Precautions

Guaiacol is harmful if swallowed. Irritating to the eyes and skin.

Experiment 6: Lignin Content

Lignin is the main structural component of secondarily thickened plant cell walls. Structurally, lignin is a heteropolymer of hydroxylated and methoxylated phenylpropane units, derived from the oxidative polymerization of different hydroxycinnamyl alcohols (*p*-coumaryl, coniferyl and sinapyl) connected by labile ether bounds and/or resistant carbon-carbon linkages. It is initially formed in the middle lamella and primary walls of cells, such as xylem vessel elements and phloem fibers. At final stages of xylem cell differentiation, lignin is deposited within the carbohydrate matrix of the cell wall by infilling of interlamellar voids and, at the same time, by the formation of chemical bonds with non-cellulosic carbohydrates. As a fundamental component of the cell wall, lignin imparts mechanical support and allows efficient conduction of water and solutes over long distances within the vascular systems (Boerjan *et al.*, 2003). Lignification has been reported as a plant response against several biotic and abiotic stresses. Root growth inhibition of maize (Devi and Prasad, 1996), cucumber (Politycka, 1999) and soybean (Santos *et al.*, 2004) under stress of allelochemicals has been associated with cell wall stiffening originated in the cross-linking among cell wall polymers and lignin production. In this experiment, the spectrophotometric determination of lignin is adopted to evaluate the effects of allelochemicals on cell wall of soybean roots.

Materials and equipments required

Seeds, dry plant roots, analytical balance, beakers, centrifuge tubes, dark bottle, germination paper, hydroponic system, mortar and pestle, oven, pH meter (or pH indicator sticks), pipettes, polypropylene flasks, rotatory shaker, screw-cap centrifuge tubes, table top clinical centrifuge capable of speed of 1,400 × g, ultraviolet light (200 to 400 nm) spectrophotometer, 1-cm cuvette; vacuum desiccator, water bath.

Reagents

Nutrient solution (see Table 1); acetone; thioglycolic acid; 50 mM potassium phosphate buffer, pH 7.0 (see Table 2); 1 M NaCl: dissolve 5.85 g of NaCl 100 mL of 50 mM potassium phosphate buffer (pH 7.0); 1% Triton X-100: dilute 1 mL of Triton X-100 in 99 mL of 50 mM potassium phosphate buffer (pH 7.0); 2 M HCl: dilute 16.6 mL of concentrated HCl (12.06 M) in 83.4 mL of distilled water and store in a dark bottle; 0.5 M NaOH: dissolve 2 g of NaOH in 100 mL of distilled water and store in a polypropylene flasks; 1.0 mM ferulic acid (see Experiment 3, *Observation* 5); 1.0 mM L-3,4-dihydroxyphenylalanine (see Experiment 4, *Observation* 2); 2% sodium hypochlorite, 5 M KOH and 5 M HCl solutions (see Experiment 1, *Reagents*).

Procedure

(i) Sterilize soybean seeds with 2% sodium hypochlorite for 5 min, and rinse extensively with deionized water.

(ii) Spread the seeds uniformly on germination paper. Roll up the paper and transfer to a beaker (or another container) with a small amount of distilled water in the bottom. Seeds must be germinated in the dark at 25°C for 72 h.

(iii) Transfer twenty-five equal-sized seedlings into the hydroponic system as described in the Introduction.

(iv) Incubate the seedlings without or with the allelochemical solution (1.0 mM ferulic acid or L-3,4-dihydroxyphenylalanine) for 24 h.

(v) Cut the roots. Determine the dry weight after oven-drying at 80°C, for 24 h.

(vi) To determine lignin, proceed as follows:

Step 1. Cell wall isolation.

Homogenize the dry roots (0.3 g) in 50 mM potassium phosphate buffer (7 mL, pH 7.0) with mortar and pestle, and transfer into a centrifuge tube. Centrifuge the pellet ($1{,}400 \times g$, 4 min). Wash the pellet by successive stirring and centrifugation as follows: twice with phosphate buffer pH 7.0 (7 mL); three times with 1% (v/v) Triton X-100 in pH 7.0 buffer (7 mL); twice with 1 M NaCl in pH 7.0 buffer (7 mL); twice with distilled water (7 mL) and twice with acetone (5 mL). Dry the pellet in an oven (60°C, 24 h). Cool the pellet in a vacuum desiccator. The dry matter obtained is defined as protein-free cell wall fraction (Ferrarese *et al.*, 2002).

Step 2. Lignin determination.

Transfer all dry protein-free tissue into a screw-cap centrifuge tube containing the reaction mixture (1.2 mL of thioglycolic acid plus 6 mL of 2 M HCl). Heat at 95°C for 4 h. After cooling at room temperature, centrifuge the sample ($1{,}400 \times g$, 5 min). Discard the supernatant. The pellet contained the complex lignin-thioglycolic acid (LTGA). Wash the pellet three times with distilled water (7 mL). Extract the product by shaking (30°C, 18 h, 115 oscillations min^{-1}) in 6 mL of 0.5 M NaOH. Centrifuge at $1{,}400 \times g$ for 5 min. Store the supernatant. Wash the pellet again with 3 mL of 0.5 M NaOH. Mix the supernatant with the supernatant obtained earlier. Add 1.8 mL of HCl in the combined alkali extracts. Recover the LTGA formed after 4 hours at 0°C by centrifugation ($1{,}400 \times g$, 5 min). Wash the pellet twice with 7 mL of distilled water. Dry the pellet at 60°C. Dissolve the pellet in 1 mL of 0.5 M NaOH. Dilute the sample (about five times) to yield an appropriate absorbance for spectrophotometric determination at

280 nm. Determine the LTGA concentration by using the molar extinction coefficient (ε = 17.87 g^{-1} LTGA L cm^{-1}), as reported by Ferrarese *et al.* (2002). Express the results as mg LTGA g^{-1} dry root weight.

Observations

Make at least three independent experiments (N = 3).

Calculations

Step 1. Calculate the sample concentration (C).

$$C = A_{280}/\varepsilon, \quad \text{where } \varepsilon = 17.87 \text{ g}^{-1} \text{ LTGA L cm}^{-1}.$$

For example, if A_{280} = 0.700 and the sample dilution before spectrophotometric reading = 5, concentration should be:

$$C = 0.700/17.87 \text{ g}^{-1} \text{ LTGA L cm}^{-1} \times 5$$
$$C = 0.196 \text{ g LTGA L}^{-1}$$

Volume of NaOH used to solubilize the LTGA pellet = 1 mL. Then, the concentration in assay is:

$$C = 0.196 \text{ g LTGA} \times 1 \text{ mL}/1000 \text{ mL}$$
$$C = 0.000196 \text{ g LTGA (or 0.196 mg LTGA)}$$

Step 2. The weight of the original dry root is = 0.3 g. Then, calculate lignin content:

Lignin content = 0.196 mg LTGA × 1 g/0.3 g
Lignin content = 0.65 mg LTGA g^{-1} dry root weight.

Note: Similar procedure should be applied to calculate lignin content in the control (without allelochemical treatment).

Statistical analysis

Difference between data of treated and untreated (control) roots may be evaluated by Student *t*-test.

Precautions

Thioglycolic acid is toxic by inhalation, contact or swallowing.

Experiment 7: Respiration

Mitochondria are the principal producers of ATP, which diffuses to all parts of the cell, providing energy for cellular work. Mitochondria are widely used for measurement of respiratory activity and the oxygen consumption is a direct measure of electron flow within of inner membrane. One of the distinguishing characteristics of the plant respiration is the presence of a cyanide-resistant pathway, alternative to the universal cytochrome pathway. In contrast to the cytochrome pathway, beyond the branch point (ubiquinone), the alternative pathway does not contribute to the generation of a proton-motive force for the synthesis of ATP. Electron flow through the alternative oxidase is insensitive to classic inhibitors of cytochrome oxidase, such as cyanide, azide, carbon monoxide and antimycin A. However, it may be specifically inhibited by *n*-propyl gallate and salicylhydroxamic acid (SHAM). Then, the contribution of the cytochrome and alternative pathways to total respiration may be determined using the inhibitors KCN and SHAM. Allelochemicals interacts with the mitochondrial membrane and strongly affects the respiratory activity of soybean radicular mitochondria (Abrahim *et al.*, 2003). Salicylic acid, a plant-derived allelochemical collapses the transmembrane electrochemical potential of pea stem mitochondria (Macri *et al.*, 1986). In this experiment, polarography is applied to evaluate the effects of salicylic acid on mitochondrial respiratory activity of excised soybean root apices.

Materials and equipments required

Seeds, fresh plant roots, analytical balance, beakers, centrifuge tubes, dark bottle, germination paper, hydroponic system, pipettes, polypropylene flasks, oxygen monitor (Fig. 7) containing a circulating water bath, a Clark-type oxygen electrode, a closed acrylic chamber, a magnetic bar, a magnetic stirrer, a polarograph and a chart recorder, pH meter (or pH indicator sticks), vacuum flask.

Reagents

Nutrient solution (see Table 1); 2.7 mM KCN: dissolve 4.4 mg of KCN in 25 mL of distilled water; 6.7 mM salicylhydroxamic acid (SHAM): dissolve 25.5 mg of SHAM in 25 mL of 2-methoxyethanol; 2 mM $Ca(NO_3)_2$: dissolve 47.2 mg of $Ca(NO_3)_2$ in 100 mL of distilled water; 2 mM KNO_3: dissolve 20.2 mg of KNO_3 in 100 mL of distilled water; 0.7 mM $MgSO_4$: dissolve 18.4 mg of $MgSO_4$ in 100 mL of distilled water; 0.4 mM NH_4Cl: dissolve 2.3 mg of NH_4Cl in 100 mL of distilled water; 20 μM NaH_2PO_4: dissolve 2.4 mg in one liter of distilled water; 1.0 mM salicylic acid (see Experiment 3, *Observation* 5); 2% sodium hypochlorite, 5 M KOH and 5 M HCl solutions (see Experiment 1, *Reagents*).

Procedure

(i) Sterilize soybean seeds with 2% sodium hypochlorite for 5 min, and rinse extensively with deionized water.

(ii) Spread the seeds uniformly on germination paper. Roll up the paper and transfer to a beaker (or another container) with a small amount of distilled water in the bottom. Seeds must be germinated in the dark at 25°C for 72 h.

(iii) Transfer twenty five equal-sized seedlings into the hydroponic system as described in the Introduction.

(iv) Incubate the seedlings with 1.0 mM salicylic acid or without it (control) for 24 h.

(v) Remove the seedlings and rinse rapidly with distilled water.

(vi) The O_2 consumption rate by mitochondria from the root apices must be monitored using a Clark-type electrode connected to the oxygen monitor, and positioned in the closed reaction chamber (Fig. 7). The chamber must be maintained on the magnetic stirrer. The temperature must be maintained at 25°C by circulating water from a temperature-controlled water bath

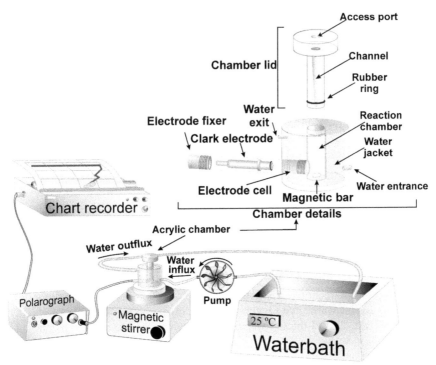

Figure 7 The oxygen monitor. The suspension to be assayed is placed in the acrylic chamber that is exposed to the surface of the Clark O_2 electrode. The movable lid of the chamber contains one access port for adding/removing materials. The reaction mixture is stirred to ensure homogeneity and to ensure that O_2 may freely diffuse into the electrode. Under O_2, the electrode delivers a current to the polarograph, which amplifies this current and converts it to a voltage output that is directly proportional to the concentration of O_2 in the chamber. The recorder moves a paper chart at constant speed, so that when the recorder pen moves in response to voltage changes, O_2 content is recorded as a function of time. For mitochondria studies (Experiment 7) the rate at which total chamber O_2 declines is referred as the O_2 consumption rate. For chloroplasts studies (Experiment 8) the rate at which total chamber O_2 increases is referred as the O_2 production rate (Source: Adapted from Estabrook, 1967).

around the Clark cell. For determination of the mitochondrial respiratory activity, proceed as follows:

Step 1. Root apices preparation.
Detach five root apices (1.0 cm each) and place immediately in the reaction chamber filled with 2 mL of respiration medium (pH 6.0) containing 2 mM $Ca(NO_3)_2$, 2 mM KNO_3, 0.75 mM $MgSO_4$, 0.43 mM NH_4Cl and 20 µM NaH_2PO_4 in a final volume of 2 mL at 25°C. This mixture must be continuously stirred by a magnetic bar inserted into the chamber.

Step 2. O_2 consumption measurement.

After addition of apices, the polarographic measurement (Fig. 8) is followed by the chart recorder. Monitor the O_2 consumption until stabilization (about 5 min). To inhibit the electron transport by the mitochondria, add 0.1 mL of 2.67 mM KCN (the inhibitor of the cytochrome oxidase) or 0.1 mL of 6.67 mM SHAM (the inhibitor of the alternative oxidase) into the reaction medium. Monitor the O_2 consumption again. Afterwards, suck the apices from the reaction chamber by using a pipette connected to a vacuum flask. Weigh the apices. Calculate the O_2 consumption rate from the polarographic record (see *Calculations*). Express the results as nmol O_2 min^{-1} mg^{-1} apices.

Observations

Make at least three independent experiments (N = 3).

Calculations

Determination of the O_2 consumption rate by the root apices: at standard temperature and pressure a buffered solution, equilibrated with room air, holds around 237 nmol O_2 mL^{-1} (at 25°C). If the medium volume into the chamber is known, the O_2 total may be calculated. A typical polarographic record may be seen in Fig. 8. The recorded traces refer to measurements of the O_2 activity. Its slopes aids to determine the O_2 consumption after addition of root apices (1), KCN or SHAM (2) and the O_2 exhaustion (3), which is obtained by grounding

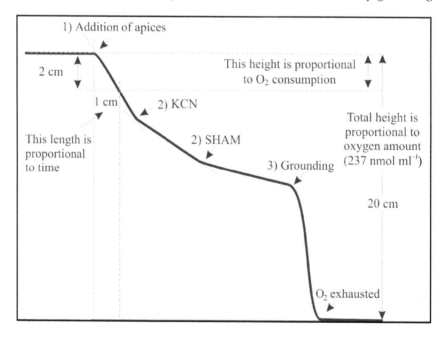

Figure 8 Typical polarographic trace recording O_2 consumption as measured using a Clark-type O_2 electrode (Source: Adapted from Estabrook, 1967).

the recorder to zero electrical potential. To monitor the O_2 consumption, the recorder pen must be firstly set until saturation of the chamber with O_2 (237 nmol O_2 mL^{-1}, i. e., 20 cm in the figure), before addition of apices (the straight line in the figure). This adjustment must be done with the chamber filled with reaction medium under stirring. From the record, it is possible to calculate exactly the amount of O_2 (in nmol) that is consumed per unit time since its concentration is proportional to height recorded (Estabrook, 1967). To do that, the dotted thin lines parallel to the slopes should be measured using a pencil.

Step 1. Calculate total the sample O_2. It is obtained by multiplying the reaction volume by the known volume of dissolved oxygen per unit volume at the temperature of the experiment (237 nmol O_2 mL^{-1}, at 25°C, for a typical respiration medium):

$$O_2 \text{ total} = \text{reaction volume} \times 237 \text{ nmol mL}^{-1}$$

For example, if the reaction volume in the chamber is 2 mL, O_2 total should be:

$$O_2 \text{ total} = 2 \text{ mL} \times 237 \text{ nmol mL}^{-1} \quad \therefore O_2 \text{ total} = 474 \text{ nmol}$$

Step 2. If the height measure of the record is 2 cm and the total height is 20 cm, the amount of O_2 consumed for a given interval is:

$$O_2 \text{ consumed} = O_2 \text{ total} \times \text{height measure/total height}$$
$$O_2 \text{ consumed} = 474 \text{ nmol } O_2 \times 2 \text{ cm/20 cm}$$
$$O_2 \text{ consumed} = 47.4 \text{ nmol}$$

Step 3. If the length of the record is 1 cm and the paper speed is 0.5 cm min^{-1}, the time variation (Δt) of the recorded sample is:

$$\Delta t = \text{length/paper speed}$$
$$\Delta t = 1 \text{ cm/0.5 cm min}^{-1}$$
$$\Delta t = 2 \text{ min}$$

Step 4. Calculate the O_2 consumption rate:

$$O_2 \text{ consumption rate} = O_2 \text{ consumed/}\Delta t$$
$$O_2 \text{ consumption rate} = 47.4 \text{ nmol/2 min}$$
$$O_2 \text{ consumption rate} = 23.7 \text{ nmol min}^{-1}$$

The weight of the soybean root apices is = 22 mg. Then, the O_2 consumption rate is:

$$O_2 \text{ consumption rate} = 23.7 \text{ nmol } O_2 \text{ min}^{-1} \times 1 \text{ mg/22 mg}$$
$$O_2 \text{ consumption rate} = 1.077 \text{ nmol } O_2 \text{ min}^{-1} \text{ mg}^{-1} \text{ apices}$$

Note 1. Similar procedure should be applied to calculate O_2 consumption rate in apices treated with KCN or SHAM, and in the control (without allelochemical treatment).

Note 2. Alternatively, an interface by generating a digital signal from the Clark's electrode may be used. Software calculates data and makes the plots. This technology furnishes a reliable measure of O_2 and simplifies calculations. It is currently applied in our laboratory.

Statistical analysis

Difference between data of treated and untreated (control) roots may be evaluated by Student t-test.

Precautions

KCN should be stored in a cool, dry, well-ventilated and locked location. It is absorbed through the skin by causing redness and irritation. In acidic environment, KCN hydrolyzes to form HCN, a severely toxic, flammable, colorless gas with almond-like odor. Make up the solution in a fume cupboard. SHAM is harmful by inhalation, in contact with skin and if swallowed. It is irritating to respiratory system.

Experiment 8: Photosynthesis

Chloroplasts are specialized organelles found in all higher plant cells. They have a double outer membrane. Within the stroma are other membrane structures-the thylakoids and grana. The photosynthetic apparatus of chloroplasts is contained within the expansive thylakoid membrane system. In photosynthesis, chloroplasts capture light energy from the sun to form ATP and NADPH. These compounds are used as energy sources to make carbohydrates and other components from CO_2 and H_2O with simultaneous release of O_2 into the atmosphere. In this context, the Hill reaction is defined as the reduction of an electron acceptor by electrons and protons from H_2O, with the evolution of O_2, when chloroplasts are exposed to light. The final electron acceptor *in vivo* is NADP. Several artificial electrons acceptors may be used to study the Hill reaction *in vitro*. The most commonly used electron acceptor is the dye 2,6-dichlorophenolindophenol (DCPI). After addition of DCPI into a chloroplast suspension, the following reaction occurs:

$$H_2O + DCPI \xrightarrow{\text{light}} DCPI \text{ (reduced)} + 1/2\ O_2$$

and the O_2 production may be measured polarographically. Quinone (juglone and sorgoleone) and phenolic (vanillic and ferulic acids) allelochemicals are powerful inhibitors of chloroplast CO_2-dependent O_2 evolution (Einhellig, 1995). In this experiment, polarographic measures are determined to evaluate the effects of ferulic acid, a cinnamic acid derivative on chloroplast O_2 production.

Materials and equipments required

Seeds, fresh plant roots, aluminum foil, analytical balance, beakers, centrifuge tubes, cheesecloth, dark bottle, fluorescent lamp (15 Watt), germination paper, hydroponic system, mortar and pestle, paintbrush, oxygen monitor (Fig. 7) containing a circulating water bath, a Clark-type oxygen electrode, a closed acrylic chamber, a magnetic bar, a magnetic stirrer, a polarograph and a chart recorder, pH meter (or pH indicator sticks), pipettes, polypropylene flasks, refrigerated preparative centrifuge capable of speed of 2,200 × g, visible light (400 to 700 nm) spectrophotometer, 1-cm cuvette.

Reagents

Nutrient solution (see Table 1); acetone; 50 mM potassium phosphate buffer, pH 6.5 (see Table 2); grinding solution: dissolve 12 g of sucrose in 100 mL of 50 mM potassium phosphate buffer (pH 6.5); 30 mM potassium phosphate buffer (pH 6.5); reaction mixture: dissolve 74.5 mg of KCl in 100 mL of 30 mM potassium phosphate buffer (pH 6.5); Prepare 1.0 mM 2,6-dichlorophenolindophenol (DCPI) before use dissolving 32.6 mg of the compound in 100 mL of reaction mixture and store in a dark bottle; 0.5 mM ferulic acid (see *Observation* 3); 2% sodium hypochlorite, 5 M KOH and 5 M HCl solutions (see Experiment 1, *Reagents*).

Procedure

(i) Sterilize soybean seeds with 2% sodium hypochlorite for 5 min, and rinse extensively with deionized water.

(ii) Spread the seeds uniformly on germination paper. Roll up the paper and transfer to a beaker (or another container) with a small amount of distilled water in the bottom. Seeds must be germinated in the dark at 25°C for 72 h.

(iii) Transfer twenty-five equal-sized seedlings into the hydroponic system as described in the Introduction.

(iv) Detach the cotyledons, and isolate the chloroplasts as follows:

Step 1. Chloroplasts isolation.

Prepare an ice bath and pre-cool all glassware before experiment. Transfer tissues (2.5 g fresh weight) to a mortar and mix thoroughly with 5 mL of grinding solution containing 50 mM potassium phosphate buffer (pH 6.5) and 350 mM sucrose. Filter the homogenate through double layered cheesecloth into a centrifuge tube, and squeeze the tissue pulp to recover all suspension. Centrifuge the filtrate at 200 × g for 2 min at 4°C to pellet the unbroken cells and fragments. Decant the supernatant into a clean centrifuge tube and re-centrifuge at 1,000 × g for 7 minutes. Discard the supernatant. The pellet contains chloroplasts. Dissolve gently the pelleted chloroplasts in 2 mL of cold grinding solution by using a soft paintbrush. The centrifuge tube containing this final suspension must be wrapped in aluminum foil and placed in an ice bucket (see *Observation* 4) for use in subsequent experiments.

Step 2. O_2 production measurement.

Measure the chloroplasts O_2 production by the oxygen monitor described in Fig. 7. The O_2 production must be monitored with a Clark-type O_2 electrode which is positioned in the closed chamber connected to polarography. The chamber is maintained on the magnetic stirrer. The temperature must be maintained at 25°C by circulating water from a temperature-controlled water bath around the Clark cell. A fluorescent lamp of white light intensity (15 Watt) is positioned (about 15 cm) on the surface of the chamber. Add chloroplasts (0.25 mL of suspension) into chamber which contains 1 mL of reaction mixture (see *Reagents*). This mixture must be continuously stirred by a magnetic bar inserted into the reaction chamber. After addition of chloroplasts, the polarographic measurement (Fig. 9) is followed in the chart recorder. After to reach the steady-state, add 0.2 mL of 1.0 mM DCPI, the artificial electron acceptor, into the reaction mixture. After to reach a new steady-state (about 2 minutes), switch on the lamp. Monitor the O_2 production until stabilization, and consider as control. To determine the effects of ferulic acid, the procedure must be restarted with a new sample of chloroplasts suspension. Undertake all steps described earlier. Then, after addition of DCPI, add 0.1 mL of 0.5 mM ferulic acid into the reaction mixture, under light. Monitor the O_2 production. Calculate the O_2 production rates from the polarographic records (see *Calculations*). Express the results as nmol O_2 minute^{-1} mg^{-1} chlorophyll.

Step 3. Determination of chlorophyll content.
Transfer a sample (0.5 mL) of chloroplast suspension into a 15 mL centrifuge tube containing 1.0 mL of 80% acetone. Mix the tube thoroughly. Centrifuge the sample at $1,000 \times g$ for 3 minutes. Decant the supernatant. Measure the absorbance at 645 and 663 nm, against blank (water). The amount of total chlorophyll is given by the equation: mg chlorophyll mL^{-1} = $20.2A_{645nm} + 8.02A_{663nm}$ (Arnon, 1949).

Observations

(i) All material must be refrigerated before chloroplasts extraction.
(ii) Make at least three independent experiments (N = 3).
(iii) Prepare 0.5 mM ferulic acid immediately before use:

   (a) dissolve 9.7 mg of ferulic acid in 25 mL of distilled water;
   (b) add drops of 5 M KOH until complete dissolution of the compound (up to pH 12);
   (c) add drops of 5 M HCl until pH 6.0 to avoid precipitation of the compound;
   (d) complete with enough water for 100 mL of solution. Finally, adjust to pH 6.0.

(iv) Chloroplasts are extremely light labile. This suspension must be kept cool and in reduced light at all times, except when actually being used for analysis of photosynthesis. Once diluted, the chloroplasts will be reasonably unstable and the experiments must be undertaken as rapidly as possible.

Calculations

Determination of the chloroplasts O_2 production rate: at standard temperature and pressure a buffered solution, equilibrated with room air, holds around 237 nmol O_2 mL^{-1} (at 25°C). If the medium volume into the chamber is known, the O_2 total may be calculated. A typical polarographic record may be seen in Fig. 9. The recorded traces refer to measurements of the O_2 activity. The slopes (Fig. 9A) aid to determine the O_2 production after addition of chloroplasts (1), DCPI (2), light (3) and ferulic acid (4). Fig. 9B shows the recorded traces of the control condition (without allelochemical treatment). For reliable measure of the O_2 production, the recorder pen must be initially set with the appropriate bottom (or by displacing to the zero electrical potential in some equipment). This adjustment must be done with the chamber filled with reaction medium under stirring. This condition refers to 237 nmol O_2 mL^{-1} (16 cm in the figure). From the record, it is possible to calculate exactly the amount of O_2 (in nmol) that is produced per unit time since its concentration is proportional to height recorded. To do that, the dotted thin lines parallel to the slopes are measured using a pencil (see Experiment 7, *Note 2*).

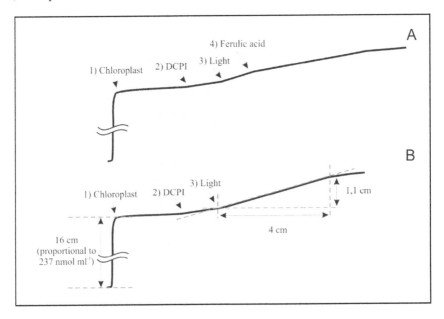

Figure 9 Typical polarographic trace recording O_2 production as measured using a Clark-type O_2 electrode. A, with ferulic acid and B, control (Source: Adapted from Estabrook, 1967).

Step 1. Calculate the chlorophyll concentration (C) in the chloroplasts suspension:

$$C = 20.2A_{645nm} + 8.02A_{663nm}$$

For example, if $A_{645nm} = 0.061$; $A_{663nm} = 0.092$ and the sample dilution before spectrophotometric reading = 3 (Note that this dilution factor is derived from the dilution of 0.5 mL of chloroplasts in 1.0 mL of 80% acetone, i.e., 1.5 mL / 0.5 mL), concentration should be:

$$C = (20.2 \times 0.061 + 8.02 \times 0.092) \times 3$$
$$C = 5.91 \text{ mg chlorophyll } mL^{-1}$$

Volume of chloroplasts suspension is 0.25 mL. Then, the chlorophyll content (Chl) in assay is:

$$\text{Chl} = 5.91 \text{ mg chlorophyll} \times 0.25 \text{ mL}/1 \text{ mL}$$
$$\text{Chl} = 1.48 \text{ mg chlorophyll}$$

Step 2. Calculate the dissolved O_2 amount in equilibrium with room air. It is obtained by multiplying reaction volume by the known volume of dissolved oxygen per unit volume at the temperature of the experiment (237 nmol O_2 mL^{-1}, at 25°C, for a typical respiration medium).

$$O_2 \text{ total} = \text{reaction volume} \times 237 \text{ nmol mL}^{-1}$$

The reaction volume into the chamber is 1.55 mL. Then, the O_2 total in assay is:

$$O_2 \text{ total} = 1.55 \text{ mL} \times 237 \text{ nmol mL}^{-1} \quad \therefore \quad O_2 \text{ total} = 367.3 \text{ nmol}$$

Step 3. If the height measure in record of the control condition (Fig. 9B) is 1.1 cm and the total height is 16 cm, the amount of O_2 produced for a given interval is:

$$O_2 \text{ produced} = O_2 \text{ total} \times \text{height measure}/\text{total height}$$
$$O_2 \text{ produced} = 367.3 \text{ nmol } O_2 \times 1.1 \text{ cm}/16 \text{ cm}$$
$$O_2 \text{ produced} = 25.2 \text{ nmol}$$

Step 4. If the length of the record is 4 cm and the paper speed is 0.5 cm minute^{-1}, the time variation (Δt) of the recorded sample is:

$$\Delta t = \text{length}/\text{paper speed}$$
$$\Delta t = 4 \text{ cm}/0.5 \text{ cm minute}^{-1}$$
$$\Delta t = 8 \text{ minutes}$$

Step 5. Calculate the O_2 production rate:

$$O_2 \text{ production rate} = O_2 \text{ produced}/\Delta t$$
$$O_2 \text{ production rate} = 25.2 \text{ nmol}/8 \text{ minutes}$$
$$O_2 \text{ production rate} = 3.15 \text{ nmol minute}^{-1}$$

The chlorophyll content (Chl) in assay is 1.48 mg chlorophyll. Then, calculate O_2 production rate:

$$O_2 \text{ production rate} = 3.15 \text{ nmol } O_2 \text{ minute}^{-1} \times 1 \text{ mg}/1.48 \text{ mg}$$
$$O_2 \text{ production rate} = 2.13 \text{ nmol } O_2 \text{ minute}^{-1} \text{ mg}^{-1} \text{ chlorophyll.}$$

Note 1. Similar procedure should be applied to calculate O_2 production rate in the sample (with allelochemical treatment).

Statistical analysis

Difference between data of treated and untreated (control) roots may be evaluated by Student t-test.

Precautions

Acetone is extremely flammable. DCPI should be stored in cool place and out of direct sunlight. Handle with caution. It is irritating to eyes, skin and to the respiratory tract. Make up the solution in a fume cupboard.

SUGGESTED READINGS

Abrahim, D., Takahashi, L., Kelmer-Bracht, A.M. and Ishii-Iwamoto, E.L. (2003). Effects of phenolic acids and monoterpenes on the mitochondrial respiration of soybean hypocotyl axes. *Allelopathy Journal* **11**: 21-30.

Arnon, D.I. (1949). Copper enzymes in isolated chloroplasts. Polyphenoloxidase in *Beta vulgaris*. *Plant Physiology* **24**: 1-15.

Baker, C.J. and Mock, N.M. (1994). An improved method for monitoring cell death in cell suspension and leaf disc assays using Evans blue. *Plant Cell, Tissue and Organ Culture* **39**: 7-12.

Baziramakenga, R., Leroux, G.D. and Simard, R.R. (1995). Effects of benzoic and cinnamic acids on membrane permeability of soybean roots. *Journal of Chemical Ecology* **21**: 1271-1285.

Boerjan, W., Ralph, J. and Baucher, M. (2003). Lignin biosynthesis. *Annual Review of Plant Biology* **54**: 519-546.

Böhm, P.A.F., Zanardo, F.M.L., Ferrarese, M.L.L. and Ferrarese-Filho, O. (2006). Peroxidase activity and lignification in soybean root growth-inhibition by juglone. *Biologia Plantarum* **50**: 315-317.

Cakmak, I. and Horst, W.J. (1991). Effect of aluminum on lipid peroxidation, superoxide dismutase, catalase, and peroxidase activities in root tips of soybean (*Glycine max*). *Physiologia Plantarum* **83**: 463-468

Deng, F., Aoki, M. and Yogo, Y. (2004). Effect of naringenin on the growth and lignin biosynthesis of gramineous plants. *Weed Biology and Management* **4**: 49-55.

Devi, S.R. and Prasad, M.N.V. (1996). Ferulic acid mediates changes in oxidative enzymes of maize seedlings: Implications in growth. *Biologia Plantarum* **38**: 387-395.

Dixon, R.A., Achnine, L., Kota, P., Liu, C.J., Srinivasa Reddy, M.S. and Wang, L. (2002). The phenylpropanoid pathway and plant defence - a genomics perspective. *Molecular Plant Pathology* **3**: 371-390.

Doblinski, P.M.F., Ferrarese, M.L.L., Huber, D.A., Scapim, C.A., Braccini, A.L. and Ferrarese-Filho, O. (2003). Peroxidase and lipid peroxidation of soybean roots in response to *p*-coumaric and *p*-hydroxybenzoic acids. *Brazilian Archives of Biology and Technology* **46**: 193-198.

Estabrook, R. (1967). Mitochondrial respiratory control and the polarographic measurement of ADP:O Ratios. In: *Methods in Enzymology* (Eds., R.W. Estabrook and M.E. Pullman). Academic Press, New York, USA.

Einhellig. F.A. (1995). Allelopathy. Mechanism of action of allelochemicals in allelopathy. In: *Allelopathy. Organisms, Processes and Applications* (Eds., Inderjit, K.M.M. Dakshini, F.A. Einhellig,). *ACS Symposium Series* **582**: 97-116. American Chemical Society, Washington, USA.

Einhellig, F.A. and Barkosky, R.R. (2003) Allelopathic interference of plant-water relationships by *p*-hydroxybenzoic acid. *Botanical Bulletin of Academia Sinica* **44**: 53-58.

Ferrarese, M.L.L., Rodrigues, J.D. and Ferrarese-Filho, O. (2000). Phenylalanine ammonia-lyase activity in soybean roots extract measured by reversed-phase high performance liquid chromatography. *Plant Biology* **2**: 152-153.

Ferrarese, M.L.L., Zottis, A. and Ferrarese-Filho, O. (2002). Protein-free lignin quantification in soybean (*Glycine max*) roots. *Biologia* **57**: 541-543.

Herrig, V., Ferrarese, M.L.L., Suzuki, L.S., Rodrigues, J.D. and Ferrarese-Filho, O. (2002). Peroxidase and phenylalanine ammonia-lyase activities, phenolic acid contents, and allelochemicals-inhibited root growth of soybean. *Biological Research* **35**: 59-66.

Hoagland, D.R. and Arnon, D.I. (1950). The water-culture method of growing plants without soil. *California Agricultural Experimental Station Circular* No. 347. USDA-ARS.

Macri, F., Vianello, A. and Pennazio, S. (1986). Salicylate-collapsed membrane potential in pea stems mitochondria. *Physiologiae Plantarum* **67**: 136-140.

McKenzie, H.A. and Dawson, R.M.C. (1969). pH, buffers, and physiological media. In: *Data for Biochemical Research* (Eds., R.M.C., Dawson, D.C., Elliot, W.H., Elliot, and K.M. Jones). Oxford University Press, Oxford, pp. 475-508.

Passardi, F., Cosio, C., Penel, C. and Dunand, C. (2005). Peroxidases have more functions than a Swiss army knife. *Plant Cell Reports* **24**: 255-265.

Politycka, B. (1996). Peroxidase activity and lipid peroxidation in roots of cucumber seedlings influenced by derivatives of cinnamic and benzoic acids. *Acta Physiologiae Plantarum* **18**: 365-370.

Politycka, B. (1999). Ethylene-dependent activity of phenylalanine ammonia-lyase and lignin formation in cucumber roots exposed to phenolic allelochemicals. *Acta Societatis Botanicorum Poloniae* **68**: 123-127.

Politycka, B., Kozlowska, M. and Mierlcarz, B. (2004). Cell wall peroxidases in cucumber roots induced by phenolic allelochemicals. *Allelopathy Journal* **13**: 29-35.

Santos, W.D., Ferrarese, M.L.L., Finger, A., Teixeira, A.C.N. and Ferrarese-Filho, O. (2004). Lignification and related enzymes in *Glycine max* root growth-inhibition by ferulic acid. *Journal of Chemical Ecology* **30**: 1199-1208.

15

Bioassays on Microorganisms: Antifungal and Antibacterial Activities

Melina A. Sgariglia, Jose R. Soberón, Diego A. Sampietro, Emma N. Quiroga, and Marta A. Vattuone*

1. INTRODUCTION

Microorganisms are broad and heterogeneous groups of living organisms, including bacteria, fungi, protozoa and viruses (Madigan *et al.*, 2002). They are everywhere in nature and have a deep impact on the biosphere. Most microorganisms are beneficial or have no apparent effect on their hosts. Nevertheless, the few pathogenic species found are very harmful for humans, plants and animals and generate important economical loses in human activities. Compounds that kill or inhibit the growth of these pathogenic microorganisms are known as antimicrobials. The search for substances with selective antimicrobial activity, without toxic effects on animals, plants and beneficial organisms is of important concern in natural products research. Microbial responses to natural products are very heterogeneous because the strong dependence on applied methods and tested microbial species. Several conditions must be fulfilled to standarize an antimicrobial method (Dey and Harborne, 1991):

(i) The tested sample must be brought in contact with the cell membrane or cell wall of the microorganism which has been selected for the test.

Authors' address: *Instituto de Estudios Vegetales "Dr. A. R. Sampietro"*, Facultad de Bioquímica, Química y Farmacia, Universidad Nacional de Tucumán, España 2903, 4000, San Miguel de Tucumán, Tucumán, Argentina.
Corresponding author: E-mail: instveg@unt.edu.ar

(ii) There must be measurable parameters that provide quantitative growth evidence of the microorganism tested.

(iii) The choice of test organisms largely depends on the purpose of the research. For general purposes, the test organisms selected should be as diverse as possible and preferably representative of all important groups of pathogenic microbials according to its physical and chemical composition and resistance pattern.

(iv) Test compounds should be dissolved and diluted in pure solvents or mixtures of them. These solvents should be innocuous for microbial growth.

(v) Mixtures of compounds (i. e. essential oils) whose components have different solubility in water must be tested after addition to the culture medium of an emulsifier or solvent that ensures proper contact between the test organisms and the bioactive products (Mann and Markham, 1998).

This chapter provides assays to measure antimicrobial activity of natural products. Assays are based on diffusion and dilution methods. Some are adequate for work with bacteria and yeasts, while others only are suitable for work with moulds.

2. GENERAL ASPECTS

The following aspects must be considered before performing the antimicrobial assays:

I. **Microbial preservation:** Primary methods for preservation of bacterial and fungal strains are continuous growth, drying and freezing. Continuous growth methods, in which cultures are grown on agar, typically are used for short-term storage. Such cultures are stored at temperatures of 4°C (i.e. fungi), or they may be frozen to increase the interval between subcultures (i.e. bacteria). Drying by lyophilization is widely used for culture preservation (i.e. bacteria, yeasts and spore-forming fungi). Freezing methods, including cryopreservation, are versatile and widely applicable. Most fungi and bacteria are preserved at –20°C, –70°C or in liquid nitrogen. Bacteria are often frozen in ssBHI (semi-solid Brain Heart Infusion) with a cryoprotectant, usually 20% glycerol.

II. **Growth re-initiation:** Re-initiation of microbial growth also requires proper temperature and medium conditions. Bacterial strains preserved in ssBHI with 20% glycerol are incubated at 37°C for 2 h. Lyophilized bacterial strains are usually re-hydrated in Mueller Hinton broth (MHB) or saline physiological solution (0.85% NaCl) before incubation. Then, bacteria are streaked on Petri plates (Fig. 1B) containing a non selective medium (e.g. Mueller Hinton Agar, MHA) to allow growth of individual colonies. After 24 hours at 37°C, bacterial colonies are isolated to prepare inoculum suspensions used for assays. Procedure for growth re-initiation of yeasts is similar to that described for bacteria, but different solid media are used such as Sabouraud, BHI, Mycobiotic (Mycosel), Solid Malt Medium (SMM), Inhibitory Mold Agar, Cottonseed Agar, Corn Meal Agar and Yeast

Nitrogen Base Agar. Fungus media are in slant tubes. For moulds, medium inoculation is achieved placing a 3-mm diameter plug of a growing mycelium of mushrooms or spore-forming fungi harvested from 8 to 10-day culture slant tubes. Moulds growth is achieved after incubation at 30°C for 48 to 120 h, according with the species.

3. DIFFUSION METHODS

Experiment 1: Disc Diffusion Test

A suspension of a bacterial strain is adjusted to a standard density and is evenly swabbed on an agar medium plate. Filter paper discs impregnated with the test compound are placed on the agar layer to allow allelochemicals diffusion into the agar. The size of the agar area impregnated with the test compound will depend on its solubility and molecular size. Susceptible microorganisms will not grow around the paper discs. These areas are known as "inhibition zones". The disc diffusion test provides a relative measure of antimicrobial activity, since biocide effect depends on the type and speed of microbial growth.

Materials and equipments required

Petri dishes (90 mm); test tubes; haemolysis tubes; empty haemolysis tubes; general laboratory glassware (flasks, bottles, graduated cylinders); 1, 2, 5 and 10 mL sterile pipettes; 10-100 µL micropipettes; pipette tips; 4 mm diameter Whatman N° 4 paper discs; loops; forceps; sterile applicator cotton swabs with wooden handles; cryogenic tubes for freezing down the working cultures; sterile Millipore filtration units with 0.22-µm-pore-size; elements for personal biosecurity (gloves, gowns, goggles or protective eye wear, chemical/biological safety hood).

Autoclave; boiling water bath for melting the solid media; balances; incubator at 30 and 37°C; horizontal laminar flow cabinet (human pathogen species should be used the horizontal and vertical laminar flow); refrigerator/freezer or liquid nitrogen tank; spectrophotometer; vortex mixer.

Reagents

Test compound; commercially available antibacterial or antifungal standards; ATCC bacterial or fungal strains (for example Table 1); MHA (Mueller Hinton Agar) (dissolve 3.7 g of commercially available MHA (Mueller Hinton Broth) powder into 100 mL of distilled water; warm in the water bath until total dissolution and fractionate 10 mL per tube; this culture medium should be used for bacterial assays); SMM (Solid Malt Medium) (add 1.5 g malt extract, 0.5 g peptone and 1.8 agar to 100 mL of distilled water, warm in a water bath until total dissolution and fractionate 20 mL per tube, this culture medium should be used for moulds assays); Media for bacterial suspensions: 3 mL of sterile distilled water, 0.85% NaCl (dissolve 0.85 g NaCl in distilled water and take up to 100 mL) or MHB (add 2.2 g of commercially available MHB powder to 100 mL of distilled water, warm in water bath until total dissolution and fractionate) and for yeasts Liquid Malt Medium (LMM), LMM (Add 1.5 g malt extract and 0.5 g peptone to 100 mL of distilled water, warm in water bath until total dissolution and fractionate); Disinfectant solution for work space (70% ethanol, v/v).

Table 1 Acceptable inhibitory zone diameter (mm) limit of control strains recommended for use in the disc diffusion test of antimicrobial sensitivity testing of bacteria isolated from animals (Source: Adapted from National Committee for Clinical Laboratory Standards, 1997)

Antimicrobial agent	Disc content	Escheri- chia coli	Staphylococ- cus aureus	Pseudomonas aeruginosa	Streptococcus pneumoniae
		ATCC 25922	ATCC 25923	ATCC 27853	ATCC 49619
Amikacin	30 µg	19-26	20-26	18-26	–
Amoxicillin- Clavulanic acid	10–20 µg	18-24	28-36	–	–
Ampicillin	10 µg	16-22	27-35	–	30-36
Cefazolin	30 µg	21-27	29-35	–	–
Cefoxitin	30 µg	23-29	23-2	–	–
Cephalothin	30 µg	15-21	29-37	–	26-32
Chloramphe- nicol 30µg	21–27 µg	19-26	–	26-32	–
Clindamycin	2 µg	–	24-3	–	19-25
Erythromycin	15 µg	–	22-30	–	25-30
Gentamicin	10 µg	19-26	19-27	16-21	–
Imipenem	10 µg	26-32	–	20-28	–
Kanamycin	30 µg	17-15	19-26	–	–
Oxacillin	1 µg	–	18-24	–	<12
Penicillin	10 units	–	26-37	–	24-30
Rifampin	5 µg	08-10	26-34	–	25-30
Tetracycline	30 µg	18-25	24-30	–	27-31
Ticarcillin	75 µg	24-30	–	21-27	–
Ticarcillin- Clavulanic acid	75/10 µg	24-30	29-37	20-28	–
Spectinomycin	100 µg	21-25	13-17	10-14	–
Sulfisoxazol	250 µg or 300 µg	15-23	–	–	–
Trimethoprim- Sulfamethoxa- zole	1.25/ 23.75 µg	23-29	24-34	–	20-28
Vancomycin	30 µg		17-21	–	20-27

(-): no established range.

Procedure

(i) **Sterilize materials as under: a)** In autoclave at 121°C (1.5 atm) for 20 minutes: culture media fractionated in test tubes; empty haemolysis tubes; haemolysis tubes with 3 mL of distilled water, 0.85% NaCl solution;

graduated pipettes (in port pipettes); open Eppendorf tubes (contained in semi-closed recipient); 0.22-µm Millipore filters (in Petri dish); port-filter (wrapped in paper and in semi-closed recipient); paper discs (in Petri dish); **b)** Petri dishes wrapped in paper at 180°C for 3 hours in an oven; **c)** Test and standard compounds by filtration through 0.22-µm filter membranes; **d)** Clean non-sterilizable materials and laminar flow cabinet with 70% ethanol, v/v.

(ii) **Preparation of Petri dishes:** Melt MHA contained in tubes in the water bath. Hold temperature of tubes at 40° to 45°C. Load the content of each tube in each Petri-dish (to achieve a thickness of 4 mm). Let dry Petri dishes in the laminar flow placing them with half-open cover. The surface of the dried medium should be smooth and should not show signs of desiccation (webbed ribbing pattern on the agar surface). Cover plates until inoculation.

(iii) **Preparation of antimicrobial discs:** Load in each paper disc the sterile test compound (filter the test solution through Millipore units and receive in a sterile tube). Let dry in laminar flow. Discs should be placed on plates so that the zones of inhibition do not overlap (Fig. 1A). Preliminary tests may be needed to determine the best number of discs and distance among discs per plate. Discs impregnated with the solvent used for dissolution of the test compound and others impregnated with standard antimicrobial compounds must be included as controls (see *Observations*).

(iv) **Microbial suspension for inoculation:**

(a) **For aerobic bacteria:** Pick up, with a sterile loop, well-isolated colonies from the bacterial subculture and suspend them into 3 mL of distilled water, 0.85% NaCl or MHB. Incubate aerobically at 37°C for 4-5 h. Read absorbance at 625 nm against non inoculated medium. Repeat this operation until obtain an absorbance of 0.08-0.1 at 625 nm (equivalent to N° 0.5 McFarland standard). This suspension contains between $1\text{-}2 \times 10^8$ CFU/mL. Dilute this suspension (1/10, with the same medium before use) and inoculate 10 µL in each plate. This aliquot contains between 1×10^5 CFU/plate. The 1/10 dilution should be used up to 15 minutes after preparation.

(b) **For anaerobic bacteria:** Proceed as indicated above but suspend bacterium colonies into a broth (i.e. MHB or BHI) and incubate at 37°C for 4-6 hours according with the species, with tight screwed cap, until obtain an absorbance of 0.08-0.1 at 625 nm (equivalent to N° 0.5 McFarland standard), dilute up to 10^6 CFU/mL. Following incubation, keep the tube at room temperature in the dark. It is necessary to use an adequate cabinet and working technique.

(c) **For yeasts:** Pick up colonies from a subculture (Read General aspects, Growth re-initiation) with a sterile loop and suspend them in 3 mL of LMM or 0.85% NaCl. To do it, vortex 20 s and add LMM or 0.85% NaCl until an 85% transmittance at 580 nm is obtained. The resulting suspension contains $1\text{-}5 \times 10^6$ CFU/mL.

(d) For spore-forming fungi: Spores harvested from 8 to 10 days cultures in slant tubes are suspended in 3 mL of LMM or 0.85% NaCl. To do it, vortex 20 s and add LMM or 0.85% NaCl until a 85% transmittance at 580 nm is obtained.

(v) Inoculation: Moist a sterile applicator (i.e. a cotton swabs with wooden handles) in the microbial cell suspension. Then, inoculate the entire surface of each plate in three different directions to ensure uniform, confluent growth (Fig. 1B). Use a distinct sterile applicator for each inoculated species.

(vi) Disc application: After the medium surface is dried, place discs (impregnated as mentioned above) on the plates tamping gently with a sterile loop or forceps to ensure contact between discs and the agar surface (Fig. 1C). All discs should be at the same distance from the edge of the plate and from each other (Fig. 1A). Allow plates to dry for 5 minutes. Invert the inoculated plates with the lid upside down (Fig. 2). Incubation conditions should be selected as follows:

- Aerobic bacteria: at 37°C for 16-24 h.
- Anaerobic bacteria: at the same conditions above, but in 5% CO_2.
- For fungi: at 30°C during 20-48 hours for yeast and during 48-72 hours for spore-forming fungi.

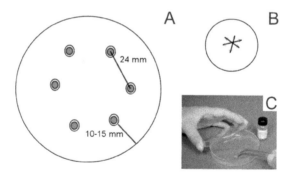

Figure 1 (A) Distribution of paper discs on a Petri dish, (B) Inoculums streak. The entire surface of the plate is inoculated in three directions to ensure uniform growth (Source: Alderman and Smith, 2001), (C) Discs application on agar surface (Source: Alderman and Smith, 2001).

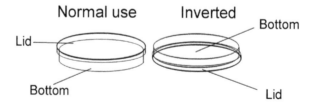

Figure 2 Normal and inverted position of a Petri dish.

Calculations

(i) Measure the inhibitory zone diameter (IZD) (mm) for each treatment in 6 distinct directions, in each Petri dish. Calculate diameter mean for each treatment expressed in mm (Figs. 3 and 4).

(ii) Calculate percentage of inhibition (% I) as follows:

$$\% \ I = [(IZD \ in \ control - IZD \ in \ sample)/IZD \ in \ control] \times 100$$

All the tests should be done in duplicate. Experiments should be twice repeated.

(iii) For quality control of the experiment, compare the measures obtained for individual antimicrobial drugs (read the preparation of antimicrobial solutions in section 5, Quality control) with the values from standard tables (Table 1). Afterward, compare the obtained values of the samples to evaluate the potency.

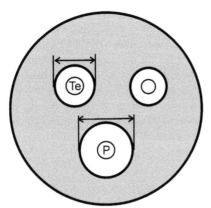

Figure 3   Zones of inhibition around paper discs. The shaded area represents the uniform growth of the strain on the plate. Zones of inhibition are represented by the white areas surrounding the discs (P and Te). Zones of inhibition are measured as indicated by the double-arrow lines.

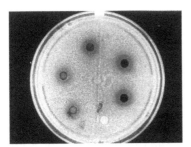

Figure 4   Effect of different amounts of *Larrea divaricata* extract on *Saccharomyces carlsbergensis* growth. Paper discs were impregnated with different amounts of *Larrea* extract, which diffuses around them. The control is observed as a white disc (Source: Quiroga *et al.*, 2001).

Observations

(i) Clean micropipette stems with 70% ethanol.

(ii) An alternative is the "E-test", which is commercially available as AB BIODISK. E-test comprises a previously defined gradient of antibiotic concentrations on a plastic strip. It is used to determine the exact minimum inhibitory concentration (MIC) of antimicrobial and antifungal agents.

Experiment 2: Inhibition of Hyphal Growth

A plug of a fungal strain is placed in the centre of an agar medium plate. After a first fungus growth, paper discs impregnated with different amounts of test compounds are placed at 0.5 cm from the mycelium border. Antifungal activity is observed as a crescent-shaped inhibitory zone at the mycelial front (Quiroga *et al.*, 2006). This method allows determining the inhibitory capacity of test compounds on mushrooms.

Materials and equipments required

Test tubes; haemolysis tubes; empty haemolysis tubes; general laboratory glassware (flasks, bottles, graduated cylinders); sterilized pipettes (1, 2, 5 and 10 mL); sterile Millipore filtration units with 0.22-μm-pore-size; paper discs (4 mm in diameter, Whatman N° 4); micropipettes (adjustable volumes up to 10 and 100 μL); sterile pipette tips; loops, forceps (sterilized to flame); sterile Petri dishes (90 mm); personal biosecurity elements (gloves, gowns, goggles or protective eye wear, chemical/biological safety hood); autoclave; horizontal laminar flow cabinet (human pathogen species requires the horizontal and vertical laminar flow); incubator oven at 30°C; boiling water bath; water bath set at 45 to 48°C; balances; vortex mixer.

Reagents

Fungal strains; SMM (dissolve 1.5 g malt extract, 0.5 g peptone and 1.8 g agar in 100 mL of distilled water, warm in water bath until total dissolution and fractionate 5 mL and 10 mL per tube); microbial suspension medium: 3 mL of fractioned sterile distilled water, 0.85% NaCl or Liquid Malt Medium, LMM (dissolve 1.5 g malt extract and 0.5 g peptone in 100 mL of distilled water, warm in water bath until total dissolution and fractionate); pure or partially purified test compound; commercially available fungicide standards; Disinfectant solution for work space (70% ethanol, v/v).

Procedure

(i) **Sterilize materials:** As indicated in Experiment 1.

(ii) **Re-initiation of mould growth:** Inoculate a SMM slant tubes with mycelium from a stored culture. Incubate at 30°C for 48-96 h. This culture can be used until 8 to 10 days after incubation, when stored at 4°C.

(iii) **Preparation and inoculation of Petri dishes:** Add 10 mL of sterile SMM to each Petri dish. A 3-mm diameter plug of a growing mycelium of a filamentous fungus harvested from a culture is placed onto the centre of the Petri plates. Incubate the Petri plates in the dark at 30°C, up to the mycelium attains 3 cm in diameter (48-72 hours).

(iv) **Preparation of antimicrobial discs:** Load in each filter paper disc a sterile test compound (filter the test compound solution, through Millipore units and take it in a sterile tube) at a given concentration and volume. Let dry in laminar flow. Discs should be placed on plates at 0.5 cm from the mycelium border. Discs impregnated with the solvent used for dissolution of the test compound and others impregnated with standard antimicrobial compounds must be included as controls.

(v) Incubate the plates at 30°C for 48-72 hours in the dark. Antifungal activity is observed as a crescent-shaped inhibitory zone at the mycelial front when compared with the solvent controls (Fig. 5).

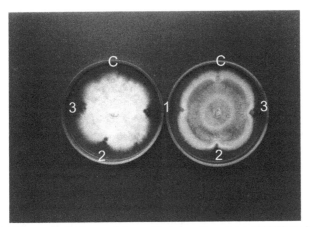

Figure 5   Inhibitory effect of a partially purified propolis extract (PPPE) on hyphal radial extent. C is the control disc (containing ethanol); 1, 2 and 3 are discs containing increased concentrations of extract. *Schizophyllum commune* (left) and *Pycnoporus sanguineus* (right) are the assayed fungi (Source: Quiroga *et al.*, 2006).

Observations

(i) The tubes containing SMM must be slanted after sterilization.

(ii) Experiments should be twice repeated and should be done in duplicate.

(iii) This method may be used to obtain an idea of the MIC's values.

Experiment 3: Direct Bioautography

Direct bioautography allows *in situ* localization of antimicrobial activity on a TLC (Thin Layer Chromatography) plate. Components of a sample (a crude extract, a partially purified or pure test compound) are separated by TLC. After evaporation of the mobile phase, the plate is placed directly in contact with a culture medium inoculated with a bacterial or a fungus strain. Antimicrobial compounds are detected as "zones of inhibition" where microbial growth is absent. This method is widely used for screening of antimicrobial activity in bacteria, yeast and fungi form spores, and bio-guided purification (Chomnawang *et al.*, 2005; Quiroga *et al.*, 2005).

Materials and equipments required

Test tubes; haemolysis tubes; empty haemolysis tubes; general laboratory glassware (flasks, bottles, graduated cylinders); sterile Petri dishes (90 mm) with three divisions; silica gel GF-254 (Merck, Darmstadt, Germany) plates (6 × 8 cm); cuvettes for chromatography; solvents for chromatography. UV lamps (366 and 254 nm); micropipettes (adjustable volumes up to 10 and 100 µL); sterile Millipore filtration units with 0.22-µm-pore-size; sterile pipette tips; loops; forceps (sterilized to flame); autoclave; horizontal laminar flow cabinet (human pathogen species require horizontal and vertical laminar flow); incubator oven at 30°C and 37°C; boiling water bath for melting the solid media; water bath set at 45-48°C; balances; vortex mixer.

Reagents

ATCC strains (Tables 1 and 3). Medium for bacteria: ssMH (dissolve 2.2 g of commercially available MHB powder and 0.8 g of agar in 100 mL of distilled water; warm in water bath until total dissolution and fractionate 5 mL per tube); medium for yeast and spore-forming fungi: ssMM (dissolve 1.5 g malt extract, 0.5 g peptone and 0.8 g agar in 100 mL of distilled water; warm in water bath until total dissolution and fractionate 5 mL per tube); pure or partially purified test compounds; 3-{4,5 -dimethylthiazol-2-yl}-2,5-diphenyltetrazolium bromide, MTT (dissolve 0.8 g of MTT in 1 L of distilled water, sterilize with 0.22 µm filter membrane); commercially available antimicrobial drugs; Disinfectant solution for work space (70% ethanol, v/v); sterile distilled water.

Table 3 Appropriate controls, depending on genera, which must be included with every batch of MIC determinations (Source: Andrews, 2001)

Organism	ATCC control strain	NCTC control strain
Escherichia coli	25922 (NCTC 12241)	10418
Staphylococcus aureus	25923 (NCTC 12981)	6571
Pseudomonas aeruginosa	27853 (NCTC 12934)	10662
Enterococcus faecalis	29212 (NCTC 12697)	
Haemophilus influenzae	49247 (NCTC 12699)	11931
S. pneumoniae	49619 (NCTC 12977)	
N. gonorrhoeae	49226 (NCTC 12700)	
B. fragilis		9343

Procedure

(i) **Sterilize materials:** As indicated in Experiment 1.

(ii) **Activation of microbial strains:** For spore-forming fungi, put 3-mm diameter plug of mycelium from a stored culture. Incubate at 30°C for 24-120 h, according to the species. For yeasts and bacteria, see *Procedure* in Experiment 1.

(iii) **TLC development:** Spot the test compound at the bottom of a TLC plate (see *Observations*). Develop the TLC plate with a suitable mobile phase.

Visualize the test compound under UV light (366 and 254 nm). Calculate and record the *Rf* of each spot. Run the plates in duplicate, one set is used as reference and the other is used for bioautography. Allow total evaporation of solvents in laminar flow.

(iv) Prepare the microbial suspension according to Experiment 1, section 4.

(v) Dilute the suspension (1/10) in Eppendorf tubes (Load 100 µL of bacterial suspension and 900 µL of dilution medium). Put an aliquot of 500 µL of this suspension in a tube with 5 mL of semi solid-medium (10^5 CFU or spore/mL), blend and immediately distribute it on the plate. Let dry in laminar flow.

(vi) Put the plate in a sterile Petri dish with three divisions; on bottom add a little amount of sterile distilled water to prevent the medium dryness. Incubate bacteria at 37°C or fungi at 30°C, during 24-96 h.

(vii) Antimicrobial activity can be visually observed or after spraying the plate with a MTT solution. Spray the plate with MTT solution and incubate at 30 or 37°C for 1 hour (Gy Horváth1 *et al.*, 2002). The tetrazolium salt solution is converted by the dehydrogenases of living microorganisms to intensely coloured, formazan. Bacterial growth inhibition was observed as pale zones (Yellow) on a coloured (blue) background of formazan.

Reading and Interpretation

The areas of inhibition should be compared with the *Rf* of the related spots on the reference TLC plate. Therefore, the spots associated to the antimicrobial activity are identified (Fig. 6).

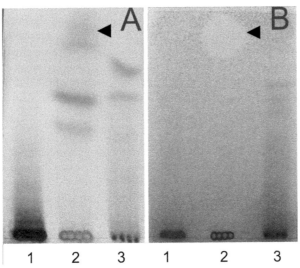

Figure 6 (A) TLC fingerprints of a crude plant extract and (B) their corresponding bioautographies against a bacterial strain. Arrows indicate an active reaction observed as a bare area in line 2 of bioautographies.

Observations

(i) The test compound in general is dissolved in organic solvents. When aqueous solvents are used, they should be sterilized by filtration.

(ii) Silica gel plates should be sterilized by immersion in 70% ethanol and let to dry in laminar flow. TLC spotting and development should be made in laminar flow too.

(iii) The bioautographic plates should be observed each day since its incubation before the addition of MTT.

4. DILUTION METHODS

Experiment 4: Macrodilution Assay

The test compound is diluted in agar contained in Petri dishes. Several bacterial or yeast strains are inoculated in punctual sowing on the surface of the agar. Bacterial growth is evaluated after incubation overnight. The MIC is interpreted as the lowest concentration of compound in agar medium, in which the growth is not observed. The method allows simultaneous estimation of MIC's for several microbial strains but it is not as accurate as the *Microdilution Assay*.

Materials and equipments required

Petri dishes (90 mm); sterile glass tubes; haemolysis tubes; empty haemolysis tubes; test tube racks; general laboratory glassware (flasks, bottles, gradated cylinders); Pipettes (1, 2, 5 and 10 mL); micropipettes (adjustable volumes up to 10 and 100 μL); pipette tips; loops, forceps; cryogenic tubes for freezing down the working cultures; sterile Millipore filtration units with 0.22-μm-pore-size; personal biosecurity elements (gloves, gowns, goggles or protective eye wear, Chemical/biological safety hood); biohazard bags; protective covers in work space; autoclave; horizontal laminar flow cabinet (human pathogen species requires the horizontal and vertical laminar flow); oven incubator at 30 and 37° C; water bath set at 43-48°C to maintain temperature of melted agar; boiling water bath; balances; spectrophotometer; refrigerator/freezer or liquid nitrogen tank; vortex mixer.

Reagents

Pure or partially purified test compounds; commercially available bactericide and fungicide standards; ATCC bacterial and fungal strains (for bacteria species, see Table 3); Culture Media: for bacteria use MHA (dissolve 3.7 g of commercially available MH agar powder in 100 mL of distilled water; warm in water bath until total dissolution and fractionate 9 mL per tube), for fungi use SMM (dissolve 1.5 g malt extract, 0.5 g peptone and 1.8 g agar in 100 mL of water, warm in water bath until total dissolution and fractionate 9 mL per tube); Media to prepare the inoculum suspensions: 3 mL sterile distilled water, 0.85% NaCl or MHB for bacteria (dissolve 2.2 g of commercially available MHB powder in 100 mL; warm in water bath until total dissolution and fractionate); 3 mL of LMM for yeast (dissolve 1.5 g malt extract and 0.5 g peptone in 100 mL of distilled water, warm in water bath until total dissolution and fractionate); commercially available antimicrobial drugs; Disinfectant solution for work space (70% ethanol, v/v).

Procedure

Chronological Scheme

Day 1	Day 2	Day 3
Sterilize materials	Prepare Petri plates	Reading
Bacterial Activation	Prepare Inoculum	final-point
(37ºC, 24 h)	Assay	(MIC determination)
	Incubation (37ºC, 20-24 h)	

(i) **Sterilize materials:** As indicated in Experiment 1.

(ii) **Activate growth of microbial strains:** As indicated in *Growth re-initiation, General Aspects*, section 2.

(iii) **Preparation of Petri dishes**: Sterilize the test compound solutions by filtration through millipore units with 0.22 μm-pore-size; receive in a sterile tube. Prepare and label dilution series of this solution (i.e. 1/2, 1/4, 1/8, 1/16, 1/32, 1/64) under sterile conditions. Melt the culture media. Hold temperature of tubes at 48-50°C. Add 1 mL of corresponding dilution, blend and immediately distribute it in each Petri dish. Let dry dishes in laminar flow placing them with half-open cover (Fig. 2). Prepare two Petri dishes per dilution. Inoculate Petri dishes containing only medium and medium plus an antibiotic solution. These will be the negative and positive controls, respectively.

(iv) **Preparation of microbial working suspension:**

    (a) Proceed as indicated in Experiment 1, *Procedure*, section 4 (Microbial suspension for inoculation) for preparation of working stock suspension.

    (b) Dilute the working stock suspension (1/10) in the same medium employed in the previous preparation. Use Eppendorf tubes (Load 50 μL of bacterial suspension and 450 μL of dilution medium). This bacterial suspension contains 5×10^6 CFU/mL, it is called working suspension and should be used up to 15 minutes after the preparation.

(v) **Sowing Inoculum:**

    (a) The surface of the medium must be dry before the punctual sowing is applied.

    (b) On the base (external) of each Petri plate identify the bacterial species writing a number or letter.

    (c) With a micropipette measure 2 μL (10^5 CFU/mL) of the working suspension and carry out a punctual sowing on the number or letter of the respective species.

    (d) Effect the sowing in all the Petri plates, including the viability and positive controls.

(e) The "ghost" control (GhC) is sown after the working stock cultures is inactivated, heating them at 100°C, for 5 minutes, seeding them on a Petri dish with medium and without other substances (see *Observations*).

(f) Inspect the inoculated plates to ensure that there is no visible liquid on the surface of the medium.

(vi) **Incubation**: Invert the inoculated plates (Fig. 2), and incubate the plates. Incubation conditions should be selected as follows:

- Aerobic bacteria: at 37°C for 16-24 h.
- Anaerobic bacteria: at 37°C for 20-24°C, but in 5% CO_2.
- For fungi: at 30°C during 20-24 hours for yeast.

Reading and Interpretation

Microbial growth is compared with viability (VC) and positive [C (+), with addition of antimicrobial drug] controls. It is hope to observe the growth of all species in the VC Petri plates and to observe no growth in the C (+) Petri plate controls (Fig. 7). The breakpoint or MIC for a microorganism given, correspond to plate that contain the highest dilution of antimicrobial in which there is no sign of growth of this microorganism.

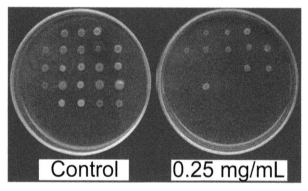

Figure 7 Inhibitory activity of methanolic extract of *Ligaria cuneifolia*. The left plate is a viability control (Control) and right plate is a dilution plate (Source: Soberón *et al.*, 2006).

Observations

(i) The selection of antibiotic substances should be done according to tested species.

(ii) The preparation of GhC-plate is very important in order to carry out the reading. This confirms the inhibition (+) and it leaves aside the accidental inactivation of the working suspension.

(iii) Microbial viability control (VC) demonstrates good growth of the inoculated species and that the culture is free of contaminating organisms.

(iv) Positive control C (+) should be free of microbial growth.

(v) For inoculum size validation, read observations (ii) in Experiment 5 (Microdilution Assay).

Experiment 5: Microdilution Assay

Dilutions of a test compound are placed in contact with a microorganism in a liquid culture medium in established conditions of temperature and time. After incubation for 16-24 h, the lowest concentration of sample resulting in no growth is registered as the MIC value. The selection of the assayed concentrations is based on the MIC's values obtained by the *Macrodilution Assay* (Experiment 4).

Materials and equipments required

The materials and equipment mentioned in Experiment 4: *Macrodilution Assay*. Sterile 96-well U-bottom microtitre plates (Fig. 8A), with cover.

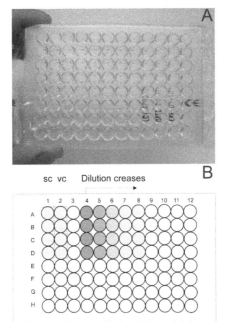

Figure 8 (A) A 96 well plate used for microdilution assay and (B) a scheme of a 96 well plate showing the sterility control (SC) at column 1, the viability control (VC) at column 2 and dilutions of a tested compound from columns 4 to 9.

Reagents

MHB (weigh 2.2 g. MHB powder, commercially available, take up to 100 mL, warm in water bath until total dissolution and fractionate 10 mL per tube) is the recommended medium by the NCCLS for sensibility testing of bacteria; LMM (weigh 1.5 g malt extract and 0.5 g peptone, take up all to 100 mL with distilled water, warm in water bath until total dissolution and fractionate 10 mL per tube) is recommended for yeasts and spore-forming fungi; 10 mg Ca^{++}/ mL (dissolve

$CaCl_2 \cdot 2H_2O$ 3.68 g in 100 mL distilled water) and 10 mg Mg^{++}/ mL (dissolve $MgCl_2 \cdot 6H_2O$ 8.36 g in 100 mL distilled water) solutions. They are sterilized by filtration through a Millipore units with 0.22-μm-pore-size; commercially available antimicrobial drugs; Disinfectant solution for work space (70% ethanol, v/v).

Procedure

Chronological Scheme

Day 1	Day 2	Day 3	Day 4

• Sterilize material	• Prepare Dilutions
	• Working Culture
	  Preparation
		• Reading of
		  final-point
		  (MIC's det.)
			• Reading of
			  final-point

• Microbial
  Activation
  (T°, 24 h)
	• Mix preparation
	• Assay
	• Incubation
		• **MBC** assay	(MBC det.)

• Incubation
  (T°, 16-24 h)
	• Incubation
	  (T°, 16-24 h)		

(i) **Sterilize materials:** As indicated in Experiment 1.

(ii) **Activate growth of microbial strains:** As indicated in *Growth re-initiation, General Aspects*, section 2.

(iii) **Dilution Preparations:** Sterilize the test compound solutions by filtration, receive in a sterile tube.

(iv) **Culture Medium Preparation:** The media must be prepared according with manufacturer instructions, control the pH at 7.2-7.4, sterilize in autoclave and leave at 4°C overnight. Supplement the culture medium with $CaCl_2$ and $MgCl_2$ solutions (1 μL of each solution in 10 mL MHB). The supplementation is done immediately after the medium is taking out from the refrigerator. This MHB medium, supplemented with cations is called CAMHB (Cations Adjusted MHB). The LMM is not supplemented. Before adding the inoculum to the bacterial or fungal culture medium, put aliquots of 50 μL of this medium in a column of wells (sterile control, Fig. 8B).

(v) **Working Suspension Preparation:**

(a) Label the tubes with the name or number of the respective species.

(b) Proceed as indicated in Experiment 1, *Procedure,* section 4 (Microbial suspension for inoculation) for preparation of working stock suspension.

(c) Take 50 μL of the working stock suspension (0.5 Mc Farland suspension, 10^8 CFU or spore/mL) and dilute in a tube containing 10 mL of CAMHB or LMM (dilution 1/20; 5×10^6 CFU or spore/mL). This is called *working suspension* and should be used up to 15 minutes after the preparation.

(vi) **Inoculation:**

(a) Load 50 μL of the *working suspension* in each well, therefore, the concentration in each well is 2.5×10^5 CFU or Spore/mL, since the final volume in each well is 100 μL.

(b) Load the sample (50 μL) in the respective well of the microtitre plate (Fig. 8B). Make the controls: Viability (add distilled water), Solvent (add the solvent in which the sample was dissolved).

(c) Carefully mix the microtitre plate.

(d) Seal each microtitre plate with the plastic covers or adhesive film to avoid dryness.

(vii) **Incubation:** Incubate at 37°C for 16-24 hours (in the case of most bacterial species). For *Haemophylus* spp., *St. pneumoniae* and *N. gonorrhoeae*, incubate between 20 and 24 hours. Incubate yeasts and spore-forming fungi at 30°C for 24-72 hours.

Reading of final point and interpretation

MIC is the lowest concentration of sample that completely inhibits the microbial growth of a particular test organism. MIC's are determined by plate visual examination after incubation for 16-24 hours for bacteria and 24-96 hours for fungi. Each well should be compared with the viability control well as shown in Table 2.

Table 2 A possible result of a microdilution assay. The MIC observed is 1/8

	SC	VC	1/2	1/4	1/8	1/16	1/32	1/64
	1	2	3	4	5	6	7	8
A	−	+	−	−	−	+	+	+
B	−	+	−	−	−	+	+	+
C	−	+	−	−	−	+	+	+
D	−	+	−	−	−	+	+	+

(+): Growth; (−): no growth.

Observations

(i) Open the trays in laminar flow, since these are inoculum for MBC (minimum bactericidal concentration) determinations.

(ii) Make CFU recount from the viability control well. This is done to confirm the final concentration of loaded inoculums:

(a) Take aliquots of 10 μL from the viability control wells immediately after the inoculation was made, and dilute it in 10 mL of broth (1:1000) and mix.

(b) Take aliquots of 100 μL of each well and distribute on agar Petri plates.

(c) Incubate over night at 37°C (bacteria), 30°C (yeasts).

(d) Count the CFU per plate. If 50 CFU/plate are observed, then the final concentration of inoculum per well is 5×10^5 CFU/mL.

(iii) Advantages of microscale method:

(a) It needs low amounts of the sample and materials to carry out the test.

(b) It is possible to test simultaneously several combinations of sample-microorganism.

Experiment 6: Minimum Bactericidal Concentration

The minimum bactericidal concentration (MBC) is the lowest concentration of a test compound needed to kill 99.9% of the original microbial inoculum in a given time. This method is possible to carry out for bacteria and yeasts.

Materials and equipments required

Test tubes; Micropipettes (adjustable volumes up to 10 and 100 µL); sterile Pipette tips; Loops; forceps (sterilized to flame); sterile Petri dishes (90 mm in diameter); autoclave; Horizontal Laminar flow cabinet (For working with Human Pathogen species should be used the Horizontal and Vertical laminar flow); oven incubator at 30°C; water bath set at 45-48°C to maintain temperature of melted agar; Boiling water bath to melt the solid media; Balances; Vortex mixer.

Reagents

Culture medium for bacteria: MHA (weigh 3.7 g of culture medium powder, commercially available, take up to 100 mL with distilled water, warm in water bath until total dissolution and fractionate in 10 mL per tube); Culture medium for fungi: SMM (weigh 1.5 g malt extract and dissolve in water, 0.5 g peptone and 1.8 g agar, take up all to 100 mL with distilled water, warm in water bath until total dissolution and fractionate in 10 mL per tube); Disinfectant solution for work space (70% ethanol, v/v) .

Procedure

  (i) **Sterilize materials:** as indicated in Experiment 1.
 (ii) **Petri dish preparation:** Melt the MHA or SMM culture media contained in tubes (10 mL per tube). Add the melted media in Petri dishes, distribute and let dry them in laminar flow with the half-open cover, then, close until the inoculation.
(iii) The necessary amount of Petri dishes is:

$$N° \text{ dilution} \times 2 + 2 \text{ (VC)}$$

i.e. 4 dilutions are calculated as under:

$$(4 \times 2) + 2 = 8 + 2 = 10$$

therefore, 10 Petri plates are needed.
 (iv) Label the Petri plates with the corresponding dilution.
  (v) **Inoculation:** Extract aliquots of the well mix that have the same or higher sample concentration than the MIC, where the bacterial growth was not observed (Table 2). Take aliquots of 10 µL from wells of the lines A and C (for example) of the microtitre plate; put it on the agar surface dispersing with a sterile loop. Let the plates open in laminar flow at room temperature until the agar absorbs the inoculum.
 (vi) **Incubation:** Invert the inoculated plates (Fig. 2) and incubate at 37°C for 16 to 24 hours (for bacteria) and 30°C for 24 to 72 hours (for yeasts and spore-forming fungi).

Reading and Interpretation: Observe if CFU are found in the plates.

(i) Presence of CFU shows non-bactericidal activity of the plant extract on microbial growth. Absence of CFU shows bactericidal activity.

(ii) The sample affects the microbial viability (it kills microorganisms) and the lowest concentration that shows this effect is the MBC (of a given sample). i.e., original inoculum was 10^5 CFU/mL, sample of 100 µL. A dilution with 10 or less CFU on plate is the CHB.

Experiment 7: Hyphal Radial Growth Inhibition

When a mould species is cultured in a medium containing a test compound (extracts, purified extracts, pure compounds or commercial drugs), hyphal length may be affected (Fig. 9). This growth inhibition is interpreted as antifungal activity (Quiroga *et al.*, 2006).

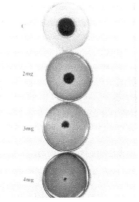

Figure 9 Hyphal Radial Growth Inhibition Test. The photo shows the control plate (C) and plates with different amounts of incorporated test sample. Increasing concentrations of test sample in the plates, produce higher inhibition (Source: Quiroga *et al.*, 2001).

Materials and equipments required

Millipore units with 0.22 µm-pore-size; Test tubes, Micropipettes (adjustable volumes up to 10 and 100 µL); Sterile Pipette tips; Loops, forceps (sterilized to flame); sterile Petri dishes (60 × 15 mm); Autoclave; Horizontal Laminar flow cabinet (For working with Human Pathogen species should be used the Horizontal and Vertical laminar flow); Oven incubator at 30°C; Boiling water bath for melting the solid media; Water bath set at 45-48°C to maintain temperature of melted agar; Balances, Vortex mixer.

Reagents

Fungal strains; Culture medium for fungi: SMM (dissolve 1.5 g malt extract and 0.5 g peptone in distilled water, add 1.8 g agar and complete to 100 mL with distilled water. Warm in water bath until total dissolution and divide properly); Sterile test compounds and pure compounds or antimicrobial chemicals available commercially; Disinfectant solution for work space (70% ethanol).

Procedure

  (i) **Sterilize materials:** as indicated in Experiment 1.
 (ii) **Activate growth of fungal strains:** Prepare the fungal cultures in SMM slant tubes, sowing a growing mycelium (See *Growth Re-initiation, General Aspects*, section 2).
(iii) **Petri dish preparation:** Add 5 mL of sterile SMM in each Petri plates (60 × 15 mm) plus increasing concentrations of sample (extracts, purified compound or commercial drug). This should be sterilized by filtration (using Millipore units with 0.22-μm-pore-size). Make the viability and positive controls. In the first, a plate without addition of sample and in the last, adding a given concentration of a commercial drug. For the solvent control, a plate is added with the solvent in which the sample was dissolved.
 (iv) **Working suspension preparation:** Proceed as indicated in Experiment 1, *Procedure*, section 4 (Microbial suspension for inoculation) for preparation of working stock suspension, this contains between 1×10^7 and 1×10^8 spore/mL.
  (v) **Inoculation:** Put onto the centre of Petri plates a 3-mm diameter plug of a growing mycelium of a filamentous fungus harvested from an 8- to 10-day culture or for spore-forming fungi, a 10 μL of spore suspensions (2.5×10^4 spore/mL).
 (vi) **Incubation:** Incubate the plates at 30°C for 4 to 5 days in a moist chamber.

Calculations

The percentage of growth inhibition is calculated (Reyes Chilpa *et al.*, 1997):

$$\% \ I = MGc - MGx/MGc \times 100$$

where, % I: Percentage Inhibition; MGc: mycelial growth in control; MGs: mycelial growth in sample.

Observations

  (i) This method is suitable for filamentous and spore-forming fungi.
 (ii) The experiment should be done per duplicate.
(iii) This method may be used for obtaining approximate data about the MIC's values.

5. QUALITY CONTROL

Quality Control (QC) is an essential part of any procedure. It consists in a series of steps conducted to ensure accurancy and reproducibility of a microbial test. The QC should be conducted before testing the samples, to adjust the test technique, taking as reference the MIC's values of tables.

Use QC strains (ATCC) according with Table 3 (For fungi search referent strains in NCCLS M38-P document, 2002). Maintain permanent stock cultures at −20°C on ssBHI (bacteria) and at 4°C on SMM slants (moulds).

5.1 Prepare Working Stock Culture

(i) Pick up a small portion of the frozen culture surface of MHA plates or SMM slants with a sterile loop and streak on the surface of a fresh culture medium.

(ii) Weekly working stock cultures are prepared by subculturing three to five colonies, with similar morphology, from the previous week's working stock plate to a fresh culture plate. Moulds are subcultured to a fresh SMM slant.

(iii) Subculture from working stock is plated to another MHA plate or SMM slant a day before use. The procedure is equal to the described experiments (1 and 5) for compounds of unknown antimicrobial activity.

5.2 Preparation of Antibiotic Stock Solutions (Andrews, 2001)

(i) Choose a suitable range of antibiotic concentrations for the organisms to be tested (see suggested ranges in Table 4).

(ii) Prepare stock solutions using the formula:

$$W = (1000/P) \times V \times C$$

where, P = potency given by the manufacturer ($\mu g/mg$), V = volume required (mL), C = final concentration of solution (multiples of 1000) (mg/L), and W = weight of antibiotic in mg to be dissolved in volume V (mL).

i.e. $\qquad (1000/980) \times 20 \times 10 = 204.08$ mg

Dissolve the powder (204.08 mg at a potency of 980 $\mu g/mg$) in 20 mL of solvent = 10,000 mg/L stock solution. Microbial contamination of powder is extremely rare. If broth methods are used, stock solution may be filter sterilized (Millipore units with 0.22 μm-pore-size, Germany); however, it must be ascertained from the manufacturer that the antibiotic does not bind to the surface of the filter. For preparation of further stock solutions, from the initial 10,000 mg/l solution, prepare the following:

1 mL of 10,000 mg/L solution + 9 mL diluent = 1000 mg/L
100 μL of 10,000 mg/L solution + 9.9 mL diluent = 100 mg/L

5.3 Preparation of Antibiotic Dilution Range

Example of dilution range: 0.25-128 mg/L

Label 1 L universal containers (containers and amounts of antibiotic and agar can be varied depending on the number of plates to be poured) as follows: 128, 64, 32, 16, 8, 4, 2, 1, 0.5, 0.25 and 0 mg/L.

From the 10,000 mg/L stock, dispense the following amounts with a micropipette:

256 μL into the container labelled 128

Table 4 Suggested ranges for MIC determination (mg/L) (Source: Andrews, 2001)

Antibiotic	Enterobacteraceae	Pseudomonas spp.	Haemophilus spp.	Neisseria spp.	B. fragilis	Staphilococci	Haemolytic streptococci	Enterococci	P. neumococci
Amikacin	0.03–128	0.06–128	0.12–16	0.5–16	–	0.008–128	1–128	1–128	1–128
Amoxycillin	0.25–128	–	0.06–128	0.004–32	1–128	0.03–128	0.008–0.12	0.12–128	0.008–4
Ampicillin	0.25–128	–	0.06–128	0.004–32	1–128	0.03–128	0.008–0.12	0.12–128	0.008–4
Azithromycin	0.25–128	–	–	–	–	–	–	–	–
Azlocillin	0.25–128	0.5–512	0.03–2	0.004–8	1–16	0.06–128	–	–	–
Aztreonam	0.004–128	0.5–128	0.015–2	0.015–2	8–128	> 128	–	–	–
Cefaclor	–	–	0.5–128	–	–	–	–	–	0.25–64
Cefixime	0.03–128	–	0.006–0.12	0.002–1	8–128	4 – 64	0.03–0.5	8–128	0.12–16
Cefotaxime	0.004–128	0.5–128	0.004–0.5	0.004–0.5	0.5–128	0.5–128	–	–	–
Cefoxitin	0.5–128	–	1–8	0.06–8	2–128	1–32	–	–	–
Cefpirome	0.008–32	0.25–128	0.008–0.5	0,001–0.12	4–128	0.06–128	0.004–0.12	1–128	0.008–1
Cefpodoxime	0.06–128	0.25–128	0.06–0.5	0.002–0.06	8–128	1–128	0.015–0.12	1–128	0.03–4
Ceftazidime	0.004–128	0.25–128	0.015–0.5	0.004–0.5	4–128	2–128	0.03–1	0.12–128	0.03–32
Ceftizoxime	0.004–128	–	0.008–0.25	0.004–0.015	0.5–128	1–128	–	–	–
Ceftriaxone	0.001–128	0.5–128	0.001–0.06	0.001–0.06	2–128	0.25–128	0.008–0.12	0.004–128	0.004–16
Cefuroxime	0.03–128	–	0.25–16	0.008–1	1–128	0.25–64	0.008–0.12	2–128	0.015–8
Cephalexin	0.25–128	–	1–128	–	4–128	0.5–128	–	–	–
Cephradine	0.25–128	–	1–128	–	1–128	0.25–128	–	–	–
Chloramphenicol	0.25–128	–	0.06–128	0.06–8	1–8	2– 16	1–16	1–128	1–16
Ciprofloxacin	0.004–128	0.015–128	0.002–0.06	0.001–0.12	2– 8	0.06–128	0.12–4	0.25–128	0.25–128
Clarithromycin	–	–	1–32	0.015–1	0.03–2	0.03–128	0.015–16	0.03–128	0.03–128
Clindamycin	–	–	–	–	0.015–2	0.03–8	–	–	–
Co-amoxyclav	0.5–28	–	0.03–128	0.004–32	0.5–128	0.008–16	0.008–0.12	0.12–16	0.008–4

Antibiotic									
Colistin	0.5–128	0.5–64	—	—	—	0.25–16	—	—	—
Doxycycline	—	—	0.03–128	0.25–16	—	0.06–128	—	—	—
Erythromycine	—	—	0.25–128	0.03–0.5	0.25–128	0.03–128	0.06–8	0.25–128	0.06–128
Fusidic acid	—	—	—	—	—	—	—	—	—
Gatifloxacin	—	—	—	0.001–0.12	—	—	—	—	—
Gemifloxacin	—	—	—	0.001–0.12	—	—	—	—	—
Gentamicin	0.03–128	0.06–128	0.12–16	0.5–16	—	0.008–128	—	0.5–2048	—
Grepafloxacin	—	—	0.002–0.06	0.001–0.12	0.015–4	—	—	—	0.002–0.25
Imipenem	0.06–4	0.06–16	0.25–4	0.004–0.25	—	0.03–128	0.002–0.25	0.25–128	—
Levofloxacin	—	—	—	0.001–0.12	1–4	—	—	—	0.5–32
Linezolid	0.03–128	0.12–16	0.007–1	—	—	0.12–8	0.25–8	0.25–8	0.5–8
Mecillinam	0.03–128	—	—	—	—	—	—	—	—
Meropenem	0.015–4	0.015–16	0.015–0.25	0.002–0.03	0.03–32	0.015–128	0.002–0.06	0.004–128	0.002–16
Methicillin	—	—	—	—	0.06–32	0.12–128	—	—	—
Metronidazole	—	—	—	—	1–128	—	—	—	—
Mezlocillin	0.25–128	0.5–512	—	—	—	0.12–128	—	—	—
Moxifloxacin	—	—	—	0.001–0.12	—	—	—	—	—
Mupirocin	—	—	—	—	—	0.06–1024	—	—	—
Nalidixic acid	1–128	32–128	—	0.5–8	32–64	16–128	—	—	—
Netilmicin	0.03–128	0.06–128	0.12–16	0.5–16	—	0.008–128	—	—	1–128
Ofloxacin	0.06–128	0.25–8	0.015–2	0.001–0.06	1–8	0.12–128	—	—	—
Oxacillin	—	—	—	—	—	0.12–128	—	—	—
Penicillin	—	—	—	0.004–32	4–128	0.015–128	0.004–0.06	0.5–128	0.5–128
Piperacillin	0.25–128	0.5–512	0.004–128	0.015–32	0.25–128	0.25–128	—	—	—
Quinupristin/Dalfopristin	—	—	—	—	—	—	—	—	0.12–32
Rifampicin	—	—	0.004–128	0.25–2	4–32	0.12–16	0.12–1	0.25–8	—
Roxithromycin	—	—	2–32	0.015–2	0.12–16	0.03–128	0.015–16	0.03–128	0.03–128

Sparfloxacin	0.008–128	0.12–16	0.004–0.03	0.001–0.12	0.12–1	0.06–0.25	0.12–1	0.25–128	0.12–128
Spectinomycin	4–128	–	–	4–64	–	–	–	–	–
Sulphamethoxazole	–	–	0.5–32	0.25–8	–	–	–	–	–
Teicoplanin	–	–	–	–	–	0.06–32	–	0.5–2048	–
Telithromycin	–	–	0.25–8	0.002–0.5	0.03–8	0.03–128	0.001–0.25	0.015–4	0.004–1
Tetracycline	0.25–128	–	0.06–128	–	0.06–128	–	–	–	–
Ticarcillin	0.25–128	0.5–512	0.06–128	–	4–128	0.5–128	–	–	–
Tobramycin	0.03–128	0.06–128	–	0.5–16	–	0.008–128–	–	–	–
Trimethoprim	0.03–128	–	0.03–8	–	0.03–8	–	–	–	–
Vancomycin	–	–	–	–	–	0.06–32	0.12–1	0.12–128	0.12–1

128 µL into the container labelled 64

64 µL into the container labelled 32

32 µL into the container labelled 16

From the 1000 mg/L stock, dispense the following amounts:

160 µL into the container labelled 8

80 µL into the container labelled 4

40 µL into the container labelled 2

From the stock 100 mg/L dispense the following amounts:

200 µL into the bottle labelled 1

100 µL into the container labelled 0.5

50 µL into the container labelled 0.25

5.4 Record Results on QC Form

(i) Positive growth control should demonstrate good growth and be free of contaminating organisms.

(ii) Negative growth control should be free of growth.

(iii) Inoculum count verification plate should show several colonies.

(iv) Solvent growth control should demonstrate good growth.

(v) Test considered in control when:

(a) The MIC for the control strain should be within one or two-fold dilution of the expected MIC (Table 5a,b).

(b) Controls show appropriate growth.

Table 5a Target MICs (mg/L) for reference strains (Source: Andrews, 2001)

Antibiotic	H. influ- enzae	H. influ enzae	E. faecalis	S. pneu moniae	B. fragilis	N. gonor- rhoeae
	NCTC 11931	ATCC 49247	ATCC 29212	ATCC 49619	NCTC 9343	NCTC 49226
Amikacin	–	–	128	–	–	–
Amoxycillin	0.5	4	0.5	0.06	32	0.5
Ampicillin	–	–	1	0.06	32	–
Azithromycin	2	2	–	0.12	–	–
Azlozillin	–	–	–	–	4	–
Aztreonam	–	–	>128	–	2	–
Cefaclor	–	128	>32	2	>128	–
Cefamandole	–	–	–	–	8	–
Cefixime	0.03	0.25	–	1	64	–
Cefotaxime	–	0.25	32	0.06	4	–

contd.

Table 5a *Contd.*

Antibiotic	H. influenzae	H. influenzae	E. faecalis	S. pneumoniae	B. fragilis	N. gonorrhoeae
	NCTC 11931	ATCC 49247	ATCC 29212	ATCC 49619	NCTC 9343	NCTC 49226
Cefoxitin	–	–	–	–	4	–
Cefpirome	0.06	0.5	16	–	16	–
Cefpodoxime	0.12	0.5	>32	0.12	32	–
Ceftazidime	0.12	–	>32	–	8	–
Ceftriaxone	–	–	>32	0.06	4	–
Cefuroxime	2	16	>32	0.25	32	–
Cephadroxil	–	–	>32	–	32	–
Cephalexin	–	–	>32	–	64	–
Cephalothin	–	–	16	–	–	–
Chloramphenicol	–	–	4	4	4	–
Ciprofloxacin	0.008	0.008	1	1	2	0.004
Clarithromycin	8	4	–	0.03	0.25	0.5
Clindamycin	–	–	8	0.12	0.5	–
Co-amoxyclav	0.5	8	0.5	0.06	0.5	0.5
Co-trimoxazole	–	1	2	4	–	–
Enoxacin	–	–	–	–	1	–
Ertapenem	0.12	0.5	–	0.12	0.25	–
Erythromycin	8	8	4	0.12	1	0.5
Faropenem	–	–	–	0.06	1	–
Fleroxacin	–	–	–	–	4	–
Flucloxacillin	–	–	–	–	16	–
Fucidic acid	–	–	2	–	–	–
Gatifloxacin	–	–	–	–	0.5	–
Gemifloxacin	0.12	–	0.03	0.03	0.25	0.002
Gentamicin	–	–	8	–	128	–
Grepafloxacin	–	0.004	–	0.25	–	–
Imipenem	–	–	0.5	–	0.06	–
Levofloxacin	–	0.015	–	0.5	0.5	–
Linezolid	–	–	–	2	4	–
Loracarbef	–	128	>32	2	>128	–
Mecillinam	–	–	>128	–	>128	–
Meropenem	–	–	2	–	0.06	–
Metronidazole	–	–	–	–	0.5	–
Moxalactam	–	–	–	–	0.25	–
Moxifloxacin	0.03	0.03	0.25	0.5	–	0.004
Nalidixic acid	–	1	–	>128	64	–
Nitrofurantoin	–	–	8	–	–	–
Norfloxacin	–	–	2	–	16	–
Ofloxacin	–	–	2	–	1	–

contd.

Table 5a *Contd.*

Antibiotic	H. influenzae	H. influenzae	E. faecalis	S. pneumoniae	B. fragilis	N. gonorrhoeae
	NCTC 11931	ATCC 49247	ATCC 29212	ATCC 49619	NCTC 9343	NCTC 49226
Oxacillin	–	–	–	1	–	–
Pefloxacin	–	–	–	–	1	–
Penicillin	–	4	2	0.5	16	–
Piperacillin	–	–	2	–	2	–
Piperacillin + 4 mg/L tazobactam	–	–	2	–	–	–
Quinupristin/ dalfopristin	–	–	1	0.5	16	–
	–	–	–	–	–	–
Rifampicin	–	–	2	0.03	–	–
Roxithromycin	16	16	–	0.12	2	–
Rufloxacin	–	–	–	–	16	–
Sparfloxacin	–	0.002	–	0.25	1	–
Teicoplanin	–	–	0.25	–	–	–
Telithromycin	1	2	0.008	0.008	–	0.03
Temocillin	–	–	–	–	–	–
Tetracycline	–	16	16	0.12	0.5	–
Ticarcillin	–	–	–	–	4	–
Tobramycin	–	–	16	–	–	–
Trimethoprim	–	–	0.25	4	16	–
Trovafloxacin	0.008	0.002	0.06	0.12	0.12	–
Vancomycin	–	–	2	0.25	16	–

Table 5b Target MICs (mg/L) for reference strains (Source: Andrews, 2001)

Antibiotic	E. coli	E. coli	P. aeruginosa	P. aeruginosa	S. aureus	S. aureus
	NCTC 10418	ATCC 25922	NCTC 10662	ATCC 27853	ATCC 25923	ATCC 29213
Amikacin	0.5	1	2	2	–	2
Amoxycillin	2	4	>128	>128	0.25	–
Ampicillin	2	4	>128	>128	–	–
Azithromycin	–	–	–	–	0.12	0.12
Azlozillin	4	–	4	–	0.25	–
Aztreonam	0.03	0.25	4	2	–	>128
Carbenicillin	2	–	32	–	–	–
Cefaclor	1	2	>128	>128	–	1
Cefamandole	0.25	–	>128	>128	–	–
Cefixime	0.06	0.25	16	–	8	16
Cefotaxime	0.03	0.06	8	8	–	1

contd.

Table 5b *Contd.*

Antibiotic	E. coli NCTC 10418	E. coli ATCC 25922	P. aeruginosa NCTC 10662	P. aeruginosa ATCC 27853	S. aureus ATCC 25923	S. aureus ATCC 29213
Cefoxitin	4	–	>128	>128	–	–
Cefpirome	0.03	0.03	4	1	–	0.5
Cefpodoxime	0.25	0.25	>128	>128	4	2
Ceftazidime	0.06	0.25	1	1	–	8
Ceftriaxone	0.03	0.06	8	8	–	2
Cefuroxime	2	4	>128	>128	1	1
Cephadroxil	8	8	>128	>128	–	2
Cephalexin	4	8	>128	>128	–	4
Cephalothin	4	8	>128	>128	–	0.25
Chloramphenicol	2	4	128	–	–	2
Ciprofloxacin	0.015	0.015	0.25	0.25	0.5	0.5
Clarithromycin	–	–	–	–	0.12	0.12
Clindamycin	–	–	–	–	0.12	0.06
Co-amoxyclav	2	4	>128	128	0.12	0.25
Colistin	0.5	–	2	–	–	–
Enoxacin	0.25	–	1	–	–	–
Ertapenem	0.008	0.015	–	–	–	–
Erythromycin	–	–	–	–	0.5	0.25
Faropenem	0.25	–	>128	>128	–	–
Fleroxacin	0.06	0.12	1	–	–	–
Fosfomycin	4	–	>128	>128	–	–
Fusidic acid	>128	–	–	–	0.12	0.06
Gatifloxacin	0.015	–	1	–	–	–
Gentamicin	0.25	0.5	1	1	0.25	0.25
Imipenem	0.6	0.12	2	1	–	0.015
Kanamycin	1	–	1	–	–	–
Levofloxacin	0.03	0.03	0.5	0.5	0.25	0.25
Methicillin	–	–	>128	>128	2	2
Mupirocin	–	–	–	–	0.25	0.12
Nalidixic acid	2	4	>128	>128	128	128
Neomycin	–	–	32	–	–	–
Nitrofurantoin	4	8	–	–	–	16
Norfloxacin	0.06	0.06	1	1	–	1
Penicillin	–	–	>128	>128	0.03	0.12
Piperacillin	0.5	2	4	2	–	1
Rifampicin	16	–	–	–	0.015	0.004
Sulphonamide	16	–	>128	>128	–	–
Telithromycin	–	–	–	–	0.06	0.06
Tetracycline	1	2	–	32	–	0.5
Trimethoprim	0.12	0.25	32	–	–	0.5
Vancomycin	–	–	–	–	0.5	1

Precautions

Sink the contaminated Petri plates, tubes, tips, Eppendorf tubes and graduated pipettes into a container with 5% sodium hypochlorite; let it act for 24 h. Afterward, carry on regular cleaning processes.

SUGGESTED READINGS

Alderman, D.J. and. Smith, P. (2001). Development of draft protocols of standard reference methods for antimicrobial agent susceptibility testing of bacteria associated with fish diseases. *Aquaculture* **196**: 211-243.

Andrews, J.M. (2001). Determination of Minimum Inhibitory Concentrations. *Journal of Antimicrobial Chemotherapy* **48**: 5-16.

Chomnawang, M.T., Surassmo, S., Nukoolkarn, V.S. and Gritsanapan, W. (2005). Antimicrobial effects of Thai medicinal plants against acne-inducing bacteria. *Journal of Ethnopharmacology* **101**: 330-333.

Dey, P.M. and Harborne, J.B. (1991). Chapter 2: Assays for Antifungal Activity/ Chapter 3: Screening Methods for Antibacterial and Antiviral Agents from Higher Plants. In: *Methods in Plant Biochemistry. Volume 6: Assays for Bioactivity.* (Ed., K. Hostettmann). Academic Press, London.

Homans, A.I. and Fuchs, A. (1970). Direct bioautography on thin-layer chromatograms as a method for detecting fungitoxic substances. *Journal of Chromatography* **51**: 327–329.

Horváth, G., Kocsis, B., Botz, L., JNémeth, J. and Szabó L.G. (2002). Antibacterial activity of Thymus phenols by direct bioautography. *Acta Biologica Szegediensis* **46**: 145-146.

Kavanagh, F. (1975). Antibiotic assays principles and precautions. *Methods in Enzymology* **43**: 55–69.

Madigan, M., Martinko, J. and Parker, J. (2002). In: *Brock Biology of Microorganisms.* 11th Edition. International Edition.

Mann, C.M. and Markham, J.L. (1998). A new method for determining the minimum inhibitory concentration of essential oils. *Journal of Applied Microbiology* **84**: 538-544.

National Committee for Clinical Laboratory Standards (1997). *Specialty Collection: Susceptibility Testing.* SC21-L. M7-A4. NCCLS, Wayne, PA.

National Committee for Clinical Laboratory Standards (NCCLS) (1997). *Reference Method for Broth Dilution Antifungal Susceptibility Testing of Yeasts.* Wayne, PA: Approved standard M27-A.

National Committee for Clinical Laboratory Standards (NCCLS) (1998). *Reference Methods for Broth Dilution Antifungal Susceptibility Testing of Conidium-Forming Filamentous Fungi.* Wayne, PA: Proposed standard M38-P.

National Committee for Clinical Laboratory Standards (NCCLS) (2002). *Performance Standards for Antimicrobial Disc and Dilution Susceptibility Tests for Bacteria Isolated from Animals*; Approved Standard— Second Edition. NCCLS document M31-A2. Wayne, Pennsylvania 19087-1898, USA.

National Committee for Clinical Laboratory Standards (NCCLS) (2003). *Methods for Dilution Antimicrobial Susceptibility Tests for Bacteria that Grow Aerobically Approved Standard*—Sixth edition. Wayne, PA: NCCLS, NCCLS document N° : M07-A6.

Quiroga, E.N., Sampietro, A.R. and Vattuone, M.A. (2001). Screening antifungal activities of selected medicinal plants. *Journal of Ethnopharmacology* **74**: 89-96.

Quiroga, E.N., Sampietro, A.R. and Vattuone, M.A. (2004). In vitro fungitoxic activity of *Larrea divaricata* cav. Extracts. *Letters in Applied Microbiology* **39**: 7–12.

Quiroga, E.N., Sampietro, D.A., Soberón, J.R., Sgariglia, M.A. and Vattuone, M.A. (2006). Propolis from the northwest of Argentina as a source of antifungal principles. *Journal of Applied Microbiology* **101**: 103-110.

Reyes Chilpa, R., Quiroz Vázquez, R.I., Jiménez Estrada, M., Navarro-Ocaña, A. and Cassani-Hernández, J. (1997). Antifungal activity of selected plant secondary metabolites against *Coriola versicolor*. *Journal of Tropical Forest Products* **3**: 110-113.

Sgariglia, M.A., Soberón, J.R., Sampietro, D.A., Quiroga, E.N., Vattuone, M.A. (2005). Fungitoxic action of propolis from Argentinian Northwest. 6° *American Latin Symposium of Food Science*. Campinas, Brazil.

Sgariglia, M.A., Soberón, J.R., Quiroga, E.N., and Vattuone, M.A. (2006). Antibacterial activity of *Caesalpinia paraguariensis* extracts. *Bio Cell* **30**: 188.

Soberón, J.R., Sgariglia, M.A., Sampietro, D.A., Quiroga, E.N., and Vattuone M.A. (2006). Antibacterial activity of plant extracts from northwestern argentina. *Journal of Applied Microbiology* (In Press). Online accessible (www.blackwell-synergy.com) from December 7th, 2006.

Zgoda, J.R. and Porter, J.R. (2001). A Convenient Microdilution Method for Screening Natural Products against Bacteria and Fungi. *Pharmaceutical Biology* **39**: 221-225.

Bioassays for Antioxidant, Genotoxic, Mutagenic and Cytotoxic Activities

José R. Soberón, Melina A. Sgariglia, Diego A. Sampietro,
Emma N. Quiroga, and Marta A. Vattuone*

1. INTRODUCTION

The plant kingdom is an important source of chemical compounds employed as agrochemicals (e.g. pesticides, herbicides, fungicides, fertilizers and growth regulators), food additives (e.g. nutraceuticals, preservatives and flavour agents), pharmacological drugs (antibiotics, anti-inflammatory drugs and hormones) and cosmetics (e.g. as components of creams and lotions). Safety application of these bioactive compounds requires characterization of their toxic effects, a main topic in natural product research. Toxicological assays allow to assess cytotoxic, genotoxic, or mutagenic hazard of natural products. Bioactive compounds are usually toxic in high doses at the cell level. However, these substances often have valuable positive biological activities on living cells at lower concentrations, being cytotoxic effect a dose-dependent response. *In vivo* lethality test using shrimp larvae allows preliminary screening for cytotoxicity of organic compounds (Rahman *et al.*, 2001). Toxicity studies also involve the effects of natural products at genomic level. Genotoxicity refers to any deleterious change in the genetic material regardless of the mechanism by which the change is induced. The *Bacillus subtilis* rec-assay and the *Allium* test are *in vivo* assays used for detection of

Authors' address: *Instituto de Estudios Vegetales "Dr. A.R. Sampietro",* Facultad de Bioquímica, Química y Farmacia, Universidad Nacional de Tucumán, España 2903, 4000, San Miguel de Tucumán, Tucumán, Argentina.
Corresponding author: E-mail: instveg@unt.edu.ar

interactions between a substance and DNA of living cells. Sometimes a toxic compound induces mutations that are changes in the amount or structure of genetic material from an organism or a cell. The Ames test with *Salmonella tiphymurium* or *Escherichia coli* strains is the most used assay for detection of potential mutagens. Genotoxic and mutagenic agents could exert their toxicity through the oxidative mechanism, which involves generation of free radicals, specially the reactive oxygen species (ROS). Oxidative damage participates in the development of degenerative diseases that involve DNA damage, carcinogenesis and cellular aging (Jaiswal *et al.*, 2002). Antioxidants substances are free radical scavengers that, either as food additives or pharmaceutical supplements, seem to play an important role in the protective effect against oxidative damage, avoiding deleterious changes. The antioxidant activity of natural products is measured according to their free radical scavenger capacity. This chapter describes the simple, rapid and economical techniques commonly used for evaluation of antioxidant, genotoxic, mutagenic and cytotoxic activities in natural product research.

2. FREE RADICAL SCAVENGING ASSAYS

Free radicals are molecules with unpaired electrons. They are highly unstable and easily react with donor-electron molecules. When the donor-molecules lose electrons, they become free radicals, beginning a chain reaction (Fig. 1). Oxygen, a bi-radical molecule, readily accepts unpaired electrons arising from reactive oxygen species (ROS), including superoxide anion ($O_2^{\bullet}$), hydrogen peroxide (H_2O_2) and hydroxyl radical ($HO^{\bullet}$) until complete reduction to water (Singh *et al.*, 2004).

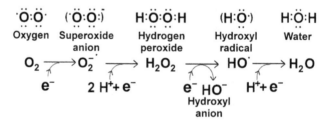

Figure 1 Reactive oxygen species.

ROS are the most prevalent, damaging free radicals *in vivo*. They are formed during the normal oxygen metabolism of living cells. Cells have an antioxidant system for ROS elimination, which involves enzymes (i.e. superoxide dismutase, catalase, and glutathione peroxidase) and natural antioxidants (i.e. glutathione, ascorbic acid, α-tocopherol, β-carotene and NADPH). However, an imbalance between free radicals production and antioxidant defenses occur when ROS overcome the protective capacity of the antioxidant cell system. This situation, known as oxidative stress, leads to significant oxidative damage of cell structures. Oxidative stress is often detected in live beings when they are exposed to a physical or a chemical environmental stress factor.

The superoxide anion is a free radical formed by addition of one electron to molecular oxygen. This oxidative species is generated as a reaction product of several enzyme systems or after a non-enzymatic electron transfer (Aruoma, 1996). In aqueous solution, superoxide can oxidize ascorbic acid or reduce iron complexes, such as cytochrome c. Hydroxyl radical is an extremely reactive oxidizing molecule with a very short half-life (Moorhouse *et al.*, 1987). It can damage purine and pyrimidine bases and deoxyribose of DNA *in vivo* (Aruoma, 1998), being a very dangerous compound for live beings. Fenton reaction (Fig. 2) is one of the main sources of hydroxyl radical *in vivo*. Hydroxyl radical cannot be eliminated by an enzymatic reaction inside the cells.

$$Fe^{+2} + H_2O_2 \longrightarrow Fe^{+3} + HO^- + HO^\bullet$$

Hydroxyl Hydroxyl
anion radical

Figure 2 Fenton reaction.

Lipid peroxidation of membranes is the most damaging effect of free radicals on living cells (Del Maestro, 1980). Lipid peroxidation is a chemical reaction between polyunsaturated fatty acids and ROS where oxygen is incorporated into the lipid molecule to form reactive hydroperoxides. These molecules easily react with cell components, such as lipids and other chemicals, generating an oxidative chain reaction. Oxidative process terminates after free radical scavenging by antioxidant cell systems. Lipid peroxidation leads to detrimental structural changes of membranes and lipoproteins.

Experiment 1: DPPH Scavenging Activity

The 1,1-diphenyl-2-picrylhydrazyl (DPPH) is a deep violet free radical. DPPH is reduced by electron transfer from a hydrogen donating substance, an antioxidant, to diphenylpycrilhydrazine which has a pale yellow colour (Fig. 3). Production of this coloured compound is spectrophotometrically measured at 514 nm (Yamaguchi *et al.*, 1998). DPPH assay is commonly used to evaluate the scavenging activity of natural products.

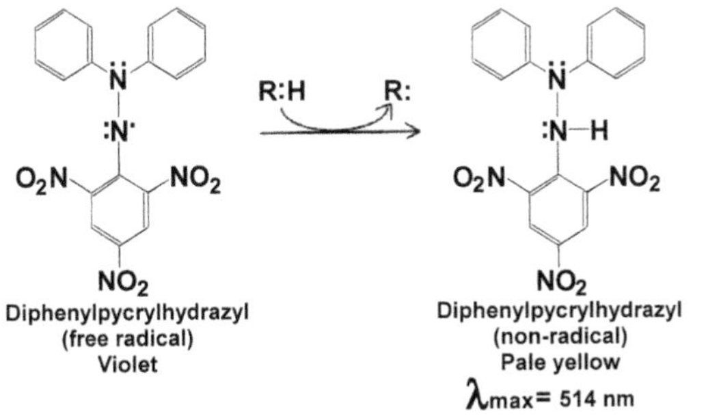

Figure 3 DPPH scavenging activity assay reaction.

Materials and equipments required

Analytical balance; glass tubes; pipettes; micropipettes; tips for micropipettes; vortex agitator; spectrophotometer; cuvettes for spectrophotometry.

Reagents

300 µM DPPH (dissolve 1.18 mg of DPPH in 5 mL of 96% ethanol, mix and make up to 10 mL with 96% ethanol (this solution can be stored in a dark glass flask at −20°C for 3 months).

Procedure

(i) Prepare labeled tubes with 96% ethanol as follows: blank (2 mL), dilution blanks (1.9 mL), reagent control (1.5 mL), and samples (1.4 mL) (see *Observations* 1 and 2).

(ii) Prepare dilutions of test compounds in 96% ethanol (see *Observations* 3 and 4).

(iii) Add 100 µL of each dilution into the corresponding dilution blank and sample tubes.

(iv) Add 0.5 mL of 300 µM DPPH into each reagent control and sample tube and swirl.

(v) Place tubes in the dark at room temperature for 20 minutes.

(vi) Measure absorbance at 514 nm against blank (see *Observation* 5).

Observations

(1) Each dilution of the test compound requires a corresponding dilution blank.

(2) Methanol can also be used as solvent instead of ethanol.

(3) Preliminary assays are needed to establish the range of dilutions where the DPPH decolouration is linear. To do it, assay a wide range of dilutions. Select the dilution where DPPH is completely converted to diphenylpycrilhydrazine (yellow pale colour). Then, prepare graded dilutions to obtain a linear response.

(4) Make at least 5 dilutions. Reference substances such as ascorbic acid (Vitamin C), α-tocopherol (Vitamin E), natural phenolic compounds such as quercetin, rutin, or hyperoside, synthetic phenolics such as butylated hydroxytoluene (BHT) or butylated hydroxyanisole (BHA) should be included as positive controls.

(5) At least three independent experiments (N=3) should be made with three replicates for each concentration of the assayed compound.

Calculations

Step 1. Correct absorbance readings at 514 nm as follows:

$$As = Asd - Adb$$

where, As is the corrected absorbance of the sample, Asd is the absorbance read for a sample dilution and Adb is the absorbance read for the corresponding dilution blank.

Step 2. Calculate mean of absorbances at each test compound dilution. The percent of inhibitory activity (IA) is calculated as:

$$IA\ (\%) = [(A_0 - As)/A_0] \times 100$$

where, A_0 is absorbance of the reagent control and As is corrected absorbance of the sample.

Step 3. Plot the percent inhibitory activity against the sample concentration, considering the final volume of 2 mL (Fig. 4). Calculate the efficient concentration of test compound needed to scavenge 50% of the available DPPH (EC_{50}). Substances with high antioxidant activity have low EC_{50} values.

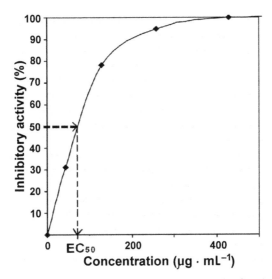

Figure 4 Percent inhibitory activity vs concentration and EC_{50} determination.

Experiment 2: Hydroxyl Radical Scavenging Activity

The deoxyribose (DR) assay is a technique for *in vitro* measure of hydroxyl radical scavenging activity. In this assay, hydroxyl radicals are generated from the Fe^{3+}/ascorbate/EDTA/H_2O_2 system (Fig. 5). The hydroxyl radical reacts with DR, forming thiobarbituric acid reactive substances (TBARS). Malondialdehyde (MDA) is the main TBARS produced. MDA forms a chromogen with 2-thiobarbituric acid (TBA), whose absorbance can be read at 532 nm.

The ability of a substance to uptake hydroxyl radicals, preventing DR oxidation, is known as **non-site-specific scavenging activity**. When this assay is performed without EDTA, unchelated iron ions are weakly associated with the DR and react with H_2O_2, releasing hydroxyl radicals (Fig. 2). The ability of a substance to chelate iron ions preventing DR oxidation is defined as **site-specific scavenging activity** (Gutteridge and Halliwell, 1988). When ascorbate is omitted from the DR reaction mixture, the ability of a compound to reduce the Fe^{3+}-EDTA complex can be tested. This is named **prooxidant activity**.

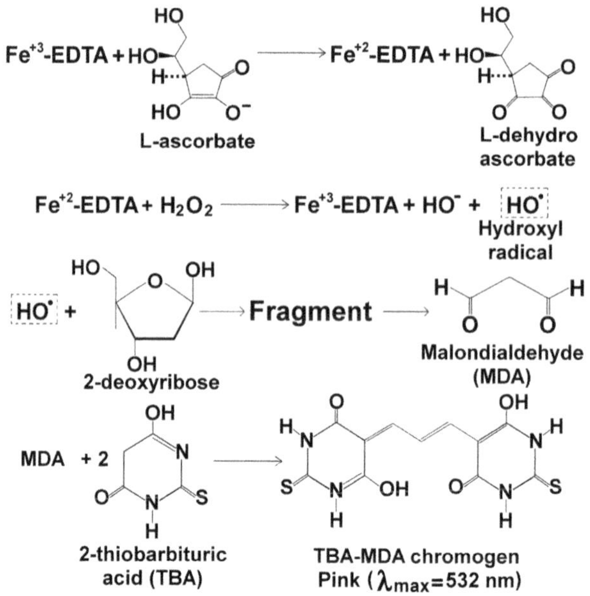

Figure 5 Deoxyribose assay reactions.

Materials and equipments required

Analytical balance; glass tubes; pipettes; micropipettes; tips for micropipettes; vortex agitator; thermostatized water bath; spectrophotometer; cuvettes for spectrophotometry.

Reagents

20 mM NaH_2PO_4 buffer (prepare 0.2 M NaH_2PO_4 dissolving 27.6 g of $NaH_2PO_4 \cdot H_2O$ in 1 L of distilled water; prepare 0.2 M NaOH dissolving 8 g NaOH in 1 L of distilled water; 20 mL of 0.2 M NaH_2PO_4 are added to 160 mL of distilled water and mixed well; adjust pH to 7.4 using 0.2 M NaOH; mix and make up to 200 mL with distilled water; dispense 100 mL aliquots in 250 mL flasks and store at 4°C). 0.1 M 2-DR solution (dissolve 134.1 mg of 2-DR in 10 mL of distilled water; dispense 1 mL aliquots in glass or Eppendorff tubes; see *Observation* 1). 4 mM $FeCl_3$ solution (dissolve 6.5 mg of $FeCl_3$ in 10 mL of distilled water; dispense 1 mL aliquots in glass or Eppendorff tubes; see *Observations* 1 and 2). 4.16 mM EDTA (dissolve 15.5 mg of EDTA disodium salt in 10 mL of distilled water; dispense 1 mL aliquots in glass or Eppendorff tubes; see *Observation* 1). 4 mM L-ascorbic acid solution (dissolve 7.04 mg of L-ascorbic acid in 10 mL of distilled water; dispense 1 mL aliquots in glass tubes). 40 mM H_2O_2 solution (dilute 40.8 µL of 30% H_2O_2 in 10 mL of distilled water; dispense in glass tubes; see *Observation* 1). 0.5% w/v 2-thiobarbituric acid (TBA) solution (dilute 12.5 mL of 0.2 M NaOH in 37.5 mL of water; add 0.5 g of TBA; mix and make up to 100 mL with distilled water; dispense in glass flask; it can be stored at 4°C for 6 months. See

Observation 4). 2.8% Trichloroacetic acid (TCA) solution (dissolve 2.8 g of TCA in 100 mL of distilled water; dispense in glass flask; see *Observation* 1).

Procedure

Non-site-specific scavenging activity
   (i) Prepare labelled tubes with 20 mM NaH_2PO_4 buffer as follows: blank (900 µL), dilution blanks (800 µL), reagent control (875 µL) and samples (775 µL) (see *Observation* 3).
  (ii) Prepare dilutions of test compounds with distilled water (see *Observations* 4, 5 and 6).
 (iii) Add 100 µL of 0.1 M DR into the reagent control and sample tubes.
 (iv) Add 100 µL of each dilution into the corresponding dilution blanks and sample tubes.
  (v) Add in the indicated order: 25 µL of 4 mM $FeCl_3$, 25 µL of 4.16 mM EDTA (see *Observation* 7) and 25 µL of 4 mM L-ascorbic acid (see *Observation* 8) to the assay tubes.
 (vi) Start the reaction by adding 25 µL of 40 mM H_2O_2 to the assay tubes and vortex.
(vii) Incubate tubes at 37°C for 1 h.
(viii) After incubation, the colorimetric reaction is performed as follows: Add 1 mL of 0.5% TBA solution and 1 mL of 2.8% w/v TCA solution and swirl tubes.
 (ix) Incubate tubes at 80°C for 30 minutes (see *Observation* 9). Cool tubes in ice.
  (x) Absorbance is measured at 532 nm against the blank (see *Observation* 10).

Observations
   (1) $FeCl_3$, EDTA and deoxyribose solutions can be stored at –20°C for 6 months. Ascorbic acid, H_2O_2 and TCA solutions can be stored at –20°C for 3 months.
   (2) As $FeCl_3$ is very hygroscopic, it could be found as a saturated solution, with 131.1 g of $FeCl_3$/100 mL of saturated solution (at 25°C). Add 5 µL of this saturated solution to 5 mL of water and proceed as indicated before.
   (3) Each dilution of test compound requires a corresponding dilution blank.
   (4) Methanol, ethanol and other organic solvents should not be used in this assay because they interfere with the reaction.
   (5) Preliminary assays are needed to establish the range of dilutions where increase of colour, due to formation of the TBA-MDA complex, is linear. Proceed as indicated in *Observation* 3 of Experiment 1.
   (6) Make at least 5 dilutions. Reference substances such as manitol, dimethyl sulphoxide, quercetin, BHT or BHA should be included as positive controls.
   (7) Add 25 µL of 20 mM NaH_2PO_4 buffer to the reaction mixture instead of 4.16 mM EDTA for **site-specific scavenging assay.**
   (8) Add 25 µL of 20 mM NaH_2PO_4 buffer instead of 4 mM L-ascorbic acid for **prooxidant activity assay.**
   (9) The temperature might not exceed 80°C.

(10) At least three independent experiments (N = 3) should be made with three replicates for each concentration assayed.

Calculations

Step 1. Correct absorbance readings at 532 nm as follows:

$$As = Asd - Adb$$

where, As is the corrected absorbance of the sample, Asd is the absorbance read for a sample dilution and Adb is the absorbance read for a corresponding dilution blank.

Step 2. Calculate means of absorbances at each test compound dilution. The percent inhibitory activity (IA) is calculated as:

$$IA\ (\%) = [(A_0 - As)/A_0] \times 100$$

where, A_0 is absorbance of the reagent control and As is the corrected absorbance of the sample.

Step 3. Plot the percent inhibitory activity against the sample concentration, considering the final volume of 1 mL (Fig. 4). Calculate the efficient concentration of test compound needed to scavenge 50% of the available hydroxyl radical (EC_{50}). Substances with high antioxidant activity have low EC_{50} values.

Precautions

(i) Hydrogen peroxide is a strong oxidant agent. Avoid contact with skin and eyes. Wear rubber gloves and goggles. Handle with care.

(ii) 2-thiobarbituric acid has a strong mercaptan odor. Do not breathe dust. Avoid contact with skin and eyes.

(iii) Trichloroacetic acid is caustic. Handle with care.

Experiment 3: Superoxide Anion Scavenging Activity

Superoxide anion has the ability to reduce the nitro-substituted aromatic salts, such as Nitroblue Tetrazolium (NBT), forming radicals (NBT$^\bullet$) and finally insoluble blue formazan (Fig. 6). The ability of a substance to uptake superoxide anions, preventing NBT reduction, is known as superoxide anion scavenging activity (Kulkarni *et al.*, 2004).

Materials and equipments required

Analytical balance; pipettes; micropipettes; tips for micropipettes; vortex agitator; spectrophotometer; 1 mL glass cuvette for spectrophotometry; ice bath.

Reagents

20 mM Tris-HCl buffer (prepare 0.2 M Hydroxymethyl aminomethane (Tris) solution dissolving 2.42 g of Tris in 100 mL of distilled water; prepare 0.2 M hydrochloric acid solution diluting 0.6 mL of concentrated HCl (36.5%) in 100 mL of distilled water; add 19.2 mL of 0.2 M HCl to 50 mL of Tris solution in 2 L flask and mix well; adjust pH to 8.3 using 0.2 M NaOH; mix and make up to 1000 mL

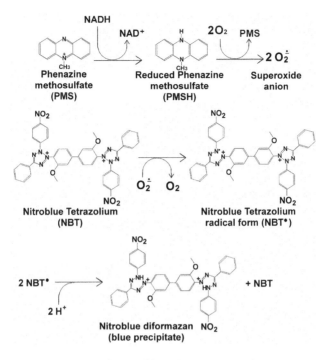

Figure 6 Superoxide anion assay reactions.

with distilled water; dispense in glass flask; it can be stored at 4°C). 1 M Nitro Blue Tetrazolium (NBT) stock solution 10X (dissolve 24 mg of NBT in 1 mL of distilled water; mix and make up to 3 mL with distilled water; it can be stored at 4°C for 3 months; see *Observation* 1). 1 M Phenazine methosulfate (PMS) stock solution 10X (dissolve 9 mg of PMS in 3 mL of distilled water; it can be stored at 4°C for 3 months; see *Observation* 2). 1 mM Reduced Nicotinamide Adenine Dinucleotide (NADH) solution (dissolve 1.42 mg of β-NADH disodium salt in 2 mL of distilled water; see *Observation* 2).

Procedure
(i) Add 20 mM Tris-HCl buffer into the following 1 ml glass cuvettes: blank (800 µL), dilution blanks (700 µL), reagent control (700 µL), and samples (600 µL) (see *Observation* 3).
(ii) Prepare dilutions of test compound with a suitable solvent (see *Observations* 4, 5 and 6).
(iii) Add 100 µL of each dilution into the dilution blank and sample cuvettes.
(iv) Add 100 µL of 1 mM NBT into the cuvettes.
(v) Add 100 µL of 1 mM NADH into the reagent control and sample cuvettes.
(vi) Start the reaction by adding 100 µL of 0.1 mM PMS into the cuvettes (see *Observation* 7).
(vii) Swirl tubes for 5 s and measure the absorbance at 560 nm.

(viii) After 1 minute incubation at room temperature (inside the spectrophoto-meter) absorbance is measured again at 560 nm (see *Observation* 8).

Observations

(1) The NBT solution must be diluted 10 times before performing the assay.
(2) The NADH solution must be prepared just before use and kept in ice during the assay.
(3) Each dilution of test compound requires a corresponding dilution blank.
(4) The most recommended solvents are water, 96% ethanol or methanol.
(5) Preliminary assays are needed to establish the range of dilutions where NBT reduction is linear. Proceed as indicated in *Observation* 3 of Experiment 1.
(6) Make at least 5 dilutions. Standard substances such as L-ascorbic acid, quercetin, BHT or BHA should be included as positive controls.
(7) All the reactives (PMS, NADH, and NBT) should be kept in ice bath during the assay.
(8) At least three independent experiments ($N = 3$) should be made with three replicates for each concentration assayed.

Calculations

Step 1. Correct absorbance readings at 560 nm as follows:

$$A = Asd - Adb$$

where, A is the corrected absorbance of the sample (at 0 or 1 minute), Asd is the absorbance read for a sample dilution (at 0 or 1 minute) and Adb is the absorbance read for a corresponding dilution blank (at 0 or 1 minute).

Step 2. Calculate mean of absorbances for each test compound dilution. The percent inhibitory activity (IA) is calculated, according with absorbances at 560 nm, as follows:

$$IA (\%) = [(A_0c - A1c) - (A_0s - A1s)] \times 100/(A_0c - A1c)$$

where, A_0c is the absorbance of reagent control at 0 minute, A1c is the absorbance of reagent control at 1 minute, A_0s is the absorbance of the sample at 0 minute, and A1s is the absorbance of the sample at 1 minute.

Step 3. Plot the percent inhibitory activity against the sample concentration, considering the final volume of 1 mL (Fig. 3). Calculate the efficient concentration of test compound needed to scavenge 50% of the available superoxide anion (EC_{50}). Substances with high antioxidant activity have low EC_{50} values.

Precautions

Hydrochloric acid is caustic. Handle with care.

Experiment 4: Inhibition of Lipid Peroxidation

The Fe^{+2}/ascorbate mix is used as an inductor of lipid peroxidation, based on the catalytic effect of ferrous ion on the oxidation of ascorbic acid to dehydroascorbic acid, with H_2O_2 generation (Fig. 7). Hydroxyl radical is generated from H_2O_2 according with the Fenton reaction (Fig. 2).

Hydroxyl radicals induce lipid peroxidation, which in turn generates toxic reactive compounds such as 4-hydroxynonenal and MDA (Fig. 8). Detection of MDA is a good indicator of lipid peroxidation. Hence, compounds with antioxidant activity inhibit lipid peroxidation and avoid MDA accumulation.

Lipid source commonly used for peroxidation process are red blood cell ghosts (RBCG), rat liver microsomes, sarcoplasmic reticulum vesicles, rat kidney microsomes and rat brain homogenates.

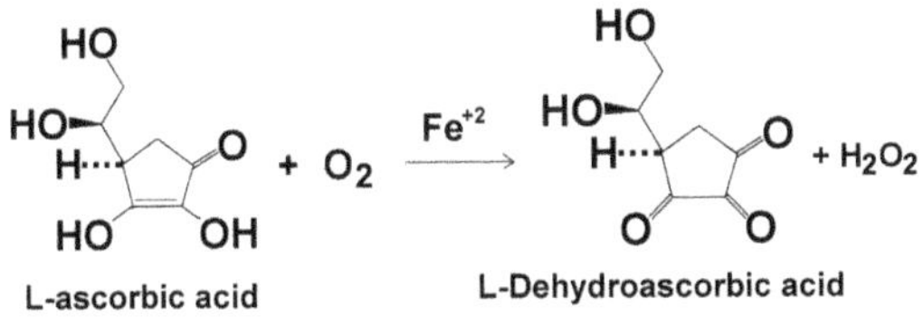

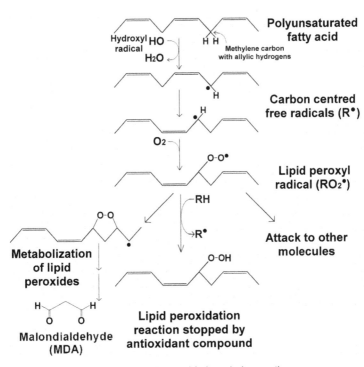

Figure 7 Generation of H_2O_2 through the oxidation of ascorbic acid.

Figure 8 Lipid peroxidation chain reaction.

Materials and equipments required

Analytical balance; pipettes; micropipettes; tips for micropipettes; vortex agitator; spectrophotometer; thermostatized water bath; cuvettes for spectrophotometry; ice bath; analytical centrifuge; refrigerated high speed centrifuge; centrifuge flasks.

Reagents

0.1 M sodium phosphate buffer (prepare 0.1 M NaH_2PO_4 dissolving 13.8 g of $NaH_2PO_4 \cdot H_2O$ in 1 L of distilled water; prepare 0.1 M Na_2HPO_4 dissolving 14.2 g of Na_2HPO_4 in 1 L of distilled water; add 120 mL of 0.1 M NaH_2PO_4 to 880 mL of 0.1 M Na_2HPO_4 and mixed well; adjust pH to 7.4 using 0.1 M Na_2HPO_4; dispense 100 mL aliquots in 250 mL screw-cap bottles; autoclave for 20 minutes at 121°C; it can be stored at 4°C; see *Observation* 1); 0.1 mM $FeSO_4$ solution (dissolve 2.78 g of $FeSO_4 \cdot 7\ H_2O$ in 100 ml of distilled water; dispense 1 mL aliquots in glass or Eppendorff tubes; it can be stored at –20°C for 6 months). 1 mM L-ascorbic acid solution (dilute 1 mL of 4 mM L-ascorbic acid solution in 3 ml of distilled water; see *reagents* in Experiment 2); 0.5% w/v TBA solution (see *reagents* in Experiment 2). 2.8% TCA solution (see *reagents* in Experiment 2).

Procedure

Preparation of red blood cell ghosts (RBCG)

(i) Collect blood from a mammalian source, into EDTA treated tubes (1 mg mL^{-1} of blood) (see *Observation* 2).

(ii) Wash the samples by mixing thoroughly with three volumes of saline phosphate buffer (10 mM sodium phosphate buffer, pH 7.4, with 0.9% NaCl, w/v) (See *Observations* 3 and 4).

(iii) Centrifugate washed samples at 600 × g for 10 minutes, to separate the red cells from plasma.

(iv) Discard supernatant.

(v) Repeat steps 2, 3 and 4.

(vi) Re-suspend the pellets (red cells) mixing thoroughly with three volumes of sodium phosphate buffer (10 mM, pH 7.4) and then washing (see *Observation* 4). The hypotonic medium disrupts most of the cells and hemoglobin is released into the buffer, turning solution to a red colour.

(vii) Centrifuge washed samples at 12,000 × g for 10 minutes.

(viii) Discard supernatant (see *Observation* 5).

(ix) Repeat steps 6, 7 and 8 until the pellet is nearly white or much lighter red colour. Five consecutive washing-centrifugation cycles are usually needed.

(x) Agitate the tubes gently to liquefy the membrane pellet, then remove the pellet to a sample tube, leaving the fibrous mass (clotted material as fibrin and platelets) stuck to the centrifuge tube.

(xi) Ghost protein concentration is determined by Bradford method (refer Chapter 4, Section 3.6, Experiment 13), using bovine serum albumin as standard. The RBCG concentration should be adjusted to 7.5 mg of proteins mL^{-1} with 10 mM sodium phosphate buffer.

(xii) Pellets, which consists of red blood cell ghosts (RBCG) can be stored into glass flasks at −20°C for 6 months.

Inhibition of lipid peroxidation (Kang *et al.*, 2003)

(i) Prepare labeled tubes with 10 mM phosphate buffer: blank (800 μL), dilution blank (700 μL), reagent control (700 μL), and sample (600 μL) (see *Observation 6*).

(ii) Prepare dilutions of test compounds with a suitable solvent (see *Observations 7, 8 and 9*).

(iii) Add 100 μL of each dilution into the dilution blank and sample tubes.

(iv) Add 100 μL of RBCG suspension (7.5 mg of proteins mL^{-1}) into the reagent control and the sample tubes.

(v) Add 100 μL of 0.1 mM $FeSO_4$ into the assay tubes.

(vi) Add 100 μL of 1 mM L-ascorbic acid into the assay tubes.

(vii) Vortex tubes. Incubate them at 37°C for 1 h.

(viii) After incubation, the colorimetric reaction is performed as follows: add 1 mL of TBA (0.5% w/v in NaOH 0.025 M) and 1 mL of TCA (2.8% w/v).

(ix) Vortex the tubes. Then, incubate at 100°C for 30 minutes. Let cool in ice.

(x) Centrifuge the tubes for 5 minutes to remove precipitated protein (see *Observation 10*).

(xi) Absorbance is measured at 532 nm against blank (see *Observation 11*).

Observations

(1) Phosphate buffer solution must be diluted 10 times for the preparation of red blood cell ghosts and for the lipid peroxidation inhibition assay. If phosphate buffer saline is needed, add 0.9 g of NaCl per 100 mL of buffer.

(2) Blood sample must be treated with anticoagulant and either used immediately or stored refrigerated until use.

(3) The whole procedure should be carried out into refrigerated centrifuge flasks.

(4) All the washing steps should be performed by pulling liquid into a pipette and ejecting it repeatedly, while keeping the tip immersed.

(5) The supernatant contains hemoglobin, so it should be removed very carefully in order to avoid the pellet resuspension.

(6) Each dilution of test compound requires a corresponding dilution blank.

(7) The most recommended are water and 96% ethanol.

(8) Preliminary assays are needed in order to establish the range of dilutions where increase of color, due to formation of the TBA-MDA complex, is linear. Proceed as indicated in *Observation 3* of Experiment 1.

(9) Make at least 5 dilutions. Reference substances such as BHT, BHA, or quercetin should be included as positive controls.

(10) Use a test tube centrifuge for this step.

(11) At least three independent experiments (N = 3) should be made with three replicates for each concentration assayed.

Calculations

Step 1. Correct absorbance readings at 532 nm as follows:

$$As = Asd - Adb$$

where, As is the corrected absorbance of the sample, Asd is the absorbance of a sample dilution and Adb is the absorbance of the corresponding dilution blank.

Step 2. Calculate means of the absorbances at each dilution of test compound. The percent of inhibitory activity (IA) is calculated as:

$$IA\ (\%) = [(A_0 - As)/A_0] \times 100$$

where, A_0 is absorbance of the control and As is absorbance of the sample.

Step 3. Plot the percent of inhibitory activity against the sample concentration, considering the final volume of 1 mL (Fig. 4). Calculate the efficient concentration of test compound needed to inhibit 50% lipid peroxidation (EC_{50}). Substances with high antioxidant activity have low EC_{50} values.

Experiment 5: Bleaching of β-carotene

The mechanism of β-carotene bleaching is a free-radical mediated phenomenon. The hydroxyl radicals, generated by thermal cleavage of H_2O_2, attack linoleic acid (see Experiment 4). Free radicals formed from linoleic acid are mainly hydroperoxides, which oxidate the double bonds of the highly unsaturated β-carotene molecules. As a consequence of this oxidative process, β-carotene molecules lose their characteristic orange color. Decoloration can be spectrophotometrically measured at 470 nm. Antioxidant substances can scavenge free radicals, protecting β-carotene from oxidation.

Materials and equipments required

Analytical balance; pipettes; micropipettes; tips for micropipettes; vortex agitator; spectrophotometer; cuvettes for spectrophotometry; rotary evaporator; round-bottom flask for rotary evaporator; thermostatized water bath.

Reagents

0.02% w/v β-carotene solution (dissolve 1 mg of β-carotene in 5 mL of chloroform; see *Observation* 1). 50 mM H_2O_2 solution (dilute 51 μL of 30% hydrogen peroxide solution in 10 mL of distilled water; dispense in glass tubes; store at –20°C for 3 months).

Procedure

According to Shon et al. (2003)

(i) Prepare labeled tubes with water or 96% ethanol as follows: blank (100 μL), and reagent control (100 μL) (see *Observation* 2).

(ii) Prepare dilutions of test compounds with a suitable solvent (see *Observations* 3 and 4).

(iii) Add 100 μL of each dilution into the corresponding dilution blank and sample tubes.

(iv) Prepare β-carotene solution.

(v) Add 1 mL of β-carotene solution into a 100 mL round-bottom flask.

(vi) Remove completely the chloroform under vacuum (see *Observation* 5).

(vii) Add 20 µL of linoleic acid and 200 µL of tween 20 and 50 mL of 50 mM H_2O_2 into the flask containing β-carotene. Shake vigorously until obtaining of an uniform emulsion (working emulsion).

(viii) Add 20 µL of linoleic acid and 200 µL of tween 20 and 50 mL of 50 mM H_2O_2 into a round-bottom flask. Shake vigorously until obtaining of an uniform emulsion (dilution blank emulsion).

(ix) Add 4.9 mL of working emulsion into the sample tubes. Vortex tubes for 1 minute.

(x) Add 4.9 mL of dilution blank emulsion into the dilution blank tubes (see *Observation* 6). Vortex tubes for 1 minute.

(xi) Read absorbance at 470 nm against blank.

(xii) Incubate the tubes at 50°C for 2 hours (until orange color of β-carotene in the control disappear).

(xiii) Read absorbance at 470 nm against blank (see *Observation* 7).

Observations

(1) β-carotene solution must be prepared before performing the β-carotene bleaching assay.

(2) The most recommended solvents for this procedure are water, 96% ethanol or methanol.

(3) Preliminary assays are needed in order to establish the range of dilutions where decoloration of β-carotene is linear. Proceed as indicated in *Observation* 3 of Experiment 1.

(4) Make at least 5 dilutions. Reference substances such as α-tocopherol, quercetin, rutin, BHT and BHA should be included as positive controls.

(5) Chloroform must be totally evaporated.

(6) Each sample dilution requires a dilution blank.

(7) At least three independent experiments (N = 3) should be made with three replicates for each concentration assayed.

Calculations

Step 1. Correct absorbance readings at 470 nm as follows:

$$A = Asd - Adb$$

where, A is the corrected absorbance of the sample (at 0 or 2 h), Asd is the absorbance read for a sample dilution (at 0 or 2 h) and Adb is the absorbance read for a corresponding dilution blank (at 0 or 2 h).

Step 2. Calculate means of the absorbances at each dilution of test compound. The percent of antioxidant activity (AA) is calculated, according with absorbances at 470 nm, as follows:

$$AA\ (\%) = (A1 - A_0) \times 100$$

where, A_0 is absorbance of the sample at 0 minute, and A1 is the absorbance of the sample at 2 h.

Step 3. Plot the antioxidant activity against the sample concentration, considering the final volume of 5 mL (Fig. 4). Calculate the efficient concentration of test compound needed to inhibit 50% β-carotene discoloration (EC_{50}).

Precautions

Hydrogen peroxide is a strong oxidant agent. Avoid contact with skin and eyes. Wear rubber gloves and goggles. Handle with care.

3. GENOTOXIC ASSAYS

Experiment 6: *Bacillus subtilis* rec-assay

The detection of DNA-damaging agents can be achieved using repair-proficient and repair-deficient bacterial strains (Leifer *et al.*, 1981). In the *Bacillus subtilis* rec-assay, a DNA repair-recombination proficient strain (H17), and a deficient strain (M45) are exposed to an allelochemical. Genotoxic compounds induce a set of DNA repairing functions in H17, called the "SOB response". Induction of SOB response is triggered by a metabolic signal that activates the synthesis of RecE protein. This protein is a 5'-3' specific exonuclease that degrades one strand of a linear DNA substrate to mononucleotide products, conducting to a *recE*-mediated homologous recombination. M45 carries a *recE* mutation (Ceglowski *et al.*, 1990), that avoid normal DNA repairing functions, which make this strain more sensitive to γ rays and genotoxic agents than H17. Growth differences between H17 and M45 strains are spectrophotometrically detected at 595 nm. The differential growth of the two strains is interpreted as a genotoxic response (Takigami *et al.*, 2002).

Compared with Ames test, the rec-assay offers several advantages (Mazza, 1982):

(i) Increase permeability to several chemicals, due to the gram-positive cell envelopes.
(ii) Easier strains maintenance.
(iii) Allows the assay of purified spores instead of vegetative cells. This increases 50 times assay sensitivity.
(iv) Strong bactericidal compounds used in other genotoxic assays are not needed for rec-assay.
(v) This assay is faster and easier than Ames test, thus constituting a powerful tool for detection of genotoxic substances.

Materials and equipments required

Analytical balance; glass tubes; magnetic bar; a magnetic stirrer; pH meter (or pH indicator sticks); pipettes; micropipettes; sterile tips for micropipettes; sterile 96 well plate (flat bottom); sterile plate cover; 96 plate well reader (with 595 nm filter); microscope slide; Bunsen flame; optical microscope; oil immersion optic; immersion oil; sterile Petri dishes (90 mm diameter); sterile cryotubes; loops for microbiology; vortex agitator with foam rubber surface; gloves; goggles or protective eye wear; autoclave; laminar flow hoot; sterile 0.22 and 0.45 μm membrane filters; sterile filter holders; stroke shaker, *B. subtilis* strains: Rec + is the H17 strain; Rec – is M45 strain; both strains are arginine dependent (Arg–); tryptophan dependent (Trp–); and uvrABC proficient (uvrABC+); Assay of test compounds should be performed taken account that some substances are genotoxics itself (direct genotoxics) or are metabolized to be genotoxics (indirect genotoxics o progenotoxics); inclusion of a metabolic activation system constituted by a mix of rat liver enzymes extract and cofactors into the *Bacillus subtilis* rec-assay allows the determination of indirect genotoxics; Reference chemicals such as $K_2Cr_2O_7$, 4-nitroquinoline-1-oxide (4-NQ), Mitomycin C (MMC), Kanamycin (KM) and dimethylsulfoxide (DMSO) should be tested before performing the *B. subtilis* rec-assay for comparative purposes.

Reagents and Media

Nutrient broth (dissolve 17.5 g of Antibiotic medium 3 (Difco) in 1 L of distilled water; dispense 5 mL in glass flasks; autoclave for 20 minutes at 121°C; store at 4°C. See *Observation* 1). Nutrient agar (dissolve 17 g of agar in 1 L of nutrient broth; autoclave for 20 minutes at 121°C; pour in sterile petri dishes, 20 mL per dish; store at 4°C). Double strength growth medium (dissolve 35 g of antibiotic medium 3 (Difco) in 1 L of distilled water, dispense 5 mL in glass flasks; autoclave for 20 minutes at 121°C; store at 4°C. See *Observation* 1). Schaeffer's agar medium (SM) (dissolve 8 g of Antibiotic medium 3 (Difco) in 500 mL of distilled water, dissolve (in the following order): 17 g of agar, 0.12 g of $MgSO_4$, 1 g of KCl, 1.98 mg of $MnCl_2 \cdot 4\ H_2O$, 0.278 mg of $FeSO_4 \cdot 7\ H_2O$, 236 mg of $Ca(NO_3)_2 \cdot 4\ H_2O$, 5 g of Glucose, 50 mg of L-Tryptophan, 50 mg of L-Arginine hydrochloride, 50 mg of L-Leucine, 50 mg of L-Methionine, 50 mg of L-Histidine monohydrochloride, and 20 mg of Adenine; make up to 1 L with distilled water; autoclave for 20 minutes at 121°C; dispense 1 mL aliquots into 2 mL sterile cryotubes; store at –20°C for 6 months). 0.021 M Arginine solution (dissolve 22.5 mg of L-Arginine hydrochloride in 5 mL of distilled water; sterilize by filtration through a 0.22 μm membrane filter, store at 4°C for 30 days. See *Observation* 2). 0.022 M Tryptophan solution (dissolve 22.5 mg of L-Tryptophan in 5 mL of distilled water; sterilize by filtration through a 0.22 μm membrane filter, store at 4°C for 30 days. See *Observation* 2). 0.1 M, pH 7.4 sodium phosphate buffer (see reagents in lipid peroxidation assay). Co-factors for metabolic activation system (dissolve in the following order: 1.6 g D-Glucose-6-phosphate dipotassium salt, 3.5 g of β-$NADP^+$ sodium salt, 1.8 g of $MgCl_2$, 2.7 g of KCl, 12.8 g of Na_2HPO_4, 2.8 g of $NaH_2PO_4 \cdot H_2O$ in 900 mL of distilled water; mix and make up to 1 L with distilled water; sterilize by filtration through a 0.45 μm membrane filter; dispense 9 or 9.5 mL aliquots in sterile glass tubes; store at –20°C for 6 months). S9 fraction (mince and mix fresh rat liver with sterile 10 mM potassium phosphate buffer added with 1.15% (w/v) of KCl in a cooled sterile mortar, 0.2-0.25 g of fresh tissue per mL of buffer); homogenize the tissue with the pylon; centrifuge at 9,000 × g for 20 minutes; separate the supernatant (S9 fraction) from the pellet; measure the protein concentration using the Bradford method (see Chapter 4, Section 3.6, Experiment 13; use bovine serum albumin as standard); adjust the protein concentration to 20 mg/mL with sterile 10 mM potassium phosphate buffer added with 1.15% (w/v) of KCl; distribute 1 mL aliquots into sterile glass tubes; store at –80°C for 1 tear or at –20°C for 5 months. See *Observations* 3, 4, and 5). Metabolic activation system (MAS) (thaw co-factors for metabolic activation system and S9 fraction during 3 minutes at room temperature); transfer to ice bath; dissolve 1 mL of S9 fraction in 9 mL of cofactors (final protein concentration of 2 mg/mL); sterilize by filtration through a 0.45 µm membrane filter; keep in ice bath as long as the assay is carried out. See *Observations* 6, 7 and 8); 10 mM potassium phosphate buffer added with 1.15% w/v of KCl (prepare 0.1 M KH_2PO_4 solution dissolving 13.6 g of KH_2PO_4 in 1 L of distilled water); 0.1 M K_2HPO_4 solution (17.4 g of K_2HPO_4 in 1 L of distilled water); 19 mL of 0.1 M KH_2PO_4 solution are added to 81 mL of K_2HPO_4 solution; mix and adjust pH to 7.4 using 0.1 M K_2HPO_4; add 11.5 g of KCl; mix and make up to 1000 mL with distilled water; dispense 100 mL aliquots in 250 mL flasks; autoclave at 121°C for 20 minutes; store at 4°C).

Strains

Bacillus subtilis spores for rec-assay may be acquired from Instituto de Estudios Vegetales "Dr. Antonio R. Sampietro" (Tucumán, Argentina). The spores should be stored in criotubes with SM at –20°C for 6 months. Before performing the assay the strains should be identified by having certain characteristics (Gram positive, red vegetative cells and green endospores by Wirtz-Conklin differential spore stain method, and auxotrophy for arginine and tryptophan. See *Observations* 9 and 10). When the purity of the strains is compromised (i.e. media contamination), an isolation and selection of the strains should be performed.

Procedure

(i) Each *B. subtilis* strain cryotube is left at room temperature till unfrozen (5 minutes approximately). Inoculate a looful into 5 mL of nutrient broth and shake at 37°C till the absorbance value at 595 nm range between 0.3-0.5 (nearly 16 h).

(ii) Inoculate 300 µL of culture into 5 mL of nutrient broth and shake at 37°C till the absorbance value at 595 nm is 0.1 (about 2 h). The cells are at an exponential growth phase at this stage.

(iii) Prepare the "working bacterial suspensions": 10 µL of exponential growth phase culture plus 990 µL of nutrient broth for H17 (the absorbance at 595 nm should be close to 0.001) and 20 µL of exponential growth phase culture plus 980 µL of nutrient broth for M45 (the absorbance at 595 nm should be close to 0.002).

(iv) For evaluation of direct genotoxics, add 10 µL of sterile 0.1 M sodium phosphate buffer into a 96 well plate with flat bottom. For the evaluation of indirect genotoxic substances, replace the buffer with 10 µL of MAS.

(v) Add 60 µL of the evaluated substance into the suitable wells. Solutions should be previously sterilized through 0.22 µm membrane filter. Each dilution of test compound requires a corresponding dilution blank. At least 3 wells per dose level should be made for each strain.

(vi) Add 30 µL of the working bacterial suspensions into the suitable wells.

(vii) Growth controls are made with 60 μL of sterile distilled water instead of the test substance. Sterility controls are made with 30 μL of broth medium instead of working bacterial suspensions.
(viii) Cover the plate with the plate cover and mix well the contents (see *Observation* 11).
(ix) The plate is held at 37°C for 1 hour under shaking on a stroke shaker (this is the interaction period).
(x) After interaction period, add 100 μL of double strength growth medium into the wells. Mix the well contents.
(xi) Uncover the plate and read the absorbance values (time = t_0) in a plate reader at 595 nm (see *Observation* 12).
(xii) Cover the plate with the plate cover and incubate at 37°C until the absorbance values for each strain growth controls reach 0.1 at 595 nm (see *Observation* 13).
(xiii) After incubation, mix the well contents.
(xiv) Uncover the plate and read the absorbance values (time = t_1) in a plate reader at 595 nm (see *Observation* 13).

Observations
(1) For each experiment use only nutrient broth from the same batch.
(2) Tryptophan solution should be kept in ice during the *B. subtilis* checking process.
(3) A number of commercial sources provides S-9 preparations. This S-9 provides information about enzyme activities and effectiveness against standard mutagens.
(4) The whole process for obtaining S9 fraction should be preformed in ice bath. The S9 fraction should be stored at –20°C as long as the protein determination is carried out.
(5) Aroclor 1254 could be used in order to induce rat liver enzymes. When using live animals, it is required that officially approved procedures must be followed for the care and use of the animals.
(6) Prepare MAS before the assay is carried out.
(7) MAS is usually prepared at 10% (v/v) concentration, though some laboratories use up to 30% v/v concentration in order to extent the detection of mutagens.
(8) Do not freeze prepared MAS for more than 2 days.
(9) Arginine requirements: streak a loopful (or 100 μL) of the overnight culture across a Cg minimal medium plate supplemented with 100 μL of 0.022 M tryptophan solution. The arginine requirement is confirmed if no growth is observed after incubation at 37°C for 48 h.
(10) Tryptophan requirements: streak a loopful (or 100 μL) of the overnight culture across a Cg minimal medium plate supplemented with 100 μL of 0.021 M arginine solution. The tryptophan requirement is confirmed if no growth is observed after incubation at 37°C for 48 h.
(11) Do not mix the well contents with the tips of the micropipettes (this could scratch the well's bottom, causing further lecture mistakes). Instead, place

the covered plate over the foam rubber surface of a vortex agitator and mix thoroughly (without creating air bubbles) for 15 seconds. Extreme care should be taken in order to avoid any liquid loses during the process.

(12) At least three independent experiments (N = 3) should be made and three measures should be performed for each plate.

(13) The incubation period is different for each strain. Approximately 7 and 8 hours for H17 and M45, respectively.

Calculations

Step 1. Correct absorbance readings at 595 nm as follows:

$$As = Asd - Adb$$

where, As is the corrected absorbance of the sample, Asd is the absorbance of a sample dilution and Adb is the absorbance of the corresponding dilution blank. This step is performed for t_0 and t_1 readings.

Step 2. Calculate means of the absorbances at each dilution of test compound. This step is performed for t_0 and t_1 readings.

Step 3. Calculate the increase in absorbance between t_1 and t_0 readings as follows:

$$Ai = At_1 - At_0$$

where, Ai is the increase in absorbance, At_1 is the absorbance of a sample dilution at t_1 and At_0 is the absorbance of a sample dilution at t_0.

Step 4. Calculate percent of survival (PS%) for each sample dilution as follows:

$$PS\% = Aisd \times 100/Aigc$$

where, Ai is the absorbance increase in a sample dilution, and Aigc is the absorbance increase in the growth control.

Step 5. Plot the percent of survival (%) against the sample concentration (logarithmic scale) (Fig. 9). Calculate the concentration of a substance giving 84%, 50%, and 16% of survival (C_{84}, C_{50}, and C_{16}) respectively, for each strain.

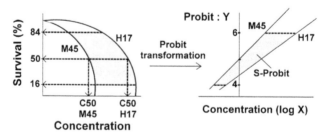

Figure 9 Survival curves for H17 and M45 strains tested with a typical genotoxic substance and Probit transformation.

Genotoxicity can be evaluated by **R50** and **S-Probit** (Matsui, 1988). **R50** is the relationship between the C_{50}s of H17 and M45:

$$(1) \qquad R50 = C_{50} \; (H17)/C_{50} \; (M45)$$

S-Probit represents the area enclosed between the survival curves of H17 and M45. Probit transformation (Finney, 1971) is performed in order to obtain a precise area. In this transformation, 84% and 16% of survival corresponds to 4 and 6 (Probit scale) respectively. After conversion to the Probit scale the survival curves are plotted (Figs. 9 and 10). With Probit transformation the two lines can be converted to linear functions, represented by the following equations:

$$(2) \qquad Y_1 = a_1 + b_1 \log x \qquad \text{for H17}$$
$$(3) \qquad Y_2 = a_2 + b_2 \log x \qquad \text{for M45}$$

where "x" is the concentration of a test substance, "a_1" and "a_2" are standard deviations for H17 and M45 respectively, and "b_1" and "b_2" are the mean of normalized populations for H17 and M45. As (2) and (3) show, "a" and "b" represent the origin ordinate and slope respectively for each survival line, so they can be easily calculated from the plot. "a" is the Probit value at log concentration = 0, and "b" can be calculated as follows:

$$(4) \qquad b_1 = \Delta Y(\text{Probit})/\Delta x_1 \; (\text{concentration log x}) \qquad \text{for H17}$$
$$(5) \qquad b_2 = \Delta Y(\text{Probit})/\Delta x_2 \; (\text{concentration log x}) \qquad \text{for M45}$$

Figure 10 shows an example of these calculations.

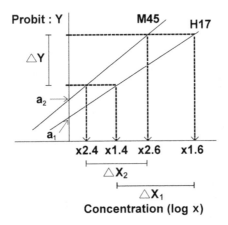

Figure 10 Survival curves for H17 and M45 strains tested with a typical genotoxic substance and examples for the origin ordinate (a) and slope calculations.

The enclosed area between the converted lines can be calculated by integration as follows:

$$(6) \qquad \text{S-Probit} = \int_4^6 \{[(a_2/b_2) - (a_1/b_1)] + (1/b_1) - (1/b_2)Y\} \, dY$$

Equation (6) can be written as follows, according to the second fundamental theorem of calculus:

$$(7) \quad \text{S-Probit} = \left(\frac{a_2}{b_2} - \frac{a_1}{b_1} \right) \cdot 2 + \left[\frac{\left(\frac{1}{b_1} - \frac{1}{b_2} \right)}{2} \right] \cdot 20$$

S-Probit is a qualitative index to evaluate genotoxic potency. Different organic and inorganic chemicals, including genotoxic and cytotoxic compounds were evaluated by *B. subtilis* rec-assay, and the criteria to assess genotoxicity were derived from those results. The criteria derived are shown in Table 1, and consist in four ranges of genotoxicity.

Table 1 Criteria for judging genotoxicity by the *B. subtilis* rec-assay

S-Probit	Classification
More than 0.593	Strong genotoxic response (++)
0.200–0.593	Genotoxic response (+)
−0.123–0.199	Non-genotoxic (−)
Less than−0.124	Reverse effect (r)

Reverse effect indicates that the growth of H17 is retarded more than that of M45. This could be a new class of genotoxicity. However, there is no clear explanation available on his effect. The results for a substance are reported according to the criteria exposed in Table 1.

Experiment 7: *Allium* Test

The cell cycle is an ordered set of events that culminate in cell growth and division into two daughter cells (Fig. 11). Four phases are often distinguished in the cell cycle: G1 is the gap between cell division and DNA synthesis, S is the phase where DNA is synthesized, G2 is the gap between DNA synthesis and cell division, and M is the mitosis, that is, the phase of cell cycle where nuclear and cytoplasmic division occur. The first three phases (G1, S and G2) are considered part of the cell interphase but only M is the cell division itself. Cell division

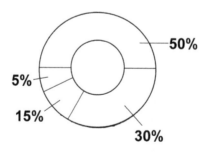

Figure 11 The mitotic cell division cycle.

comprises four steps: prophase, metaphase, anaphase and telophase. These steps are often observed in plant meristematic tissues. In the *Allium* test, roots of onion (*Allium cepa*) are subjected to solutions with and without (control) an organic compound for a proper time. Mitotic cells are observed after fixation, hydrolysis and staining of meristematic root tips (Fig. 12). According with the sample

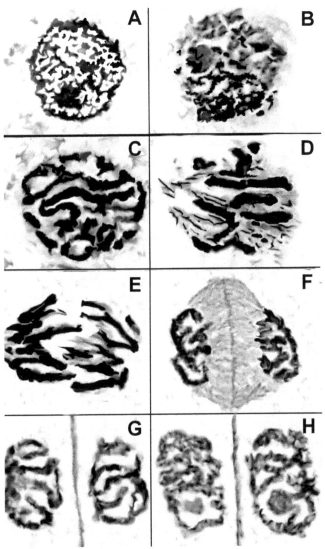

Figure 12 Light microscopy micrographs of cells from root tips of *Allium cepa* observed at different steps of the cell cycle: (A) Interphase, (B) Early Prophase, (C) Late Prophase, (D) Metaphase, (E) Early Anaphase, (F) Late Anaphase, (G) Telophase and (H) daughter cells.

preparation and the staining procedures, chromosomes can be examined under optical microscopy (Ruzin, 1999; Dayan *et al.*, 2000). Percentage of cells under division are registered. Root tips with and without exposure to the allelochemical are observed to establish if the compound affects cell division at DNA level. The *Allium* test is a fast and inexpensive method allowing the investigation of universal mechanisms of toxicity in plant cells.

Materials and equipments required
Containers (10 mL volume); onion bulbs with uniform size; thermostatic Water bath set at 60°C; immersion oil; optical microscopy; oil immersion optic; coverslips; microscope slide; glass tubes; razor blade; filter paper.

Reagents
Carnoy fixer solution (dilute 250 mL of glacial acetic acid in 500 mL of absolute ethanol; mix and make up to 1 L with absolute ethanol; store at 4°C). Schiff's reagent (prepare 0.5% w/v basic fuchsin solution: dissolve 0.5 g of basic fuchsin in 80 mL of boiling distilled water; mix and make up to 100 mL with distilled water; cool up to 50°C; filter through Whatman N° 1 paper; prepare 1 N HCl solution: dissolve 100 mL of 36.5% HCl in 900 mL of distilled water; see *Observation* 1; dissolve 1.5 g of $K_2S_2O_5$ in 100 mL of basic fuchsin solution; add 15 mL of 1 N HCl and mix well; allow to stand overnight with cap on tightly; if colour is observed, add 500 mg of charcoal; shake for 2 minutes; filter through Whatman N° 1 paper; store at 4°C in a dark flask). 45% v/v glacial acetic acid (dilute 45 mL of glacial acetic acid in 55 mL of distilled water; store at 4°C). 40% v/v ethanol (dilute 417 mL of 96% ethanol in 500 mL of distilled water; mix and make up to 1 L with distilled water; store at 4°C). 70% v/v ethanol (dilute 271 mL of water in 500 mL of 96% ethanol; mix and make up to 1 L with 96% ethanol; store at 4°C).

Procedures
Root tips production

(i) Clean onion bulbs with water.
(ii) Add 9 mL of water to each container. Put the bulbs on containers (Fig. 13). Leave at room temperature for 24 h.
(iii) Select bulbs with uniform root length.
(iv) Prepare dilutions of test compounds dissolved in distilled water (see *Observation* 2).
(v) Add the dilutions into the assay containers (see *Observation* 3)
(vi) Put the bulbs over the containers in contact with the sample solution at room temperature for 24 h.
(vii) Transfer bulbs to containers filled with distilled water. Leave bulbs at room temperature for 24 h.
(viii) Cut several roots of each bulb (approximately 1.5 cm of length).
(ix) Add roots to labelled tubes containing a known amount of the Carnoy fixer solution.
(x) Leave the roots in contact with the fixer solution for 2 hours at 4°C.

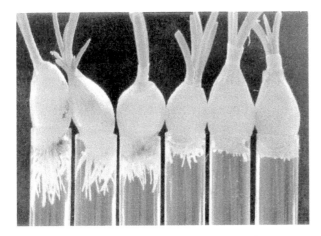

Figure 13 Dose-response relationship between toxin concentrations and growth inhibition of *A. cepa* roots (Source: Rodrigues and Salabert de Campos, 2006).

(xi) Replace the fixer solution with 40% ethanol and leave in contact for 24 hours at 4°C.

(xii) Replace the 40% ethanol with 70% ethanol and store at 4°C (see *Observation* 4)

Squashing cells and tissues

(i) Add the roots into a flask containing 1 N HCl. Incubate in water bath at 60°C for 10 minutes (hydrolysis).

(ii) Transfer the roots to microscope slides using a paintbrush.

(iii) Add drops of Schiff's reagent to the root tips. Leave for 1 hour in the dark. Absorbs excess of Schiff´s reagent with a little piece of filter paper. Separate root apex by cutting at 2 mm from the root end. Discard the remaining parts of the root tips.

(iv) Add drops of 45% glacial acetic acid to the root apexes. Squash them on the slide applying pressure over a coverslip to get a single cell layer preparation.

(v) Put an immersion oil drop and observe in bright-field microscopy. Watch at least 1000 cells per slide at a magnification level of 40 × (Fig. 14).

Observations

(1) HCl is caustic. Handle with care.

(2) Make at least 3 dilutions. The dilutions should be prepared in order to reach the desired concentration in the final volume added to the recipient.

(3) At least three independent experiments should be made (N = 3) and three slides should be prepared for each concentration assayed. Root apexes grown in water serve as controls. If test compounds are dissolved in an organic solvent before preparation of aqueous dilutions, control solvents must also be prepared.

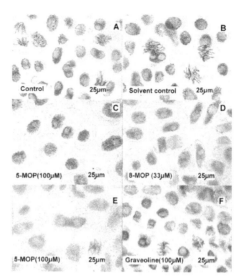

Figure 14 Representative light micrographs of cells from root tips of *A. cepa* treated with the furanocoumarins 5-methoxypsoralen (5-MOP) and 8-methoxypsoralen (8-MOP) and the quinolone alkaloid graveoline. Note that controls (A and B) are under an active mitotic activity while allelochemical treatments almost inhibit cell division. Bars represent 25 μm (Source: Hall *et al.*, 2004).

(4) The low temperature helps the chromosome condensation process, in order to improve post-stain observation.

(5) The advantage of *Allium* test respect to other genotoxic methods is that genotoxicity is tested over eukaryotic cells, and sterile conditions are not needed.

Calculations

Mitotic index is calculated as follows:

$$\text{Mitotic index} = [(\text{Pph} + \text{Mph} + \text{Aph} + \text{Tph}) \times 100]/\text{Tc}$$

where, Pph are cells in prophase, Mph are cells in metaphase, Aph are cells in anaphase, Tph are cells in telophase and Tc are total cells counted.

In addition to quantitative data, visual observation of the root squashes may provide evidences of abnormal mitotic arrangement or a typical cell wall formation that would suggest either a disruption of the microtubule-organizing centres or alteration of processes involved in cell wall biosynthesis.

4. MUTAGENIC ASSAY

Experiment 8: Ames test

The Ames test allows screening of compounds with potential biological activity at the nucleotide level (Ames *et al.*, 1973). This test evaluates if an organic compound is able to induce reverse mutations at a specific loci in bacterial strains. The assay is performed with mutant *Salmonella typhymurium* or *Escherichia coli* strains which are turned off for histidine or tryptophan biosynthesis (Green and Muriel, 1976). Because of these mutations, the bacteria require exogenous supply of histidine or tryptophan to survive. Exposed to a mutagen agent, some bacteria undergo a reverse mutation turning the histidine or tryptophan synthesis back on. The assay quantifies the number of revertant strains induced (Alldrick *et al.*, 1985). Different strains provide specific information about mutation mechanisms (i.e. oxidative DNA damage), because each strain was created by a specific type of mutation - either a base-pair substitution or frame-shift mutation. Ames test is sensitive to a wide range of mutagenic and carcinogenic chemicals (Ames *et al.*, 1975). Mutagenic agents are classified into four groups according to their action mechanism:

(i) **Base pair substitution mutagens** are agents that cause a base change in DNA. In a reversion test, this change may occur at the site of the original mutation or at a second site in the bacterial genome.

(ii) **Frameshift mutagens** are agents that cause the addition or deletion of one or more base pairs in the DNA, thus changing the reading frame in the RNA (Hartman *et al.*, 1984).

(iii) **Cross-linking mutagens** are chemicals that cause significant distortion of the secondary structure of DNA. Cross-linking can be intra-strand or inter-strand; inter-strand links block DNA replication.

(iv) **Oxidative mutagens** are compounds that generate reactive oxygen species (ROS). Mutations may be induced as a consequence of oxidative DNA damage by ROS. Damage to DNA by ROS has been suggested to result in the mutations associated with cancer initiation and progression (Halliwell and Aruoma, 1991).

Materials and equipments required

Analytical balance; glass tubes; magnetic bar; a magnetic stirrer; pH meter (or pH indicator sticks); pipettes; micropipettes; sterile tips for micropipettes; Bunsen flame; sterile petri dishes (90 mm diameter); sterile cryotubes; loops for microbiology; vortex agitator; gloves; goggles or protective eye wear; autoclave; thermostatized water bath; ice bath; laminar flow hoot; sterile 0.22 and 0.45 μm membrane filters; sterile filter holders; stroke shaker; marker pen; colony counter (not essential).

Strains

There are several international guidelines referred to the strains that should be employed for this assay. Although a single strain is sufficient to demonstrate a mutagenic response, a negative result should be defined using 4 or 5 strains. The most recommended strains battery are listed below (Food and Drug Administration, 2000):

(i) *S. typhimurium* TA98
(ii) *S. typhimurium* TA100
(iii) *S. typhimurium* TA1535

(iv) *S. typhimurium* TA1537 or TA97 or TA97a
(v) *E. coli* WP2 *uvrA* (pKM101) or *S. typhimurium* TA102
(vi) *E. coli* WP2 *uvrA/OxyR* (pKM101)

The strains are commercially available from ATCC (*American Type Culture Collection*). All *Salmonella* strains are histidine-biotin dependent (His-), and *E. coli* strains are tryptophan dependent (Trp-). Strains TA98, TA100, and TA97 (or TA97a or TA1537) are always considered necessary. These strains have G-C base pairs at the primary reversion site. Some DNA crosslinking mutagens are not detected with these strains, but they may be detected by *E. coli* WP2 or *S. typhimurium* TA102 strains, which have an A-T base pair at the primary reversion site (Levin *et al.*, 1982). Some mutations may be induced as a consequence of oxidative DNA damage, therefore, *E. coli* WP2 *uvrA/OxyR* (pKM101) strain is proposed as specific indicator of cell mutability induced by reactive oxygen species (ROS) that could attack DNA (Blanco *et al.*, 1985). Some strains have additional genetic markers that increase sensitivity to certain types of mutagens. The additional genetic markers and strain characteristics are listed below:

(i) uvrA mutation in *E. coli* or *uvrB* mutation in *S. typhimurium*, allows sensitivity to UV light. The *uvrB* deletion extends through the biotin gene, hence the mutant strain needs biotin to grow.
(ii) A mutation (*rfa*) in all strains leads to a defective lipolysaccharide (LPS) layer that coats the bacterial surface, making the strains more permeable to bulky chemicals, as crystal violet.
(iii) Plasmid pKM101 confers ampicillin resistance in strains TA98, TA100, TA97a, TA97, and WP2 and ampicillin + tetracycline resistance in strain TA102, which is a convenient marker to detect the presence of the plasmid.
(iv) Mutation *hisG428* on the plasmid pAQ1 (*S. typhimurium* TA102) amplifies the number of target sites. To enhance the ability of this strain to detect DNA cross-linking agents, the *uvrB* gene was retained making the bacterium DNA repair proficient.
(v) OxyR is a DNA-binding transcription factor that activates the expression of several genes encoding antioxidant enzymes. The OxyR deficiency increases the strains mutability by oxidative mutagens.

Table 2 shows the mutagenicity event detected by each strain.

Table 2 Mutagenicity event detected by each strain in the Ames test

Strain	Mutagenicity event
S. typhimurium TA97 *S. typhimurium* TA98 *S. typhimurium* TA1537	Frameshifts
S. typhimurium TA100 *S. typhimurium* TA1535	Base-pair substitution
E. coli WP2 *uvrA* (pKM101) *S. typhimurium* TA102	Cross-linking mutagens
E. coli WP2 *uvrA/OxyR* (pKM101)	Oxidative mutagens

Test chemicals

Many chemicals are not mutagenic in their native forms, but they are converted into mutagenic substances after liver metabolization; these substances are referred as "promutagens" or "indirect mutagens". Since bacteria do not have the same metabolic capabilities as mammals, some test protocols utilize a mix added with rat liver enzyme extract (S9) in order to promote metabolic conversion of the tested chemical. The inclusion of metabolic activation system into the Ames test allows the determination of indirect mutagens. On the other hand, there are substances with mutagenic activity without metabolic activation system, usually referred as "direct mutagens". Table 3 shows positive control chemicals frequently used in Ames test.

Table 3 Chemicals often used as positive controls in Ames test

Strain	In the absence of MAS (µg/plate)*	In the presence of MAS (µg/plate)*
S. typhimurium TA97	9-aminoacridine (50)	2-aminoanthracene (1-5)
S. typhimurium TA98	4-nitro-o-phenylene-diamine (2.5)	2-aminoanthracene (1-5)
S. typhimurium TA100	sodium azide (5)	2-aminoanthracene (1-5)
S. typhimurium TA102	mitomycin C (0.5)	2-aminoanthracene (5-10)
S. typhimurium TA1535	sodium azide (5)	2-aminoanthracene (2-10)
S. typhimurium TA1537	9-aminoacridine (50)	2-aminoanthracene (2-10)
E. coli WP2 uvrA pKM101	methyl methane-sulfonate (250)	Terbutyl hydroperoxide (100)
E. coli WP2 uvrA/*OxyR* pKM101	methyl methane-sulfonate (250)	Terbutyl hydroperoxide (100)

*Concentration based on 100 × 15 mm petri plate containing 20 to 25 ml of GM agar.

Reagents and media

Vogel–Bonner (VB salts) medium E (50×). Dissolve in the following order: 10 g of $MgSO_4$, 100 g of Citric acid monohydrate, 500 g of K_2HPO_4, and 175 g of $NaNH_5PO_4 \cdot 4 H_2O$ in 500 mL of warm distilled water; mix and make up to 1 L with distilled water; dispense 20 mL in glass flasks; autoclave for 20 minutes at 121°C; store at room temperature in the dark). 10% v/v Glucose solution (dissolve 100 g of glucose in 500 mL of distilled water; mix and make up to 1 L with distilled water; autoclave for 20 minutes at 121°C; store at 4°C for 3 months). Glucose medium (GM) agar plates (dissolve 17 g of agar in 500 mL of boiling distilled water; mix and make up to 1 L with distilled water; autoclave for 20 minutes at 121°C; let cool for about 65°C; add 20 mL of sterile VB salts and 50 mL of sterile 10% v/v glucose solution (see *Observations* 1 and 2); pour in sterile petri dishes (20 mL per plate), store at 4°C for 7 days (see *Observations* 3, and 4). Nutrient Broth (dissolve 25 g of Oxoid nutrient broth N° 2 in 500 mL of boiling distilled water; mix and make up to 1 L with distilled water; dispense 50 mL in flasks; autoclave for 20 minutes at 121°C; store at 4°C for 30 days). Luria Bertrani

(LB) agar medium (dissolve 17 g of agar, 10 g of bacto-tryptone, 5 g of bacto-yeast extract, and 10 g of NaCl in 500 mL of boiling distilled water; mix and make up to 1 L with distilled water; autoclave for 20 minutes at 121°C; pour in sterile petri dishes (20 mL per plate), store at 4°C for 7 days). Top agar (dissolve 6 g of agar, and 5 g of NaCl in 500 mL of boiling distilled water; mix and make up to 1 L with distilled water; autoclave for 20 minutes at 121°C; store at 4°C for 15 days). 0.5 mM Biotin/Histidine solution (dissolve 30.5 mg of D-biotin in 100 mL of boiling distilled; separately dissolve 26.2 mg of L-histidine monohydrochloride in 100 mL of distilled water; mix both solutions and make up to 250 mL with distilled water; sterilize by filtration through a 0.22 μm membrane filter, store at 4°C for 30 days. See *Observation* 5). 0.5 mM Tryptophan solution (dissolve 3 mg of L-Tryptophan in 10 mL of distilled water; sterilize by filtration through a 0.22 μm membrane filter, store at 4°C for 30 days. See *Observation* 5). 0.1% w/v Crystal violet solution (dissolve 0.1 g of Crystal violet in 100 mL of distilled water sterilize by filtration through a 0.22 μm membrane filter, store at 4°C. See *Observation* 6). 0.8%, w/v Ampicillin solution (dissolve 8 mg of Ampicillin sodium salt in 100 mL of warm distilled water (65°C) sterilize by filtration through a 0.22 μm membrane filter, store at 4°C for 30 days). 0.1 M histidine solution (dissolve 209.6 mg of L-Histidine monohydrochloride in 10 mL of distilled water; sterilize by filtration through a 0.22 μm membrane filter, store at 4°C for 30 days. See *Observation* 5). 0.5 mM biotin solution (dissolve 2.4 mg of D-biotin in 10 mL of boiling distilled water; mix and make up to 20 mL with distilled water; sterilize by filtration through a 0.22 μm membrane filter, store at 4°C for 30 days. See *Observation* 5). 0.1 M Tryptophan solution (dissolve 204.2 mg of L-Tryptophan in 10 mL of distilled water; sterilize by filtration through a 0.22 μm membrane filter, store at 4°C for 30 days. See *Observation* 5). 0.1 M, pH 7.4 sodium phosphate buffer (see reagents in lipid peroxidation assay). Co-factors for metabolic activation system, S9 fraction, metabolic activation system (MAS) (see reagents and media in *Bacillus subtilis* rec-assay)

Procedures
Preparation of Frozen Permanent Cultures
Strains are usually received as stab cultures. Using a sterile platinum wire, inoculate 10 mL of nutrient broth (supplemented with 100 μL of 0.8% w/v ampicillin solution for strains carrying plasmid pKM101) with a loopful from the stab culture. Incubate at 37°C for 24 h. Distribute 1 mL samples into sterile 2 mL cryotubes. Add 90 μL of DMSO to each tube, mix gently and store the tube at –80°C for 1 year or at –20°C for 6 months (see *Observation* 7).

Identification of Test System
Strains will be identified by having certain characteristics (Mortelmans and Zeiger, 2000). The strains also yield spontaneous revertant colony plate counts within the frequency ranges stated in the historical control data.

(i) **Histidine requirements:** (for *Salmonella* strains) streak a loopful (or 100 μL) of the overnight culture across a GM agar plate supplemented with 100 μL of 0.5 mM D-biotin solution. The histidine requirement is confirmed if no growth is observed after incubation at 37°C for 24 h.

(ii) **Biotin requirements:** (for *Salmonella* strains) streak a loopful (or 100 µL) of the overnight culture across a GM agar plate supplemented with 100 µL of 0.1 M L-histidine solution. The biotin requirement is confirmed if no growth is observed after incubation at 37°C for 24 h. This assay would not be necessary for TA102 (it is biotin independent).

(iii) **Tryptophan requirements:** (for *E. coli* strains) streak a loopful (or 100 µL) of the overnight culture across the surface of GM agar plates, one without and one with 100 µL of 0.1 M L-tryptophan solution. The tryptophan requirement is confirmed if growth is observed on the GM agar plate supplemented with tryptophan and no growth is observed on the unsupplemented GM agar plate after incubation at 37°C for 24 h.

(iv) **Rfa mutation:** Add 100 µL of a fresh overnight culture ($1–2 \times 10^8$ viable cells) to a test tube containing 2.2 mL of molten top agar (see *Observation* 8). After vigorously vortex, the mixture is poured on LB plates. Cover the plates and leave till reach complete top agar solidification (2-3 minutes). Add 10 µL of 0.1% w/v crystal violet solution to 5 mm sterile filter paper disk (Whatman N° 1). Place the paper disks in the centre of the plates. Rfa mutation is confirmed if all strains show a zone of growth inhibition surrounding the disk after incubation at 37°C for 24 h.

(v) **Uvr mutation:** Set 4 LB plates for each strain. Streak 100 µL of a fresh overnight culture ($1–2 \times 10^8$ viable cells) on LB plates surface. Expose the surface of the first plate to UV radiation (254 nm, 3.5 J/m^2) for 5 seconds. Increase the exposition time to 10 seconds, 15 seconds and 30 seconds for the following plates. Uvr mutation is confirmed if an inversal relationship between growth and exposition time is observed after incubation at 37°C for 24 h.

(vi) **Presence of plasmid pKM101 (ampicillin resistance):** Streak a loopful of the pKM101-carrying strain culture across a LB plate supplemented with 100 µL of 0.8% w/v Ampicillin solution. A control strain (which does not carry plasmid pKM101) should be included. Growth should be observed after incubation at 37°C for 24 h, only with strains carrying pKM101 (all listed strains but TA 1535 and 1537).

(vii) **Oxy-R mutation:** this phenotype can be verified by standard plate incorporation assay. As *E. coli* WP2 uvrA/*OxyR* pKM101 carries mutations in *oxyR* gen, the sensibility is increased for the same dose of a mutagenic agent at least two times compared with the OxyR proficient strain (*E.coli* WP2 uvrA/pKM101) (Blanco *et al.*, 1998).

Spontaneous Mutant Rate

The number of spontaneous revertants varies between experiments (Storz *et al.*, 1987), therefore it is recommended inclusion of at least three control plates for each strain in a mutagenicity assay. This is especially important when the mutagen is weak. Any deviation from a laboratories historical range indicates that genetic characteristics of the strain may be altered and this should be tested. Each laboratory should establish its own historical range as these ranges are likely to vary from laboratory to laboratory. Abnormally high spontaneous reversion may indicate contamination or accumulation of back mutations by repeated sub-culturing. If this occurs, the strain should be recovered by re-isolation from the frozen permanent copy. Table 4 shows a guide for a typical spontaneous mutant rate for each strain, but it is recommended that each laboratory establish its own historical data base.

Table 4 Spontaneous revertant control values (number of revertants/plate)

Strain	Without MAS	With MAS
Salmonella typhimurium		
TA97	75–200	100–200
TA98	20-50	20-50
TA100	60-220	60-220
TA1535	5-50	5-50
TA1537	5-25	5-25
TA102	100-300	200-400
Escherichia coli		
WP2 uvrA pKM101	45-151	45-151
WP2 uvrA/OxyR pKM101	60-175	75-200

Standard Plate Incorporation Assay

(i) Leave a frozen permanent tube at room temperature till unfrozen (5 minutes approximately). Inoculate 100 µL into 10 mL of nutrient broth and incubate at 37°C for 24 h.

(ii) Label an appropriate number of GM agar plates and sterile test tubes for each test chemical.

(iii) Prepare MAS (metabolic activation system) if needed (see *Observation* 5).

(iv) Bring 0.5 mM tryptophan solution and/or 0.5 mM histidine-biotin solution into ice bath.

(v) Add 100 µL of a fresh overnight culture ($1–2 \times 10^8$ viable cells) to a test tube containing 2.2 mL of molten top agar (see *Observation* 8), 100 µL of a suitable dilution of the test compound, and 0.5 mL of MAS (if metabolic activation is needed) or 0.5 mL of 0.1 M, pH 7.4 sodium phosphate buffer (when MAS is not used). Make up to 3 mL with 0.1 M, pH 7.4 sodium phosphate buffer. Mix and pour on GM plates. Cover the plates and leave till complete top agar solidification (2-3 minutes).

(vi) Incubate the plates inverted at 37°C for 48 h.

(vii) Count the colonies from the bottom of the Petri plates by marking each colony with a marker pen, and record the number of colonies (see *Observation* 9).

(viii) To evaluate the number of pre-plating mutants originated during the overnight growth ("spontaneous revertants control"), the assay is performed as indicated in step 5, with or without MAS, without test compound added. This is the "negative control" of the assay.

(ix) "Growth controls" are prepared as indicated in step 5, plus 100 µL of 0.5 mM tryptophan solution (for *E. coli* strains), and 100 µL of 0.5 mM histidine-biotin solution (for *Salmonella* strains), with or without MAS added, without test compound. (See *Observation* 10.)

(x) "Antimutagenic activity" can also be evaluated. The assay is performed as indicated in step 5, plus 100 µL of mutagenic agent (see *Observation* 11). "Growth controls" for this assay should include 100 µL of 0.5 mM histidine-biotin solution (for *Salmonella* strains) or 100 µL of 0.5 mM tryptophan solution (for *E. coli* strains), and 100 µL of mutagenic agent (with or without MAS added).

Pre-incubation Assay

This is a modification of the standard plate incorporation assay, suitable for those mutagenic agents that could not have good chances to affect the strains by the standard plate incorporation assay (e.g. short-lived mutagenic metabolites). The small volume of the preincubation mixture allows better chances for reaction between the chemical agent and the strain.

(i) Leave a frozen permanent tube at room temperature till unfrozen (5 minutes approximately). Inoculate 100 µL into 10 mL of nutrient broth and incubate at 37°C for 24 h.

(ii) Label an appropriate number of GM agar plates and sterile test tubes for each test chemical.

(iii) Prepare MAS (metabolic activation system), if needed and keep it on ice until use.

(iv) Bring 0.5 mM tryptophan solution and/or 0.5 mM histidine-biotin solution into ice bath.

(v) Add 100 µL of test substance, and 100 µL of a fresh overnight culture (1-2 $\times 10^8$ viable cells) into a test tube in a small volume (0.5 mL) of either phosphate buffer (for direct mutagens evaluation) or MAS (for indirect mutagens evaluation); aerate the tubes by using a stroke shaker (for 20 to 30 minutes) at 30-37°C.

(vi) After incubation add the mixture to 2.2 mL of molten top agar (see *Observation* 8), make up to 3 mL with 0.1 M, pH 7.4 sodium phosphate buffer. Mix and pour on GM plates. Cover the plates and leave till complete top agar solidification (2-3 minutes).

(vii) Incubate the plates inverted at 37°C for 48 h.

(viii) Count the colonies from the bottom of the Petri plates by marking each colony with a marker pen, and record the number of colonies (see *Observation* 9).

(ix) To evaluate the number of pre-plating mutants originated during the overnight growth ("spontaneous revertants control"), the assay is performed as indicated in step 5., with or without MAS, without test compound added. This is the "negative control" of the assay.

(x) "Growth controls" are prepared as indicated in step 5, plus 100 µL of 0.5 mM tryptophan solution (for *E. coli* strains), and 100 µL of 0.5 mM histidine-biotin solution (for *Salmonella* strains), with or without MAS added, without test compound. (See *Observation* 10.)

(xi) "Antimutagenic activity" can also be evaluated. The assay is performed as indicated in step 5, plus 100 µL of mutagenic agent (see *Observation* 11). "Growth controls" for this assay should include 100 µL of 0.5 mM histidine-biotin solution (for *Salmonella* strains) or 100 µL of 0.5 mM tryptophan solution (for *E. coli* strains), and 100 µL of mutagenic agent (with or without MAS added).

Assay Design Protocol

For broad screening aims, use TA98 and TA100 strains for a row approach (with and without MAS). The assay should be repeated with the specific strain only if a positive response is obtained. When negative results are obtained, TA1535, TA1537 (or TA97 or TA97a) strains are used, with and without MAS. For cross-linking mutagens detection TA102 or *E. coli* WP2 (pKM101) should be included. TA102 is recommended when the chemical is suspected to cause oxidative DNA damage (Mortelmans and Riccio, 2000), though *E. coli* WP2 (pKM101)/OxyR is more sensitive for detecting these kind of mutagens.

Observations

(1) The preparation should be performed into laminar flow hood.

(2) Sterile VB salts and sterile glucose solutions should be added slowly, and the agar flask should be agitated to avoid precipitation.

(3) Prior to use, the stored plates are incubated overnight at 37°C to check sterility.

(4) Do not autoclave the agar added with VB salts and glucose.

(5) The solution should be kept in ice as the mutagenicity assay is carried out.

(6) Warning! Crystal violet causes severe eye irritation.

(7) One permanent frozen culture tube should be used for each experiment, and the unused portion might be discarded.

(8) Top agar should be kept between 43°C and 48°C to minimize prolonged heat exposure, which may kill the bacteria.

(9) The counting process could be carried out with the aid of a colony counter, or by hand with a marker pen.

(10) At least two independent experiments should be made for each strain, with four concentration levels per experiment, and three replicates for each concentration (Good Laboratory Practice, 2001).

(11) Be sure to include MAS if the mutagenic agent needs it.

(10) A preliminary toxicity determination is recommended before start the mutagenicity assay. *S. typhimurium* TA100 is generally suitable for that purpose. The chemical should be tested for toxicity over a broad range of concentrations, with the highest dose limited by solubility, or by an arbitrary value (usually 5000 or 10,000 µg/plate).

(11) Doses for definitive assay: For toxic chemicals, only the highest dose used should exhibit toxicity. For non-toxic chemicals, higher values than 5000 or 10000 µg/plate are acceptable.

(12) This assay is generally used in the evaluation of solids substances, but there are many types of substances, as liquids or gases that can also be evaluated by this assay with some modifications.

Calculations

Calculate mutagenic index as follows:

$$MI = Rst/Rgc$$

where, MI is the mutagenic index, Rst is the number of revertants in the test sample and Rgc is the number of revertants in the growth control. MI and the means of revertant colonies per plate ± the S.D (of at least three replicates) should be presented. The concentration of the test chemical is expressed as mg/plate or µg/plate. A positive result (mutagenic activity) is defined as a reproducible, dose-related increase in the number of revertants. The increase should be three times the number of spontaneous revertants (i.e. MI > 3) or two times for strain TA100. A compound is considered a non-mutagen (negative result) if all quotients range between 1.0 (or lower) and 1.6. A non-existent dose effect relationship could underline this conclusion. A positive result indicates that the test compound induces mutations in *Salmonella typhimurium* and/or *E. coli* cells (Mahon *et al.*, 1989). A mutagenic response should be judged oxidative in nature when it was higher in *E. coli* WP2 uvrA/*OxyR* pKM101 than in *E. coli* WP2 uvrA pKM101.

5. CYTOTOXICITY

Experiment 9: Brine Shrimp Lethality Assay

The brine shrimp lethality assay evaluates cytotoxicity of organic compounds. Larvae suspensions of the brine shrimp *Artemia salina* (Leach) hatch are exposed to different concentrations of an allelochemical and the number of survivors are counted and registered after incubation at specific conditions. Compounds that kill 50% of a brine shrimp population (LD_{50}) at concentrations less than 1000 ppm are considered cytotoxics (Meyer *et al.*, 1982).

Materials and equipments required

Artemia salina (Leach) brine shrimp eggs (commercially available at pet shops); analytical balance; glass tubes; pipettes; micropipettes; tips for micropipettes; Pasteur pipettes; vortex agitator; 100 W lamp; thermometer; 50 mm Petri dishes; small half-covered container (7 × 5 × 2 cm); aluminum foil; magnifying glass.

Reagents

Artificial sea water (dissolve 3.8 g of NaCl in 1 L of distilled water).

Procedures

(i) Add the sea water to the half covered container.

(ii) Add shrimp eggs to the covered part of the container (see *Observations* 1 and 2).

(iii) Turn on the lamp and place over the container (see *Observation* 3).

(iv) Leave shrimps at room temperature (22-29°C) for 17 h, to hatch and mature (see *Observation* 4).

(v) Prepare dilutions of test compounds (see *Observations* 5 and 6).

(vi) Add 100 µL of each dilution and 4.9 mL of sea water into the Petri dishes (see *Observation* 7).

(vii) Add 10 matured shrimps into the Petri dishes (see *Observation* 8).

(viii) Place the dishes under the lamp light at room temperature for 18 h.

(ix) Count and record the number of surviving shrimps with the aid of a 3× magnifying glass (see *Observation* 9).

Observations

(1) Brine shrimp eggs are available as fish food in pet shops.

(2) The container should be half covered with aluminum foil.

(3) The lamp should be placed at 50 cm above the container surface.

(4) The incubation period is variable (24-48 h), and depends of many factors (temperature, illumination and *A. salina* batch).

(5) The most recommended is water. However, 96% ethanol can be used. Alternatively, DMSO could be employed as solvent considering that 50 µL may be added per 5 mL of sea water.

(6) Make at least 5 serial dilutions. It is recommended to test initial concentrations from 10-1000 µg mL^{-1}.

(7) Control vials should be made with the solvent employed for sample dilutions.

(8) Hatched shrimps should be used within 3 h. Use a Pasteur pipette or a micropipette. Count the shrimps in contrast with a lamp light.

(9) At least three independent experiments (N = 3) should be made with three replicates for each assayed concentration.

Calculations

Calculate mean of number of surviving shrimps for each concentration of test compound assayed, considering control as 100% of survival. Lethal concentration (LC$_{50}$) is determined using Probit analysis with a 95% confidence intervals. LC$_{50}$ is the concentration of the test compound needed to kill 50% of the shrimps. Graphical determination of LC$_{50}$ is also possible plotting % of shrimp survival vs. sample concentration.

SUGGESTED READINGS

Alldrick, A.J. and Rowland, I.R. (1985). Activation of the food mutagens IQ and MeIQ by hepatic S-9 fractions derived from various species. *Mutation Research* **144**: 59-62.

Ames, B.N., Lee, F.D. and Durston, W.E. (1973). An improved bacterial test system for the detection and classification of mutagens and carcinogens. *Proceedings of the National Academy of Sciences of the United States of America* **70**: 782-786.

Ames, B.N., McCann, J. and Yamasaki, E. (1975.) Methods for detecting carcinogens and mutagens with *Salmonella*/mammalian-microsome mutagenicity test. *Mutation Research* **31**: 347-364.

Aruoma, O.I. (1996). Assessment of potential prooxidant and antioxidant actions. *Journal of the American Oil Chemists' Society (JAOCS)* **73** (12): 1617-1625.

Aruoma, O.I. (1998). Free radicals, oxidative stress and antioxidants in human health and disease. *Journal of the American Oil Chemists' Society (JAOCS)* **75** (2): 199-212.

Blanco, M., Herrera, G. and Urios, A. (1995). Increased mutability by oxidative stress in OxyR-deficient *Escherichia coli* and *Salmonella typhimurium* cells: clonal occurrence of the mutants during growth on nonselective media. *Mutation Research* **346**: 215-220.

Blanco, M., Urios, A. and Martínez, A. (1998). A New *Escherichia coli* WP2 tester strains highly sensitive to reversion by oxidative mutagens. *Mutation Research* **413**: 95-101.

Ceglowski, P., Luder, G. and Alonso, J.C. (1990). Genetic analysis of rec E activities in *Bacillus subtilis*. *Molecular & General Genetics* **222** (2-3): 441-445.

Chang, W.H. and Julin, D.A. (2001). Structure and Function of the *Escherichia coli* RecE Protein, a Member of the RecB Nuclease Domain Family. *The Journal of Biological Chemistry* **276** (49): 46004-46010.

Clain, E. and Brulfert, A. (1980). Hydroxyurea-induced mitotic synchronization in *Allium sativum* root meristems. *Planta* **150**: 26-31.

Dayan, F.E., Romagni, J.G. and Duke, S.O. (2000). Investigating the mode of action of natural phytotoxins. *Journal of Chemical Ecology* **26**: 2079-2094.

Del Maestro, R.F. (1980). An approach to free radicals in medicine and biology. *Acta Physiologica Scandinavica supplement* **492**: 153-168.

Dunkel, V.C., Zeiger, E., Brusick, D., McCoy, E., McGregor, D., Mortelmans, K., Rosenkranz, H.S. and Simmon, V.F. (1984). Reproducibility of microbial mutagenicity assay: 1. Test with *Salmonella typhimurium* and *Escherichia coli* using a standardized protocol. *Environmental Mutagenesis* **6**: 1-254.

Finney, D.I. (1971). *Probit Analysis*. 3rd Ed. Cambridge University Press, London.

Food and Drug Administration (FDA). Center for Veterinary Medicine. (2000). *Safety Studies for Veterinary Drug Residues in Human Food: Genotoxicity Studies* — Draft guidance. U.S. Department of Health and Human Services.

Good Laboratory Practice. *Regulations of the EC Enacted in Germany in the Chemikaliengesetz.* (Chemicals Act), dated July 25, 1994, BGBl. I, p. 1703 (1994). Modifications of the Appendix I of the Chemicals Act dated May 22, 1997, BGBl. I p. 1060 (1997) and May 8, 2001, BGBl. I, p. 843 (2001).

Green, M.H.L. and Muriel, W.J. (1976). Mutagen testing using trp+ reversion in *Escherichia coli*. *Mutation Research* **38**: 3-32.

Gutteridge, J.M.C. and Halliwell, B. (1988). The deoxyribose assay: an assay both for 'free' hydroxyl radical and for site-specific hydroxyl radical production. *The Biochemical Journal* **253**: 932-933.

Hale, A.L., Meepagala, K.M., Giovanni Aliotta, A.O. and Duke, S.O. (2004). Phytotoxins from the leaves of *Ruta graveolens*. *Journal of Agricultural and Food Chemistry* **52**: 3345-3349.

Halliwell, B. and Aruoma, O.I. (1991). DNA damage by oxygen-derived species. Its mechanism and measurement in mammalian systems. *FEBS Letters* **281**: 9-19.

Halliwell, B., Gutteridge, J.M.C. and Cross, C.E. (1992). Free radicals, antioxidants and human disease: Where are we now?. *The Journal of Laboratory and Clinical Medicine* **119**: 598-620.

Hamouda, T., Shih, A.Y. and Baker Jr., J.R. (2002). A rapid staining technique for the detection of the initiation of germination of bacterial spores. *Letters in Applied Microbiology* **34**: 86-90.

Hartman, Z., Hartman, P.E., Barnes, W.M. and Tuley, E. (1984). Spontaneous mutation frequencies in *Salmonella*: enhancement of G/C to A/T transitions and depression of deletion and frameshift mutation frequencies afforded by anoxic incubation. *Environmental and Molecular Mutagenesis* **6**: 633-650.

Hirano, K., Hagiwara, T., Ohta, Y., Matsumoto, H. and Kada, T. (1982). Rec-Assay with spores of *Bacillus subtilis* with and without metabolic activation. *Mutation Research* **97**: 339-347.

Jaiswal, R., Khan, M.A. and Musarrat, J. (2002). Photosensitized paraquat-induced structural alterations and free radical mediated fragmentation of serum albumin. *Journal of Photochemistry and Photobiology B, Biology* **67**: 163-170.

Kada, T., Tutikawa, T. and Sadaie, Y. (1972). In vitro and host-mediated "rec assay" procedures for screening chemical mutagens and phloxine, a mutagenic red dye detected. *Mutation Research* **16**: 165-174.

Kang, D.G., Yun, C.K. and Lee, H.S. (2003). Screening and comparison of antioxidant activity of solvent extracts of herbal medicines used in Korea. *Journal of Ethnopharmacology* **87**: 231-236.

Kaur, C. and Kapoor, H.C. (2001). Antioxidants in fruits and vegetables — the millennium's health. *International Journal of Food Science and Technology* **36**: 703-725.

Kulkarni, A.P., Aradhya, S.M. and Divakar, S. (2004). Isolation and identification of a radical scavenging antioxidant punicalagin from pith and carpellary membrane of pomegranate fruit. *Food Chemistry* **87**: 551-557.

Kusano, K., Sunohara, Y., Takahashi, N., Yoshikura, H. and Kobayashi, I. (1994). DNA double-strand break repair: Genetic determinants of flanking crossing-over. *Proceedings of the National Academy of Sciences of the United States of America* **91**: 1173-1177.

Leifer, Z., Kada, T., Mandel, M., Zeiger, E., Stafford, R. and Rosenkranz, H.S. (1981). An evaluation of tests using DNA repair-deficient bacteria for predicting genotoxicity and carcinogenicity. A report of the U.S. EPA's Gene-TOX Program. *Mutation Research* **87** : 211-97.

Levin, D.E., Hollstein, M.C., Christman, M.F., Schwiers, E.A. and Ames, B.N. (1982). A new *Salmonella* tester strain (TA102) with A:T base pairs at the site of mutation detects oxidative mutagens. *Proceedings of the National Academy of Sciences of the United States of America* **79**: 7445-7449.

Mahon, G.A.T., Green, M.H.L., Middleton, B., Mitchell, I. de G., Robinson, W.D. and Tweats, D.J. (1989). Analysis of data from microbial colony assays. In: *UKEMS Sub-committee on Guidelines for Mutagenicity Testing. Report: Part III. Statistical Evaluation of Mutagenicity Test Data.* (Ed., D.J. Kirkland). Cambridge University Press, Cambridge, UK.

Matsui, S. (1988). The *Bacillus subtilis*/microsome rec-assay for the detection of DNA damaging substances in waters of night soil treatment. *Toxicity Assessment: An International Journal* **3**: 173-193.

Mazza, G. (1982). *Bacillus subtilis* "rec assay" test with isogenic strains. *Applied and Environmental Microbiology* **43**: 177-184.

Meyer, B.N., Ferrigni, N.R., Putnam, J.E., Jacobsen, L.B., Nichols, D.E. and McLaughlin, J.L. (1982). Brine Shrimp: A convenient general bioassay for active plant constituents. *Planta Medica* **45**: 31-34.

Moorhouse, P.C., Grootveld, M., Halliwell, B., Quinlan, G. and Guttetidge, J.M.C. (1987). Allopurinol and oxypurinol are hydroxyl radical scavengers. *FEBS Letters* **213**: 23-28.

Mortelmans, K. and Riccio, E.S. (2000). The bacterial tryptophan reverse mutation assay with *Escherichia coli* WP2. *Mutation Research: Fundamental and Molecular Mechanisms of Mutagenesis* **455**: 61-69.

Mortelmans, K. and Zeiger, E. (2000). The Ames Salmonella Microsome Mutagenicity Assay. *Mutation Research: Fundamental and Molecular Mechanisms of Mutagenesis* **455**: 29-60.

Rahman, A.U., Choudhary, M.I. and Thomsen, W.J. (2001). *Bioassay Techniques for Drug Development.* Harwood Academic Publishers, Amsterdam, Netherlands.

Rodríguez, G.S. and Salabert de Campos, J.M. (2006). Análise de mutagenicidade em lodo de esgoto. *Workshop on Plant Bioassays for the Detection of Mutagens and Carcinogens in the Environment.* May 22 to 26, 2006. CENA-USP. Piracicaba. Brazil. Availavble online at: www.sbmcta.org.br/workshop-plant/Download-br/S01.pdf (15-08-06).

Ruzin, S.E. (1999). Staining. In: *Plant Microtechnique and Microscopy.* (Ed., S. Ruzin). Oxford University Press, New York, USA.

Sanchez-Moreno, C. (2002). Review: methods used to evaluate the free radical scavenging activity in foods and biological systems. *Food Science and Technology International* **8**: 121-137.

Shon, M.Y., Kim, H.K. and Sung, N.J. (2003). Antioxidants and free radical scavenging activity of *Phellinus baumii* (Phellinus of Hymenochaetaceae) extracts. *Food Chemistry* **82**: 593-597.

Singh, R. P., Sharad, S. and Kapur, S. (2004). Free radicals and oxidative stress in neurodegenerative diseases: relevance of dietary antioxidants. *Journal, Indian Academy of Clinical Medicine* **5**: 218-225.

Storz, G., Christman, M.F., Sies, H. and Ames, B.N. (1987). Spontaneous mutagenesis and oxidative damage to DNA in *Salmonella typhimurium*. *Proceedings of the National Academy of Sciences of the United States of America* **84**: 8917-8921.

Takigami, H., Matsui, S., Matsuda, T. and Shimizu, Y. (2002). The *Bacillus subtilis* rec-assay: a powerful tool for the detection of genotoxic substances in the water environment. Prospect for assessing potential impact of pollutants from stabilized wastes. *Waste Management* **22**: 209-213.

Wisniewska, H. and Chelkowski, J. (1994). Influence of deoxynivalenol on mitosis of root tip cells of wheat seedlings. *Acta Physiologiae Plantarum* **16**: 159-162.

Wonisch, A., Tausz, M., Müller, M., Weidner, W., De Kok, L.J. and Grill, D. (1999). Treatment of young spruce shoots with SO_2 and H_2S effects on fine root chromosomes in relation to changes in the thiol content and redox state. *Water, Air and Soil Pollution* **116**: 423-428.

Yamaguchi, T., Takamura, H., Matoba, T. and Terao, Y. (1998). HPLC method for evaluation of the free radical scavenging activity of food by using 1,1-diphenyl-2-picrylhydrazil. *Bioscience, Biotechnology and Biochemistry* **62**: 1201-1204.

Zetterberg, A. and Larsson, O. (1985). Kinetic analysis of regulatory events in G1 leading to proliferation or quiescence of Swiss 3T3 cells. *Proceedings of the National Academy of Sciences of the United States of America* **82**: 5365-5369.

Bioassays: Inhibitors of Insect Chitin-degrading Enzymes

Teruhiko Nitoda* and Hiroshi Kanzaki

1. INTRODUCTION

Chitin is a structural component of the cuticle and the peritrophic membrane in the mid-gut of insects. Strict regulation of its metabolism is essential for normal insect growth. Chitinase and β-N-acetylglucosaminidase (GlcNAcase) are key enzymes in chitin degradation. Hence, their inhibitors are expected to be biorational insect growth regulators (IGRs). Several compounds have been reported so far as inhibitors of chitinase or GlcNAcase (Sakuda *et al.*, 1987; Aoyagi *et al.*, 1992; Spindler and Spindler-Barth, 1999). However, most of them show inhibitory activity not only against insect enzymes but also against the enzymes of other organisms such as mammals, plants and microorganisms. Therefore, they are hardly developed as IGRs. To screen inhibitors specific for insect chitin-degrading enzymes, we have developed assays for inhibitors of *Spodoptera litura* chitinase and GlcNA case (Kawazu *et al.*, 1996; Nitoda *et al.*, 1999; Usuki *et al.*, 2006), which are applicable to enzymes of other *Spodoptera* species. This chapter describes the methods applied for these assays.

Experiment 1: Preparation of Colloidal Chitin Powder

Chitin, a natural substrate for chitinase, is insoluble in water and common organic solvents, and it takes long time to digest by chitinase. Hence, chitin is not suitable for direct use as a substrate for a rapid chitinase assay. Colloidal chitin obtained

Authors' address: *The Graduate School of Natural Science and Technology, Okayama University, Okayama 700-8530, Japan.
Corresponding author: nitoda@cc.okayama-u.ac.jp

by the dissolution of chitin and the reprecipitation from an acid is readily digested by chitinase. In this experiment, colloidal chitin is prepared for chitinase assay (Shimahara and Takiguchi, 1988; Nitoda *et al.*, 1999).

Materials and equipments required

Chitin powder (crab shell); analytical balance; crushed ice; measuring cylinders; beakers; magnetic stirrer; water bath (30°C, 37°C); glass filter (pore size 40-100 μm); universal pH indicator paper; dialysis tubes (seamless cellulose tubing available from Viskase Companies, Inc., Illinois, USA); dialysis tube closures (available from Spectrum Laboratories, Inc., California, USA); freeze dryer; ultrasonic cleaner; autopipettes; Pasteur pipettes; spectrophotometer; photometer cell with a path length of 10 mm and a width of 2 mm.

Reagents: NaCl; 12 M HCl; 0.5 N NaOH.

Procedure

Chitin powder (2.5 g) was slowly added over 5 minutes into 100 mL of 12 M HCl below 5°C (in an ice bath) with stirring. After 10 minutes, the suspension was heated gently up to 30°C in a water bath and kept for 60 minutes with stirring. The suspension will become clearer during the stirring and finally a clear solution is obtained. The solution is filtered quickly through a glass filter (pore size 40-100 μm) to remove insoluble residues. The solution is poured into 1000 mL of deionized water below 5°C with vigorous stirring. Within a few minutes, the solution will become turbid because of chitin reprecipitation. After 30 minutes, stirring is stopped and then the suspension is kept overnight below 5°C. The supernatant is removed by decantation and ice-cold deionized water (800 mL) was added to the precipitate. The suspension is stirred and kept until the precipitate has almost settled. This precipitation and decantation process is repeated until the pH rise above 2. Then, suspension is neutralized with 0.5 N NaOH and kept for 30 minutes at room temperature. The supernatant is removed by decantation and the precipitate is resuspended in a small amount of tap water and dialyzed. Dialysis tubes must be boiled in deionized water for 10 minutes and washed with running tap water to remove glycerol coating before use. One end of the tube is closed tightly either by putting a knot or using a dialysis tube closure so that no liquid leaks. The suspension of the precipitate above is carefully poured into each tube. The tube must not be completely filled (about 3/4 full). Any excess air is squeezed out and the top end of each tube is closed tightly either by putting a knot or using a dialysis tube closure. The dialysis bags are immersed in running tap water overnight at room temperature. The dialyzed suspension is freeze-dried to obtain a white powder (1.0-1.5 g).

Observations

Dispersibility of the obtained colloidal chitin powder must be evaluated to determine whether it can be used or not as a substrate for chitinase assay (Experiment 3). To do it, a colloidal chitin powder (30-50 mg) is added to distilled water (5 mL) and completely dispersed using an ultrasonic cleaner to obtain a homogeneous suspension with a milky white appearance, which is used as "substrate suspension" in the chitinase assay. This suspension (20 µL) is mixed with buffer A (380 µL) and transferred into a photometer cell (path length of 10 mm, width of 2 mm), gently and thoroughly mixed in the cell by sucking and blowing with a Pasteur pipette and the absorbance at 610 nm of the diluted suspension is measured with a spectrophotometer. This absorbance value must be higher than 0.5 to ensure a sensitive detection of chitinase activity. Absorbance must not fluctuate (difference before and after 1 h-incubation at 37°C should be less than 0.05).

Precautions
The yield and dispersibility of colloidal chitin powder depend on purity, powder size and other characteristics of chitin used. Therefore, several trials and modifications of experimental conditions would be required to obtain a colloidal chitin powder suitable for chitinase assay.

Experiment 2: Preparation of a Chitin-degrading Enzyme Solution
A crude chitin-degrading enzyme solution is obtained by ammonium sulfate precipitation. Ultracentrifugation makes the solution more stable by removing factors that would lead to enzyme inactivation (Kawazu *et al.*, 1996). The obtained enzyme solution shows both chitinase and GlcNAcase activity.

Materials and equipments required
Young pupae of *S. litura*: the larvae of *S. litura* were reared on an artificial diet (Insecta LFS, Nihon-Nousan Kogyo, Yokohama, Japan) at 25°C under 16 hours light and 8 hours dark conditions, collected just after pupation and stored at −80°C; ultrafreezer (−80°C); analytical balance; crushed ice; measuring cylinders; beakers; Pasteur pipettes; centrifuge tubes; mortar and pestle; pH meter; gauze; magnetic stirrer; refrigerated preparative centrifuge capable of 20,000 × g; dialysis tubes (see Experiment 1, *Materials required*); dialysis tube closures (see Experiment 1, *Materials required*); autopipettes; 1.5 mL or 0.6 mL microcentrifuge tubes; preparative ultracentrifuge capable of 100,000 × g.

Reagents
Ammonium sulfate, 14.3 mM buffer A/PTU (prepare a stock solution dissolving 12.0 g of citric acid monohydrate, 7.79 g of KH_2PO_4 and 3.54 g of boric acid in 2000 mL of distilled water; prepare 14.3 mM buffer A adding 0.2 M NaOH (ca. 500 mL) to 1000 mL of stock solution until pH 7.0; then, fill up to 2000 mL with distilled water; prepare 14.3 mM buffer A/PTU adding 2.0 g of phenylthiourea to 2000 mL of 14.3 mM buffer A; stir vigourously and kept at 5°C overnight).

Procedure
(i) Young pupae (50 g) are homogenized with a mortar and pestle on ice in 50 mL of 14.3 mM buffer A/PTU. The homogenate is filtered through two layers of gauze to remove debris and centrifuged at 20,000 × g at 4°C for 30 minutes.

(ii) The clarified homogenate is centrifuged again at 100,000 × g at 4°C for 60 minutes. Supernatant of second centrifugation is transferred into a beaker placed in an ice bath over a magnetic stirrer.

(iii) Finely ground (with a mortar and pestle) ammonia sulfate is slowly added with constant stirring to make a 60% saturated solution, and the solution is kept on ice with stirring for 60 minutes. The solution is centrifuged at 20,000 × g at 4°C for 15 minutes.

(iv) The precipitate is suspended (with a Pasteur pipette) in a small volume (up to 10 mL) of 14.3 mM buffer A/PTU and dialyzed against 10-fold diluted 14.3 mM buffer A/PTU for 3 h, and then against 14.3 mM buffer A/PTU overnight. The dialyzate is centrifuged at 20,000 × g at 4°C for 15 minutes, and the supernatant is dispensed into 1.5 mL or 0.6 mL microcentrifuge tubes, and stored at −80°C.

(v) The frozen enzyme solution is thawn on ice just before use as an enzyme solution.

Precautions

All procedure should be carried out at temperature between 0 and 4°C. Vigorous agitation, foaming, and repeated freeze and thaw cycles should be avoided. The ultracentrifugation step (100,000 × g at 4°C for 60 minutes) is not essential, but recommended to avoid enzyme inactivation.

Experiment 3: Chitinase Assay

The chitinase activity of a *S. litura* chitin-degrading enzyme solution is measured by a turbidimetric method, in which the degradation of colloidal chitin is monitored by a decrease in light scattering at 610 nm (Yabuki *et al.*, 1986; Kawazu *et al.*, 1996; Nitoda *et al.*, 1999).

Materials and equipments required

Colloidal chitin powder (see Experiment 1), *S. litura* chitin-degrading enzyme solution (see Experiment 2), analytical balance, crushed ice, measuring cylinders, beakers, autopipettes, Pasteur pipettes, pH meter, water bath (37°C), 1.5 mL microcentrifuge tubes, spectrophotometer, photometer cell with 10 mm path length and a width of 2 mm, ultrasonic cleaner.

Reagents

14.3 mM Buffer A (see Experiment 2, *Reagents*), 232 mM buffer A (prepare a stock solution dissolving 4.88 g of citric acid monohydrate, 3.16 g of KH_2PO_4, 1.44 g of boric acid in 50 mL of distilled water; prepare 232 mM buffer A adding 3.2 M NaOH (ca.12.5 mL) to 25 mL of stock solution until pH 7.0; fill up to 50 mL with distilled water).

Procedure

(i) Three kinds of reaction mixtures, namely ES (enzyme-substrate mixture), E (enzyme mixture without substrate addition or substrate blank), and S (substrate mixture without enzyme addition or enzyme blank), are prepared in 1.5 mL-microcentrifuge tubes on ice.

(ii) ES-mixtures are prepared as follow: 20 µL of a substrate suspension (see Experiment 1, *Observations*), 80 µL of 232 mM buffer A, and 200 µL of distilled water are added into 5 tubes. Five different volumes of 14.3 mM buffer A, comprised between 0 and 100 µL (for example, 50, 60, 70, 80, 90 µL), are added into the tubes. A *S. litura* chitin-degrading enzyme solution is then added into each tube to make a total volume of 400 µL per tube. The volumes of reagents used in preparing ES-mixtures are shown in Table 1.

Table 1   Volumes of reagents in preparing ES-mixtures for *Spodoptera litura* chitinase assay

Reagents	Volume (µL) in each tube
Substrate suspension	20
232 mM buffer A	80
Distilled water	200
14.3 mM buffer A	0–100*
Enzyme solution	0–100*
Total reaction volume	400

* Five different volumes of 14.3 mM buffer A, comprised between 0 and 100 µL are added into the 5 tubes. A *Spodoptera litura* chitin-degrading enzyme solution is then added into each tube to make a total volume of 400 µL per tube.

(iii) E-mixtures (5 tubes) are prepared in the same manner as ES-mixtures except for using distilled water instead of a substrate suspension.

(iv) S-mixture (1 tube) is prepared as follows: 20 µL of a substrate suspension, 80 µL of 232 mM buffer A, 200 µL of distilled water, and 100 µL of 14.3 mM buffer A were added into a tube. All reaction mixtures are prepared at least in duplicate.

(v) Each mixture is transferred into a photometer cell (path length of 10 mm, width of 2 mm), mixed thoroughly but gently in the cell by sucking and blowing with a Pasteur pipette, and the absorbance is measured at 610 nm immediately with a spectrophotometer.

(vi) The mixtures are transferred back into 1.5 mL-microcentrifuge tubes and incubated in a water bath at 37°C for 1 hour. Then, tubes are placed back on ice. The absorbance of each mixture is measured at 610 nm just after mixing by sucking and blowing with a Pasteur pipette.

Calculations

The chitinase activity is calculated as follows:

$$\text{Chitinase activity } (\Delta A_{610}) = \Delta A_{610} \text{ (ES)} - \Delta A_{610} \text{ (E)} - \Delta A_{610} \text{ (S)}$$

$$\Delta A_{610} \text{ (ES)} = A_{610} \text{ (ES, 0 minute)} - A_{610} \text{ (ES, 60 minutes)}$$
(for each volume of an enzyme solution)

$$\Delta A_{610} \text{ (E)} = A_{610} \text{ (E, 0 minute)} - A_{610} \text{ (E, 60 minutes)}$$
(for each volume of an enzyme solution)

$$\Delta A_{610} \text{ (S)} = A_{610} \text{ (S, 0 minute)} - A_{610} \text{ (S, 60 minutes)}$$

where A_{610} is absorbance at 610 nm. ES, E and S are reaction mixtures whose absorbances are measured just after preparation (0 minute) and after 1 hour (60 minutes).

Plot the chitinase activity versus the volume of an enzyme solution. A typical plot is shown in Fig. 1.

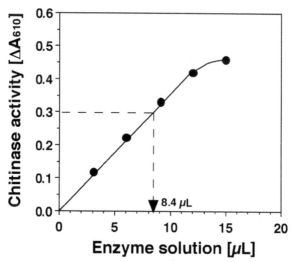

Figure 1 Typical plot of *Spodoptera litura* chitinase activity versus the volume of the enzyme solution. Each closed circle represents the mean of 2 determinations. The Y-axis denotes the decrease in absorbance at 610 nm of the ES-mixture after incubation at 37°C for 1h subtracted by the decrease in absorbance of the E-mixture and S-mixture.

This curve shows a linear relationship in the range of 0 to 12 μl with the volume of an enzyme solution. In this range, find out the volume of an enzyme solution to show activity 0.2-0.3 (ΔA_{610}), which is the amount of the enzyme used in each reaction for the chitinase inhibitory assay (Experiment 3). In Fig. 1, the volume of an enzyme solution to achieve an activity (ΔA_{610}) of 0.3 is 8.4 μL.

Precautions

Spodoptera litura chitin-degrading enzyme solution, 14.3 mM buffer A, 232 mM buffer A and distilled water must be kept on ice. Substrate suspension must be kept at room temperature. Vigorous agitation, foaming should be avoided during the mixing in a photometer cell. *Spodoptera litura* chitin-degrading enzyme solution must be added to the corresponding reaction mixture.

Experiment 4: Chitinase Inhibitory Assay

The inhibitory activity of a sample solution against *S. litura* chitinase is measured by adding a sample solution to the reaction mixture for the chitinase assay (Yabuki *et al.*, 1986; Kawazu *et al.*, 1996; Nitoda *et al.*, 1999).

Materials and equipments required

Sample solution; colloidal chitin powder (see Experiment 1); *S. litura* chitin-degrading enzyme solution (see Experiment 2); analytical balance; crushed ice; measuring cylinders; beakers; autopipettes; Pasteur pipettes; pH meter; water bath (37°C); 1.5 mL microcentrifuge tubes; spectrophotometer; photometer cell with a path length of 10 mm and a width of 2 mm; ultrasonic cleaner.

Reagents

Buffer A (see Experiment 2, *Reagents*); 232 mM buffer A (see Experiment 3, *Reagents*).

Procedure

(i) Six kinds of reaction mixtures, namely ESI (enzyme-substrate-sample mixture), EI (enzyme-sample mixture), and SI (substrate-sample mixture), ES (enzyme-substrate mixture), E (enzyme mixture), and S (substrate mixture), are prepared in 1.5 mL-microcentrifuge tubes on ice.

(ii) ESI-mixtures are prepared as follow: 20 μl of a substrate suspension (see Experiment 1, *Observations*) and 80 μL of 232 mM buffer A are added into 5 tubes. Five different volumes of distilled water, comprised between 0 and 200 μL (for example, 20, 40, 60, 80, 100 μL), are added into the tubes. An aqueous sample solution is then added into each tube to make a total volume of 300 μL per tube. If the test sample is insoluble in water, it is dissolved in methanol or dimethyl sulfoxide (less than 10 μL) and the solution is diluted with distilled water. In this case, all reaction mixtures must be prepared to contain the same amount of solvent. Finally, 100 μL of a diluted *Spodoptera litura* chitin-degrading enzyme solution is added. The diluted enzyme solution is prepared by adding 14.3 mM buffer A to achieve an activity (ΔA_{610}) of 0.2-0.3 (determined in Experiment 2). The volumes of reagents used in preparing ESI-mixtures are shown in Table 2.

(iii) EI-mixtures (5 tubes) are prepared in the same manner as ESI-mixtures except for using distilled water instead of a substrate suspension.

(iv) SI-mixtures (5 tubes) are prepared in the same manner as ESI-mixtures except for using 14.3 mM buffer A instead of a diluted enzyme solution.

(v) ES-mixture (1 tube) is prepared in the same manner as ESI-mixture except for using distilled water instead of sample solution.

(vi) E-mixture (1 tube) is prepared in the same manner as ES-mixture except for using distilled water instead of a substrate suspension.

(vii) S-mixture (1 tube) is prepared in the same manner as ES-mixture except for using 14.3 mM buffer A instead of a diluted enzyme solution. All reaction mixtures are prepared at least in duplicate.

(viii) Each mixture is transferred into a photometer cell (path length of 10 mm, width of 2 mm), mixed thoroughly but gently in the cell by sucking and blowing with a Pasteur pipette, and the absorbance is measured at 610 nm immediately with a spectrophotometer.

(ix) The mixtures are transferred back into 1.5 mL-microcentrifuge tubes and incubated in a water bath (37°C) for 1 hour. Then tubes are placed back on ice. The absorbance of each mixture is measured at 610 nm after mixing by sucking and blowing with a Pasteur pipette.

Table 2 Volumes of reagents in preparing ESI-mixtures for *Spodoptera litura* chitinase inhibitory assay

Reagents	Volume (µL) per tube
Substrate suspension	20
232 mM buffer A	80
Distilled water*	0–200*
Sample solution*	0–200*
Diluted enzyme solution*	100
Total reaction volume	400

* Five different volumes of distilled water, comprised between 0 and 200 µL are added into the 5 tubes. An aqueous sample solution is then added into each tube to make a total volume of 300 µL per tube.

Calculations

The chitinase inhibitory activity is calculated as follows:

$$\% \text{ Inhibition} = 100 - [\text{Activity } \Delta A_{610} \text{ (with a sample solution)}]/$$
$$[\text{Activity } \Delta A_{610} \text{ (without a sample solution)}] \times 100$$
$$= 100 - [\Delta A_{610} \text{ (ESI)} - \Delta A_{610} \text{ (EI)} - \Delta A_{610} \text{ (SI)}]/$$
$$[\Delta A_{610} \text{ (ES)} - \Delta A_{610} \text{ (E)} - \Delta A_{610} \text{ (S)}] \times 100$$

$$\Delta A_{610} \text{ (ES)} = A_{610} \text{ (ES, 0 minute)} - A_{610} \text{ (ES, 60 minutes)}$$
$$\Delta A_{610} \text{ (E)} = A_{610} \text{ (E, 0 minute)} - A_{610} \text{ (E, 60 minutes)}$$
$$\Delta A_{610} \text{ (S)} = A_{610} \text{ (S, 0 minute)} - A_{610} \text{ (S, 60 minutes)}$$
$$\Delta A_{610} \text{ (ESI)} = A_{610} \text{ (ESI, 0 minute)} - A_{610} \text{ (ESI, 60 minutes)}$$
(for each volume of a sample solution)
$$\Delta A_{610} \text{ (EI)} = A_{610} \text{ (EI, 0 minute)} - A_{610} \text{ (EI, 60 minutes)}$$
(for each volume of a sample solution)
$$\Delta A_{610} \text{ (SI)} = A_{610} \text{ (SI, 0 minute)} - A_{610} \text{ (SI, 60 minutes)}$$
(for each volume of a sample solution)

where A_{610} is the absorbance at 610 nm. ESI, EI, SI, ES, E and S are reaction mixtures whose absorbances are measured just after preparation (0 minute) and after 1 hour (60 minutes).

Plot the chitinase inhibitory activity versus the concentration of an aqueous solution of a sample solution. A typical plot is shown in Fig. 2.

The inhibitory activity is expressed as ID_{50} or IC_{50} defined as the weight or concentration required to achieve 50% inhibition of the enzyme activity. The inhibitory activity of a solution can also be expressed tentatively as ID_{50} in volume. In Fig. 2, the ID_{50} of the culture supernatant of a fungal strain is 16 µl.

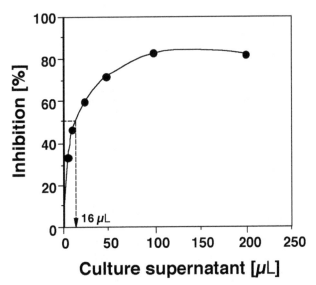

Figure 2 Typical plot of *Spodoptera litura* chitinase inhibitory activity versus volumes of a sample solution. Culture supernatant from a soil fungi strain was used as source of inhibition. Each closed circle represents the mean of 2 determinations. Eight point four microliter of the *Spodoptera litura* chitin-degrading enzyme solution used in Fig. 1 are used for each reaction.

Precautions

Spodoptera litura chitin-degrading enzyme solution, 14.3 mM buffer A, 232 mM buffer A and distilled water must be kept on ice. Substrate suspension must be kept at room temperature. Vigorous agitation and foaming should be avoided during mixing in the photometer cell. The pH of a sample solution must be near neutral, otherwise inactivation of the enzyme may occur. *Spodoptera litura* chitin-degrading enzyme solution must be added to the corresponding reaction mixture.

Experiment 5: GlcNAcase Assay

The GlcNAcase activity of a *S. litura* chitin-degrading enzyme solution is measured by a colorimetric method using *p*-nitrophenyl β-*N*-acetylglucosaminide (pNP-GlcNAc) as a substrate. The release of *p*-nitrophenol from the substrate is determined by measuring the absorbance at 415 nm (Usuki *et al*, 2006).

Materials and equipments required

Spodoptera litura chitin-degrading enzyme solution (see Experiment 2); analytical balance; crushed ice; measuring cylinders; beakers; autopipettes; pH meter; incubator (37°C); 96-well microtiter plate (flat bottom); microplate shaker; 96-well microplate reader equipped with a 415 nm filter; 8-channel autopipette (volume range 10-100 µL).

Reagents

p-Nitrophenyl β-*N*-acetylglucosaminide (pNP-GlcNAc); 14.3 mM buffer B (prepare a stock solution dissolving 1.20 g of citric acid monohydrate, 0.78 g of KH_2PO_4 and 0.35 g of boric acid in 200 mL of distilled water; prepare 14.3 mM buffer B adding 0.2 M NaOH (ca. 40 mL) to 100 mL of stock solution until pH 6.0; then, fill up to 200 mL with distilled water); 648 mM buffer B (prepare a stock solution dissolving 13.62 g of citric acid monohydrate, 8.82 g of KH_2PO_4 and 4.01 g of boric acid in 50 mL of distilled water; prepare 648 mM buffer B adding 8.9 M NaOH (ca. 10 mL) to 25 mL of stock solution until pH 6.0; then, fill up to 50 mL with distilled water).

Procedure

(i) Three kinds of reaction mixtures, namely ES (enzyme-substrate mixture), E (enzyme mixture without substrate addition or substrate blank), and S (substrate mixture without enzyme addition or enzyme blank), are prepared in 96-well microplate wells on ice.

(ii) ES-mixtures are prepared as follow: 16 µL of 5 mM aqueous solution of pNP-GlcNAc, 24 µL of 648 mM buffer B and 80 µL of distilled water are added into 5 wells. Five different volumes of 14.3 mM buffer B, comprised between 0 and 40 µL (for example, 30, 32, 34, 36, 38 µL), are added into the wells. A *Spodoptera litura* chitin-degrading enzyme solution is then added to reach a total volume of 160 µL per well. The volumes of reagents used in preparing ES-mixtures are shown in Table 3.

(iii) E-mixtures (5 wells) are prepared in the same manner as ES-mixtures except for using distilled water instead of 5 mM pNP-GlcNAc.

(iv) S-mixture (1 well) is prepared as follows: 16 µL of a 5 mM pNP-GlcNAc, 24 µL of 648 mM buffer B, 80 µL of distilled water, and 40 µL of 14.3 mM buffer B were added into a well.

(v) All reaction mixtures are prepared at least in duplicate. The microplate is gently shaken with a microplate shaker for 1 minute followed by incubation at 37°C for 1 h. 1.3 M NaOH (100 µL) is added to each well. After gentle shaking for 10 s, absorbance of each well is read at 415 nm on a microplate reader.

Table 3 Volumes of reagents in preparing ES-mixtures for *Spodoptera litura* GlcNAcase assay

Reagents	Volume (µL) per well
5 mM pNP-GlcNAc	16
648 mM buffer B	24
Distilled water	80
14.3 mM buffer B	0–40*
Enzyme solution*	0–40*
Total reaction volume	160

* Five different volumes of 14.3 mM buffer B, comprised between 0 and 40 µl are added into the 5 wells. A *Spodoptera litura* chitin-degrading enzyme solution is then added to reach a total volume of 160 µL per well.

Calculations

The GlcNAcase activity is calculated as follows:

$$\text{GlcNAcase activity } (\Delta A_{415}) = A_{415} \text{ (ES)} - A_{415} \text{ (E)} - A_{415} \text{ (S)}$$
(for each volume of an enzyme solution)

where A_{415} is absorbance at 415 nm. ES, E and S denote the corresponding reaction mixtures.

Plot the GlcNAcase activity versus the volume of an enzyme solution. A typical plot is shown in Fig. 3.

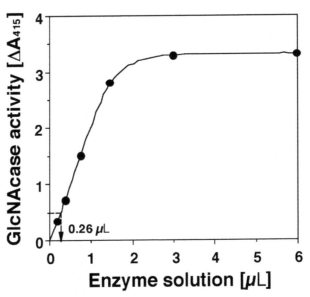

Figure 3 Typical plot of *Spodoptera litura* GlcNAcase activity versus the volume of the enzyme solution. Each closed circle represents the mean of 2 determinations. The Y-axis denotes the absorbance at 415 nm of the ES-mixture subtracted by the absorbance of the E-mixture and S-mixture.

This curve shows a linear relationship in the range of 0 to 1.5 μl with the volume of an enzyme solution. In this range, find out the volume of an enzyme solution to show activity 0.5 (ΔA_{415}), which is the amount of the enzyme used in each reaction for the GlcNAcase inhibitory assay (Experiment 5). In Fig. 3, the volume of an enzyme solution to achieve an activity of 0.5 (ΔA_{415}) is 0.26 μL.

Precautions

Spodoptera litura chitin-degrading enzyme solution, aqueous solution of pNP-GlcNAc, 14.3 mM buffer B, 648 mM buffer B and distilled water must be kept on ice. *Spodoptera litura* chitin-degrading enzyme solution must be added to the corresponding reaction mixture.

Experiment 6: GlcNAcase Inhibitory Assay

The inhibitory activity of a sample solution against *S. litura* GlcNAcase is measured by adding a sample solution to the reaction mixture for the GlcNAcase assay (Usuki *et al*, 2006).

Materials and equipments required

Sample solution; *Spodoptera litura* chitin-degrading enzyme solution (see Experiment 2); analytical balance, crushed ice; measuring cylinders; beakers; autopipettes; pH meter; incubator (37°C); 96-well microtiter plate (flat bottom); microplate shaker; 96-well microplate reader equipped with a 415 nm filter; 8-channel autopipette (volume range 10-100 µL).

Reagents

p-Nitrophenyl β-*N*-acetylglucosaminide (pNP-GlcNAc); 14.3 mM buffer B (see Experiment 4, *Reagents*); 648 mM buffer B (see Experiment 4, *Reagents*).

Procedure

(i) Six kinds of reaction mixtures, namely ESI (enzyme-substrate-sample mixture), EI (enzyme-sample mixture), SI (substrate-sample mixture), ES (enzyme-substrate mixture), E (enzyme mixture) and S (substrate mixture), are prepared in 96-well microplate wells on ice.

(ii) ESI-mixtures are prepared as follow: 16 µL of 5 mM pNP-GlcNAc and 24 µL of 648 mM buffer B are added into 5 wells. Five different volumes of distilled water, comprised between 0 and 80 µL (for example, 40, 50, 60, 70, 80 µL), are added into the wells. An aqueous sample solution is then added to reach a total volume of 120 µL per well. If the test sample is insoluble in water, it is dissolved in dimethyl sulfoxide (less than 8 µL) and the solution is diluted with distilled water. In this case, all reaction mixtures must be prepared to contain the same amount of the solvent. Finally, 40 µL of a diluted *S. litura* chitin-degrading enzyme solution is added. The diluted enzyme solution is prepared by adding 14.3 mM buffer B to achieve an activity (ΔA_{415}) of 0.5 (determined in Experiment 4). The volumes of reagents used in preparing ESI-mixtures are shown in Table 4.

Table 4 Volumes of reagents in preparing ESI-mixtures for *Spodoptera litura* GlcNAcase inhibitory assay

Reagents	Volume (µL) per well
5 mM pNP-GlcNAc	16
648 mM buffer B	24
Distilled water	0–80
Sample solution*	0–80
Diluted enzyme solution*	40
Total reaction volume	160

* Five different volumes of distilled water, comprised between 0 and 80 µL are added into the 5 wells. An aqueous sample solution is then added to reach a total volume of 120 µL per well.

(iii) EI-mixtures (5 wells) are prepared in the same manner as ESI-mixtures except for using distilled water instead of 5 mM pNP-GlcNAc.

(iv) SI-mixtures (5 wells) are prepared in the same manner as ESI-mixtures except for using 14.3 mM buffer B instead of a diluted enzyme solution.

(v) ES-mixture (1 well) is prepared in the same manner as ESI-mixture except for using distilled water instead of sample solution.

(vi) E-mixture (1 well) is prepared in the same manner as ES-mixture except that distilled water is used instead of a substrate suspension.

(vii) S-mixture (1 well) is prepared in the same manner as ES-mixture except that 14.3 mM buffer B is used instead of a diluted enzyme solution.

(viii) All reaction mixtures are prepared at least in duplicate. The microplate is gently shaken with a microplate shaker for 1 minute and incubated at 37°C for 1 h. 1.3 M NaOH (100 μL) is added to each well. After gentle shaking for 10 sec, absorbance of each well is measured at 415 nm on a microplate reader.

Calculations

The GlcNAcase inhibitory activity is calculated as follows:

$$\% \text{ Inhibition} = 100 - [\text{Activity } \Delta A_{415} \text{ (with a sample solution)}/$$
$$\text{Activity } \Delta A_{415} \text{ (without a sample solution)}] \times 100$$
$$= 100 - [A_{415} \text{ (ESI)} - A_{415} \text{ (EI)} - A_{415} \text{ (SI)}]/[A_{415} \text{ (ES)} - A_{415} \text{ (E)}$$
$$- A_{415} \text{ (S)}] \times 100 \quad \text{(for each volume of a sample solution)}$$

where A_{415} (ESI) is absorbance at 415 nm. ESI, EI, SI, ES, E and S denote the corresponding reaction mixtures.

Plot the GlcNAcase inhibitory activity versus the concentration of an aqueous solution of a test material. A typical plot is shown in Fig. 4.

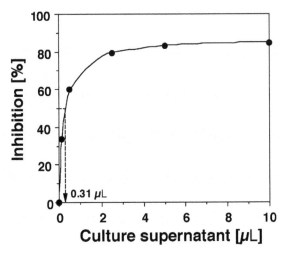

Figure 4 Typical plot of *Spodoptera litura* GlcNAcase inhibitory activity versus volume of a sample solution. The culture supernatant from a soil actinomycete strain was used as source of inhibition. Each closed circle represents the mean of 2 determinations.

The inhibitory activity is expressed as ID_{50} or IC_{50} defined as the weight or concentration required to achieve 50% inhibition of the enzyme activity. The inhibitory activity of a solution can also be expressed as ID_{50} in volume. In Fig. 4, the ID_{50} of the culture supernatant of an actinomycete strain is 0.31 µL.

Precautions

Spodoptera litura chitin-degrading enzyme solution, aqueous solution of pNP-GlcNAc, 14.3 mM buffer B, 648 mM buffer B and distilled water must be kept on ice. *Spodoptera litura* chitin-degrading enzyme solution must be added to the corresponding reaction mixture.

SUGGESTED READINGS

Aoyagi, T., Suda H., Uotani, K., Kojima, F., Aoyama, T., Horiguchi, K., Hamada, M. and Takeuchi, T. (1992). Nagstatin, a new inhibitor of *N*-acetyl-β-D-glucosaininidase, produced by *Streptomyces amakusaensis* MG846-fF3. Taxonomy, production, isolation, physico-chemical properties and biological activities. *Journal of Antibiotics* **45**: 1404-1408.

Kawazu K., Ohnishi S., Kanzaki H. and Kobayashi A. (1996). A stable crude chitinase solution from *Spodoptera litura* pupae and a search for its inhibitors. *Zeitschrift für Naturforschung* **51c**: 738-742.

Nitoda, T., Kurumatani, H., Kanzaki, H. and Kawazu, K. (1999). Improved bioassay method for *Spodoptera litura* chitinase inhibitors using a colloidal chitin powder with a uniform particle size as substrate. *Pesticide Science* **55**: 563-565.

Sakuda, S., Isogai, A., Matsumoto, S. and Suzuki, A. (1987). Search for microbial insect growth regulators II. allosamidin, a novel insect chitinase inhibitor. *Journal of Antibiotics* **40**: 296-300.

Shimahara, K. and Takiguchi, Y. (1988). Preparation of crustacean chitin. *Methods in Enzymology* **61**: 417-423.

Spindler, K.-D. and Spindler-Barth M. (1999). Inhibitors of chitinases. In: *Chitin and Chitinases* (Eds., P. Jollés and R.A.A. Muzzarelli). Birkhäuser Verlag, Basel, Germany.

Usuki, H., Nitoda, T., Okuda, T. and Kanzaki, H. (2006). Screening and partial characterization of inhibitors of insect β-*N*-acetylglucosaminidase. *Journal of Pesticide Science* **31**: 41-46.

Yabuki, M., Mizushima, K., Amatatsu, T., Ando, A., Fujji, T., Shimada, M. and Yamashida, M. (1986). Purification and characterization of chitinase and chitobiase produced by *Aeromonas hydrophila* subsp. *anaerogenes* A52. *Journal of General and Applied Microbiology* **32**: 25-38.

Bioassays: Insect Behaviour, Development and Survival

María C. Carpinella[1], María T. Defagó[2],
Graciela Valladares[2] and Sara M. Palacios[1*]

1. INTRODUCTION

Many natural products affect insect behaviour and physiology and could be used for crop protection. Among them, compounds that inhibit insect feeding, or antifeedants, have received increasing attention (Mordue and Blackwell, 1993). Plant allelochemicals can affect feeding behaviour of insects by attracting or repelling them from relatively long distances, whereas, others act only upon direct contact (Bernays and Chapman, 1994; Mordue and Blackwell, 1993). Antifeedant effects can result from a sensory response of chemoreceptors on the insect mouthparts or other body appendices (Mordue et al., 1998). This represents a primary feeding inhibition, which may culminate in insect starvation and death by feeding deterrency (Mordue and Nisbet, 2000). A secondary mechanism of feeding inhibition observed in most species of phytophagous insects involves a combination of antifeedant and physiological effects after ingestion of plant allelochemicals. In this case, food consumption and digestive efficiency can be reduced not only after ingestion but also by application or injection of antifeedants (Schmutterer, 1985). Such secondary deterrent effects result from

Authors' addresses: [1]Laboratorio de Química Fina y Productos Naturales, Facultad de Ciencias Químicas, Universidad Católica de Córdoba, Camino a Alta Gracia Km 10, (5000) Córdoba, Argentina.
[2]Centro de Investigaciones Entomológicas de Córdoba, Facultad de Ciencias Exactas, Físicas y Naturales, Universidad Nacional de Córdoba, Av. Velez Sarsfield 299, X5000JJC Córdoba, Argentina.
*Corresponding author: E-mail: sarapalacios@campus1.uccor.edu.ar

the disturbance of hormonal and/or other physiological systems (Bernays and Chapman, 1973; Schmidt *et al.*, 1997; Mordue and Nisbet, 2000). In other cases, many natural products act as growth regulators or cause toxicity and mortality, with no disturbance in feeding, affecting physiological processes that are essential for insect life. This chapter describes several methods for evaluation of primary and secondary effects of plant allelochemicals on insect feeding behaviour, development and survival.

2. EXPERIMENTAL METHODS

Experiment 1: Orientational Responses in Y-tube Olfactometers

Olfactometers are devices often used for measuring sensory odour impact. In this experiment, test insects are exposed simultaneously to a solvent and to an allelochemical selected as odour sources, to detect long distance effects on insect feeding behaviour. This technique allows detection of primary antifeedant effects acting at relatively long distance.

Materials and equipments required

Y-tube olfactometer, mortar, fresh host plant leaves, white fluorescent light, filter paper, allelochemical, detergent, air pump, camel hair brush, organic solvent, thermometer, stopwatch.

Procedure

(i) **Preparation of test solution:** The test substance (extracts or pure compounds) is dissolved in water or a proper organic solvent at the desired concentrations.

(ii) **Bioassay set-up:** A strip of filter paper (1.5 by 0.5 cm) is held within an acetate tube (2.5 cm, 1 cm i.d.) (Fig. 1) fitted at the apical end of each diverging arm of the olfactometer (Riddick *et al.*, 2000). A pump air output of 1000 mL minute^{-1} is used to ensure air circulation within the olfactometer (Fig. 1). The latter is positioned horizontally on a countertop in the laboratory at 23-25°C with bright fluorescent illumination. One filter paper strip is painted with the active principle and/or crude extract solution (treatment) and another strip with the corresponding solvent (control). The impregnated strips are placed inside each arm of the Y-tube olfactometer, and an insect is added to the base of the apparatus by removing the tube plug. The plug is repositioned and the air flow is restored through the apparatus. Arms used as treatment and control are alternated after every trial, after carefully washing them. At least 15 replicates should be used for each treatment.

(iii) **Fasting time:** Time elapsed after a meal can affect insect response to food stimuli. Thus, previous to allelochemical testing with olfactometer, it is advisable to assess the threshold time, after a meal, required by the studied species for respond to food presence. This can be achieved by separating

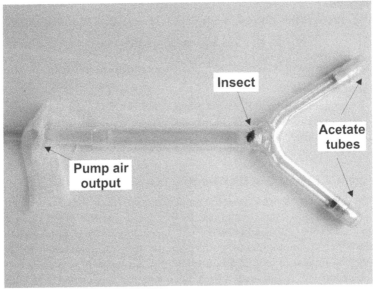

Figure 1 Y-Tube olfactometer

insects in groups and subjecting such groups to different fasting regimes. Water should be provided on a piece of damp material such as cotton wool. A leaf from the host plant is grounded in a mortar, the resulting material being used to impregnate a strip of filter paper (treatment). Another strip is impregnated with water to act as control. The olfactometer is then used as indicated before, and the same variables are measured. The fasting period at which the insect begins to actively respond to its host plant by choosing the treatment arm, should be used in allelochemical tests.

(iv) **Variables:** The amount of time (in seconds) that each insect spends in test versus control arms during a 10 minutes period is recorded with a stopwatch (Riddick *et al.*, 2000). Total time spent by each insect in the treatment arm, number of switches from one arm to the other, and number of individuals first choosing the treatment arm are recorded.

Alternatively, the insect activities can be recorded using the Observer Software System (Noldus, 1991).

Observations

Identical brushes, properly labelled, should be used to impregnate treatment and control filter papers. Experiments should be conducted within the main activity period of the studied species (Acar *et al.*, 2001). The olfactometer must be washed (in water or a suitable solvent) and dried between trials.

Calculations

Repellency Index (RI) is calculated as follows (Hassanali and Lwande, 1989):

$$RI = (Bc - Bt/Bc + Bt)*100$$

where Bc = time spent in control arm and Bt = time spent in treatment arm.

Statistical analysis

Data are compared using the paired t-test. Data not normally distributed are compared using Wilcoxon signed-rank test. When more than two treatments need to be compared, Analysis of Variance or its non-parametrical equivalent (Kruskal Wallis) can be used.

Experiment 2: Choice Test for Chewing Insects

A feeding choice test allows to scan the antifeedant activity of either crude extracts or pure compounds at short distances. In this assay, the test insect can choose between food treated either with the test substance or with the dissolution solvent (control), which are simultaneously offered in the same recipient where the insect is placed. The substance is considered an antifeedant when insects preferentially choose the control over the treated food. Such tests are suitable for Coleopterans, Lepidopterans, grasshoppers, locusts, and other chewing insects.

Materials and equipments required

Leaves, seedlings or whole plants that are usually food of herbivore insects. Immature or adult insects (same age). Petri dishes, plastic containers or entomological cages, tissue paper, filter paper or cotton disks, round glasses (with the same diameter as Petri dishes) with two holes of 1 cm^2 in diameter (optional), distilled water, high quality solvents, high-precision balance, Hamilton syringe and paint brush or atomizer.

Procedure

(i) **Preparation of test solution:** The test substance (extracts or pure compounds) is dissolved in water or in a solvent according with its solubility. The solvents should be selected among those that preserve leaf turgor. Chlorinated solvents should be discarded because they damage leaf structure.

(ii) **Application of the test solution on leaves:** A given concentration of test solution is applied on leaves, whole or cut into 1 cm^2 circles (or in other food sources), in the quantity needed to reach the final desired concentrations. The volume of solvent applied on control leaves is equal to the maximum volume of test solution, used in treated leaves. Both solution and solvent could be applied with a volumetric microsyringe. This procedure is advisable since concentration units of the test principle, in µg/cm^2 of leaf could be measured; thus, effectiveness between different allelochemicals could be compared. A 5 to 20 µl of volume solution is the best for totally covering whithout drowning of the 1 cm^2 leaf disk. Recommended concentrations for complete extracts solution are 0.5 to 10% or a dosage of 100 to 2,000 $\mu g/cm^2$. Test solutions could also be applied to leaves using a paint brush or by dipping leaves in the solution for a certain time. In these forms of application, the amount of allelochemical applied to the leaves is unknown. Alternatively, a known volume of test solution can be sprayed using an atomizer.

(iii) **Bioassay set-up:** Once solvent has been evaporated, both treated and control leaves (or other food source) are placed in the same recipient, as distant as possible, on wet paper or cotton disks in order to maintain moisture and preserve leaf integrity (Fig. 2). A glass disk with two 1 cm^2 holes is placed on leaves (Carpinella *et al.*, 2003), or food is directly offered to the insect. Then, the insect is placed equidistant from both leaves (or other food source) and allowed to eat. Recipients are kept under 27-28°C and 70% relative humidity (Carpinella *et al.*, 2002; 2003; Valladares *et al.*, 1997; 2003). Insects could be deprived of food for at least two hours, depending on the species, prior to the experiment. Ten to 20 replicates should be made for each treatment.

(iv) **Optimal time for the assay:** The optimal time for coleopterans is 24 h, for locusts is about 2 hours and for lepidopterans the time is fixed when the insect has consumed about 50% of the total amount of food available.

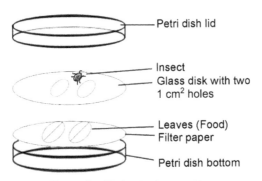

Petri dish lid

Insect
Glass disk with two
1 cm^2 holes

Leaves (Food)
Filter paper

Petri dish bottom

Figure 2 Choice test assembly.

Variables: After a previously determined time of exposure, the relative amounts of the treated and untreated substrate areas eaten in both leaves in each replicate are visually estimated (visual ranking system: recorded in percentage from 0 to 100), by dividing the food area in imaginary quarters. The consumed area could also be digitally estimated. This procedure is carried out by comparing the contrast between the eaten or uneaten surface under a video-image software after taking a photograph of the treated and untreated groups. Feeding could also be evaluated by placing both treated and untreated leaves on graph paper (8.2 divisions per cm) under a stereo microscope and counting the number of units consumed (Landis and Gould, 1989)

Observations

If larvae are near to molting or pupating they should be discarded. Each specimen should be used only once.

Calculations

Antifeedant Index (AI%) is calculated as follows:

$$AI\% = (1 - T/C) \times 100$$

where T and C are the consumption of treated and control leaves, respectively (Carpinella *et al.*, 2003). Values of AI % vary between 0 and 100%; Higher AI % value indicates higher antifeedant effect.

Feeding Ratio Index (FR) is calculated as follows:

$$FR = (T/C)$$

where T are units consumed in treatments and C are units consumed in control leaves (Landis and Gould, 1989). Values of FR vary between 0 and 1; important antifeedant effects are represented by small FR values.

Feeding deterrence (FD%) is calculated as follows:

$$FD\% = (C-T)/(C+T) \times 100$$

where T and C are the mean area eaten in the treated and control leaves, respectively (Koul *et al.*, 1990). This index ranges from 0 to 100% similar to AI%, however the resulting values are lower than those obtained for AI % in the same experiment. Higher FD% value indicates higher antifeedant effect.

Statistical analysis

Consumption data are compared between treated and control leaves by using the Wilcoxon signed paired rank test, $p < 0.05$.

Experiment 3: Choice Test for Granivorous Insects

This test is a variation of the previous assay. It is designed to assess the antifeedant effect of either extracts or pure compounds on granivorous beetles (weevils). The test insect can choose between the food treated either with the test substance or with the dissolution solvent, which are simultaneously offered in the same recipient where the insect is placed. The substance is considered an antifeedant when insects preferentially choose the control instead of the treatment.

Materials and equipments required

Flour, grains or wafers, larvae or adult insects, Petri dishes or plastic containers, distilled water, high quality solvents and high-precision balance are needed. Hamilton syringe, pipette, paint brush or atomizer are used for applying test products.

Procedure

(i) **Preparation of test solution:** The test substance (extracts or pure compounds) is dissolved in water or a proper organic solvent at the desired concentrations.

(ii) **Application of the test solution on flour, wafers or grains:** The test solution is applied at a given concentration on the whole surface of wafers or grains with a volumetric microsyringe or a pipette. In the case of flour, substances are applied with any volumetric device, providing a homogeneous mixing. The volume of solvent applied on control wafers, grains or flour is equal to the maximum volume of test solution, applied to the treatment.

(iii) **Set-up of the bioassay:** Once solvent is evaporated, both treated and control wafers or grains, are placed at a considerable distance in the same recipient. Flour containing the test products is placed in different containers.

Then, the larvae or adult insect is placed equisdistant from both treated and untreated food and allowed to eat. Larvae or adult insect should be starved for a determined time before placing in the dishes. Recipients are kept at 27-28°C and 70% relative humidity (Carpinella *et al.*, 2002; 2003; Valladares *et al.*, 1997; 2003). When using very small insects, several individuals (e.g. 5) can be used simultaneously within a recipient, as one replication. At least 10-20 replications should be made.

(iv) **Optimal time for the assay:** The optimal time for weevils, is 24 h or more, in case the insect has approximately eaten less than 20% of total offered food.

(v) **Variables:** After a previously determined time of exposure, the relative amounts of treated and untreated substrate areas eaten are visually estimated (visual ranking system: recorded in percentage from 0 to 100), by dividing the wafer area in imaginary quarters, or digitally. This last procedure is carried out comparing the contrast between the eaten or uneaten surface with a proper software program after taking a photograph of the treated and control wafers. When using very small insects, several individuals (e.g. 5) can be used simultaneously within a recipient, as one replication. At least 10-20 replications should be made.

In case of using grains or flour as food, either treated or untreated substrate must be weighed before and after the assay.

Calculations

Antifeedant Index (AI%) is calculated as follows:

$$AI\% = (1 - T/C) \times 100$$

where T and C are the consumption in the treated and control wafers, respectively. In the experiment with flour and grains, T and C are the weight difference of treated and control food matrix before and after the bioassay, respectively (Carpinella *et al.*, 2003). The AI% value ranges from 0 to 100%; Higher AI% value indicates higher antifeedant effect.

Feeding Ratio Index (FR) is calculated as follows:

$$FR = (T/C)$$

where T are units consumed in treated wafers and C are units consumed in control wafers. The value of FR varies between 0 and 1; important antifeedant effect is represented by small values.

In the experiment with flour or grains, T and C are the weight difference of food matrix before and after the bioassay, in treated and untreated food, respectively.

Feeding deterrence (FD%) is calculated as follows:

$$FD\% = (C - T)/(C + T) \times 100$$

where T and C are the areas eaten in the treated and control wafers, respectively (Koul *et al.*, 1990). In experiments carried out with flour or grains, T and C are the weight differences of treated and control food matrix before and after the bioassay, respectively. This index ranges from 0 to 100% similar to AI%. However the resulting values are lower than those obtained for AI% in the same experiment. Higher FD% value indicates higher antifeedant effect.

Statistical analysis

Data are compared using the Wilcoxon signed paired rank test, $p < 0.05$.

Experiment 4: Topic Application on Mouth Parts

Topic application of allelochemicals on mouth parts would affect insect feeding behaviour if there is a primary antifeedant action.

Materials and equipments required

Hamilton Microsyringes, CO_2 current, Petri dish, high precision balance, allelochemicals, solvent used in allelochemical solution, stereomicroscope, fresh host plant leaves.

Procedure

(i) **Preparation of test solution:** The test substance (extracts or pure compounds) is dissolved, in the desired concentrations, preferentially in water or in the solvent where the substance has the best solubility.

(ii) **Bioassay set-up:** Insects are anesthetized by a CO_2 current and placed under magnification on a stereomicroscope. The test solution is then applied using a microsyringe, on all or specific mouth parts. Controls receive the same treatment but only the solvent used to prepare the test solution is applied. The insects are then placed in Petri dishes and fed with fresh plant material for 24 hs. After that, food consumption is measured. At least 15 replications are advised, always using insects of the same age and nutritional status.

(iii) **Variables:** Amount of food consumed by test and control insects after 24 hours is determined.

Observation

Insects must be kept under a CO_2 current until the solvent is fully evaporated.

Statistical analyses

The data are analysed using Mann-Whitney test. Analysis of variance (ANOVA) and Tukey honestly significant difference test or their nonparametric equivalent can be used if different concentrations are tested.

Experiment 5: Ablation of Insect Mouthparts or other Appendices

The sensory structures (chemoreceptors) that enable insects to identify feeding attractants and deterrents can be located in different body parts such as maxillary palps, galea, lacinia, labrum-epipharynx, antennae and first tarsi. In deterrence-detection experiments (feeding choice tests), ablation of particular appendices can therefore affect insect response to allelochemicals in food, thus allowing identification of the receptors involved.

Materials and equipments required

Freezer or CO_2 current, 1 mL Eppendorf plastic pipette, compressed air, Dumont #5 forceps or microintraocular scissors, stereomicroscope, allelochemicals, solvent, Petri dish, fresh host plant leaves.

Procedure

(i) **Ablation experiments:** The insects must be chilled at 5°C for at least 30 minutes or anesthetized by a CO_2 current (Branson and Ortman, 1969) before ablation. Insects are introduced in an Eppendorf pipette with its end cut off, so that the insect's head would emerge from tip when a flow of compressed air is applied from above (De Boer and Hanson, 1987; Chyb *et al.*, 1995; Eichenseet and Mullin, 1996). The air stream over the insect's head spreads the mouthparts apart, facilitating their identification and removal.

Appendices (one pair at a time or combinations of them) are selectively pulled off at their articulation with a fine pair of forceps (Eichenseet and Mullin, 1996) or microintraocular scissors. Non-ablated insects are used as controls. Operated insects are fed and allowed to recover for 24 or 48 h (Eichenseet and Mullin, 1996). A Choice test (24 h) as previously described can be used to check the ability of ablated insects to detect antifeedants. Fifteen replicates should be used for each treatment.

(ii) **Variables:** Percentage area eaten on treated and control leaves, estimated from choice tests.

Observations

Care must be taken to ensure that the proper mouthparts were ablated by checking the remaining pieces under a dissecting microscope after each experiment. Insects without the proper complement of mouthparts for the respective ablation treatment should be removed from statistical analysis.

Statistical analyses

An Antifeedant Index (see Experiment 2) is calculated and then compared between ablated and non-ablated insects by Mann-Whitney test. ANOVA or regression analysis can be used when different allelochemical concentrations are considered.

Experiment 6: No-choice Test

No-choice test can be carried out In order to determine the effects of complete extracts or pure compounds on insect feeding, development and survival. In this assay, the insects are forced to feed on palatable food treated with the substance of interest, throughout its complete or partial life cycle (Mordue and Blackwell 1993; Rembold, 1989). The obtained results are compared with those obtained with insects forced to eat food without treatment. This assay is frequently used to check the strength of antifeedant compounds, to confirm primary or secondary antifeedant effects and to detect growth regulator compounds or toxicants.

Materials and equipments required

Leaves, seedlings or complete plants of the habitual food for herbivore insects, dry food e.g. flour, wafers in the case of weevils, grains for granivorous species or artificial substrates which contain needed nutrients. Pots, Petri dishes or plastic containers, tissue paper, filter paper or cotton disks, round glasses with two 1 cm^2 holes (optional), distilled water, high quality solvents and high-precision balance are also needed. Hamilton syringe, paint brush or atomizer are used for applying test products.

Procedure

(i) **Preparation of test solution:** The test substance (extracts or pure compounds) is dissolved in water or a proper organic solvent at the desired concentrations. Chlorinated solvents should be discarded because they damage leaf structure.

(ii) **Bioassay set-up:** The experiment is carried out as indicated for Experiment 2 or 3, but insect larvae should be fed only with food matrix treated with known quantities of the test compounds or with the solvent. Food can be complete leaves, seedlings or plants, 1 cm^2 leaf disks, flour, grains, wafers or artificial diet. Substances should be applied as explained in choice test. In case of the artificial diet assay, neonate larvae should be exposed to the test substance or control solvent mixed with the diet. Two larvae should be placed by recipient (total of 20 replicates are recommended) in order to leave one larvae/recipient in case of natural death. In all cases, food should be periodically renewed to ensure a constant provision. A similar set of larvae should not be fed at all, then acting as starved controls (Carpinella *et al.*, 2003). All recipients should be kept under controlled atmosphere till end of study which could be defined by adult emergency or total death of tested individuals.

Alternatively, each larva can be transferred to another diet which is free of tested products after being exposed to treatment for a determined time. Insects are allowed to complete their development.

(iii) **Variables:** Food consumption (consumed area and/or weight of remaining food matrix), body and faeces weight, larval and pupal development, abnormalities, adult emergency and mortality in any stage should be recorded every 24 hours till the end of the experience. If the measured parameter is weight, then larvae or food must be weighed prior to set up of the experiment.

Observations

Nutritional indices need to be calculated on the basis of dry weights. Thus, at the onset of the experiments, 10 food samples (ca. 0.5 g each) and 20 individual insects should be dried to constant weight at 60°C to determine fresh weight/ dry weight ratios. At the end of the experiments, dry weights are determined for remaining diet, faeces and insects.

Calculations

Nutritional Indices are calculated as follows:

$$\text{Consumption Index (CI)} = E/T \times A$$
$$\text{Relative Growth Rate (RGR)} = P/T \times A$$
$$\text{Relative Consumptiom Rate (RCR)} = E/(T \times A)$$
$$\text{Relative Metabolic Rate (RMR)} = M/(A \times T) \; M = (E-F)-P$$
$$\text{Approximate digestibility (AD)} = 100(E-F)/E$$
$$\text{Efficiency of conversion of ingested food (ECI)} = 100 \, P/E$$
$$\text{Efficiency of conversion of digested food (ECD)} = 100 \, P/(E-F)$$
$$\text{Metabolic cost (MC)} = 100-ECD$$

where T is the duration of experimental period, A is mean dry weight of insect during T, E is the dry weight of food eaten, F is the dry weight of faeces produced and P is the dry weight gain of insect during feeding period (Schmidt *et al.*, 1997; Rodriguez Hernández and Vendramim, 1998; Senthil Nathan and Schoon, 2006). Mortality data are used for the determination of the lethal dosage 50 (LD_{50}) calculated by Probit analysis. The time required for 50% mortality (LT_{50}) is also calculated by Probit analysis.

Statistical analysis

Data is subjected to analysis of variance (ANOVA of arcsine square root transformed percentages) and Tukey honestly significant difference test or their equivalent no parametric tests.

Experiment 7: Topic Application of Allelochemicals on Insect Tegument

Topic application of allelochemicals on insect tegument could affect feeding behaviour, revealing secondary antifeedant effects. Topic application is the only way to prove secondary deterrence in small insects when injection is not possible. Toxic effects can also be assessed through these tests. The tested compounds are usually applied to abdominal intersegmental membranes, avoiding contact with gustatory sensilla.

Materials and equipments required

Hamilton microsyringes, CO_2 current, Petri dishes, high-precision balance, stove, allelochemicals, solvent, stereomicroscope, fresh host plant leaves.

Procedure

(i) **Preparation of test solution:** The test substance (extracts or pure compounds) is dissolved in water or a proper organic solvent at the desired concentrations.

(ii) **Bioassay set-up:** Insects, with uniform weight, are individually placed in Petri dishes and anesthetized with a CO_2 current. A droplet of the studied substance solution is then applied upon the intersegmental membranes of the abdomen and solvent is left to evaporate. The insects are kept under observation, fed with untreated leaves or the corresponding food (periodically renewed), and weighed every 48 hours post-treatment, in order to detect possible changes in food consumption and growth rate due to the allelochemical application (Schmidt *et al.*, 1997; Rodriguez Hernández and Vendramim, 1998; Senthil Nathan and Schoon, 2006). Fifteen to 20 replications should be made.

(iii) **Variables:** Amount of food consumed (e.g. percentage of leaf area eaten, visual estimation, or through weight of food pre- and post-consumption), insect and faeces weight, percent survival. Nutritional indices can be calculated as above (see Experiment 6).

Observations

Nutritional indices are calculated on the basis of dry weights. Thus, at the onset of experiments, 10 food samples (ca. 0.5 g each) and 20 individual insects should be dried to constant weight at 60°C to determine fresh weight/dry weight ratios. After 4 days of topic application, dry weights are determined for remaining diet, faeces and insects.

Statistical analysis

The data are compared between treatment and control through Wilcoxon test, or among concentrations and their controls via Analysis of Variance (ANOVA) and Tukey honestly significant difference test or their non-parametrical equivalents. Dose-response relationships and DL_{50}, TL_{50} y CE_{50} values are determined by Probit analysis.

Experiment 8: Injection Assays

When some allelochemicals are ingested by the insects, they may inhibit food consumption, or induce changes in growth rate or even cause death. In this test, insects are injected with allelochemical or solvent to assess post-ingestive secondary antifeedant or toxic effects.

Materials and equipments required

Hamilton microsyringes, CO_2 current, Petri dishes, high-precision balance, stove, allelochemicals, organic solvent, stereomicroscope, fresh host plant leaves, controlled temperature, humidity and photoperiod conditions.

Procedure

 (i) **Test solution preparation:** The test substance (extracts or pure compounds) is dissolved in water or a proper organic solvent at the desired concentrations.

 (ii) **Set-up of bioassay:** Insects (adults, larvae or nymphs) are injected into the haemolymph, using a microsyringe (5 or 10 μL Hamilton). Different amounts of allelochemicals can be obtained by varying the volume of the solution (Wheeler and Isman, 2001). Control insects are just punctured with the syringe, without actual injection, or the same volume of the corresponding solvent is injected.

   A brief anaesthesia with CO_2 or low temperature is used, when needed, to facilitate injection. The syringe tip should be inserted longitudinally with the body, on the distal lateral dorsal section of the abdomen, close to the lateral margin, in order to avoid injuries to internal organs (Kabaru and Mwangi, 2000; Wheeler and Isman, 2001).

   The insects are then placed individually in Petri dishes containing fresh untreated host plant leaves, renewed every 48 hours (Liu *et al.*, 1991). Fifteen to 20 insects of the same age should be used as replications.

 (ii) **Variables:** Amount of food consumed (% leaf area eaten), pre-treatment and 96 hours post-treatment insect weight, faeces weight, weight of offered and discarded food. See above for calculation of nutritional indices.

Statistical analysis

Data can be analysed by Wilcoxon paired signed rank test or ANOVA if several allelochemical concentrations are tested.

SUGGESTED READINGS

Acar, E.B., Medina, J.C., Lee, M.L. and Booth, G.M. (2001). Olfactory behavior of convergent lady beetles (Coleoptera: Coccinellidae) to alarm pheromone of green peach aphid (Hemiptera: Aphididae). *The Canadian Entomologist* **133**: 189-197.

Bernays, E.A. and Chapman, R.F. (1973). The regulation of feeding in *Locusta migratoria*: internal inhibitory mechanisms. *Entomologia Experimentalis et Applicata* **16**: 329-342.

Bernays, E.A. and Chapman, R.F. (1994). *Host-Plant Selection by Phytophagous Insects*. Chapman and Hall, New York, USA.

Branson, T.F. and Ortman, E.E. (1969). Feeding behavior of larvae of the western corn rootworm: normal larvae and maxillectomized with laser radiation. *Annals of the Entomological Society of America* **62**: 808-812.

Carpinella, M.C., Defago, M.T., Valladares, G. and Palacios, S.M. (2003). Antifeedant and insecticide properties of a limonoid from *Melia azedarach* (Meliaceae) with potential use for pest management. *Journal of Agricultural and Food Chemistry* **51**: 369-374.

Carpinella, M.C., Ferrayoli, G.C., Valladares, G., Defago, M. and Palacios, S.M. (2002). Potent insect antifeedant limonoid from *Melia azedarach*. *Bioscience Biotechnology and Biochemistry* **66**: 1731-1736.

Chyb, S., Eichenseer, H., Hollister, B., Mullin, C. and Frazier, J.L. (1995). Identification of sensilla involved in taste mediation in adult western corn rootworm (*Diabrotica virgifera virgifera* LaConte). *Journal of Chemical Ecology* **21**: 313-329.

De Boer, G. and Hanson, F.E. (1987). Differentiation of roles of chemosensory organs in food discrimination among host and non-host plants by larvae of tobacco hornworm, *Manduca sexta*. *Physiological Entomology* **12**: 387-398.

Eichnseer, H. and Mullin, C.A. (1996). Maxillary appendages used by western corn rootworms, *Diabrotica virgifera virgifera*, to discriminate between a phagostimulant and deterrent. *Entomologia Experimentalis et Applicata* **78**: 237-242.

Hassanali, A. and Lwande, W. (1989). Antipest secondary metabolites from African plants. In: *Insecticides of Plant Origin* (Eds., J.T. Arnason, B.J.R. Philogene and P. Morand). *ACS Symposium Series* **387**: 78-94. American Chemical Society, Washington, DC.

Kabaru, J.M. and Mwangi, R.W. (2000). Effect of post-treatment temperature on the insecticidal activity of neem, *Azadirachta indica* A. Juss. Seed extract on *Schistocerca gregaria* (Forskal): a preliminary report. *Insect Science and Its Applications* **20**: 77-79.

Koul, O., Smirle, M.J. and Isman, M.B. (1990). Asarones from *Acorus calamus* L. oil: their effect on feeding behaviour and dietary utilization in *Peridroma saucia*. *Journal of Chemical Ecology* **16**: 1911-1920.

Landis, D.A. and Gould, F. (1989) Investigating the effectiveness of feeding deterrents against the southern corn rootworm, using behavioural bioassay and toxicity testing. *Entomologia Experimentalis et Applicata* **51**: 163-174.

Liu, Y., Alford, A.R. and Bentley, M.D. (1991). A study on mode of antifeedant effects of epilimonol against *Leptinotarsa decemlineata*. *Entomologia Experimentalis et Applicata* **60**: 13-18.

Mordue A.J. and Blackwell A. (1993). Azadirachtin: an update. *Journal of Insect Physiology* **39**: 903-924.

Mordue A.J., Simmonds, M.N.S., Ley, S.V., Blaney, W.M., Mordue, W., Nasiruddin, M. and Nisbet, A.J. (1998). Actions of azadirachtin, a plant allelochemical, against insects. *Pesticide Science* **54**: 277-284.

Mordue A.J. and Nisbet, A.J. (2000). Azadirachtin from the neem tree *Azadirachta indica*: its actions against insects. *Annais da Sociedade Entomológica do Brasil* **29**: 615-132.

Noldus, L.P. (1991). The observer: a software system form collection and analysis of observational data. *Behavior Research Methods Instruments & Computers* **23**: 415-429.

Rembold, H. (1989). Azadirachtins: their structure and mode of action. In: *Insecticides of Plant Origin* (Eds., J.T. Arnason, B.J.R. Philogene and P. Morand), *ACS Symposium Series* **387**: 150-163. American Chemical Society, Washigton DC, USA.

Riddick, E.W., Aldrich, J.R., De Miro, A. and Davis, J.C. (2000). Potential for modifying the behavior of the multicolored Asian lady beetle (Coleoptera: Coccinellidae) with plant-derived natural products. *Annals of the Entomological Society of America* **93**: 1314-1321.

Rodriguez Hernandez, C. and Vendramim, J.D. (1998). Use of Indexes to measure insecticidal effect of extracts from Meliaceae on *Spodoptera frugiperda*. *Manejo Integrado de Plagas* **48**: 11-18. (In Spanish).

Schmidt, G.H., Ahmed, A.A. I. and Breuer, M. (1997). Effect of *Melia azedarach* extract on larval development and reproduction parameters of *Spodoptera littoralis* (Boisd.) and *Agrotis ipsilon* (Hufn.) (Lep. Noctuidae). *Anz. Schadl.* **70**: 4-12.

Schmutterer, H. (1985). Which insect pests can be controlled by application of neem seed kernel extracts under field conditions. *Zeitschrift Fur Angewandte Entomologie* **100**: 468-475.

Senthil Nathan, S. and Schoon, K. (2006). Effects of *Melia azedarach* extract on the teak defoliator *Hyblaea puera* Cramer (Lepidoptera: Hyblaeidae). *Crop Protection* **25**: 287-291.

Valladares, G., Defagó, M.T., Palacios, S.M. and Carpinella, M.C. (1997). Laboratory evaluation of *Melia azedarach* (Meliaceae) extracts against the Elm Leaf Beetle (Coleoptera: Chrysomelidae). *Journal of Economical Entomology* **90**: 747-750.

Valladares, G., Garbin, L., Defagó, M.T., Carpinella , M.C. and Palacios, S.M. (2003). Evaluation of insecticide and antifeedant activity of senescent leaf extracts from *Melia azedarach*. *Revista de la Sociedad de Entomología Argentina* **62**: 49-57. (In Spanish).

Wheeler, D.A. and Isman, M.B. (2001). Antifeedant and toxic activity of *Trichilia americana* extract against the larvae of *Spodoptera litura*. *Entomologia Experimentalis et Applicata* **98**: 9-16.

SECTION

IV

Appendices

I

Abbreviations

ANOVA: Analysis of the variance
ATCC: American Type Culture Collection
ATP: Adenosine triphosphate
AS: Anisaldehyde-sulfuric acid reagent
BAW: Butanol-acetic acid-water
BHT: Butylated hydroxytoluene
BIRD: Bilinear Rotation pulse
BSA: Bis(trimethylsilyl)-acetamide
BSTFA: Bis (trimethylsilyl) trifluoroacetamide
CC: Column chromatography
CCC: Counter-current chromatography
CCD: Charged coupled device
CE: Capillary electrophoresis
CFU: Colony forming units
COLOC: Correlation spectroscopy for long-range couplings
COSY: Correlation spectroscopy
CZE: Capillary zone electrophoresis
DAD: Diode array detection system
DC: Dielectric constant
DMDS: Dimethyldisulfide
DMP: 2,2-Dimethoxypropane
DMSO: Anhydrous dimethysulfoxide
DPPH: 1,1-diphenyl-2-picrylhydrazyl
DQC: Dichloroquinonechlorimide reagent
DQF: Double quantum frequency
DR: 2-deoxyribose
EC: Electrical conductivity
ECD: Efficiency of conversion of digested food
EDTA: Disodium ethylenediamine tetraacetic acid
EI: Electron impact

ELSD: Evaporative Light-Scattering Detector
EM: Electromagnetic radiation
EO: Essential oil
ESI: Electrospray ionization
EtOAc: Ethyl acetate
FAB: Fast atom bombardment
FAMEs: Fatty acids methyl esters
FID: Fame ionization detection
FT-IR: Fourier transformed IR
GC: Gas chromatography
GLC: Gas-liquid chromatography
GlcNAcase: β-N-acetylglucosaminidase
GPC: Gel permeation chromatography
GSC: Gas-solid chromatography
HDPE: High-density polyethylene
HETCOR: Heteronuclear correlation spectroscopy
HMBC: Heteronuclear multiple bond correlation
HMQC: Heteronuclear multiple-quantum coherence experiment
HPLC: High performance liquid chromatography
HPTLC: High Performance Thin Layer Chromatography
IEC: Ion exchange chromatography
IGRs: Insect growth regulators
INEPT: Insensitive nucleus enhanced by polarization transfer
IR: Infrared
LC: Liquid chromatography
LLC: Liquid-liquid chromatography
LLE: Liquid-liquid extraction
LMM: Liquid Malt Medium
LSC: Liquid-solid chromatography
LSD: Least significant difference
LSIMS: Liquid secondary ions mass spectrometry
LTGA: Lignothioglycolic acid
MALDI: Matrix assisted laser desorption ionization
MC: Metabolic cost
MDA: Malondialdehyde
MEKC: Micellar electrokinetic chromatography
MHA: Mueller Hinton Agar
MHB: Mueller Hinton Broth
MS: Mass spectrometry
MSTFA: N-methyl-N-trimethylsilyltrifluoracetamide
MTBE: Methyl *ter*-butyl ether
MTT: 3-{4,5-dimethylthiazol-2-yl}-2,5-diphenyltetrazolium bromide
NACE: Non-aqueous capillary electrophoresis
NADPH: Nicotinamide adenine dinucleotide phosphate
NMR: Nuclear magnetic resonance

NOESY: Nuclear overhauser enhancement spectroscopy
NP/PEG: Natural products/polyethylene glycol
NPD: Nitrogen-phosphorus detection
PAL: Phenylalanine Ammonia-lyase
PC: Paper chromatography
pNP-GlcNAc: p-Nitrophenyl β-N-acetylglucosaminide
POD: Peroxidase
PTFE: Polytetrafluoroethylene
PVP: Polyvinylpyrrolidone
PVPP: Polyvinylpolypyrrolidone
Rf: Retardation factor
RI: Refractive Index
ROESY: Rotating-frame overhauser enhancement spectroscopy
RP: Reverse Phase
SDE: Simultaneous distillation and extraction
SFE: Supercritical fluid extraction
SMM: Solid Malt Medium
SPE: Solid phase extraction
SPME: Solid phase microextraction
ssBHI: Semi solid Brain Heart Infusion
TAG: Triacylglycerols
TBA: 2-thiobarbituric acid
TCA: Trichloroacetic acid
TeBA: Tertiary butanol-glacial acetic acid-water
TFA: Trifluoroacetamide
TLC: Thin layer chromatography
TMCS: Trimethylchlorosilane
TMS: Trimethylsilyl
TOCSY: Total correlated spectroscopy
UV-Vis: Ultraviolet-visible
VOCs: Volatile organic compounds
XHCORR: Heteronuclear (X, H) shift correlation Spectroscopy

Chemical Formulae and Molecular Weight of Solvents and Reagents

Compound Name	Chemical Formulae	Molecular Weight
Acetic acid	CH_3CO_2H	60.1
	HOAc	
Aluminium chloride (anhydrous)	$AlCl_3$	133.3
Aluminium chloride hexahydrate	$AlCl_3 \cdot 6\ H_2O$	241.4
Ammonia	NH_3	17
Ammonium chloride	NH_4Cl	53.5
Ammonium sulfate	$(NH_4)_2SO_4$	132.1
Boric acid	H_3BO_3	61.8
Calcium chloride dihydrate	$CaCl_2 \cdot 2H_2O$	147
Carbon dioxide	CO_2	44
Carbon disulfide	CS_2	76.1
Calcium nitrate	$Ca(NO_3)_2$	236.2
Carbon tetrachloride	CCl_4	153.2
Chloroform	$CHCl_3$	119.4
Copper sulfate pentahydrate	$CuSO_4 \cdot H_2O$	249.7
Dibasic sodium phosphate	Na_2HPO_4	141.9
Dichloromethane	Cl_2CH_2	86.9
Ethanol	EtOH	46.1
Ferric chloride	$FeCl_3$	162.2
Ferrous sulfate	$FeSO_4 \cdot 7\ H_2O$	278.0
Hydrochloric acid	HCl	36.5 (d = 1.18 g /mL)
Hydrogen cyanide	HCN	27
Hydrogen peroxide	H_2O_2	34
Magnesium chloride	$MgCl_2$	95.2
Magnesium chloride hexahydrate	$MgCl_2 \cdot 6H_2O$	230.3
Magnesium sulfate	$MgSO_4$	120.4
Manganese chloride tetrahydrate	$MnCl_2 \cdot 4\ H_2O$	197.8
Methanol	MeOH	32

Methyl iodide	CH_3I	141.9
Potassium bromide	KBr	119
Potassium chloride	KCl	74.6
Potassium cyanide	KCN	65.1
Potassium dihydrogen phosphate	KH_2PO_4	136.1
Potassium hydroxide	KOH	56.1
Potassium metabisulfite	$K_2S_2O_5$	222.3
Potassium monohydrogen phosphate	K_2HPO_4	174.2
Potassium nitrate	KNO_3	101.1
Potassium permanganate	$KMnO_4$	158
Silver chloride	AgCl	143.3
Sodium acetate trihydrate	$NaAcO \cdot 3\ H_2O$	136.8
Sodium ammonium phosphate tetrahydrate	$NaNH_5PO_4 \cdot 4\ H_2O$	209.1
Sodium borate	$Na_2B_4O_7$	381.4
Sodium borohydride	$NaBH_4$	37.8
Sodium carbonate (anhydrous)	Na_2CO_3	105.9
Sodium chloride	NaCl	58.4
Sodium dihydrogen phosphate monohydrate	$NaH_2PO_4 \cdot H_2O$	138
Sodium hydrogen phosphate	Na_2HPO_4	142
Sodium hydroxide	NaOH	40
Sodium methoxide	NaMeO	54
Sodium oxide sulfide	$Na_2S_2O_3$	158.1
Sodium sulfate (anhydrous)	Na_2SO_4	142
Sulfuric acid	H_2SO_4	98.1 (d = 1.84 g/mL)
Trifluoroacetic acid	CF_3CO_2H	114

III

Molecular Weight of Organic Compounds

Compound Name	Molecular Weight
Acetylsenkirkine	407
Adenine	135.1
Ampicillin sodium salt	371.4
Angelic acid	100
Angustifoline	234
Apotropic acid	148
L-Arginine hydrochloride	210.7
L-ascorbic acid	176.1
Berberine	336
D-biotin	244.3
Caffeine	194
β-carotene	536.9
Castanospermine	190
α-chaconine	852
Cinnamic acid	148
trans-cinnamic acid	148.2
Citric acid monohydrate	210.1
Cocaine	303
Codeine	299
Colchicine	399
p-coumaric acid	164.2
Crystal violet	408
1-deoxynojirimycin	164
2-deoxyribose	134.1
Desacetyldoronine	417
2,6-dichlorophenolindophenol	326.1
L-3,4-dihydroxyphenylalanine	197.2
N,N-dimethylformamide	73.1
3-{4,5-dimethylthiazol-2-yl}-2,5-diphenyltetrazolium Bromide (MTT)	414.3

1,1-diphenyl-2-picrylhydrazyl (DPPH)	394.3
Doronine	459
Ephedrine $C_{10}H_{15}NO$	165
Ethylenediaminetetraacetic acid disodium salt dihydrate (EDTA)	372.2
Evans blue	960.8
Ferulic acid	194.2
Fuchsin, basic	305.4
Gallic acid	
Glucose	180.2
D-Glucose-6-phosphate dipotassium salt	336.3
L-glutamic acid	147.1
Florosenine	423
Heliotridine	155
L-Histidine monohydrochloride	209.6
Hordenine	165
p-hydroxybenzoic acid	138.1
13α-hydroxylupanine	264
Hyoscyamine	289
Integerrimine	335
Isoretronecanol	157
Isovaleric acid	102
L-Leucine	131.2
Lupanine	248
Malachite green oxalate salt	927
L-Methionine	149.2
Morphine	285
β-NADH disodium salt	709.4
β-NADP+ sodium salt	765.4
Naringenin	272.3
Nicotine	162
Nitro Blue Tetrazolium (NBT)	817.6
p-Nitrophenyl β-N-acetylglucosaminide	301.3
Noscapine	413
Otosenine	381
Papaverine	339
Phenazine Methosulfate (PMS)	306.3
L-phenylalanine	165.2
Platynecine	157
Retronecine	153
Retrorsine	351
Safranin T	350.9
Salicylic acid	138.1
Salicylhydroxamic acid	153.1
Scopolamine	303
Senecionine	335
Seneciphylline	333
Senecivernine	335
Senkirkine	365
Sodium hypochlorite	74.4
Sodium citrate · 2 H_2O	272.1
Solanidine	397

Solanine	868
Sucrose	342.3
Swainsonine	174
Tebaine	311
2-thiobarbituric acid	144.2
Tiglic acid	100
Tigloyloxylupanine	346
Tomatidine	415
Tricloroacetic acid (TCA)	163.4
Tris	121.1
Tropic acid	166
Tropine	141
α-truxillic acid	296
L-Tryptophan	204.2
Usaramine	351
Vanillic acid	168.1

Index

About the Editors

Prof. S.S. Narwal, Department of Agronomy, CCS Haryana Agricultural University, Hisar, India, is an internationally renowned allelopathy scientist and prossess 37 years teaching and research experience. He is the first National Fellow (Allelopathy) of the Indian Council of Agricultural Research, New Delhi from 1995-2005. Currently he is the President, of the Indian Society of Allelopathy and Chief Editor, of the Allelopathy Journal. Till now, he has authored/edited 18 Books on allelopathy and more books are under preparation.

He has travelled to more than 35 countries in all the six continents, to deliver lectures in International Allelopathy Conferences. In 1989, when he was a Visiting Professor, Department of Allelopathy, Ukrainian Academy of Sciences, Kiev, he attended all Soviet allelopathy conferences and also read allelopathy reprints from Prof. A.M. Grodzinsky's Collection. Prof. Narwal established the Indian Society of Allelopathy in 1990. Prof. E.L. Rice, USA wrote *'Congratulations, Prof. Narwal for establishing the First National Society in the World'*. He has organized, four International Allelopathy Conferences in India: 1992 (Hisar), 1994 (New Delhi), 1998 (Dharwad) and 2004 (Hisar). He is also Founder of the International Allelopathy Society (1994) during the II International Allelopathy Conference, New Delhi.

Diego A. Sampietro is Assistant Professor (Phytochemistry and Plant Biotechnology), National University of Tucumán (Tucumán, Argentina). He completed his Ph.D. (Biochemistry) in 2005, from National University of Tucumán. He possesses 10 years experience of teaching and research at National University of Tucumán (Argentina) and is Assistant Researcher, National Council of Scientific and Technological Research (CONICET-Argentina). Currently he is English and

Regional Editor of the Allelopathy Journal. He is author or co-author of more than 30 scientific publications, including four book chapters. He is member of the research group of Dr. Catalán and is actively working on allelopathy and secondary compounds involved in plant defence.

César A. N. Catalán is Professor of Organic Chemistry since 1983. He had been Head of the Organic Chemistry Group of the Faculty of Chemistry, Biochemistry and Pharmacy (National University of Tucumán, Tucumán, Argentina) since 1985. He completed his Ph.D. (Chemistry) in 1977, from National University of Tucumán and in 1982, he was Postdoc at Stanford University (California,

USA). He possesses 40 years of teaching and research experience at National University of Tucumán (Argentina) and is Senior Researcher, National Council of Scientific and Technological Research (CONICET-Argentina). He was Director of CONICET, on national universities, Dean of Faculty of Chemistry, Biochemistry and Pharmacy, National University of Tucumán, and Rector of this University. He has received the award "Dr. Venancio Deulofeu 2005" from the Argentinian Chemical Association (ACA) for his major contributions to organic chemistry. He is the author or co-author of 150 journal articles, six book chapters and has supervised over 20 M.Sc. and Ph.D. students in the fields of chemotaxonomy, bioactive metabolites and organic synthesis, using readily available natural products. Currently his research group is actively working on natural products, especially on sesquiterpene lactones from Argentine Asteraceae and bioactive constituents from aromatic and medicinal plants native of South America.

Marta A. Vattuone is Professor of Phytochemistry. Her main research interests are Plant Biochemistry and Natural Products. She has been head of the Phytochemistry Group of the Faculty of Biochemistry, Chemistry and Pharmacy (National University of Tucumán, Tucumán, Argentina) since 1999. She completed Ph.D. (Chemistry) in 1980 and was Postdoc in France, Spain, Sweden and Belgium. She has 40 years of teaching experience at the National University of

Tucumán (Argentina), and is Senior Researcher from the National Council of Scientific and Technological Research (CONICET-Argentina). Dr. Vattuone is working on plant biochemistry, secondary metabolites involved in allelopathic processes and antimicrobial activity of propolis and medicinal plants. She has published more than 100 papers in international journals and written seven book chapters.